BIRKHÄUSER

Progress in Mathematics
Volume 283

Arrangements, Local Systems and Singularities

CIMPA Summer School, Galatasaray University, Istanbul, 2007

Fouad El Zein
Alexander I. Suciu
Meral Tosun
A. Muhammed Uludağ
Sergey Yuzvinsky
Editors

Birkhäuser
Basel · Boston · Berlin

Editors:

Fouad El Zein
Institute of Mathematics
Case 7012, 2 place Jussieu
75251 Paris Cedex 05
France
e-mail: elzein@math.jussieu.fr

and

Center for Advanced Mathematical
Sciences (CAMS)
American University of Beirut
P.O. Box 11-236
Beirut
Lebanon

Alexandru I. Suciu
Northeastern University
Department of Mathematics
360 Huntington Avenue
Boston, MA 02115
USA
e-mail: a.suciu@neu.edu

Meral Tosun
Galatasaray University
Mathematics Department
Ciragan Cad. No 36
Besiktas, 34357 Istanbul
Turkey
e-mail: mrltosun@gmail.com

A. Muhammed Uludağ
Galatasaray University
Mathematics Department
Ciragan Cad. No 36
Besiktas, 34357 Istanbul
Turkey
e-mail: muhammed.uludag@gmail.com

Sergey Yuzvinsky
Department of Mathematics
University of Oregon
Eugene, OR 97405
USA
e-mail: yuz@uoregon.edu

2000 Mathematics Subject Classification: Primary 52C35, 32S22, 55N25, 14C21, 32J25, 32S20, 32S35; Secondary 20F34, 05E25, 06A07, 14F35, 55R80, 13D02

Library of Congress Control Number: 2009938996

Bibliographic information published by Die Deutsche Bibliothek.
Die Deutsche Bibliothek lists this publication in the Deutsche Nationalbibliografie; detailed bibliographic data is available in the Internet at http://dnb.ddb.de

ISBN 978-3-0346-0208-2 Birkhäuser Verlag AG, Basel · Boston · Berlin

Basel · Boston · Berlin
P.O. Box 133, CH-4010 Basel, Switzerland
Part of Springer Science+Business Media
Printed on acid-free paper produced from chlorine-free pulp. TCF ∞

ISBN 978-3-0346-0208-2

ISBN 978-3-0346-0209-9 (eBook)

9 8 7 6 5 4 3 2 1

www.birkhauser.ch

Contents

Preface

This volume comprises a set of lecture notes from the CIMPA Summer School, *Arrangements and Local systems and Singularities*, held at Galatasaray University, Istanbul, during June 11-22, 2007. The school was attended by 68 mathematicians, 35 of them from 19 countries outside Turkey. The Summer School was made up of eleven short courses and five seminars presented by an outstanding group of lecturers who covered a wide range of topics related to the concepts of arrangements, local systems and singularities. The list of lectures of the workshop appears below. Most members of the audience were graduate students or young researchers and the primary purpose of the school was to introduce them, in particular those from developing countries, to the many advances and opportunities in this widely applicable field.

Historically, research on arrangements of hyperplanes, starting with the work of V.I. A'rnold and E. Brieskorn and later Peter Orlik, was presented at meetings devoted mainly to discussions of singularities. Some of the articles in this volume demonstrate that the interaction between these two subjects is ongoing and particularly productive. Beyond this, the volume is intended for a large audience in pure mathematics, including researchers and graduate students working in algebraic geometry, singularity theory, topology and related fields. The reader will find in the Problem Section at the end of the volume a variety of open problems proposed by the lecturers at the end of the School's sessions as directions ripe for further study and development.

We would like to thank the *Centre des Mathématiques Pures et Appliquées (CIMPA)* for sponsoring the School and Professor Richard Grin for organizational help. We would also like to thank the *Scientific and Technological Research Council of Turkey (TUBITAK)* for their financial help which made it possible to invite 16 young mathematicians from Turkey as well as some lecturers from abroad. We were able to support participants from across the region, thanks to the generous financial help provided by the *International Center for Theoretical Physics (ICTP)* and the *International Mathematical Union (IMU)*.

During the long preparatory process and also during the school, Gülay Kaya, Ayşe Altıntaş and Celal Cem Sarıoğlu contributed at various levels to the organization. We are grateful to them.

F. El Zein, A. Suciu, M. Tosun, A.M. Uludağ and S. Yuzvinsky

The fourth named editor was supported by Tübitak/Kariyer 103T136 and Tuba/Gebip during the summer school and during the preparation of this volume. The third and fourth named authors were supported by the Galatasaray University Research Fund during the preparation of this volume.

List of Lectures of the School

T. Abe
Addition-deletion theorems for multiarrangements and graphic deformation of braids.

D. Cohen
Fundamental groups of complements of hyperplane arrangements.

G. Denham
Homological algebra of hyperplane arrangements.

F. El Zein
Local systems, constructible sheaves and Geometry.

M. Falk *Geometry and combinatorics of resonance weights.*

G. Kaya
Moduli of surfaces admitting non-smooth genus two fibrations over elliptic curves.

A.U.O. Kisisel
Toric varieties and the diagonal properties.

Lê D.T.
Simple singularities and simple Lie algebras.

D. Matei
Artin groups, Bestvina-Brady groups and arrangements of hypersurfaces.

D. Mond
(1) *Introduction to singularities of mappings*
(2) *Conservation of multiplicity. Cohen-Macaulay rings. Free divisors (from groups).*

M. Oka
Introduction to plane curve singularities.

H. Terao
(1) *The characteristic polynomial of a multiarrangement*
(2) *Periodicity of arrangements with integral coefficients modulo positive integers.*

M. Yoshida
Hyperbolic Schwarz map for the hypergeometric differential equation.

M. Yoshinaga
Minimal CW decomposition of the complement of hyperplane arrangements.

S. Yuzvinsky
Orlik-Solomon algebras of hyperplane arrangements.

List of Participants

Name	Institution
Aase Feragen	University of Helsinki
Abdigappur Narmanov	National University of Uzbekistan
Ayberk Zeytin	Middle East Technical University
Ayşe Altıntaş	University of Warwick
Ayşegül Özgüner	Bilkent University
Barbu Berceanu	IMAR Bucharest
Bayat Farshad Merrikh	Sharif University of Technology
Bedia Akyar Møller	Dokuz Eylül University
Begüm Uçar	Yıldız Technical University
Bülent Tosun	Middle East Technical University
Celal Cem Sarıoğlu	Dokuz Eylül University
Daniel Cohen	Louisiana State University
Daniel Matei	IMAR Bucharest
David Mond	University of Warwick
Duygu Irmak	Yıldız Technical University
Emanuele Delucchi	Universita Di Pisa
Emel Bilgin	İstanbul Technical University
Emel Coşkun	İstanbul Technical University
Fatma Şengüler	İstanbul Technical University
Fouad El Zein	Universite de Nantes
Graham Denham	University of Western Ortario
Guljakhan Kaypnazarova	National University of Uzbekistan
Gülay Kaya	Galatasaray University
Gülçin Gökçe	Ondokuz Mayıs University
Hakan Güntürkün	Middle East Technical University
Hani Shaker	Abdus Salam School of Mathematical Sciences GCU
Hiraoki Terao	Hokkaido University
Javier Bobadilla	CSIC
Kaan Esin	İstanbul Technical University
Khurram Shabbir	Abdus Salam School of Mathematical Sciences GCU

Kouider Djerfi	Centre de Universitaire de Saida
Kürşat Hakan Oral	Yıldız Technical University
Lê Dũng Tráng	ICTP
Mamuka Shubladze	Tbilisi State University
Maria Pe Pereira	Universidad Complutense de Madrid
Masaaki Yoshida	Kyushu University
Masahiko Yoshinaga	ICTP
Meral Tosun	Galatasaray University
Mesut Arslandok	Yıldız Technical University
Michael Falk	Northern Arizona University
Mikhail Mazin	University of Toronto
Mohammed Salim Jbara	University of Baghdad
Mohan Bhupal	Middle East Technical University
Muhammed Uludağ	Galatasaray University
Mustafa Topkara	Middle East Technical University
Mutsuo Oka	Universite de Tokyo Science
Münevver Çelik	Middle East Technical University
Ömer Faruk Koç	Yıldız Technical University
Özer Öztürk	Middle East Technical University
Özgür Kişisel	Middle East Technical University
Paul Cadman	University of Warwick
Pınar Mete	Balıkesir University
Richard Grin	CIMPA
Sabri Kaan Gürbüzer	Dokuz Eylül University
Selma Altınok	Adnan Menderes University
Sergei Yuzvinsky	University of Oregon
Shaheen Nazir	Abdus Salam School of Mathematical Sciences GCU
Shiv Datt Kumar	Motilal Nehru National Institute of Technology
Sultan Erdoğan	Bilkent University
Takeshi Sasaki	Kobe University
Takuro Abe	Hokkaido University
Tevfik Şahin	Ondokuz Mayıs University
Thi Anh Thu Dinh	Laboratoire Ja Dieudonne
Yasemin Güllük	Yıldız Technical University

Progress in Mathematics, Vol. 283, 1–38

Combinatorics of Covers of Complexified Hyperplane Arrangements

Emanuele Delucchi

Abstract. This is a survey of combinatorial models for covering spaces of the complement of a complexified hyperplane arrangement. We obtain a unified picture of the subject, and a generalization of various known results, by exploiting the toolkit of homotopy colimits for combinatorial applications developed by Welker, Ziegler and Živaljević.

Mathematics Subject Classification (2000). 32C35, 52C40, 05B35, 55P20, 57Q05.

Keywords. Arrangements of hyperplanes, oriented matroids, order complexes of posets, Salvetti complex, homotopy colimits, combinatorial topology.

Introduction

A *cover of an arrangement* is a topological cover of the space obtained by removing a finite set of hyperplanes from a complex, finite-dimensional vector space.

The study of combinatorially defined complexes modeling covers of arrangements has a story that goes back to the beginnings of the topological theory of hyperplane arrangements, and arises in the context of finite real reflection groups, where one can consider the set of hyperplanes ('mirrors') fixed by the reflections in the group. In 1971 Brieskorn [16] conjectured the complement of the complexification of this set of hyperplanes to be an aspherical space (we then say that this is a $K(\pi,1)$-*arrangement*).

Brieskorn's conjecture was settled by a general theorem of Deligne [29], who proved that the complexification of any real arrangement of linear hyperplanes whose chambers are simplicial cones is $K(\pi,1)$. The idea was to prove contractibility of the universal covering space of the arrangement's complement, and the method involved designing a cell complex that, under certain conditions, models the universal cover of the arrangement's complement.

The $K(\pi,1)$ problem for hyperplane arrangements, i.e., the problem of deciding whether the property of being $K(\pi,1)$ is determined by the combinatorics of the lattice of intersections of the hyperplanes, is still open and in the focus of active research. The construction and the study of different models for the universal covering space of arrangement complements has been one of the main strategies used to attack this problem. Alternative approaches have been successfully exploited, most notably the idea to reduce the problem to a lower dimensional situation by linear fibrations (that led to the concept of supersolvable arrangements [76, 37]), the use of fibrations onto the complex torus [54], or a mix of the different techniques [21]. For a general reading on the $K(\pi, 1)$ problem for arrangements we point to the survey of Falk and Randell [38, 39].

Among more recent topics in arrangement theory are the study of local system cohomology on arrangement complements and of the topology of the Milnor fiber. In both these subjects, the homology of certain covering spaces plays an important role (see e.g. [27] and [30]).

For general complex arrangements not much is known. Björner and Ziegler [11] described a simplicial model for the complement of a complex arrangement, but no description of the covering space is at hand. After previous partial results of Nakamura [53], the case of finite complex reflection arrangements was recently settled by Bessis [3], who described a model for the universal cover of the orbit space and showed its contractibility using the theory of Garside groups and Garside categories, thus proving the $K(\pi, 1)$ conjecture for this class of arrangements (for more details see Remark 5.15 and Section 6).

In this survey we present a unified view on the different combinatorial models for covers of complexified real arrangements.

We put the subject into the framework of the theory of diagrams of spaces and homotopy colimits for combinatorial applications, as developed by Welker, Ziegler and Živaljević in [81]. Diagrams of spaces have already been fruitfully exploited to study the *link* of hyperplane arrangements (i.e., the space defined by the union of the hyperplanes) [82, 77, 46]. In our context, these techniques allow us for instance to link the two main classes of complexes we will be dealing with, namely the Salvetti-type models W_ρ (Definition 3.1) and the Garside-type models U_ρ (Definition 5.1). Each of these types of models generalizes some known constructions, that we will explain. We thus obtain a unified picture of the subject. Moreover, this language allows us to apply the known techniques for the study of the homotopy type of diagram of spaces.

We will use some facts from the covering theory of groupoids. Also, we will meet along our way the notion of oriented systems (with a corresponding covering theory) as introduced by Paris [59]. We hope that the chosen notation and the explanations will succeed in clarifying the interplay among the different notions of "cover", nevertheless avoiding confusion.

We will begin our exposition by recalling some definitions and facts that are nowadays standard in arrangement theory. In Section 2 we introduce diagrams of spaces and their homotopy colimits and state some basic facts about them.

Then, in Section 3 we will present a first type of diagram models and study their homotopy colimits. For every topological cover of the given arrangement we construct a diagram of which homotopy colimit is isomorphic to the given covering space, and can be written as the order complex of an explicitly described poset. These models are called of *Salvetti type* because the model of the identical cover is actually isomorphic to the complex introduced by Salvetti in [67]. Specializing to the universal covering of arrangements of linear hyperplanes we recover naturally the simplicial complex obtained by Luis Paris in [60]. Moreover, we will mention here the work of Charney and Davis on Artin groups [22, 23], also pointing to an application of it given by Charney and Peifer [24] in the context of affine reflection arrangements.

In fact, Paris constructed topological models for arbitrary covers of linear arrangements [59]. In Section 4 we first explain this construction. Then, we describe a stratification of it whose nerve is isomorphic to the poset obtained from the diagram model of the corresponding cover, thereby showing that the diagram models offer a compact and handy description (in fact, as order complexes of posets) of Paris' models.

Restricting our attention from affine to linear real arrangements, Section 5 introduces another type of diagram models generalizing a construction that arose in the context of Garside groups [14, 5, 25]. We call them therefore of *Garside type*. As an application, we then explain how Deligne's argument can be reformulated in view of this type of models. The closing section is about possible further applications and directions of work.

Acknowledgements

The work reported here is part of the Ph.D. thesis of the author. A great expression of sincere gratitude goes to the supervisor, Prof. Eva-Maria Feichtner. Part of the redaction was carried out during a postdoctoral fellowship at the Mathematical Sciences Research Institute of Berkeley, CA. The author also thanks the anonymous referees for their valuable help.

1. Definitions and background

1.1. Arrangements

We will denote by $\mathcal{A}$ a collection of affine hyperplanes in $\mathbb{R}^d$, also called a *real arrangement.* Our considerations will be restricted to the case where the arrangement is locally finite (i.e., every point of $\mathbb{R}^d$ is contained in at most finitely many hyperplanes) and essential (i.e., the minimal intersections of hyperplanes are points).

The classical reference on arrangements of hyperplanes is the textbook of Orlik and Terao [56], and for the combinatorics of real arrangements in terms of oriented matroids we point to [10]. Let us here only recall the facts that we will need.

The closed strata that are determined by $\mathcal{A}$ in $\mathbb{R}^d$ are the *faces* of $\mathcal{A}$. The *support* of a face F is the set $\operatorname{supp}(F)$ of all hyperplanes containing F. The set of faces of $\mathcal{A}$ is partially ordered by reverse inclusion, so that for any two faces F_1, F_2 we have $F_1 \geq F_2$ if and only if $F_1 \subseteq F_2$: this defines the poset of faces of the arrangement, denoted by $\mathcal{F}(\mathcal{A})$.

The minimal elements of $\mathcal{F}(\mathcal{A})$ are the connected components that are cut out in $\mathbb{R}^d$ by $\mathcal{A}$ and are usually called *chambers*, or *regions*. Given two chambers C_1, C_2 of $\mathcal{A}$, one may choose a point in the interior of each chamber and consider the line segment spanned by these points. The hyperplanes that are met by this segment *separate* C_1 from C_2; the set of all hyperplanes separating C_1 from C_2 is denoted by $S(C_1, C_2)$. Two chambers are said to be *adjacent* if they are separated by only one hyperplane. If the arrangement is linear, we define the *opposite* of a chamber C to be the unique chamber $-C$ so that $S(C, -C) = \mathcal{A}$. If C is any chamber and F any face of $\mathcal{A}$, we will denote by C_F the unique chamber that contains F in its closure, and that is not separated from C by any of the hyperplanes that contain F – i.e, such that $C_F < F$ and $S(C_F, C) \cap \operatorname{supp}(F) = \emptyset$. The set of all regions of $\mathcal{A}$ will be written $\mathcal{T}(\mathcal{A})$ and can be given different partial orderings, depending on the choice of a base element. Once a *base chamber* $B \in \mathcal{T}$ is fixed, an associated partial order $\prec_B$ can be defined by setting $C_1 \preccurlyeq_B C_2$ if and only if $S(B, C_1) \subseteq S(B, C_2)$. This gives rise to the *poset of regions of* $\mathcal{A}$ *with base* B (introduced in [35]), that we will denote by $\mathcal{T}_B(\mathcal{A})$.

The *arrangement graph* $G(\mathcal{A})$ has $\mathcal{T}(\mathcal{A})$ as its set of vertices, and it is constructed by putting two opposite oriented edges between each pair of vertices that represent adjacent chambers. As an example, see the left side of Figure 1 for a picture of $G(\mathcal{A})$ when $\mathcal{A}$ consists of two lines in the plane. A directed path in the arrangement graph is called *positive*; it is called also *minimal* if it does not "cross" twice any hyperplane.

The *complexification* of the arrangement $\mathcal{A}$ is the set $\mathcal{A}_{\mathbb{C}}$ of the complex hyperplanes obtained by considering the same (real) defining equations as for the elements of $\mathcal{A}$. We will be interested in the topology of the complement of the complexification (sometimes just called *the arrangement's complement*)

$$\mathcal{M}(\mathcal{A}) := \mathbb{C}^d \setminus \bigcup \mathcal{A}_{\mathbb{C}}.$$

1.2. Posets

We give a short review of some basic facts and notations about partially ordered sets (or, for short, *posets*). For a careful exposition of the subject see [75]. Given two elements x, y in a poset $\mathcal{P}$ we denote by $x \vee y$ their unique least upper bound (or *join*) and by $x \wedge y$ their unique maximal lower bound (or *meet*), if these exist. A poset where the meet and the join exist for every pair of elements is called a *lattice*. Given two posets $\mathcal{P}$ and $\mathcal{Q}$, we will partially order their disjoint union $\mathcal{P} \coprod \mathcal{Q}$ by letting $x \geq y$ if and only if either both $x, y \in \mathcal{P}$ and $x \leq y$ in $\mathcal{P}$, or $x, y \in \mathcal{Q}$ and $x \leq y$ in $\mathcal{Q}$. The main topological object associated to a poset $\mathcal{P}$ is its *order complex* $\Delta(\mathcal{P})$, that is the simplicial complex of the totally ordered subsets of $\mathcal{P}$. It is clear that if $\mathcal{P}$ has a unique minimal element $\hat{0}$, then the order complex $\Delta(\mathcal{P})$ will be a cone with apex $\hat{0}$, and thus in particular contractible. The analogous statement holds of course when $\mathcal{P}$ has a unique maximal element.

1.3. The Salvetti complex

We introduce the tool that will allow us to link the topology of $\mathcal{M}(\mathcal{A})$ to the combinatorics of the real arrangement. Let us begin with the abstract definition.

Definition 1.1. Let $\mathcal{S}(\mathcal{A})$ be the set of all pairs (F, C) with $F \in \mathcal{F}$, $C \in \mathcal{T}$ and $C < F$. We give this set a partial order by setting $(F_1, C_1) > (F_2, C_2)$ if and only if $F_1 > F_2$ in $\mathcal{F}$ and $C_2 = (C_1)_F$. The (simplicial version of the) *Salvetti complex* is

$$\mathrm{Sal}(\mathcal{A}) := \Delta(\mathcal{S}(\mathcal{A})).$$

The importance of this object lies in the following fundamental theorem, that was proved by Mario Salvetti.

Theorem 1.2 (Theorem 1 of [67]). *For every real arrangement $\mathcal{A}$, the geometrical realization of* Sal($\mathcal{A}$) *can be embedded into the arrangement's complement, and is a strong deformation retract of $\mathcal{M}(\mathcal{A})$.*

There is another way to look at this complex. Indeed, the poset $\mathcal{S}(\mathcal{A})$ satisfies the conditions given in [6] for a general poset to be actually the poset of cells of a regular CW-complex. Thus, $\Delta(\mathcal{S}(\mathcal{A}))$ is the barycentric subdivision of a regular CW-complex that we will call Sal($\mathcal{A}$) as well.

Remark 1.3. An explicit construction of the CW-version of the Salvetti complex is the following. Start with a geometric realization of the arrangement graph, and take it as the 1-skeleton of the CW-complex. The attaching of the higher dimensional cells $[F, C]$ is defined recursively by saying that the 1-skeleton of $[F, C]$ consists of the positive minimal paths that start at C and end at the chamber opposite to C with respect to supp(F); a cell $[G, K]$ is then contained in the boundary of $[F, C]$ if and only if the 1-skeleton of $[G, K]$ is a directed subgraph of the 1-skeleton of $[F, C]$ (see [67]).

Example 1.4. As an example we consider the arrangement of two lines passing through the origin of $\mathbb{R}^2$. The picture illustrates the arrangement graph and two 2-cells with their boundary.

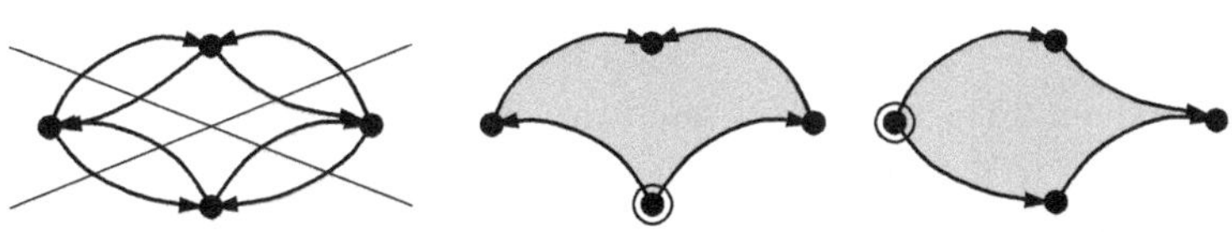

FIGURE 1. The arrangement of two lines in the plane with its arrangement graph, and two 2-cells of the cellular version of the associated Salvetti complex.

Those are the cells $[F, C]$ where F is the only codimension-2 face, namely the origin, and C is the chamber associated to the marked vertex. Note the boundary consisting of the positive minimal paths. The full complex has one such 2-cell for every chamber. △

1.4. The arrangement groupoid

A *groupoid* is a category where every arrow is invertible. This notion was first introduced by Brandt [15] as a generalization of the concept of group that he developed in his study of quadratic forms. According to [19], the use of groupoids in topology goes back to Reidemeister [65]. Let us mention also the work of Gabriel and Zisman [40], who explain and exploit the functorial relations between topological spaces, simplicial sets, and the associated groupoids. More recent textbooks exploiting groupoids in topology were written by Higgins [44] and Brown [17].

One of the classical features of groupoids is their nice covering theory, that parallels the theory of topological spaces. As this is a very classical topic, we will sketch the definitions and state the results we need; proofs and complements can be found in the elementary approach to the topic by Brown [18], while the book by Gabriel and Zisman [40] provides a more advanced treatment of the subject, together with its homological implications. For connections with homotopy of diagrams of spaces, see [19, 20]. For the basics about categories we refer to [49].

Let $\mathcal{Q}$ be a groupoid and consider an $x \in \mathrm{Ob}(\mathcal{Q})$ (we will use latin lowercase letter for objects, and Greek lowercase letters for morphisms). The set of endomorphisms $\mathrm{End}(x)$ has a natural group structure that does not depend on the choice of x; this group is called the *object group* of $\mathcal{Q}$ and will be denoted by $\pi\mathcal{Q}$ for reasons that will become clear later. The source and target object of a morphism ω will be indicated by $\mathrm{start}(\omega)$ and $\mathrm{end}(\omega)$, respectively. The *star* of the object x is the set

$$\mathrm{St}(x) := \{\omega \in \mathrm{Mor}(\mathcal{Q}) \mid \mathrm{start}(\omega) = x\}$$

of all morphisms of $\mathcal{Q}$ that start in x. The groupoid $\mathcal{Q}$ is called *connected* if for every $x, y \in \mathrm{Ob}(\mathcal{Q})$ there is a morphism ω with $x = \mathrm{start}(\omega)$ and $y = \mathrm{end}(\omega)$.

Definition 1.5. A *morphism of groupoids* is a functor

$$\rho : \mathcal{Q}' \to \mathcal{Q}$$

between groupoids. If $\mathcal{Q}$ is connected, then ρ is called a *covering* if, for every $z \in \mathrm{Ob}(\mathcal{Q}')$, the induced map

$$\rho_z : \mathrm{St}(z) \to \mathrm{St}(\rho(z))$$

is bijective. Given an $\alpha \in \mathrm{Mor}(\mathcal{Q})$ and any $z \in \rho^{-1}(\mathrm{start}(\alpha))$, the *lift of* α *at* z is the morphism $\rho_z^{-1}(\alpha)$, and will be written $\alpha^{\langle z \rangle}$ when the covering ρ is understood.

Example 1.6. The groupoid described in Example 1.14 is a cover of the groupoid of Example 1.11: the bijection can be checked directly. At the end of Example 1.15 we sketch the proof that the groupoid of Example 1.15 is a cover of the one defined in Example 1.12. △

If ρ is a covering of groupoids as above, the object group $\mathrm{End}(z) = \pi\mathcal{Q}'$ is mapped isomorphically by ρ to a subgroup of $\mathrm{End}(\rho(z)) = \pi\mathcal{Q}$ that is called the *characteristic group* of the covering.

The following result is classical.

Theorem 1.7 (see 9.4.3 of [18]). *Let $\mathcal{Q}$ be a connected groupoid, H a subgroup of $\pi\mathcal{Q}$, and choose a base object $x \in \mathrm{Ob}(\mathcal{Q})$. Consider the groupoid $\mathcal{Q}'$ defined by setting $\mathrm{Ob}(\mathcal{Q}') := \{H\omega \mid \omega \in \mathrm{St}(x)\}$ and where the morphisms between $H\omega_1$ and $H\omega_2$ correspond to morphisms α from $\mathrm{end}(\omega_1)$ to $\mathrm{end}(\omega_2)$ in $\mathcal{Q}$ such that $H\omega_1\alpha = H\omega_2$.*

The functor $\rho : \mathcal{Q}' \to \mathcal{Q}$ mapping $H\omega$ to $\mathrm{end}(\omega)$ is a covering of groupoids with characteristic group H.

Definition of the arrangement groupoid. Consider the free category on the arrangement graph (see [49, section II.4] for the definition), whose morphisms correspond to directed paths in $G(\mathcal{A})$.

Example 1.8. Take as an example the 1-dimensional arrangement given by the zero point inside the real line, that we will call $\mathcal{A}_1$. This arrangement has clearly two chambers A, B, and its arrangement graph consists of two vertices joined by two directed edges: the edge a directed from A to B, and the edge b directed from B to A (see Figure 5). The free category on it has two objects A, B, and the sets of morphisms are

$$\mathrm{Mor}(A, A) = \{(ab)^n \mid n \in \mathbb{N}_{\geq 0}\},$$
$$\mathrm{Mor}(A, B) = \{(ab)^n a(ba)^m \mid m, n \in \mathbb{N}_{\geq 0}\} = \{(ab)^n a \mid n \in \mathbb{N}_{\geq 0}\},$$

and analogously for $\mathrm{Mor}(B, B)$ and $\mathrm{Mor}(B, A)$. △

Returning to the general situation, let R denote the smallest equivalence relation compatible with morphism composition and that identifies every two morphisms that come from positive minimal paths with the same beginning and target. We might then build the quotient category $\mathcal{G}^+ := \mathtt{Free}(G(\mathcal{A}))/R$, called the category of positive paths.

It is clear that $\mathrm{Ob}(\mathcal{G}^+(\mathcal{A})) = \mathcal{T}(\mathcal{A})$. In general, the equivalence relation is such that any two chambers C_1, C_2 determine an equivalence class of positive minimal paths starting at C_1 and ending at C_2; we will write $(C_1 \to C_2)$ for any morphism representing this class.

Example 1.9. In the previous example, the relation is empty. To see a case where it actually plays a role, let $\mathcal{A}_2$ be the arrangement of two lines considered in Example 1.4 and depicted in Figure 1 together with its arrangement graph. The vertex set of $G(\mathcal{A}_2)$ is $\{C_0, C_1, C_2, C_3\}$ (say, in counterclockwise order in Figure 1) and we may label the edges $e_{i,i\pm1}$, where the edge $e_{i,j}$ is directed from the vertex C_i to the vertex C_j (the indexing is taken modulo 4). The set of morphisms from C_i to C_j in the free category $\mathtt{Free}(G)$ is the set of directed paths in G starting at C_i

and ending in C_j. The positive minimal paths in G are either single-edge paths or paths of length 2 of the form

$$e_{i,i+1}e_{i+1,i+2} \quad \text{or} \quad e_{i,i-1}e_{i-1,i-2}.$$

For any fixed i, the two above paths share the same start and the same target. Let us represent a directed path in G with the corresponding word in the alphabet $\{e_{i,i\pm1}\}_{i=0,\ldots,3}$. The set of morphisms from C_i to C_j in $\mathcal{G}^+$ is obtained from the set of directed paths from C_i to C_j after identification of any two paths represented by words that can be transformed into one another by a sequence of substitutions of the form $e_{i,i+1}e_{i+1,i+2} \leftrightarrow e_{i,i-1}e_{i-1,i-2}$.

To describe the set of morphisms of $\mathcal{G}^+$ in this case, consider any directed path in the arrangement graph starting, say, at C_0, and let us parse it following the orientation of the edges. The first two letters of ω either describe a loop (in which case the third letter represents an edge starting at C_0) or a positive minimal path. In the latter case, two situations may occur. If the second and third letter define a directed loop in the graph, then we can apply two "substitutions" as in Figure 2 to see that this three-edges path is equivalent to one which makes a loop based at C_0, and thus ω is equivalent to a path whose third vertex is again C_0.

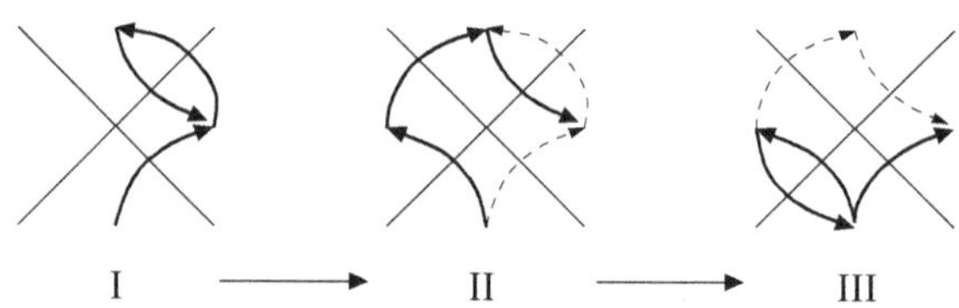

FIGURE 2

If the second and third letter define a positive minimal path, then we are already in the situation of Figure 2.II and one substitution suffices to show that ω is equivalent to a path whose third vertex is C_0. So in any case we know that ω is equivalent to a loop followed by a directed path ω', still starting at C_0 but two edges shorter than ω. By induction we see that $\mathrm{Mor}_{\mathcal{G}^+}(C_0, C_0)$ is the free commutative monoid with generators $e_{0,1}e_{1,0}$ and $e_{0,3}e_{3,0}$, and

$$\mathrm{Mor}_{\mathcal{G}^+}(C_0, C_j) = \{\alpha(C_0 \to C_j) \mid \alpha \in \mathrm{Mor}_{\mathcal{G}^+}(C_0, C_0)\}. \qquad \triangle$$

Let us again return to the general construction. We can now state the definition of the arrangement groupoid.

Definition 1.10. The *arrangement groupoid* $\mathcal{G}(\mathcal{A})$ is obtained from the category of positive paths $\mathcal{G}^+(\mathcal{A})$ by groupoid completion, i.e., adding formal inverses to every morphism.

The arrangement $\mathcal{A}$ being often understood, we will sometimes just write $\mathcal{G}$ for $\mathcal{G}(\mathcal{A})$.

Example 1.11 (Example 1.8 continued). We already described the objects and morphisms of $\mathcal{G}_1^+ := \mathcal{G}^+(\mathcal{A}_1)$ for the arrangement of one point in the real line. The associated arrangement groupoid is obtained by formally adding an element $a^{-1} \in \mathrm{Mor}(B, A)$ such that $a^{-1}a = aa^{-1} = id$ in $\mathrm{Mor}(A, A)$, and an analogous element $b^{-1} \in \mathrm{Mor}(A, B)$. Thus we have $abb^{-1}a^{-1} = id$ in $\mathrm{Mor}(A, A)$, which justifies the notation $(ab)^{-1} := b^{-1}a^{-1}$. Then, in the arrangement groupoid $\mathcal{G}_1 := \mathcal{G}(\mathcal{A}_1)$, we have

$$\mathrm{Mor}_{\mathcal{G}_1}(A, A) = \{(ab)^n \mid n \in \mathbb{Z}\} \text{ (i.e., the group } \mathbb{Z}),$$
$$\mathrm{Mor}_{\mathcal{G}_1}(A, B) = \{(ab)^n x \mid n \in \mathbb{Z},\, x = a \text{ or } b^{-1}\} \qquad \triangle$$

Example 1.12 (Example 1.9 continued). Let us now consider again the arrangemet $\mathcal{A}_2$. As in the previous example, adding formal inverses to every $e_{i,\pm 1}$ in the positive category $\mathcal{G}^+(\mathcal{A}_2)$ that we described in Example 1.9, we can for instance see that, in the arrangement groupoid $\mathcal{G}_2 := \mathcal{G}(\mathcal{A}_2)$, $\mathrm{Mor}_{\mathcal{G}_2}(C_0, C_0)$ is the free abelian group on two generators $e_{o,1}e_{1,0}$ and $e_{o,3}e_{3,0}$. △

Remark 1.13. The arrangement groupoid was first defined by Deligne [29, (1.25)]. See also the work of Paris [61] for more on the construction. As a word of caution it has to be pointed out that in [29] this object is defined under the assumption that the arrangement is simplicial, thereby obtaining 'by default' some properties that are not granted in the general case, such as the faithfulness of the natural functor $\mathcal{G}^+ \to \mathcal{G}$ that turns out to be a crucial property in view of asphericity of the complement (see [29, 69, 60] and our Section 5.1). Note that our two examples indeed enjoy this property.

Coverings of the arrangement groupoid. From the definition of $\mathcal{G}(\mathcal{A})$ and from Remark 1.3 we see that indeed $\pi\mathcal{G}(\mathcal{A}) = \pi_1(\mathcal{M}(\mathcal{A}))$. So the same subgroups characterize the coverings of $\mathcal{M}(\mathcal{A})$ and the coverings of $\mathcal{G}(\mathcal{A})$.

If we apply Theorem 1.7 to the arrangement groupoid, we obtain coverings $\rho : \mathcal{G}_\rho \to \mathcal{G}$. The objects of $\mathcal{G}_\rho$ represent (right cosets of) paths on the arrangement graph. Therefore we will freely switch between the interpretation of them as objects (written with latin letters) or as morphisms in $\mathcal{G}$ (written with Greek letters). Moreover, universal covering will be denoted by a hat on the corresponding object. So $\hat{\rho}$ for the universal covering morphism and $\widehat{\mathcal{G}}$ for $\mathcal{G}_{\hat{\rho}}$.

Example 1.14. Consider the arrangement groupoid $\mathcal{G}_1$ of Examples 1.8 and 1.11. Choose A as the base point and let $\widehat{\mathcal{G}}_1$ be the groupoid given by

$$\mathrm{Ob}(\widehat{\mathcal{G}}_1) := \{v_k\}_{k\in\mathbb{Z}}, \quad \mathrm{Mor}_{\widehat{\mathcal{G}}_1}(v_i, v_j) := \{\mu_{i,j}\},$$

$$\hat{\rho}: \quad v_i \mapsto \begin{cases} A & i \text{ even} \\ B & i \text{ odd} \end{cases}$$
$$\mu_{i,j} \mapsto \begin{cases} (ab)^{q(i,j)} a^{p(i,j)} & i \text{ even, } i < j \\ (ab)^{q(i,j)+1} b^{-p(i,j)} & i \text{ even, } i > j \\ (ba)^{q(i,j)} b^{p(i,j)} & i \text{ odd, } i < j \\ (ba)^{q(i,j)+1} a^{-p(i,j)} & i \text{ odd, } i > j \end{cases}$$

where we used the following notation:

$$p(i,j) = \begin{cases} 0 & i-j \text{ even} \\ 1 & i-j \text{ odd} \end{cases} \qquad q(i,j) := \frac{j-i-p(i,j)}{2}.$$

This groupoid can also be obtained from the free category over the directed infinite path by identifying every two morphisms with the same beginning and endpoint.

It is easy now to see that this is a covering of the groupoid $\mathcal{G}_1$. Indeed, $\mu_{i,j}$ ranges clearly over the morphisms exiting from v_i, and conversely every such morphism can be written as $\mu_{i,j}$ for an adequate choice of i and j. △

Example 1.15. The universal covering groupoid $\widehat{\mathcal{G}}_2 := \widehat{\mathcal{G}}(\mathcal{A}_2)$ of the arrangement groupoid of Example 1.9 and 1.12 is

$$\mathrm{Ob}(\widehat{\mathcal{G}}_2) = \{v_{i,j} \mid i,j \in \mathbb{Z}\}, \quad \mathrm{Mor}_{\widehat{\mathcal{G}}_2}(v_{i,j}, v_{k,l}) = \{\mu_{i,j,k,l}\} \text{ (a singleton)}.$$

The covering map is defined on objects as

$$\hat{\rho}(v_{i,j}) := C_{r(i,j)}$$

where $0 \le r(i,j) \le 3$ and $r(i,j) \equiv p(i) - p(j) + 2p(i)p(j) \pmod 4$ with $p(i) := 0$ if i is even, $p(i) := 1$ if i is odd, for every $i,j \in \mathbb{Z}$. To define $\hat{\rho}$ on morphisms it is useful to think of every $v_{i,j}$ as corresponding to the integer point (i,j) in the real plane. Let us also think of every edge of the integer grid in $\mathbb{R}^2$ as being directed following the increase of the coordinates value. So every node (i,j) is the source of two edges: one oriented in the x-direction, that we will label by $e_{r(i,j),r(i+1,j)}$, and one in the y-direction, labeled $e_{r(i,j),r(i,j+1)}$ (remember from Example 1.9 that also the labeling $e_{i,j}$ is taken modulo 4).

Given any morphism $\mu_{i,j,k,l}$, consider any path of minimal length in the integer grid joining (i,j) to (k,l). Then $\hat{\rho}(\mu_{i,j,k,l})$ is the morphism of $\mathcal{G}_2$ represented by the word read along this path, where labels of edges that are traversed against their orientation should be taken with a negative exponent.

Stated otherwise, this groupoid can be constructed from the free category on the graph obtained by directing "north" and "south" the edges of the 2-dimensional integer grid, by identifying every two morphisms with the same beginning and endpoint.

To see that this is indeed the universal covering groupoid of $\mathcal{G}_2$, we have to look, for every pair $i,j \in \mathbb{Z}$, at the stars of $v_{i,j}$ and $C_{r(i,j)}$. We have clearly

$$\mathrm{St}_{\widehat{\mathcal{G}}_2}(v_{i,j}) = \{\mu_{i,j,k,l} \mid k,l \in \mathbb{Z}\}, \quad \mathrm{St}_{\mathcal{G}_2}(C_{r(i,j)}) = \bigcup_{h=1}^{3} \mathrm{Mor}(C_{r(i,j)}, C_{r(i,j)+h}).$$

It is now clear that $\hat{\rho}$ induces a bijection between these sets. For instance, any $\mu_{i,j,k,l}$ is mapped to an element of $\mathrm{Mor}(C_{r(i,j)}, C_{r(i,j)+h})$ for $h = r(k,l) - r(i,j)$. Conversely, by comparing Example 1.12 we see that every morphism of $\mathcal{G}_2$ starting at $C_{r(i,j)}$ can be 'unwrapped' as a path on the integer grid starting at (i,j). The given morphism is then the image of $\mu_{i,j,k,l}$, where (k,l) is the endpoint of the unwrapped path. △

2. Homotopy colimits for combinatorial applications

The theory of homotopy colimits of diagrams of spaces comes from homological algebra and category theory. It was developed by Quillen, Bousfield, Kan and others (see for example [63], [12]), and has now achieved a remarkable extension and depth. Some orientation (and a good introduction to the general subject) may be found in the paper by Hollender and Vogt [45] and in the book by Goerss and Jardine [42] and its bibliography.

The work of Rainer M. Vogt [78] leaves the complete generality of diagrams over categories and focuses on what can be said if one makes more and more assumptions on the target category (never going beyond the properties satisfied by the category of topological spaces) and on the index category, requiring it to be *small* and, as a further restriction, *directed* (for example, Vogt derives the explicit form of Definition 2.2).

In our work we will take the latter and more combinatorial point of view, which was adopted by Welker, Ziegler and Živaljević in [81], where a useful toolkit for applications of homotopy colimits in discrete mathematics is developed. We will recall the main definitions and some basic results; for a more complete account of the theory we refer to [81] and the recent textbook by Kozlov [47, Chapter 15], which provides a readable and self-contained introduction to the subject.

Definition 2.1. A *diagram of spaces* is a covariant functor $\mathscr{D} : \mathcal{I} \to \texttt{Top}$ from a small index category to the category of topological spaces and continuous maps.

In our setting, $\mathcal{I}$ will always be given by some poset $\mathcal{P}$. Indeed, a poset can be thought of as a small category with at most one arrow between any two objects, where an arrow from $p \in \mathcal{P}$ to $q \in \mathcal{P}$ actually exists if and only if $p \geq q$ in $\mathcal{P}$. From now on we shall only consider diagrams over posets. In order to simplify notation, it is common to write $\mathscr{D}_p$ for the space $\mathscr{D}(p)$, and $f_{p,q}$ for the map $\mathscr{D}(p > q)$, if the diagram is understood.

A *morphism* from a diagram $\mathscr{D}$ over the poset $\mathcal{P}$ to a diagram $\mathscr{E}$ over $\mathcal{Q}$ is a pair $(\mu, (\alpha_p)_{p\in\mathcal{P}})$, where $\mu : \mathcal{P} \to \mathcal{Q}$ is a morphism of posets, and $(\alpha_p)_{p\in\mathcal{P}}$ is a family of continuous maps $\alpha_p : \mathscr{D}(p) \to \mathscr{E}(\mu(p))$ (indexed by elements of $\mathcal{P}$) that commute with the diagram maps.

Definition 2.2 (Compare (5.10) of [78]). Given a diagram of spaces $\mathscr{D} : \mathcal{P} \to \texttt{Top}$, the *homotopy colimit* of $\mathscr{D}$ is defined by

$$\mathrm{hocolim}\mathscr{D} := \coprod_{p\in\mathcal{P}} \Delta(\mathcal{P}_{\leq p}) \times D_p \Big/ \sim$$

where the relation $\sim$ is given, for $p > q$, by identifying along the maps

$$\Delta(\mathcal{P}_{\leq q}) \times D_p \hookrightarrow \Delta(\mathcal{P}_{\leq p}) \times D_p, \quad \Delta(\mathcal{P}_{\leq q}) \times D_p \overset{(id\times f_{p,q})}{\longrightarrow} \Delta(\mathcal{P}_{\leq q}) \times D_q.$$

Example 2.3. Consider the poset $\mathcal{P}$ with three elements a, b, c ordered by $a > b$, $a > c$ and b, c incomparable, as in Figure 3.

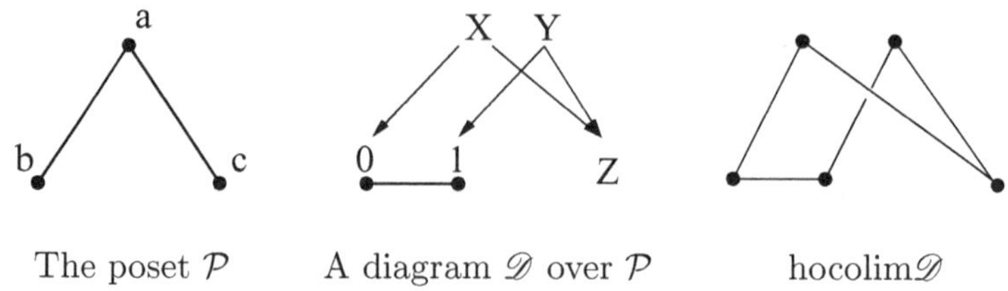

FIGURE 3

Consider now the diagram $\mathscr{D}$ with $D_a := \{X, Y\}$, two points, $D_b = [0,1]$, the unit interval, and $D_c = \{Z\}$, a single point. The map $f_{a,b}$ sends X and Y to the extremes 0 and 1 in $[0,1]$, while $f_{a,c}$ maps everything to Z.

Let us consider the three terms of the disjoint union appearing in the definition of hocolim$\mathscr{D}$. The order complexes $\Delta(\mathcal{P}_{\leq b})$ and $\Delta(\mathcal{P}_{\leq c})$ each consist of a single point. Thus, $\Delta(\mathcal{P}_{\leq b}) \times D_b = b \times [0,1]$ and $\Delta(\mathcal{P}_{\leq c}) \times D_c$ is still a single point. On the other hand, $\Delta(\mathcal{P}_{\leq a})$ consists of the segments $b - a$ and $a - c$, joined at a, so that the corresponding term in the distinct union consists of two disjoint segments $b \times X - a \times X - c \times X$ and $b \times Y - a \times Y - c \times Y$. The equivalence relation identifies $b \times X$ with $b \times 0 \subset b \times [0,1]$, $b \times Y$ with $b \times 1 \subset b \times [0,1]$, and both the points $c \times X$, $c \times Y$ with Z. The result is a subdivision of the circle S^1 with five vertices, as can be seen in the Figure 3. △

As mentioned before, homotopy colimits were designed to enjoy many naturality properties. The following functorial property is particularly useful.

Lemma 2.4 (The Homotopy Lemma, see Proposition 3.7 of [81]). *Consider a morphism $\phi := (id, (\alpha_p)_{p \in \mathcal{P}}) : \mathscr{D} \to \mathscr{E}$ between two diagrams over the same poset $\mathcal{P}$. If every α_p is a (weak) homotopy equivalence, then the induced map* hocolim$\mathscr{D} \to$ hocolim$\mathscr{E}$ *is a (weak) homotopy equivalence.*

That this is not granted with usual colimits is precisely the reason why one has to introduce homotopy colimits.

Example 2.5. Consider the diagram $\mathscr{D}$ of the previous example. By taking its colimit, the points 0, 1 and Z are identified, so that colim$\mathscr{D}$ is obtained by identifying the endpoints of the unit interval, and thus is homotopy equivalent to a circle. Now

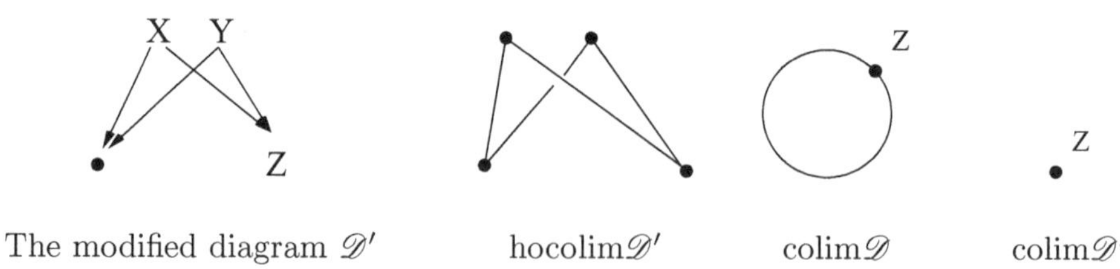

FIGURE 4

consider the diagram $\mathscr{D}'$ that is defined the same way as $\mathscr{D}$ except for the fact that D'_b is now a single point as well, so that the contraction $D_b \to D'_b$ is clearly a homotopy equivalence. Now it easy to see that colim$\mathscr{D}'$ is a single point. On the other hand hocolim$\mathscr{D}'$ is, as the reader will verify, still a subdivision of S^1. △

However, the colimit and the homotopy colimit of a diagram of space do agree in some cases, as stated in the following lemma.

Lemma 2.6 (The Projection Lemma, see Lemma 4.5 of [81]**).** *Let $\mathscr{D}$ denote a diagram of spaces on a poset $\mathcal{P}$. If all diagram maps $\mathscr{D}(p > q)$ are closed cofibrations, then the natural map* hocolim$\mathscr{D} \to$ colim$\mathscr{D}$ *induces a homotopy equivalence.*

A *diagram of posets* is a diagram $\mathscr{D} : \mathcal{P} \to \texttt{Pos}$ from a small index category (that in our work will be given as above by a poset $\mathcal{P}$) to the category $\texttt{Pos}$ of partially ordered sets and order-preserving maps.

In this situation, we also can define the *poset limit* Plim$\mathscr{D}$ of the diagram of posets $\mathscr{D}$. This is a poset whose underlying set of elements is

$$\mathrm{Plim}\mathscr{D} := \bigcup_{p \in \mathcal{P}} \{p\} \times \mathscr{D}(p)$$

and whose order relation is defined by

$$(p_1, q_1) \geq (p_2, q_2) :\Leftrightarrow \begin{cases} p_1 \geq p_2 \quad \text{and} \\ f_{p_1,p_2}(q_1) \geq q_2 \text{ in } \mathscr{D}(p_2) \end{cases}$$

where f_{p_1,p_2} as usual stands for the diagram map associated to the order relation $p_1 \geq p_2$.

To such a diagram of posets one can associate a diagram of spaces $\Delta(\mathscr{D})$: $\mathcal{P} \to \texttt{Top}$ with spaces $\Delta(\mathscr{D})_p$ defined to be the order complex $\Delta(\mathscr{D}(p))$, and maps $\Delta(f_{p,q}) : \Delta(\mathscr{D}(p)) \to \Delta(\mathscr{D}(q))$ induced by $f_{p,q}$ for all $p \geq q$.

Lemma 2.7 (The Simplicial Model Lemma). *Let $\mathscr{D}$ be a diagram of posets. Then the homotopy colimit of $\Delta(\mathscr{D})$ is homotopy equivalent to the order complex of the poset limit of $\mathscr{D}$:*

$$\mathrm{hocolim}\Delta(\mathscr{D}) \simeq \Delta(\mathrm{Plim}\mathscr{D}).$$

Proof. See [1], note after Corollary 2.11. □

3. Salvetti-type diagram models

We now introduce the first type of diagram models. The result of this section is summarized in Theorem 3.7, where it is proved that the homotopy colimit of the diagrams that we are going to introduce indeed model every covering space of the complement of a complexified arrangement. The fact that we will deal with diagrams of posets will allow us to write the covering spaces as order complexes of posets.

Definition 3.1. Given a cover of the arrangement groupoid $\rho : \mathcal{G}_\rho \to \mathcal{G}(\mathcal{A})$, we define a diagram of posets $\mathscr{D}_\rho : \mathcal{F} \longrightarrow \mathsf{Pos}$ with

$$\mathscr{D}_\rho(F) := \{v \in \mathrm{Ob}\,(\mathcal{G}_\rho) \mid \rho(v) < F\},$$

endowed with the trivial order relation defined by setting $v_1 \leq v_2$ if and only if $v_1 = v_2$, and maps being inclusions

$$f^\rho_{F_1,F_2} := \mathscr{D}_\rho(F_1 > F_2) : \begin{array}{ccc} \mathscr{D}_\rho(F_1) & \longrightarrow & \mathscr{D}_\rho(F_2) \\ v & \longmapsto & \mathrm{end}\,(\rho(v) \to \rho(v)_{F_2})^{\langle v \rangle}. \end{array}$$

The Simplicial Model Lemma 2.7 allows us to write $\mathrm{hocolim}\Delta(\mathscr{D}_\rho)$ as $\Delta(\mathrm{Plim}\mathscr{D}_\rho)$. Because this will be the main object of our attention for this section, let us from now on set $W_\rho := \Delta(\mathrm{Plim}\mathscr{D}_\rho)$. The simplicial complex W_ρ has vertex set

$$\{(F, v) \in \mathcal{F} \times \mathrm{Ob}(\mathcal{G}_\rho) \mid \rho(v) < F\}$$

and the simplexes are chains with respect to the partial order

$$(F_1, v_1) \geq (F_2, v_2) \Leftrightarrow \left\{ \begin{array}{ll} 1) & F_1 \geq F_2, \\ 2) & v_2 = \mathrm{end}(\rho(v_1) \to \rho(v_1)_{F_2})^{\langle v_1 \rangle}. \end{array} \right.$$

Remark 3.2. A chain in $\mathrm{Plim}\mathscr{D}_\rho$ is given by a chain ϕ in $\mathcal{F}$ and an object v of the groupoid such that $\rho(v) \leq \max\phi$ in $\mathcal{F}$. Everything else can be reconstructed as above. Since all chains are of this form, we can encode each simplex of W_ρ by $\Delta(\phi, v)$.

Remark 3.3. If ρ is the identical cover, then $W := W_{id}$ is exactly the simplicial version of the Salvetti complex (the proof is carried out in [31, Proposition 4.1.2]).

Remark 3.4. For any covering $\rho : \mathcal{G}_\rho \to \mathcal{G}$, W_ρ is the barycentric subdivision of a CW-complex ${W_\rho}^{CW}$ having a d-cell $[F, v]$ for every $v \in \mathrm{Ob}(\mathcal{G}_\rho)$ and every $F \in \mathcal{F}$, $F \geq \rho(v)$, with $\mathrm{codim}(F) = d$. In fact,

$$[F, v] := \bigcup_{\max(\phi)=F} \Delta(\phi, v)$$

is the barycentric subdivision of a closed $\mathrm{codim}(F)$-ball.

The vertices of this complex are of the form $[\rho(v), v]$, and thus correspond bijectively to elements of $\mathrm{Ob}(\mathcal{G}_\rho)$. We will therefore identify vertices of ${W_\rho}^{CW}$ with objects of $\mathcal{G}_\rho$.

The cell $[F, v]$ is attached to those vertices v' that can be written as $v' = \mathrm{end}(\rho(v) \to \rho(v)_{F'})^{\langle v \rangle}$ with $F' < F$.

Note also that if $\mathcal{A}$ is central and P is the maximal element of $\mathcal{F}$, then any $[P, v]$ is the barycentric subdivision of the zonotope of the arrangement.

3.1. Covering maps

The functor $\rho : \mathcal{G}_\rho \to \mathcal{G}$ naturally induces a morphism of diagrams $\lambda : \mathscr{D}_\rho \longrightarrow \mathscr{D}$. In turn, by functoriality the morphism λ induces a map between the homotopy colimits, and thus a map of simplicial complexes

$$\Lambda_\rho : W_\rho \longrightarrow W.$$

In fact, Λ_ρ is a simplicial extension of λ: the simplex $\Delta(\phi, v)$ of W_ρ is mapped to $\Delta(\phi, \rho(v))$ in W. The previous considerations can be followed step by step to see that a morphism $\eta : \mathcal{G}_{\rho_1} \to \mathcal{G}_{\rho_2}$ between two covers $\rho_i : \mathcal{G}_{\rho_i} \to \mathcal{G}$ induces a map $\Lambda_\eta : W_{\rho_1} \to W_{\rho_2}$.

We now prove that groupoid coverings indeed induce topological coverings. In the remainder of this section we shall slightly abuse notation and write W_ρ for the *geometric realization* of the simplicial complex $\Delta(\mathrm{Plim}\mathscr{D}_\rho)$ (for the definition of the geometric realization of simplicial complexes see e.g. Spanier [73, Chapter 3]). This shall not cause confusion because the topological properties of a simplicial complex are indeed defined via its geometrical realization.

Proposition 3.5. *Let $\mathcal{A}$ be a locally finite real arrangement. For every covering of groupoids $\rho : \mathcal{G}_\rho \to \mathcal{G}(\mathcal{A})$, the induced map $\Lambda_\rho : W_\rho \to W$ is a topological cover of W.*

Proof. First, one sees that the base space is connected and locally arcwise connected because W is finite dimensional and locally finite.

Now take $P \in W$ and Let X be an open neighborhood of P that does not contain any vertex of W (except P if P happens to be a vertex). Let σ be the smallest dimensional simplex of W containing P, and let U be the star of σ. We have to show that each component of the preimage ${\Lambda_\rho}^{-1}(X \cap W)$ is mapped homeomorphically to $X \cap W$. For this, it is enough to show that ${\Lambda_\rho}^{-1}(U)$ is a disjoint union of copies of U, each of which is mapped identically to U by Λ_ρ.

In view of Remark 3.2, $\sigma = \Delta(\tilde{\phi}, \widetilde{C})$ for a chain $\tilde{\phi} \subset \mathcal{F}$ and a chamber $\widetilde{C} < \max(\tilde{\phi})$. Defining $\tilde{F} := \max(\tilde{\phi})$, we can write U and its preimage as

$$U = \bigcup_{\substack{\phi \supseteq \tilde{\phi} \\ C \in R(\phi)}} \Delta(\phi, C), \qquad {\Lambda_\rho}^{-1}(U) = \bigcup_{\substack{\phi \supseteq \tilde{\phi} \\ \rho(v) \in R(\phi)}} \Delta(\phi, v),$$

where $R(\phi)$ is the set of all chambers $C < \max(\phi)$ such that $C_{\tilde{F}} = \widetilde{C}$.

For every vertex $v' \in \rho^{-1}(\widetilde{C})$ we want to distinguish the subcomplex of U spanned by all vertices that can be attained from v' by the lift of a positive minimal path. To every such v' we thus associate the subcomplex

$$W_{v'} := \bigcup_{\substack{\phi \supseteq \tilde{\phi} \\ C \in R(\phi)}} \Delta(\phi, v(C, v'))$$

where, for $w \in \mathrm{Ob}(\mathcal{G}_\rho)$ and $C \in \mathrm{Ob}(\mathcal{G})$, we denote by $v(C, w)$ the object of $\mathcal{G}_\rho$ corresponding to the morphism $w(C \to \rho(w))^{-1}$ of $\mathcal{G}$, i.e., "the object mapping to C from which w can be reached by a positive minimal path".

The proof now consists in the following three facts that amount to easy calculations with the morphisms of the involved groupoids. We list them and refer to [31, Proposition 4.3.2] for the complete argument.[1]

Fact 1. *For any $v_1 \neq v_2 \in \rho^{-1}(\widetilde{C})$, $W_{v_1} \cap W_{v_2} = \emptyset$.*

Fact 2.

$$\coprod_{\tilde{v} \in \rho^{-1}(\widetilde{C})} W_{\tilde{v}} = \Lambda_\rho{}^{-1}(U).$$

Fact 3. *Fix $\tilde{v} \in \rho^{-1}(\widetilde{C})$. Then $\Lambda_\rho : W_{\tilde{v}} \to U$ is a homeomorphism.* (Note that it is enough to prove bijectivity on the vertex set). □

3.2. The fundamental group

We want to compare the fundamental group of $W_\rho \simeq \mathrm{hocolim}\Delta(\mathscr{D}_\rho)$ with the object group of the corresponding groupoid. We begin by studying the structure of the low dimensional skeleta of the CW-complex $W_\rho{}^{CW}$ of which W_ρ is the barycentric subdivision (see Remark 3.4).

1-skeleton. Between two vertices $[\rho(v_1), v_1]$ and $[\rho(v_2), v_2]$ there is an edge if and only if the two corresponding chambers $\rho(v_i)$, $\rho(v_j)$ are separated by only one face F. Thus, there must be representatives of v_i and v_j that differ only by an edge that "crosses" F: either v_2 represents the concatenation of v_1 with the lift of $(\rho(v_1) \to \rho(v_2))$ at v_1, or v_1 represents the concatenation of a representative of v_2 with the lift $(\rho(v_2) \to \rho(v_1))^{\langle v_2 \rangle}$. In the first case, we have $(F, v_1) > (\rho(v_1), v_1)$ and $(F, v_1) > (\rho(v_2), v_2)$ in $\mathrm{Plim}\mathscr{D}_\rho$, and thus we take the element (F, v_1) of $\mathrm{Plim}\mathscr{D}_\rho$ to represent an edge $[F, v_1]$ directed from $[\rho(v_1), v_1]$ to $[\rho(v_2), v_2]$ in the 1-skeleton of $W_\rho{}^{CW}$. The second case is treated analogously and produces an edge $[F, v_2]$ 'directed' away from $[\rho(v_2), v_2]$.

2-skeleton. Fix $v \in \mathrm{Ob}(\mathcal{G}_\rho)$ and $F \in \mathcal{F}$ with $\mathrm{codim}(F) = 2$ (w.l.o.g. $\rho(v) < F$). The vertices in the boundary $\partial[F, v]$ are those of the form $\left[C, \mathrm{end}(\rho(v) \to C)^{\langle v \rangle}\right]$, where C is any chamber adjacent to F. Let us label the vertices in circular order as $[\rho(v_i), v_i]$, $i = 0, \ldots, 2k-1$, and assume w.l.o.g. $v = v_0$, $C_0 := \rho(v_0)$.

Now consider $v_i \neq v_j$, and suppose that in $\partial[F, v]$ an edge between $[\rho(v_i), v_i]$ and $[\rho(v_j), v_j]$ exists. This means that $\rho(v_i)$ is adjacent to $\rho(v_j)$, and that there is an F_1 with $F > F_1 > \rho(v_i)$ and $F > F_1 > \rho(v_j)$, such that this edge can be written as $[F_1, \tilde{v}]$. To determine whether $\tilde{v} = v_i$ or v_j (which gives the 'direction' of the edge as above), recall that the fact that $(F, v_0) > (F_1, \tilde{v})$ in $\mathrm{Plim}\mathscr{D}_\rho$ implies $\rho(\tilde{v}) = \rho(v_0)_{F_1} = (C_0)_{F_1}$ – so $\tilde{v} = v_i$ if $\rho(v_i) = (C_0)_{F_1}$, i.e., if $\rho(v_i)$ is on the same side of F_1 as C_0.

---

[1]Actually, the proof of [31] uses only the properties of *positive paths*. This makes the argument somewhat more involved, but makes sure that the arguments remain valid even restricting to the so-called "positive complexes" that will become important later.

Summarizing, in $\partial[F,v]$ we have then one edge for each codimension 1 face F_1 incident to F, and this edge is oriented away from the vertex that projects to the chamber on the same side as $\rho(v)$ w.r.t. F_1. We may also view $\partial[F,v]$ as a subgraph of G_ρ: it consists of the lift at v of the two positive minimal paths from $\rho(v)$ to its opposite chamber with respect to F. It follows that the cover $\mathcal{G}_\rho$ can be obtained from the graph G_ρ in the same way as $\mathcal{G}$ from G. In other words, $[F,v]$ provides a homotopy between two positive minimal paths in G_ρ.

It is not difficult now to compare the relations given by cells in ${W_\rho}^{CW}$ and the relation defining $\mathcal{G}_\rho$ and see that the following result holds. The proof is carried out in detail in [31, Section 4.4.3].

Theorem 3.6 (Proposition 4.4.3 of [31]). *Let $\mathcal{A}$ be a locally finite real arrangement. For every covering $\rho : \mathcal{G}_\rho \to \mathcal{G}(\mathcal{A})$ of the arrangement groupoid, we have an isomorphism*

$$\pi\mathcal{G}_\rho \simeq \pi_1(\text{hocolim}\Delta(\mathscr{D}_\rho)).$$

Higher skeleta and MH-complexes. We have seen that the 2-cells are attached to the 1-skeleton in the very same way as in the original construction of the Salvetti complex from the arrangement graph. The reader is invited to check that also every higher dimensional cell $[F,v]$ is attached so that its 1-dimensional skeleton consists of the lift at v of all positive minimal paths from $\rho(v)$ to its opposite chamber with respect to F.

This is exactly the structure of the *Metrical Hemisphere complexes* (or MH-complexes) studied by Salvetti in [69], where the influence of the graph structure on the homotopical properties of these complexes is carefully explained. In particular the classical work of Gabriel and Zisman [40] is applied to give a very deep insight into the link between the homotopy of the complexes and the category of paths on the graph. The treatment starts from the full generality, and proceeds adding more and more restrictions as the proofs require them. For the case in which the MH-complex models the universal cover of a central arrangement (and thus agrees with $W_{\hat{\rho}}^{CW}$), Salvetti recovers, and puts into this broader context, Deligne's theorem about asphericity of simplicial arrangements [29].

3.3. Classification of the covers

Theorem 3.7. *For any topological cover $r : X \to \text{Sal}(\mathcal{A})$ of the Salvetti complex of a locally finite real arrangement $\mathcal{A}$, there exists a cover of the arrangement groupoid $\rho : \mathcal{G}_\rho \to \mathcal{G}(\mathcal{A})$ such that the homotopy colimit of the associated diagram of spaces $\Delta(\mathscr{D}_\rho)$ is isomorphic to X as a covering space of* $\text{Sal}(\mathcal{A})$.

Proof. Let $\varphi : \pi\mathcal{G}(\mathcal{A}) \to \pi_1(\text{hocolim}\Delta(\mathscr{D}))$ be the isomorphism of Theorem 3.6. Since $\text{hocolim}\Delta(\mathscr{D}) \simeq \text{Sal}(\mathcal{A})$, we can consider the preimage $U := \varphi^{-1}(r_\star(\pi_1(X)))$ of the fundamental group of X in $\pi\mathcal{G}(\mathcal{A})$.

Theorem 1.7 gives a cover $\rho : \mathcal{G}_\rho \to \mathcal{G}(\mathcal{A})$ with $\rho_\star(\pi\mathcal{G}_\rho) = U$. Moreover, by Theorem 3.6 we have an isomorphism $\varphi_\rho : \pi\mathcal{G}_\rho \to \pi_1(\mathrm{hocolim}\Delta(\mathscr{D}_\rho))$. These isomorphisms come naturally from the inclusion ι of the graphs G_ρ as 1-skeletons of the CW-version of the homotopy colimits. Therefore the following diagrams commute

$$\begin{array}{ccc} G_\rho & \xrightarrow{\iota} & W_\rho{}^{CW} \\ \downarrow{\scriptstyle \rho} & & \downarrow{\scriptstyle \Lambda_\rho} \\ G & \xrightarrow{\iota} & W^{CW} \end{array} \qquad \begin{array}{ccc} \pi\mathcal{G}_\rho & \xrightarrow{\varphi_\rho} & \pi_1(W_\rho) \\ \downarrow{\scriptstyle \rho_\star} & & \downarrow{\scriptstyle (\Lambda_\rho)_\star} \\ \pi\mathcal{G} & \xrightarrow{\varphi} & \pi_1(W) \end{array}$$

and $r_\star(\pi_1(X)) = \varphi\rho_\star(\pi\mathcal{G}_\rho) = (\Lambda_\rho)_\star\varphi_\rho(\pi_1(W_\rho)) \cong (\Lambda_\rho)_\star(\pi_1(W_\rho))$. Hence, the cover $\Lambda_\rho : W_\rho \to \mathrm{Sal}(\mathcal{A})$ is isomorphic to $r : X \to \mathrm{Sal}(\mathcal{A})$. □

Corollary 3.8. *Any cover $\rho : X \to \mathrm{Sal}(\mathcal{A})$ of the Salvetti complex can be written as the order complex of a poset, namely $\mathrm{Plim}\mathscr{D}_\rho$. The poset $\mathrm{Plim}\mathscr{D}_{id}$ is naturally isomorphic to the poset $\mathcal{S}(\mathcal{A})$ of cells of the Salvetti complex.*

Proof. Apply the Simplicial Model Lemma 2.7. □

The following corollary generalizes [60, Theorem 3.7] (see also the definitions on [60, p. 164]) to affine arrangements.

Corollary 3.9. *Let $\hat\rho : \hat{\mathcal{G}} \to \mathcal{G}(\mathcal{A})$ denote the universal cover of $\mathcal{G}(\mathcal{A})$. Then $W_{\hat\rho} = \Delta(\mathrm{Plim}\mathscr{D}_{\hat\rho})$ is the universal cover of $\mathrm{Sal}(\mathcal{A})$.*

Proof. We prove universality. Take any cover $r : X \to \mathrm{Sal}(\mathcal{A})$; we have to show that there is a morphism of covers $m : W_{\hat\rho} \to X$. By the theorem, we know that there is a cover $\rho : \mathcal{G}_\rho \to \mathcal{G}(\mathcal{A})$ with $W_\rho \cong X$ as a cover. Universality of $\hat{\mathcal{G}}$ implies the existence of a cover $\mu : \hat{\mathcal{G}} \to \mathcal{G}_\rho$, and this induces a morphism of diagrams $\lambda_\mu : \mathscr{D}_{\hat\rho} \to \mathscr{D}_\rho$. By functoriality, we have $\Lambda_\mu : W_{\hat\rho} \to W_\rho$, which gives the required morphism, as in the following diagram.

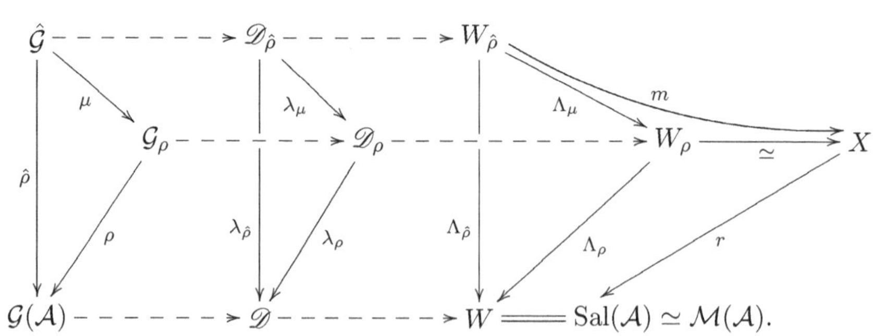

□

Example 3.10. Consider the arrangement given by one point $P \in \mathbb{R}$ (i.e., the arrangement $\mathcal{A}_1$ of Examples 1.8, 1.11, 1.14). The space $\mathbb{R}$ is divided by P in two chambers A and B. It is easy to write down the face poset $\mathcal{F}_1 := \mathcal{F}(\mathcal{A}_1)$ and the arrangement graph $G(\mathcal{A}_1)$ as in Figure 5.

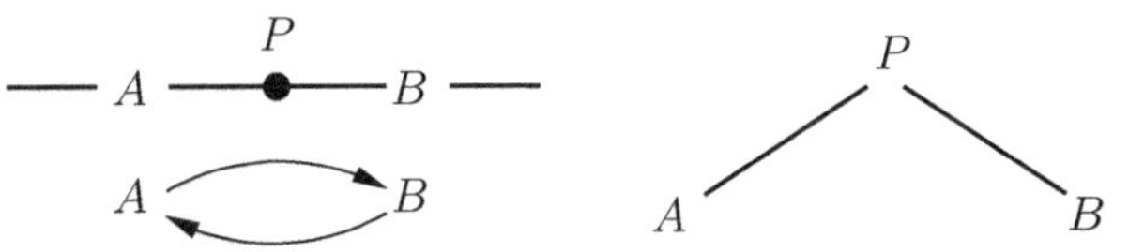

$\mathcal{A}_1$ and $G(\mathcal{A}_1)$ The poset of faces $\mathcal{F}(\mathcal{A}_1)$

FIGURE 5

The complexification of $\mathcal{A}_1$ is the arrangement given by a point in the complex plane. The complement $\mathcal{M}(\mathcal{A}_1)$ is then homotopy equivalent to S^1, hence its universal cover is $\mathbb{R}$. We will now see how the diagram models come to this conclusion.

First consider the diagram $\mathscr{D}$ on the poset $\mathcal{F}_1$. We have $\mathscr{D}(A) = \{A\}$, $\mathscr{D}(B) = \{B\}$, $\mathscr{D}(P) = \{A, B\}$. The diagram maps are in this case trivial, but let us explain where they come from:

$$\begin{aligned} \mathscr{D}(P > A):\quad & A \mapsto A_A = A \\ & B \mapsto B_A = A \\ \mathscr{D}(P > B):\quad & A \mapsto A_B = B \\ & B \mapsto B_B = B \end{aligned}$$

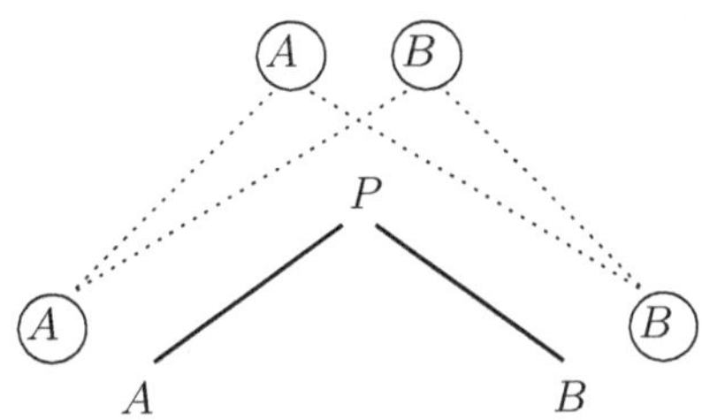

Note that the associated diagram of spaces $\Delta(\mathscr{D})$ is exactly the diagram of Example 3, where it is shown that $\mathrm{hocolim}\Delta(\mathscr{D}) \simeq S^1$.

We now have to look at the universal cover of $\mathcal{G}(\mathcal{A}_1)$. As we already pointed out, since $\mathrm{Sal}(\mathcal{A})$ has no 2-cells, the identification on $\texttt{Free}(G(\mathcal{A}_1))$ is empty. Indeed, $\mathcal{G}(\mathcal{A}_1)$ is described in Example 1.11, and in Example 1.14 we computed its universal covering groupoid $\widehat{\mathcal{G}}_1$. Here we will slightly change notation and write $A_i := v_{2i}$ and $B_i := v_{2i+1}$, so that $\hat{\rho} : \hat{\mathcal{G}}_1 \to \mathcal{G}(\mathcal{A}_1)$ is defined by $\hat{\rho}(A_i) = A$, $\hat{\rho}(B_i) = B$ for all i and $G_{\hat{\rho}}(\mathcal{A}_1)$ is an infinite path $\dots B_{-1} \to A_0 \to B_0 \to A_1 \to B_1 \dots$

Writing down the diagram $\mathscr{D}_{\hat{\rho}}$ we have to keep in mind that the poset associated to an element $F \in \mathcal{F}_1$ has as many incomparable elements as there are objects in $\hat{\mathcal{G}}$ that project to a chamber adjacent to F. So we have $\mathscr{D}_{\hat{\rho}}(A) = \{A_i | \, i \in \mathbb{Z}\}$, $\mathscr{D}_{\hat{\rho}}(B) = \{B_i | \, i \in \mathbb{Z}\}$, $\mathscr{D}_{\hat{\rho}}(P) = \mathscr{D}_{\hat{\rho}}(A) \cup \mathscr{D}_{\hat{\rho}}(B)$. For the maps one has to take care of how paths are lifted. Let us work out a special case and write down the diagram in the same fashion as above:

$$\begin{aligned}
&\mathscr{D}_{\hat\rho}(P > B):\\
&A_i \mapsto \operatorname{end}((A \to B)^{\langle A_i\rangle})\\
&\qquad = B_i\\
&B_i \mapsto \operatorname{end}((B \to B)^{\langle B_i\rangle})\\
&\qquad = B_i
\end{aligned}$$

$$\begin{aligned}
&\mathscr{D}_{\hat\rho}(P > A):\\
&A_i \mapsto \operatorname{end}((A \to A)^{\langle A_i\rangle})\\
&\qquad = A_i\\
&B_i \mapsto \operatorname{end}((B \to A)^{\langle B_i\rangle})\\
&\qquad = A_{i+1}
\end{aligned}$$

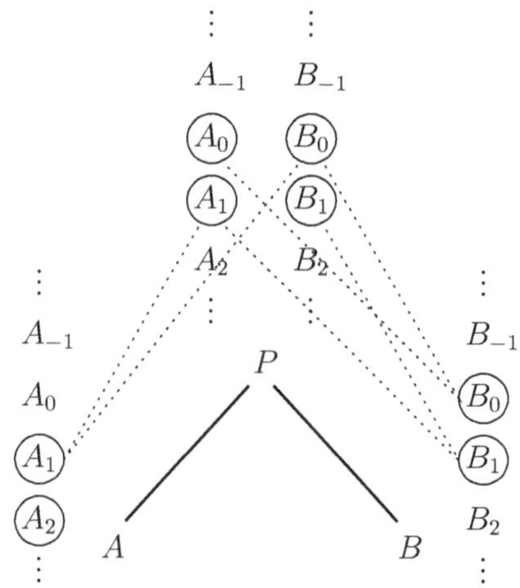

By Lemma 2.7, we now only have to write down the poset $\mathrm{Plim}\mathscr{D}_{\hat\rho}$. The order relation is such that for $F_i \in \mathcal{F}_1$ and $v_i \in \mathrm{Ob}(\hat{\mathcal{G}}_1)$ we have $(F_1, v_1) \geq (F_2, v_2)$ if and only if $F_1 \geq F_2$ and $v_2 = \operatorname{end}(\rho(v_1) \to \rho(v_1)_{F_2})^{\langle v_1\rangle}$.

In our case, this means that the dotted lines in the above picture are yet a piece of the Hasse diagram of $\mathrm{Plim}\mathscr{D}_{\hat\rho}$, which we can redraw in a more readable way as

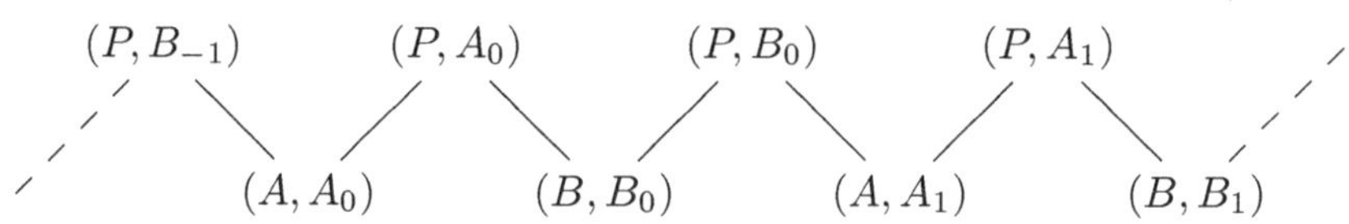

It is now clear that $\mathrm{hocolim}\mathscr{D}_\rho \cong \Delta(\mathrm{Plim}\mathscr{D}_{\hat\rho}) \simeq \mathbb{R}$, as required. △

3.4. Reflection arrangements and Charney-Davis models

Suppose that $\mathcal{A}$ is the set of reflecting hyperplanes for a finite real reflection group W; then W acts on $\mathcal{M}(\mathcal{A})$. The fundamental group $\pi_1(\mathcal{M}(\mathcal{A})/W)$ is the associated Artin group, as was proved by Brieskorn [16], and $\mathcal{M}(\mathcal{A})$ is aspherical (i.e., its homotopy groups are trivial in degree greater than or equal to 2, as proved in [29]). Among other things, this means that the Salvetti complex $\mathrm{Sal}(\mathcal{A})$, and its "quotiented" version presented in [68] are finite $K(\pi, 1)$s for the Artin groups of finite type.

Ruth Charney and Michael W. Davis showed in [23] that this situation generalizes to many infinite Coxeter groups. The argument builds on previous work [22] of the same authors, who introduced a *"modified Deligne complex"* Φ ([22, (1.5)]) in order to describe, via the theory of complexes of groups, the universal cover of a space M associated to any linear reflection group W [22, see Theorem 1.5.1, Corollary 3.2.2, Proposition 3.2.3]. The space M can be obtained as the quotient (by the action of W) of the complement of the "reflection hyperplanes" associated to the action of W on a certain space (see [22, Section 2]). Also, M is conjecturally a $K(\pi,1)$ space for the Artin group associated with W; in [22] this conjecture is proved true for two classes of reflection groups ("2-dimensional Artin groups" and "Artin groups of FC type"), by showing that Φ is contractible for these classes. In both cases this is achieved by proving that a suitable piecewise euclidean metric on Φ is CAT(0) [22, Section 4]. Then, in [23] the same authors describe a finite complex that is homotopy equivalent to M, thus providing finite $K(\pi,1)$ complexes for the Artin groups for which the above conjecture holds, and therefore making the situation for finite reflection groups part of a more general picture. The finite complex is called a *Salvetti complex* in [23, (1.2)].

One of the questions raised by Charney and Davis' work in our context is whether this similarity can lead to any generalization of their technique – i.e., whether techniques of CAT(0) geometry used on Φ can be applied to the diagram models in the general case.

The first candidates for such a program could be the affine reflection arrangements. In this context, and building on the results of [22, 23], Charney and Peifer proved the $K(\pi,1)$ conjecture for the affine braid arrangements by realizing [24, Section 3] the universal cover of these arrangements as the nerve of a certain covering by contractible subcomplexes of the "Bestvina Normal Form Complex" for the Artin group of (finite) type B_n (see [5, 25]) that we will encounter later in this survey (see Definition 5.12). This complex is contractible by either [29] (i.e., because it models the universal cover of the B_n arrangement) or [25] (i.e., using the Garside structure of the associated Artin group). At present, the $K(\pi,1)$ conjecture for affine real reflection arrangements is solved for the arrangements of type $\tilde{A}_n$ and $\tilde{C}_n$ (first by Okonek [54], type $\tilde{A}_n$ also by Charney and Peifer [24]) as well as for type $\tilde{B}_n$ (as proved by Callegaro, Moroni and Salvetti [21]).

4. Paris' topological models

We explain a construction of topological models for covers of complexified arrangements that is due to Luis Paris [59]. In this construction, the information on the fundamental group is encoded in so-called *oriented systems* rather than in groupoids, as is the case in our treatment. In later work [60], Paris himself gave a combinatorial stratification of his models for the universal cover of a linear arrangement.

We will start by giving the definition of Paris' oriented systems and outlining the parallels with the theory of groupoids. Then we will explain the construction of the topological models and conclude by giving a combinatorial stratification of them in the most general case, providing an explicit homotopy equivalence with the Salvetti-type diagram models that works for any cover of a complexified arrangement.

4.1. Oriented systems and their covers

Paris introduced the notion of *oriented system*, that we briefly recall.

Definition 4.1. An *oriented system* is a pair $(\Gamma, \sim)$ where Γ is an oriented graph and $\sim$ is an identification between paths of Γ such that:

(1) $\alpha \sim \beta$ implies $\mathrm{start}(\alpha) = \mathrm{start}(\beta)$ and $\mathrm{end}(\alpha) = \mathrm{end}(\beta)$,
(2) $\alpha\alpha^{-1} \sim \mathrm{start}(\alpha)$ for every α,
(3) $\alpha \sim \beta$ implies $\alpha^{-1} \sim \beta^{-1}$,
(4) $\alpha \sim \beta$ implies $\gamma_1\alpha\gamma_2 \sim \gamma_1\beta\gamma_2$ for any γ_1, γ_2 with $\mathrm{end}(\gamma_1) = \mathrm{start}(\alpha)$ and $\mathrm{start}(\gamma_2) = \mathrm{end}(\alpha)$,

where the 'inverse' of an oriented path α is obtained by going along α in the reverse direction.

Given a real arrangement of hyperplanes $\mathcal{A}$, the natural way to associate to it an oriented system $(\Gamma(\mathcal{A}), \sim)$ is of course to take $\Gamma(\mathcal{A}) = G(\mathcal{A})$, the arrangement graph, and identify two paths if they are both positive minimal and they start at the same point v and end at the same point w.

Forgetting orientation of edges, one can view Γ as a 1-complex, and therefore consider its fundamental group $\pi_1(\Gamma)$. The conditions that were required in the definition of the equivalence relation on paths ensure that $\sim$ induces an equivalence relation on $\pi_1(\Gamma)$; therefore we can consider the quotient $\pi(\Gamma, \sim) := \pi_1(\Gamma)/\sim$, which is called by Paris the *fundamental group* of the oriented system $(\Gamma, \sim)$.

For oriented systems, Paris [59] introduced the following concept of a cover:

Definition 4.2. Given two oriented systems $(\Theta, \sim_\Theta)$ and $(\Psi, \sim_\Psi)$, a morphism of oriented graphs $\rho : \Theta \to \Psi$ is said to be a *cover* of $(\Psi, \sim_\Psi)$ if:

(1) for every vertex v of Θ and every path α in Ψ with $\mathrm{start}(\alpha) = \rho(v)$ there is a unique path $\hat{\alpha}_v$ in Θ with $\rho(\hat{\alpha}_v) = \alpha$ and $\mathrm{start}(\hat{\alpha}_v) = v$. This path is called the *lift* of α at v.
(2) for any two paths α, β in Ψ with $\mathrm{start}(\alpha) = \mathrm{start}(\beta) = \rho(v)$, if $\alpha \sim \beta$ then $\hat{\alpha}_v \sim \hat{\beta}_v$.

At this point, the similarity with the theory of groupoids as sketched in the prologue is clear, and we summarize it.

Corollary 4.3. *Let $\mathcal{G}(\mathcal{A})$ denote the arrangement groupoid. We have immediately* $\mathrm{Ob}(\mathcal{G}(\mathcal{A})) = V(\Gamma(\mathcal{A}))$*, the set of vertices of* Γ*. Any path γ on Γ identifies an equivalence class of morphisms $[\gamma]$ of $\mathcal{G}$; moreover,*

$$\gamma_1 \sim \gamma_2 \text{ in } (\Gamma, \sim) \text{ if and only if } [\gamma_1] = [\gamma_2] \in \mathrm{Mor}_{\mathcal{G}}(\mathrm{start}(\gamma_1), \mathrm{end}(\gamma_2)).$$

In particular, observe that $\pi(\Gamma(\mathcal{A}), \sim) \cong \pi\mathcal{G}(\mathcal{A})$.

Furthermore, the requirement on the star $\mathrm{St}(v)$ *in the definition of covering of a groupoid translates naturally into condition* (1) *of Definition* 4.2*, where condition* (2) *takes account of the quotient to which we pass in defining the arrangement groupoid. Therefore, for any choice of a subgroup of $\pi(\Gamma(\mathcal{A}), \sim) \simeq \pi\mathcal{G}(\mathcal{A})$ we obtain a covering $(\Theta, \sim)$ (by Theorem* 4.4*) and a covering $\mathcal{G}_\rho$ (by Theorem* 1.7*) that may be compared as follows:*

$$V(\Theta) = \mathrm{Ob}(\mathcal{G}_\rho),$$
$$(v, w) \in V(\Theta) \Leftrightarrow \text{there is } \alpha \in \mathrm{Mor}_{\mathcal{G}_\rho}(v, w) \text{ lifting an edge of } G(\mathcal{A}).$$

The following result is now evident.

Theorem 4.4 ([59]). *Consider an oriented system* $(\Psi, \sim)$. *For each subgroup* H *of* $\pi(\Psi, \sim)$ *there exists a cover of oriented systems* $\rho : (\Theta, \sim) \to (\Psi, \sim)$ *with* $\rho_\star(\pi(\Theta, \sim)) = H$.

4.2. Topological covers associated to oriented systems

We fix from now on a cover of oriented systems $\rho : (\Theta, \sim) \to (\Gamma, \sim)$, the latter being the oriented system naturally associated to the arrangement graph, and describe the construction developed by Paris in [59] that associates a topological space to $(\Theta, \sim)$.

Remark 4.5. In this section the faces F have to be considered as convex, relatively open subsets of $\mathbb{R}^d$, and we will write $|F|$ for the affine span of any face F. Also, given a chamber C and a face F, we will denote by $C_{|F|}$ the unique chamber of the arrangement $\mathcal{A}_{|F|} := \operatorname{supp}(F)$ containing C. Thus, $C \subseteq C_{|F|}$ for every chamber C and every face F.

The key objects out of which the space is built are pieces of the form

$$M(C) := \bigcup_{F \in \mathcal{F}} (F + iC_{|F|}) \subset \mathbb{R}^d \oplus i\mathbb{R}^d = \mathbb{C}^d.$$

For every vertex v of Θ we define $N(v) := M(\rho(v))$, and the topological space associated to $(\Theta, \sim)$ is given by

$$N_\Theta := \coprod_{v \in V(\Theta)} N(v) \Big/ \approx \ .$$

The relation $\approx$ is defined pointwise by identifying two points $z \in N(v)$, $z' \in N(v')$ whenever (v, v') is an edge of Θ and they correspond to a unique point $z = z'$ in $M(\rho(v)) \cap M(\rho(v'))$ whose real part lies on the same side as $\rho(v')$ with respect to the hyperplane separating $\rho(v)$ from $\rho(v')$.

Theorem 4.6 ([59]). *Given any cover* $\rho : (\Theta, \sim) \to (\Gamma, \sim)$ *of the oriented system associated to a linear arrangement* $\mathcal{A}$, *the space* N_Θ *with the map induced by* ρ *is a topological cover of* $\mathcal{M}(\mathcal{A})$ *with characteristic group* $\pi(\Theta, \sim)$.

Paris defines for any two vertices v, w of Θ a topological space $Z(v, w)$ as the interior of $\overline{\bigcup \rho(u)}$, where the overbar denotes topological closure in $\mathbb{R}^d$ and the union is over all $u \in V(\Theta)$ that are reachable from both v and w by lifts of positive minimal paths of Γ. If (v, w) is an edge of Θ, then $Z(v, w)$ contains the real part of the set that is identified between $N(v)$ and $N(w)$. For general v and w, Paris shows the following fact that we state for later reference.

Lemma 4.7 (see Lemma 3.5 of [59]**).** *If* $N(v) \cap N(w) \neq \emptyset$ *in* N_Θ, *then* $Z(v, w)$ *is not empty. Indeed,*

$$N(v) \cap N(w) = M(\rho(v)) \cap M(\rho(w)) \cap (Z(v, w) + iV).$$

4.3. Combinatorial stratifications and diagram models

We now draw the link between Paris' construction and the diagram models. For this, we describe a stratification of N_Θ by contractible subsets with contractible intersections, and show that its nerve is isomorphic to $\mathrm{Plim}\mathscr{D}_\rho$. By the Nerve Lemma (see [7, 10.6(ii)] or [47, Theorem 15.21]) this proves directly $\mathrm{hocolim}\Delta(\mathscr{D}_\rho) \simeq N_\Theta$.

Remark 4.8. In comparing the two constructions one has to bear in mind that a vertex of an oriented system corresponds to an object of the corresponding covering groupoid, which is an equivalence class of paths in the arrangement graph. We will offer adequate explanations unless the context already makes clear which point of view is taken on the objects denoted by $v, w, \dots$

Let us define, for any $F \in \mathcal{F}$ and any $C \in \mathcal{T}$, a set

$$M_C^F := \bigcup_{F' \leq F} F' + iC_{|F'|}.$$

Accordingly, for any $v \in V(\Theta)$ let $N_v^F := M_{\rho(v)}^F \subset N(v)$.

It is then clear that

$$M(C) = \bigcup_{F \in \mathcal{F}} M_C^F, \qquad N(v) = \bigcup_{F \in \mathcal{F}} N_v^F.$$

Lemma 4.9. *For any $F \in \mathcal{F}(\mathcal{A})$ and any $C \in \mathcal{T}(\mathcal{A})$, the space M_C^F is contractible. This implies contractibility of N_v^F for all $v \in V(\Theta)$, $F \in \mathcal{F}$.*

Proof. Fix $F \in \mathcal{F}$, $C \in \mathcal{T}$ and $z \in F + iC_{|F|}$. We show that the segment connecting z to any $x \in M_C^F$ lies fully in M_C^F.

If $x \in F + iC_{|F|}$ the claim is trivial by convexity. Choose therefore $F' \leq F \in \mathcal{F}$ such that $x \in F' + iC_{|F'|}$, and consider the segment

$$\gamma(t) := tz + (1-t)x, \qquad 0 \leq t \leq 1.$$

It is clear that $\Re(\gamma(t)) \in F'$ for $0 \leq t < 1$, and by assumption $\Re(\gamma(1)) \in F$.

For the imaginary part, note that $\Im(x) \in C_{|F'|}$ and $\Im(z) \in C_{|F|} \subset C_{|F'|}$ (see Remark 4.5), so $\Im(\gamma(t))$ is a straight line between two points of the convex set $iC_{|F'|}$. □

Lemma 4.10. *Given $v_1, v_2 \in V(\Theta)$ and $F_1, F_2 \in \mathcal{F}$, $N_{v_1}^{F_1} \subset N_{v_2}^{F_2}$ if and only if $(F_1, v_1) < (F_2, v_2)$ in $\mathrm{Plim}\mathscr{D}_\rho$.*

Proof. For the "only if"-part, suppose that $N_{v_1}^{F_1} \subset N_{v_2}^{F_2}$. Then clearly $N_{v_1}^{F_1} \subset N(v_1) \cap N(v_2)$ and by Lemma 4.7 we have that $N_{v_1}^{F_1}$, which by definition is a copy of $M_{\rho(v_1)}^{F_1}$, is indeed contained in $Z(v_1, v_2)$. In particular, we thus have that $\rho(v_1) \subset Z(v_1, v_2)$, and by the very definition of $Z(v_1, v_2)$, in the oriented system v_1 can be reached from v_2 by the lift of a minimal path $(\rho(v_2) \to \rho(v_1))$. In the corresponding covering groupoid every object is an equivalence class of paths, and these objects are the vertices of G_ρ. Thus we may state the above conclusion as a composition of (equivalence classes of) paths as

$$\text{(i) } v_1 = v_2(\rho(v_2) \to \rho(v_1))^{\langle v_2 \rangle}$$

(recall that the v_i correspond to homotopy classes of paths, and as such may be concatenated with other paths). On the other hand, $N_{v_1}^{F_1} \subset N_{v_2}^{F_2}$ implies $M_{\rho(v_1)}^{F_1} \subset M_{\rho(v_2)}^{F_2}$, which is clearly equivalent to:

$$\text{(ii) } F_1 < F_2 \text{ and } \rho(v_1) = \rho(v_2)_{F_1}.$$

The sentences (i) and (ii) above constitute the definition of the order relation $(F_1, v_1) < (F_2, v_2)$ in $\mathrm{Plim}\mathscr{D}_\rho$.

For the "if"-part, suppose $(F_1, v_1) < (F_2, v_2)$ in $\mathrm{Plim}\mathscr{D}_\rho$, that means

$$\text{(a) } F_1 < F_2, \qquad \text{(b) } \rho(v_1) = \rho(v_2)_{F_1}, \qquad \text{(c) } v_1 = v_2(\rho(v_2) \to \rho(v_1))^{\langle v_2 \rangle}.$$

As above, (c) means that $N(v_1)$ and $N(v_2)$ intersect, and Lemma 4.7 describes this intersection. Since (a) and (b) above ensure that $M_{\rho(v_1)}^{F_1} \subseteq M_{\rho(v_2)}^{F_2}$, to prove the inclusion $N_{v_1}^{F_1} \subset N_{v_2}^{F_2}$ we only have to show that $Z(v_1, v_2)$ contains all faces $F < F_1$.

The latter assertion means that no chamber C such that $C < F$ (i.e., $F \subset \overline{C}$) is separated from $\rho(v_1)$ by any hyperplane that separates $\rho(v_1)$ from $\rho(v_2)$. But $C < F$ implies $C < F_1$, and so C can be separated from $\rho(v_1)$ only by hyperplanes $H \in \mathrm{supp}(F_1)$ and, by construction, the set of hyperplanes separating $\rho(v_1)$ from $\rho(v_2)$ is contained in $\mathrm{supp}(F_2) \setminus \mathrm{supp}(F_1)$. This concludes the proof. □

We end by proving the announced proposition.

Proposition 4.11. *Let $\Theta \to \Gamma(\mathcal{A})$ be a cover of oriented systems, $\rho : \mathcal{G}_\rho \to \mathcal{G}$ the associated cover of the arrangement groupoid, and $\mathscr{D}_\rho$ the corresponding diagram. Then*

$$N_\Theta = \bigcup_{\substack{F \in \max(\mathcal{F}) \\ v \in \mathrm{Ob}(\mathcal{G}_\rho)}} N_v^F$$

is a covering by open, contractible subsets with empty or contractible intersections. Moreover, the nerve of this covering is the poset $\mathrm{Plim}\mathscr{D}_\rho$.

Proof. After the above preparations, we only have to show that $N_{v_1}^{F_1} \cap N_{v_2}^{F_2}$ is not empty if and only if it equals $N_{\tilde{v}}^{F_1 \wedge F_2}$, where $\tilde{v}$ represents the path obtained by concatenating any representative of v_1 with the lift $(\rho(v_1) \to \rho(v_1)_{F_1 \wedge F_2})^{\langle v_1 \rangle}$ (where the wedge is taken in $\mathcal{F}$).

For this, recall Lemma 4.7. It implies that if $z \in N(v_1) \cap N(v_2)$ and, say, $\Re(z) \in F$, then all pieces of $N(v_1)$ of the form $F' + iC'_F$ with $F' < F$ are also in the intersection. Therefore, if an intersection is nonempty, it contains some $N_{v'}^{F'}$. Now, by Lemma 4.10, this is equivalent with $(F', v') < (F_i, v_i)$ for $i = 1, 2$.

Again by Lemma 4.10, it follows that $N_{\tilde{v}}^{F_1 \wedge F_2} \subset N_{v_1}^{F_1} \cap N_{v_2}^{F_2}$ because in $\mathrm{Plim}\mathscr{D}_\rho$ we have $(F_1 \wedge F_2, \tilde{v}) = (F_1, v_1) \wedge (F_2, v_2)$. The reverse inclusion follows because with some $z' \in N^{F_1}_{v_1} \cap N^{F_2}_{v_2} \setminus N^{F_1 \wedge F_2}_{\tilde{v}}$ we would have a whole $N^{F'}_{v'}$ included in both sets, but for which $(F', v') \not\leq (F_1, v_1) \wedge (F_2, v_2)$, contradicting Lemma 4.10. □

Corollary 4.12. *Given any cover $\rho : (\Theta, \sim) \to (\Gamma(\mathcal{A}), \sim)$, the spaces N_Θ and W_ρ are homotopy equivalent and isomorphic as covers of $\mathcal{M}(\mathcal{A})$.*

Proof. Homotopy equivalence is obtained from Proposition 4.11 via the Nerve Lemma ([7, 10.6(ii)] or [47, Theorem 15.21]). By naturality we obtain the isomorphism as covering spaces. □

5. Garside-type diagram models

The introduction of diagrams of spaces into the picture allows us to take the next step, modifying the diagram spaces so as to get a new type of combinatorial models, with different features. The name indicates that this construction is inspired by the theory of Garside groups, in a way that will be made precise in Section 5.2.

Definition 5.1. Let $\mathcal{A}$ be a linear arrangement of real hyperplanes. Given a covering $\mathcal{G}_\rho \to \mathcal{G}(\mathcal{A})$ of the arrangement groupoid, let U_ρ be the flag complex on the vertex set $\mathrm{Ob}(\mathcal{G}_\rho)$ such that a set $\{v_0, v_1, \dots, v_d\} \subset \mathrm{Ob}(\mathcal{G}_\rho)$ is a simplex if and only if, for all $i < j$ and given representatives γ_i of v_i, γ_j of v_j, the path $\gamma_j \gamma_i^{-1}$ is positive minimal.

Remark 5.2. In the previous definition, as in what follows, we need to consider specific representatives of the classes of paths that are given by objects of the groupoids. As agreed in Section 1.4, we will write $v, w, \dots$ for the objects of the groupoids (i.e., classes of paths) and use Greek lowercase letters for specific paths in the arrangement graph.

For the sake of this survey it will be enough to give a slightly different version of the diagram models than the original one in [31], also in order to reduce to a minimum the required new definitions. Also, note that throughout the whole section $\mathcal{A}$ *denotes a linear arrangement*, even if the construction can easily be modified to hold also for affine arrangements. For instance, in the case of affine braid arrangements the construction would specialize to the complex used by Charney and Peifer in [24] (see the discussion in Section 3.4). Indeed part of the motivation in introducing the Garside-type models was the goal of finding a (still missing) possible generalization of the methods of [24].

Definition 5.3. Let a cover $\mathcal{G}_\rho$ of the arrangement groupoid be given, and fix a face $F \in \mathcal{F}$. For every path γ representing a $v \in \mathrm{Ob}(\mathcal{G}_\rho)$ such that $\mathrm{end}(\gamma) < F$ we consider the set

$$\{\gamma(C \to \mathrm{end}(\gamma))^{-1} \mid C \in \mathcal{T}(\mathcal{A}),\ S(C, \mathrm{end}(\gamma)) \cap \mathrm{supp}(F) = \emptyset\} \subset \mathrm{Ob}(\mathcal{G}_\rho).$$

There is a natural partial order on this set that is induced by $\preccurlyeq_{\mathrm{end}(\gamma)}$ (for the definition see Section 1); we call this poset $\mathcal{Q}^F_\rho(\gamma)$.

Moreover, if C is a chamber, then $\text{supp}(C) = \emptyset$ and thus every $\mathcal{Q}^C_\rho(\gamma)$ is naturally isomorphic to the poset $\mathcal{T}_C(\mathcal{A})$. To emphasize this structure we will write $\mathcal{T}_\rho(\gamma)$ for $\mathcal{Q}^C_\rho(\gamma)$.

Definition 5.4. For any covering of groupoids $\rho : \mathcal{G}_\rho \to \mathcal{G}(\mathcal{A})$ we define a diagram of posets $\mathscr{G}_\rho : \mathcal{F} \to \texttt{Pos}$ by setting

$$\mathscr{G}_\rho(F) := \coprod_{\text{end}(v)<F} \mathcal{Q}^F_\rho,$$

$$\begin{array}{rccc} \mathscr{G}_\rho(F_1 > F_2) : & \mathscr{G}_\rho(F_1) & \hookrightarrow & \mathscr{G}_\rho(F_2) \\ & \gamma' \in \mathcal{Q}_{F_1}(\gamma) & \mapsto & \gamma' \in \mathcal{Q}_{F_2}(\gamma\alpha) \end{array}$$

where α is defined as the positive minimal path from $\text{end}(\gamma)$ to $\text{end}(\gamma)_{F_2}$. The right side is well defined because if $\gamma' = \gamma(C' \to C)^{-1}$, where $C = \text{end}(\gamma)$ and $C' = \text{end}(\gamma')$, then $\gamma' = (\gamma\alpha)((C' \to C)\alpha)^{-1}$.

An easy check shows that the maps are well-defined. Let us go on to the following theorem, that is now easy to prove in the context of diagram of spaces.

Theorem 5.5. *For every cover $\mathcal{G}_\rho \to \mathcal{G}$ we have a homotopy equivalence*

$$\text{hocolim}\Delta(\mathscr{D}_\rho) \simeq \text{hocolim}\Delta(\mathscr{G}_\rho).$$

Proof. It is evident that the minimal elements of $\mathcal{Q}^F_\rho(\gamma)$ are the v such that $\rho(v) < F$ in $\mathcal{F}$. Every one of these vertices belongs to a different connected component of $\Delta(\mathcal{Q}^F_\rho(\gamma))$ and is a cone point for that component, so that clearly $\Delta(\mathscr{G}_\rho(F))$ is homotopy equivalent to $\Delta(\mathscr{D}_\rho(F))$. It is easy to check that the map that sends every $\gamma \in \mathcal{Q}^F_\rho \subset \mathscr{G}_\rho(F)$ to $v := \min \mathcal{Q}^F_\rho \in \mathscr{D}_\rho(F)$ induces a morphism of diagrams. We then conclude the proof by applying the Homotopy Lemma 2.4. □

Now we can prove that the flag complexes defined at the beginning of this section indeed model the arrangement covers.

Theorem 5.6. *For every cover $\mathcal{G}_\rho \to \mathcal{G}$, we have*

$$U_\rho \simeq \mathcal{M}_\rho.$$

Proof. After Theorem 5.5 it suffices to show that $\text{hocolim}\Delta(\mathscr{G}_\rho) \simeq U_\rho$. For this, note that for any morphism $g = \mathscr{G}_\rho(F_1 > F_2)$, the image $g(\Delta(\mathscr{G}_\rho(F_1)))$ is a simplicial subcomplex of $\Delta(\mathscr{G}_\rho(F_2))$ (compare Definition 5.4).

In particular, $(\Delta(\mathscr{G}_\rho(F_2)), \Delta(\mathscr{G}_\rho(F_1)))$ is a NDR-pair and g is a closed cofibration.

So we are in the situation to apply the Projection Lemma 2.6, obtaining a homotopy equivalence $\text{hocolim}\Delta(\mathscr{G}_\rho) \simeq \text{colim}\Delta(\mathscr{G}_\rho)$. We are left with showing that the right-hand side is the complex U_ρ. Indeed, every simplex is contained in (maybe more than) a $\Delta(\mathcal{T}_\rho(\gamma))$. The maps of the diagram are inclusions, so

$$\text{colim}\Delta(\mathscr{G}_\rho) = \Big(\coprod_{\gamma \in Ob(\mathcal{G}_\rho)} \Delta(\mathcal{T}_\rho(\gamma)) \Big) \Big/ \sim$$

and we only have to check the identifications.

Of course $\omega_1 \in \mathcal{T}_\rho(\gamma_1)$ and $\omega_2 \in \mathcal{T}_\rho(\gamma_2)$ are identified if and only if $\omega_1 = \omega_2$ in $\mathcal{G}_\rho$. Given a chain $\sigma \in \Delta(\mathcal{T}_\rho(\gamma_1)) \cap \Delta(\mathcal{T}_\rho(\gamma_2))$ (which is then automatically of length $< n$), let us write $C_1 := \operatorname{end}(\gamma_1)$, $C_2 := \operatorname{end}(\gamma_2)$, $\gamma' := \min \sigma$, $A := \operatorname{end}(\gamma')$, $B := \operatorname{end}(\max(\sigma))$, as in Figure 6.

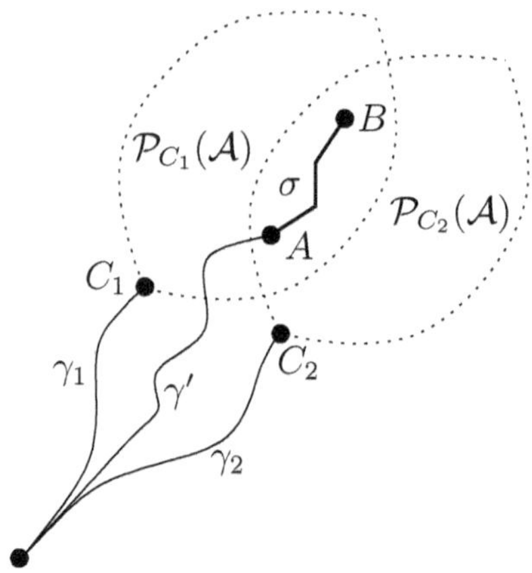

FIGURE 6. Figure for the proof of Theorem 5.5.

The case $C_1 = A = C_2$ is trivial, and if both $C_1 \neq A$ and $C_2 \neq B$ one may consider the poset $\mathcal{T}_\rho(\gamma')$ that contains the chain σ. Then it is enough to show the claim for the case $C_1 = A$, $C_2 \neq A$. In this case we may suppose that $\gamma_1 = \gamma_2\alpha$ with $\alpha = (A \to C_2)$ (all equalities of paths here are in fact equivalences in $\mathcal{G}_\rho$). We will argue by induction on the length of α, the case where $\ell(\alpha) = 0$ being trivial. If $\ell(\alpha) > 0$, let F denote the first face that is crossed by α and let $\epsilon = (C_2 \to C)$ denote the edge of α crossing F, so that $\alpha = \epsilon\tau$ for a positive minimal path τ.

Then clearly $F > C_2$, and by definition the hyperplane H supporting F does not separate A from B. Thus, we have that

$$\sigma \subset \mathcal{Q}_\rho^F(\gamma_2\epsilon\tau),$$

and σ is mapped identically to $\sigma' \subset \mathcal{Q}_\rho^C(\gamma_2\epsilon)$, the latter being equal to $\mathcal{T}_\rho(\gamma_2')$, where $\gamma_2' = \gamma_2\epsilon = \gamma_1(C \to A)^{-1}$. Therefore σ is identified with σ' in the colimit, and it remains to show that $\sigma' \subset \mathcal{T}_\rho(\gamma_2')$ is identified with $\sigma \subset \mathcal{T}_\rho(\gamma_1)$: but this follows now by induction, since $\ell((C \to A)) = \ell(\alpha) - 1$. □

Example 5.7. Consider the arrangement $\mathcal{A}_2$ of Example 1.4. The universal covering groupoid is computed in Example 1.15, and keeping those notations we may describe the complex $U_{\hat{\rho}}$ by identifying every vertex $v_{i,j}$ with the corresponding point (i, j) of the Cartesian plane, agreeing that $v_{0,0}$ projects to the base chamber we choose for the construction. Then it is clear that the morphisms that actually give edges of $U_{\hat{\rho}}$ are the $\mu_{i,j,k,l}$ of the form $\mu_{i,j,i+1,j}$, $\mu_{i,j,i,j+1}$, $\mu_{i,j,i+1,j+1}$. The

2-simplices are of the form $\{v_{i,j}, v_{i,j+1}, v_{i+1,j+1}\}$ or $\{v_{i,j}, v_{i+,j}, v_{i+1,j+1}\}$. Figure 7 shows a piece of this complex which is, in fact, a triangulation of the real plane. △

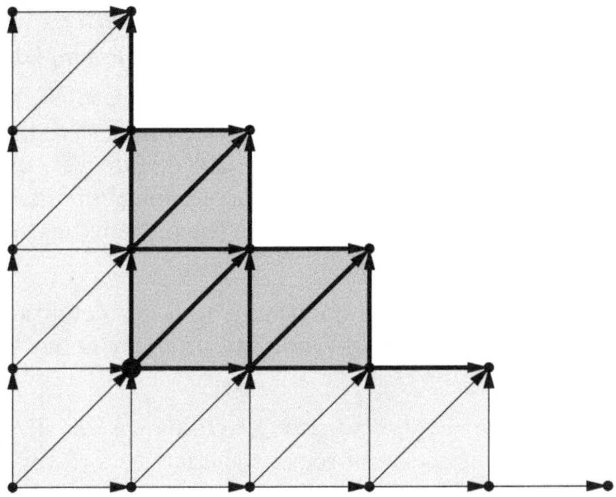

FIGURE 7. A part of the Garside-type universal cover complex for the arrangement $\mathcal{A}_2$ of Example 5.7, with the vertex $v_{0,0}$ represented by the thicker dot. The shaded part is a piece of the positive complex, namely $\widehat{U}_3$.

5.1. The 'Strong Lattice Property'

A linear real arrangement is called *simplicial* if its chambers are cones over simplexes. Brieskorn's conjecture was settled by Deligne, who showed that the complexification of every simplicial arrangement is $K(\pi, 1)$ [29].

As an application of our construction, let us recast the proof of this result in view of the Garside-type diagram models.

Deligne's strategy has been to construct a contractible simplicial complex and then to show that under some technical assumptions this complex models the universal cover of the arrangement's complement. See also [61] for a reformulation of the argument. The first part of Deligne's proof establishes a crucial property of positive paths of simplicial arrangements. This property was given the name "property D" by Paris, who proved that it is indeed *equivalent* to the arrangement being simplicial (see [62] and, for an alternative proof, [31, Chapter 6]).

Lemma 5.8 ("Property D", see [60]). *Let $\mathcal{A}$ be a simplicial arrangement. For every $v \in \mathrm{Ob}(\hat{\mathcal{G}})$ representing a positive path on the arrangement graph, the following holds: there is a unique chamber C_v such that $\beta(C \to C_0)$ represents v for a positive path β if and only if $C \preccurlyeq_{C_0} C_v$.*

Once this is established, one can look at the part of the Garside-type universal cover complex $U_{\hat{\rho}}$ that is generated by the positive paths of length at most n; let us call it $\widehat{U}_n$ (see Figure 7). It is easy to see that Lemma 5.8 implies:

If the arrangement $\mathcal{A}$ is simplicial, then for every vertex v of $\widehat{U}_n \setminus \widehat{U}_{n-1}$,
(∗) v is the apex of a cone over a contractible subcomplex of $\widehat{U}_{n-1}$.

Thus, it is clear that $\widehat{U}_{n-1}$ is homotopy equivalent to $\widehat{U}_n$ by simply successively "pushing in" the cones. By induction, we obtain that the subcomplex of U generated by the positive paths is contractible. Moreover, one can show [60, Lemma 4.16] (but see also [29, 69]) that contractibility of the positive complex implies contractibility of the whole universal cover, and thus asphericity of the arrangement.

Question I. The natural (and open) question is to find a condition on the positive paths of a real arrangement that is weaker than property D but keeps the validity of statement (∗) above.

Remark 5.9 (On the Lattice Property). It is a well-known fact that an arrangement is simplicial if and only if its poset of regions is a lattice for every choice of a base region (see [9]; let us call this the "Strong Lattice Property"). In an attempt to generalize Deligne's argument, one might consider a weakening of this condition, i.e., requiring that the poset of regions be a lattice for at least one choice of base chamber ("Weak Lattice Condition"). Indeed, this condition is satisfied by all simplicial and all supersolvable arrangements, i.e., the two major known classes of aspherical arrangements, and by all hyperfactored arrangements (which are conjecturally $K(\pi, 1)$; see [48]). However, an example exists that satisfies the Weak Lattice Condition but is *not* $K(\pi, 1)$: it is the arrangement $\mathcal{A}_2$ of [36].

The Weak Lattice Property has however nice consequences with respect to the structure of the Garside type models, e.g. leading to a coarsening of the stratification by order complexes of the $\mathcal{T}_C$'s; see [31, Chapter 7] for further details.

5.2. Garside groups and Bestvina's complex

The name "Garside-type" comes from the analogy with the following construction that can be carried out when $\mathcal{A}$ is the reflection arrangement associated to a finite reflection group W. In this situation, W acts on $\mathcal{M}(\mathcal{A})$ and the fundamental group of $\mathcal{M}(\mathcal{A})/W$ is the associated *Artin group*. Among other nice properties, one has that Artin groups are *Garside groups* (see e.g. [41, 51, 28]).

Definition 5.10. A group G is a *Garside group* if there is a bounded, graded lattice $\mathcal{L}$ of finite height, with a labeling of the edges of its Hasse diagram in some alphabet S, so that G is the group of fractions of the monoid generated by S with relations that identify any two words that can be read along saturated chains of $\mathcal{L}$ with same begin- and endpoint. The labeling must satisfy some very important technical conditions, to ensure that the monoid actually embeds into its group of fractions. For every pair $x < y$ of the lattice $\mathcal{L}$, let $\lambda(x, y)$ denote the set of all words in S that can be read along any saturated chain starting at x and ending at y. All of these words are equal, and thus $\lambda(x,y)$ represents a single element, in the generated monoid. Then the conditions are that the labeling be:

- *Balanced:* The sets $\{\lambda(\hat{0},x) \mid x \in \mathcal{L}\}$ and $\{\lambda(x,\hat{1}) \mid x \in \mathcal{L}\}$ are equal.
- *Group-like:* For every two triples $x \leq y \leq z$, $x' \leq y' \leq z'$ of elements $\mathcal{L}$, if two of the corresponding pairs of labelings are equal, so is the third.

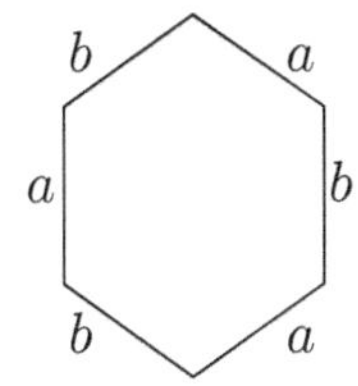

FIGURE 8.
$\{\lambda(\hat{0},x) \mid x \in \mathcal{L}\} = \{a,\ b,\ ab,\ ba,\ aba,\ bab\} = \{\lambda(x,\hat{1}) \mid x \in \mathcal{L}\}$

Example 5.11. An instructive example is one of the Garside structures that lead to the Artin group of type A_2. In this case (as for every finite-type Artin group) the poset is the weak order of the corresponding Coxeter Group, with the natural labeling by simple roots. We depict the poset with an equivalent labeling in Figure 8. For details on the weak order and the labeling see [8]. △

We then introduce the following simplicial (flag) complex that was defined by Brady for braid groups [13], by Brady and Watt [14] and Bestvina [5] for finite type Artin groups, and was extended by Charney, Meier and Whittlesey to the more general context of Garside groups [25].

Definition 5.12 (Compare [25] and Section 2.2 of [5]). Given a Garside group G, let $X(G)$ denote the simplicial complex on the vertex set G obtained by declaring a subset $\{g_0,\ldots,g_d\} \subset G$ to be a simplex if for any $0 \leq i < j \leq d$ the element $g_i^{-1}g_j$ is an *atom* (i.e., any word that can be read (bottom-to-top) along some saturated chain in $\mathcal{L}$).

The following fact is proved as Theorem 3.1 of [25], but see also Theorem 3.6 of [5] and Theorem 6.9 of [13].

Theorem 5.13 ([25]). *For any Garside group G, the complex $X(G)$ is contractible.*

Remark 5.14. Returning to real reflection arrangements and Artin groups, the lattice $\mathcal{L}$ is the so-called *weak order* on the associated Coxeter group, with the natural labeling by standard generators (see [8] for definitions). Bestvina remarked that in this case $X(G)$ has a covering with contractible intersections whose nerve is the universal cover of the Salvetti complex of $\mathcal{M}(\mathcal{A})$, and by the Nerve Lemma [7, 10.6(ii)] one concludes homotopy equivalence [5, Section 2.2].

It is well-known that for any finite reflection arrangement the weak order of the associated reflection group is isomorphic to the poset of regions (by symmetry the choice of base chamber does not matter). Thus, one sees that the Garside-type universal cover models specialize to Bestvina's complex when the arrangement is the reflection arrangement of a finite real reflection group.

Remark 5.15. In addition to Remark 5.14, it has to be mentioned that all finite-type Artin groups can be presented as Garside groups also by partially ordering the associated reflection group by *reflection length* (see [13, 14]). This gives rise to the so-called *dual Garside structure* for these Artin groups (the word "dual" coming, so far, only from some enumerative properties of the two structures). One of the many interesting things about these orderings is that they can be defined also for finite groups of *unitary* reflections as classified by Shephard and Todd [72] (see [3, 4]). The lattices associated to the dual structures can all be described as posets of *generalized noncrossing partitions* ordered by refinement [66, 4], and appear in many different contexts. Bessis [3] shows that, in general, the Garside group obtained in this way turns out to be the fundamental group of the quotient of the complement of the associated reflection arrangement by the action of the reflection group (see Orlik and Terao [56] and Orlik and Solomon [55] for a combinatorial study of arrangements defined by unitary reflection groups). In the same work, Bessis was able to exploit this structure and give a proof of asphericity of all (real and unitary) finite reflection arrangements. Since every central arrangement in $\mathbb{C}^2$ is $K(\pi, 1)$ [56, Proposition 5.6], only the case of dimension strictly greater than 2 must be handled. In all but one of those cases, Bessis shows that the universal covering space of the arrangement's complement is indeed homotopic to the associated complex $X(G)$, and thus contractible by Theorem 5.13. This holds for all *well-generated* groups, i.e., for all groups that can be generated by d reflections, where d is the dimension of an irreducible representation of the group (the dimension of the ambient space of the arrangement we are interested in). In the only remaining case the above argument must be refined, using the more general notion of Garside groupoid, also developed by Bessis in [2]. We will speak again about this topic in Section 6.

The natural question to ask is now whether these structures, and in particular the combinatorics of noncrossing partitions, can give rise to a corresponding presentation of the fundamental group of the arrangement's complement. Note that, for real reflection groups, this is the corresponding pure Artin group. For the type A_n some work was done by McCammond and Margalit [50] who gave presentations for the pure braid group that are inspired by the pictures of 'classical' noncrossing partitions. But the question is open and much work still needs to be done.

6. Applications and open ends

Spectral sequences and homology of covers. In the context of local system homology of arrangements, attention has been paid to the computation of the homology of cyclic covers of arrangement complements, as they generalize in many ways the

Milnor fibre (see the work of Cohen and Orlik [27] and, for a survey and the relevant bibliography, the paper by Suciu [74]). Therefore we want to point out that there exist spectral sequences that calculate the homology and cohomology of homotopy colimits of diagrams of spaces, thus offering an alternative to the spectral sequence approach described by Denham [30] and later generalized by Papadima and Suciu [58]. The original idea goes back to Segal [71], and a formulation for our combinatorial setting is given in [81]. The complexes resulting from the application of this spectral sequence to the general covers are described in detail in [31, Chapter 5], where also a direct derivation of this spectral sequence starting by a filtration by the skeleta of the underlying order complex can be found.

Minimality. The question whether $\mathcal{M}(\mathcal{A})$ has the homotopy type of a minimal CW-complex (i.e., one that has as many k-cells as there are generators of the k-th homology) was raised by Papadima and Suciu in [57]. An affirmative answer to this question was given by Dimca and Papadima [34] and, independently, Randell [64]. In the case of complexified arrangements, the question of minimality was studied by Yoshinaga [79] who in particular attempted to describe the attaching maps of the lift of a minimal CW-structure to the universal cover. Indeed, one point of interest of minimal complexes in this context is that the linearization of the equivariant chain complex obtained by their lift to the universal covering space is equivalent to the Aomoto complex – a well-known complex associated to arrangements and defined from the cohomology ring of the complement. This was first proved by Cohen and Orlik [26] and subsequently in increasing generality by other authors [34, 80, 58].

Recently, explicit constructions were described for such a minimal CW-complex, at least when the arrangement is complexified. Both these works exploit discrete Morse theory in order to describe a collapsing of all 'superfluous' cells of the Salvetti complex.

Salvetti and Settepanella [70] introduce a new total ordering of the faces of the arrangement that they call *polar ordering* because it is obtained by lexicographically ordering polar coordinates of a distinguished point on every face. Given this ordering, an algorithm allows us to construct the required discrete Morse vector field, and so to describe the collapsing that leads to the minimal CW-model. A closed formula for the boundary maps of the minimal complex is also given in [70]. In view of an easier computation of the polar ordering, it has to be pointed out that in fact it is not necessary to actually determine the polar coordinates. It was shown in [33] that the construction works also with a more general type of orderings, called *combinatorial polar orderings*, that are constructed from any valid sequence of "flippings" (see [10, Chapter 5]) along which a general position hyperplane can be "swept" through a generic section of the given arrangement.

The method used in [32] can be carried out entirely in terms of the intrinsic combinatorics of the associated oriented matroid. The data needed to construct the discrete Morse vector field is given by a maximal chain in the poset of regions of the arrangement (the tope poset of the associated oriented matroid). This construction exhibits a very direct correspondence between the no-broken-circuit sets of the matroid and the corresponding cells of the minimal complex. However, an explicit description of the attaching maps is still missing.

Question II. Does any interesting simplification appear by 'lifting' one of the described collapsings of the cells to the Salvetti-type universal cover complex?

Complex reflection groups. The most recent achievement about asphericity of arrangements is the result of David Bessis, who proved that every complex finite reflection arrangement is $K(\pi, 1)$ [3]. As we explained in Remark 5.15, the method of Bessis involves the technique of Garside groups, and in particular, when the group is *well generated*, the universal covering space is modeled by a "complex analogue" of $X(G)$ (Definition 5.12). Among the complex reflection arrangements associated to well-generated groups we find the complexification of the real reflection arrangements, and the translation of the argument of Bessis to that case turns out to use the so-called "dual Garside structure" instead of the "standard" one. It is then natural to ask whether, in the complexified case, a combinatorial way exists to prove the equivalence between Bessis' complex and $U_{\hat{\rho}}$. This problem amounts to a better understanding of the combinatorial relationship between the two Garside structures of finite type Artin groups, which still lacks a satisfactory explanation.

References

[1] E. Babson, D. N. Kozlov; *Diagrams of classifying spaces and k-fold boolean algebras.* ArXiv `math.CO/9704227`.

[2] D. Bessis; *Garside categories, periodic loops and cyclic sets.* ArXiv `math/0610778v1`

[3] D. Bessis; *Finite complex reflection arrangements are $K(\pi, 1)$.* ArXiv `math.AG/0610777`.

[4] D. Bessis, R. Corran; *Non-crossing partitions of type* (e, e, r). Adv. Math. **202** (2006), no. 1, 1–49.

[5] M. Bestvina; *Non-positively curved aspects of Artin groups of finite type.* Geom. Topol. **3** (1999), 269–302 (electronic).

[6] A. Björner; *Posets, regular CW complexes and Bruhat order.* European J. Combin. 5 (1984), no. 1, 7–16.

[7] A. Björner; *Topological methods.* In *Handbook of combinatorics*, vol.2, pp. 1819–1872, Elsevier, Amsterdam, 1995.

[8] A. Björner, F. Brenti; *Combinatorics of Coxeter groups.* Graduate Texts in Mathematics, **231**. Springer, New York, 2005.

[9] A. Björner, P. Edelman, G. M. Ziegler; *Hyperplane arrangements with a lattice of regions.* Discrete Comput. Geom. **5** (1990), no. 3, 263–288.

[10] A. Björner, M. Las Vergnas, B. Sturmfels, N. White, G. M. Ziegler; *Oriented matroids.* Second edition. Encyclopedia of Mathematics and its Applications **46**. Cambridge University Press, Cambridge, 1999.

[11] A. Björner, G. M. Ziegler; *Combinatorial stratification of complex arrangements.* J. Amer. Math. Soc. **5** (1992), no.1, 105–149.

[12] A. K. Bousfield, D. M. Kan; *Homotopy limits, completions and localizations.* Lecture Notes in Mathematics **304**, Springer, Berlin–New York 1972.

[13] T. Brady; *A partial order on the symmetric group and new* $K(\pi, 1)$*'s for the braid groups.* Adv. Math. **161** (2001), no. 1, 20–40.

[14] T. Brady, C. Watts; $K(\pi, 1)$*'s for Artin groups of finite type.* Geom. Dedicata **94** (2002), 225–250.

[15] H. Brandt; *Über eine Verallgemeinerung des Gruppenbegriffes.* Math. Ann. **96** (1927), no. 1, 360–366.

[16] E. Brieskorn; *Sur les groupes de tresses.* Séminaire Bourbaki (1971/1972), Exp. No. 401, pp. 21–44. Lecture Notes in Mathematics **317**, Springer, Berlin, 1973.

[17] R. Brown; *Elements of modern topology.* McGraw-Hill, New York–Toronto 1968.

[18] R. Brown; *Topology. A geometric account of general topology, homotopy types and the fundamental groupoid.* Second edition. Ellis Horwood Series: Mathematics and its Applications. John Wiley and Sons, New York, 1988.

[19] R. Brown; *Groupoids and crossed objects in algebraic topology.* Homology Homotopy Appl. **1** (1999), 1–78 (electronic).

[20] R. Brown, J. L. Loday; *Van Kampen theorems for diagrams of spaces.* Topology **26** (1987), no. 3, 311–335.

[21] F. Callegaro, D. Moroni, M. Salvetti; *The* $K(\pi, 1)$ *problem for the affine Artin group of type* $\widetilde{B}_n$ *and its cohomology.* arXiv:0705.2830.

[22] R. Charney, M. Davis; *The* $K(\pi, 1)$*-problem for hyperplane complements associated to infinite reflection groups.* J. Amer. Math. Soc. **8** (1995), no. 3, 597–627.

[23] R. Charney, M. W. Davis; *Finite* $K(\pi, 1)$*'s for Artin groups.* Prospects in topology (Princeton, NJ, 1994), 110–124, Ann. of Math. Stud. **138**, Princeton Univ. Press, Princeton, NJ, 1995.

[24] R. Charney, D. Peifer; *The* $K(\pi, 1)$*-conjecture for the affine braid groups.* Comment. Math. Helv. **78** (2003), no. 3, 584–600.

[25] R. Charney, J. Meier, K. Whittlesey; *Bestvina's normal form complex and the homology of Garside groups.* Geom. Dedicata **105** (2004), 171–188.

[26] D. C. Cohen, P. Orlik; *Arrangements and local systems.* Math. Res. Lett. **7** (2000), no. 2–3, 299–316.

[27] D. C. Cohen, P. Orlik; *Some cyclic covers of complements of arrangements.* Topology Appl. **118** (2002), no. 1–2, 3–15.

[28] P. Dehornoy; *Groupes de Garside.* Ann. Sci. école Norm. Sup. (4) **35** (2002), no. 2, 267–306.

[29] P. Deligne; *Les immeubles des groupes de tresses généralisés.* Invent. Math. 17 (1972), 273–302.

[30] G. Denham; *The Orlik-Solomon complex and Milnor fibre homology.* Topology Appl. **118** (2002), no. 1–2, 45–63.

[31] E. Delucchi; *Topology and combinatorics of arrangement covers and of nested set complexes.* Ph.D. thesis, ETH Zurich, 2006.

[32] E. Delucchi; *Shelling-type ordering of regular CW-complexes and acyclic matchings for the Salvetti complex.* Int. Math. Res. Not. 2008, no. 6, Art. ID rnm167, 39 pp.

[33] E. Delucchi, S. Settepanella; *Combinatorial polar orderings and follow-up arrangements.* ArXiv:0711.1517v1. To appear in Advances in Applied Mathematics.

[34] A. Dimca, S. Papadima; *Hypersurface complements, Milnor fibers and higher homotopy groups of arrangements.* Ann. of Math. (2) **158** (2003), no. 2, 473–507.

[35] P. H. Edelman; *A partial order on the regions of* $\mathbb{R}^n$ *dissected by hyperplanes.* Trans. Amer. Math. Soc. **283** (1984), no. 2, 617–631.

[36] P. Edelman, V. Reiner; *Not all free arrangements are* $K(\pi,1)$. Bull. Amer. Math. Soc. (N.S.) **32** (1995), no. 1, 61–65.

[37] M. Falk, R. Randell; *The lower central series of a fiber-type arrangement.* Invent. Math. **82** (1985), no. 1, 77–88.

[38] M. Falk, R. Randell; *On the homotopy theory of arrangements.* Complex analytic singularities, 101–124. In: *Adv. Stud. Pure Math.*, **8**, North-Holland, Amsterdam, 1987.

[39] M. Falk, R. Randell; *On the homotopy theory of arrangements. II.* Arrangements Tokyo 1998, 93–125. In: *Adv. Stud. Pure Math.*, **27**, Kinokuniya, Tokyo, 2000.

[40] P. Gabriel, M. Zisman; *Calculus of fractions and homotopy theory.* Ergebnisse der Mathematik und ihrer Grenzgebiete, Band 35. Springer Verlag, New York 1967.

[41] F. A. Garside; *The braid group and other groups.* Quart. J. Math. Oxford Ser. (2) **20** (1969), 235–254.

[42] P. G. Goerss, J. F. Jardine; *Simplicial homotopy theory.* Progress in Mathematics **174**. Birkhäuser Verlag, Basel, 1999.

[43] M. Gromov; *Hyperbolic groups,* in "Essays in Group Theory", ed. by S. M. Gersten, M. S. R. I. Publ. **8**, Springer Verlag, New York 1987, 75–264.

[44] P. J. Higgins; *Notes on categories and groupoids.* Van Nostrand Reinhold Mathematical Studies **32**, Van Nostrand Reinhold, London–New York–Melbourne 1971.

[45] J. Hollender, R. M. Vogt; *Modules of topological spaces, applications to homotopy limits and* E_∞ *structures.* Arch. Math. **59** (1992), no. 2, 115–129.

[46] D. N. Kozlov; *A comparison of Vassiliev and Ziegler-Živaljević models for homotopy types of subspace arrangements.* Topology Appl. **126** (2002), no. 1–2, 119–129.

[47] D. N. Kozlov; *Combinatorial algebraic topology.* Algorithms and Computation in Mathematics **21**. Springer, Berlin, 2008.

[48] M. Jambu, L. Paris; *Combinatorics of inductively factored arrangements.* European J. Combin. **16** (1995), no. 3, 267–292.

[49] S. Mac Lane; *Categories for the working mathematician.* Graduate Texts in Mathematics **5**. Springer Verlag, New York, 1998.

[50] D. Margalit, J. McCammond; *Geometric presentation s for the pure braid group.* To appear in Journal of Knot Theory and its Ramifications. ArXiv:`math/0603204v1`

[51] J. McCammond; *An introduction to Garside structures.* Preprint available at `http://www.math.ucsb.edu/%7Ejon.mccammond/papers/index.html`

[52] G. Moussong; *Hyperbolic Coxeter groups.* Ph.D. thesis, The Ohio State University, 1988.

[53] T. Nakamura; *A note on the $K(\pi, 1)$-property of the orbit space of the unitary reflection group G(m,l,n).* Sci. Papers College of Arts and Sciences, Univ. Tokio **33** (1983), 1–6.

[54] C. Okonek; *Das $K(\pi, 1)$-Problem für die affinen Wurzelsysteme vom Typ A_n,C_n.* Math. Z. **168** (1979), no. 2, 143–148.

[55] P. Orlik, L. Solomon; *Unitary reflection groups and cohomology.* Invent. Math. **59** (1980), no. 1, 77–94.

[56] P. Orlik, H. Terao; *Arrangements of Hyperplanes.* Grundlehren der Mathematischen Wissenschaften **300**, Springer, Berlin 1992.

[57] S. Papadima, A. Suciu; *Higher homotopy groups of complements of hyperplane arrangements.* Advances in Math. **165** (2002), no. 1, 71–100.

[58] S. Papadima, A. Suciu; *The spectral sequence of an equivariant chain complex and homology with local coefficients.* ArXiv:0708.4262.

[59] L. Paris; *The covers of a complexified real arrangement of hyperplanes and their fundamental groups.* Topology and its Applications **53** (1993), 75–178.

[60] L. Paris; *Universal cover of Salvetti's complex and topology of simplicial arrangements of hyperplanes.* Trans. Amer. Math. Soc. **340** (1993), no. 1, 149–178.

[61] L. Paris; *The Deligne complex of a real arrangement of hyperplanes.* Nagoya Math. J. **131** (1993), 39–65.

[62] L. Paris; *Arrangements of hyperplanes with property* D. Geom. Dedicata **45** (1993), no. 2, 171–176.

[63] D. Quillen; *Higher algebraic K-theory.* Lecture Notes in Mathematics **341**, Springer, Berlin 1973.

[64] R. Randell; *Morse theory, Milnor fibers and minimality of hyperplane arrangements.* Proc. Amer. Math. Soc. **130** (2002), no. 9, 2737–2743 (electronic).

[65] K. Reidemeister; *Einführung in die kombinatorische Topologie.* Vieweg, Braunschweig (1932). Reprint Chelsea, New York (1950).

[66] V. Reiner; *Non-crossing partitions for classical reflection groups.* Discrete Math. **177** (1997), no. 1–3, 195–222.

[67] M. Salvetti; *Topology of the complement of real hyperplanes in $\mathbb{C}^n$.* Invent. math. **88** (1987), 603–608.

[68] M. Salvetti; *The homotopy type of Artin groups.* Math. Res. Lett. **1** (1994), no. 5, 565–577.

[69] M. Salvetti; *On the homotopy theory of complexes associated to metrical-hemisphere complexes.* Discrete Math. **113** (1993), no. 1–3, 155–177.

[70] M. Salvetti, S. Settepanella; *Discrete Morse theory and minimality of arrangements.* Geom. Topol. **11** (2007), 1733–1766.

[71] G. Segal; *Classifying spaces and spectral sequences.* Inst. Hautes Études Sci. Publ. Math. **34** (1968), 105–112.

[72] G. C. Shephard, J. A. Todd; *Finite unitary reflection groups.* Can. J. Math. **6** (1954), 274–302.

[73] E. H. Spanier; *Algebraic topology.* Springer Verlag, New York–Berlin, 1981.

[74] A. Suciu; *Fundamental groups of line arrangements: Enumerative aspects.* In *Advances in algebraic geometry motivated by physics.* Contemporary Math., vol. 276, A.M.S., Providence, RI, 2001.

[75] R. Stanley; *Enumerative combinatorics, Vol. 1.* Wadsworth and Brooks/Cole, Monterey, CA, 1986; reprinted as Cambridge Studies in Advanced Mathematics, Vol. 49, Cambridge University Press, Cambridge, 1997.

[76] H. Terao; *Modular elements of lattices and topological fibration.* Adv. in Math. **62** (1986) no. 2, 135–154.

[77] V. A. Vassiliev; *Complements of Discriminants of Smooth Maps: Topology and Applications.* Transl. Math. Monographs **98**, Amer. Math. Soc., Providence, RI, 1994.

[78] R. M. Vogt; *Homotopy limits and colimits.* Math. Z. **134** (1973), 11–52.

[79] M. Yoshinaga; *Hyperplane arrangements and Lefschetz's hyperplane section theorem.* Kodai Math. J. **30** (2007), no. 2, 157–194.

[80] M. Yoshinaga; *Chamber basis of the Orlik-Solomon algebra and Aomoto complex.* ArXiv: math/0703733

[81] V. Welker, G. M. Ziegler, R. T. Živaljević; *Homotopy colimits – comparison lemmas for combinatorial applications.* J. reine angew. Math. **509** (1999), 117–149.

[82] G. M. Ziegler, R. Živaljević; *Homotopy types of subspace arrangements via diagrams of spaces.* Math. Ann. **295** (1993), no. 3, 527–548.

Emanuele Delucchi
Department of Mathematical Sciences
State University of New York
Binghamton, NY 13902-6000
USA
e-mail: delucchi@math.binghamton.edu

Progress in Mathematics, Vol. 283, 39–58

Homological Aspects of Hyperplane Arrangements

Graham Denham

Abstract. The purpose of this expository article is to survey some results and applications of free resolutions related to hyperplane arrangements. We include some computational examples and open problems.

Mathematics Subject Classification (2000). Primary 53B30; Secondary 17B55, 16S37, 20F14.

Keywords. Hyperplane arrangement, lower central series, homotopy Lie algebra, resonance.

1. Introduction

One facet of recent work on hyperplane arrangements is the influence of rational homotopy theory and the appearance of some interesting homological algebra. Arrangement complements are formal in the sense of Sullivan, and their cohomology rings are well-understood. On the other hand, the combinatorics of the cohomology ring is quite intricate (see [14]), which leads to some interesting and unsolved problems. Providing a complete survey of the rational homotopy theory of hyperplane arrangements is beyond the scope of these lecture notes; the objective here is instead to use two related topics to give some idea of the existing literature and future directions. Koszul duality plays a somewhat unifying role. Some of the topics here are also discussed in the surveys [21, 47, 19, 14].

These notes are organized as follows. The rest of this section defines graded free resolutions and introduces some Lie algebras associated with a discrete group, in this case the fundamental group of an arrangement complement.

Section §2 considers free resolutions over the arrangement's cohomology ring. If M is the complement of n hyperplanes in $\mathbb{C}^\ell$, let $A = H^\bullet(M,\mathbb{Q})$. We recall a "classical" interpretation of the linear strand of the resolution of the trivial A-module in terms of the lower central series of the fundamental group. Both can be understood in combinatorial terms in two interesting cases: when A is a

Koszul algebra (§2.1) and, more recently, for decomposable arrangements (§2.2). Beyond the linear strand, one encounters the homotopy Lie algebra. Roos [39] has recently provided examples of arrangements for which this Lie algebra is not finitely generated, and for which its Hilbert series is transcendental (§2.3).

Section §3 focusses on resolutions over an exterior algebra. This is a more recent inquiry that begins with work in [41, 18, 42]. Via a standard linear inclusion of an arrangement complement in a torus $T = (\mathbb{C}^*)^n$, the exterior algebra $E = H^\bullet(T, \mathbb{Q})$ acts on A and its vector space dual, $A^* = H_\bullet(M, \mathbb{Q})$. The homology of the arrangement complement, A^*, has a remarkable property: its free resolution as an E-module is linear (§3.3). We see this provides a link with resonance and cohomology of local systems on arrangement complements, which ties in this second topic with the lectures notes of Falk [19] and Dimca-Yuzvinsky[14].

1.1. Free resolutions

Let $\mathbf{k}$ be a field, and let R be a Noetherian $\mathbf{k}$-algebra, graded by the natural numbers. Let $|x|$ denote the degree of a homogeneous element $x \in R$. The algebra R is *graded-commutative* if $xy = (-1)^{|x||y|}yx$ for all homogeneous elements $x, y \in R$. If M is a finitely-generated, graded left R-module, for $j \in \mathbb{Z}$, let $M(j)$ denote the module with degrees shifted down by j: that is, $M(j)_i = M_{i+j}$, for all $i \in \mathbb{Z}$. A graded free resolution of M is a resolution $(F_\bullet, \partial)$ by graded, free modules in which the differential ∂ has degree 0. One can write such a resolution as follows:

$$0 \longleftarrow M \longleftarrow \bigoplus_j R(-j)^{b_{0j}} \xleftarrow{\partial_1} \bigoplus_j R(-j)^{b_{1j}} \cdots \xleftarrow{\partial_i} \bigoplus_j R(-j)^{b_{ij}} \xleftarrow{\partial_{i+1}} \cdots$$

for some integers $\{b_{ij} \colon i, j \in \mathbb{Z}\}$. The resolution is *minimal* if the entries of each ∂_i are contained in the maximal homogeneous ideal $R_+ = \bigoplus_{j>0} R_j$. In this case, the number b_{0j} is just the number of generators of M in degree j in a minimal generating set. Minimal resolutions exist, but they are not unique. However, the numbers $\{b_{ij} \colon i \geq 0, j \in \mathbb{Z}\}$ are the same for every minimal resolution of a module M, and are called the bigraded Betti numbers of M. Let

$$P(M, s, t) = \sum_{i \geq 0, j} b_{ij} s^i t^j, \tag{1.1}$$

the Poincaré-Betti polynomial of M. Computing the Euler characteristic of the resolution in each degree gives an expression for the Hilbert series of M:

$$H(M, t)/H(R, t) = P(M, -1, t). \tag{1.2}$$

Suppose $(F_\bullet, \partial)$ is a minimal resolution of M. Since $\mathbf{k} = R/R_+$, we get

$$\mathbf{k} \otimes_R F_\bullet = \bigoplus_j \Big(\mathbf{k}(-j)^{b_{0j}} \xleftarrow{0} \bigoplus_j \mathbf{k}(-j)^{b_{1j}} \cdots \xleftarrow{0} \bigoplus_j \mathbf{k}(-j)^{b_{ij}} \xleftarrow{0} \cdots \Big),$$

where the differential is zero. In homology, then, $\dim_{\mathbf{k}} \operatorname{Tor}^R_i(\mathbf{k}, M)_j = b_{ij}$, for all i, j. Similarly, $\dim_{\mathbf{k}} \operatorname{Ext}^i_R(M, \mathbf{k})_{-j} = \dim_{\mathbf{k}} \operatorname{Hom}_R(F_i, \mathbf{k})_{-j} = b_{ij}$ for all i, j.

Suppose that $R(-j)$ is a summand of F_i, for some j, $i > 0$. Then since R is nonnegatively graded, its image $\partial_i(R(-j))$ in F_{i-1} is contained in summands

$R(-j')$, where $j' < j$. It follows that, for any strictly increasing sequence of integers $c_0 < c_1 < c_2 < \cdots$, the differential in $F_\bullet$ restricts to the subcomplex

$$\bigoplus_{j \le c_0} R(-j)^{b_{0j}} \longleftarrow \bigoplus_{j \le c_1} R(-j)^{b_{1j}} \longleftarrow \cdots \longleftarrow \bigoplus_{j \le c_i} R(-j)^{b_{ij}} \longleftarrow \cdots .$$

Definition 1.1. If $F_\bullet$ is a minimal resolution of M, let d_0 be the smallest degree of a generator of M. Since $b_{0j} = 0$ for $j < d_0$, the smallest nonzero subcomplex of the form above occurs using the sequence $c_i = d_0 + i$ for $i \geq 0$. This subcomplex is called the *linear strand of* $F_\bullet$. If a minimal resolution $F_\bullet$ is equal to its linear strand, it is called a *linear resolution.*

Notice that, if a left module M has a linear resolution, then its Hilbert series determines the Betti numbers in its resolution: from (1.2),

$$\begin{aligned} H(M,t)/H(R,t) &= P(M,-1,t) \\ &= (-t)^{d_0} \sum_{i \geq 0} b_{i,d_0+i}(-t)^i. \end{aligned} \tag{1.3}$$

1.2. Fundamental groups and the lower central series

From this point onward, we will take our scalars to be the field $\mathbb{Q}$. Let $\mathcal{A} = \{H_1, H_2, \ldots, H_n\}$ be a central arrangement in $\mathbb{C}^\ell$, and let $M = M(\mathcal{A}) = \mathbb{C}^\ell - \bigcup_{H \in \mathcal{A}} H$ denote the complement of the hyperplanes in affine space. Let $G(\mathcal{A}) = \pi_1(M)$ denote its fundamental group. Such groups have been of interest since the 1960s: in particular, if $\mathcal{A}$ is the set of reflecting hyperplanes of a reflection group, then $G(\mathcal{A})$ is a (generalized) pure braid group [8, 22]. A modern survey may be found in [44].

For any group G, let $G^{(1)} = G$ and $G^{(i)} = [G, G^{(i-1)}]$ for $i \geq 1$, the lower central series of G. The quotient $G^{(i)}/G^{(i+1)}$ is abelian, and their sum

$$\operatorname{gr}_{\mathbb{Q}} G = \bigoplus_{i \geq 1} (G^{(i)}/G^{(i+1)}) \otimes_{\mathbb{Z}} \mathbb{Q} \tag{1.4}$$

forms a Lie algebra over $\mathbb{Q}$ with bracket imposed by the commutator in G, called *the rational lower central series Lie algebra* of G.

One should regard the Lie algebra $\operatorname{gr}_{\mathbb{Q}} G$ as being a simplified approximation to G. Let $G'' = [G', G']$, the second derived subgroup of G. The maximal metabelian quotient, G/G'', is another approximation of the group G. Its lower central series ranks were first considered by Chen [9]. Its lower central series Lie algebra, $\operatorname{gr}_{\mathbb{Q}}(G/G'')$, is called the (rational) *Chen Lie algebra* of G: for a modern treatment, see [32].

Returning to the case where G is the fundamental group of a hyperplane complement $M(\mathcal{A})$, we will call these the LCS and Chen Lie algebras of $\mathcal{A}$, respectively.

1.3. A third Lie algebra

Now let X be a finite-type CW complex. Dualizing the cohomology product $\cup\colon H^1(X,\mathbb{Q})\otimes H^1(X,\mathbb{Q})\to H^2(X,\mathbb{Q})$ gives a map

$$\cup^*\colon H_2(X,\mathbb{Q})\to H_1(X,\mathbb{Q})\otimes H_1(X,\mathbb{Q}). \tag{1.5}$$

Let $V=H_1(X,\mathbb{Q})$. Since $\cup$ is skew-commutative, for any $z\in H_2(X,\mathbb{Q})$,

$$\cup^*(z)=\sum_i x_i\otimes y_i-y_i\otimes x_i,$$

for some elements $\{x_i\},\{y_i\}$ of V. One can use the image of $\cup^*$ to make a (non-commutative) algebra generated by the vector space V: let

$$U=\mathbb{T}(V)/(\mathrm{im}(\cup^*)), \tag{1.6}$$

the quotient of the tensor algebra on V by the two-sided ideal generated by the image of $\cup^*$. Let $[x,y]=x\otimes y-y\otimes x$ for $x,y\in V$. Then since its relations are generated by brackets, U is the universal enveloping algebra of the Lie algebra generated by V. This Lie algebra, denoted by $\mathfrak{h}(X)$, is called the *holonomy Lie algebra of* X, introduced in [10]. We will write $U=U(\mathfrak{h}(X))$. If the space X is understood, we will write $\mathfrak{h}$ in place of $\mathfrak{h}(X)$.

More precisely, we define an algebra homomorphism $\nabla\colon U\to U\otimes U$ by letting $\nabla(x)=x\otimes 1+1\otimes x$ for $x\in V$, and extending it multiplicatively. An element $x\in U$ is *primitive* if $\nabla(x)=x\otimes 1+1\otimes x$. Let $\mathcal{P}U$ denote the set of primitive elements of U. It is not hard to check that $\mathcal{P}U$ is closed under the bracket operation, so $\mathcal{P}U$ is a Lie algebra. We define $\mathfrak{h}(X)=\mathcal{P}U$. (This is an instance of a more general fact: U has the structure of a cocommutative Hopf algebra over $\mathbb{Q}$ with coproduct ∇. In such a situation, U is always the enveloping algebra of its Lie algebra of primitive elements: see [30, 4])

In the case of hyperplane arrangement complements, $\mathfrak{h}(\mathcal{A})=\mathfrak{h}(M(\mathcal{A}))$ is called the holonomy Lie algebra of $\mathcal{A}$. Recall that the cohomology algebra of the complement $M(\mathcal{A})$ has a combinatorial presentation as the Orlik-Solomon algebra, $A=E/I$, where E is an exterior algebra on n generators, and I an ideal of relations indexed by circuits.

Let $V^*=A^1$, a $\mathbb{Q}$-vector space with basis $\{e_H\colon H\in\mathcal{A}\}$. Let $\{f_H\colon H\in\mathcal{A}\}$ be the dual basis in V. Kohno [26] showed that

$$\mathfrak{h}(\mathcal{A})=\Big\langle f_H\colon H\in\mathcal{A}\,\Big|\,\big[f_H,\sum_{H'\colon H'<X}f_{H'}\big]\colon H\in\mathcal{A},X\in L_2(\mathcal{A}),H<X\Big\rangle. \tag{1.7}$$

In particular, the holonomy Lie algebra depends only on A_i for $i=0,1,2$, which is to say that it is determined completely by the number of hyperplanes and their codimension-2 intersections.

Example 1. Let $\mathcal{A}$ be an arrangement of n lines through the origin in $\mathbb{C}^2$. Then $A=E/I$, where $I=((e_ie_j-e_ie_k+e_je_k)\colon 1\le i<j<k\le n)$. We can identify A_2 with $V^*\otimes V^*/W$, where W is the subspace generated by the elements $e_i\otimes e_j+e_j\otimes e_i$ and $e_i\otimes e_j-e_i\otimes e_k+e_j\otimes e_k$ for all i,j,k. Then the

cup product $A_1 \otimes A_1 \to A_2$ is the quotient map, and the image of its dual is $W^\perp = \{g \in V \otimes V : g(x) = 0 \text{ for all } x \in W\}$. In this case, it can be checked directly that the elements $[f_i, \sum_{j=1}^n f_j]$ span $W^\perp$, for $1 \leq i \leq n$, which recovers the presentation of $\mathfrak{h}(\mathcal{A})$ for this arrangement given by (1.7).

1.4. Relating the Lie algebras

The example above motivates a key observation due to Shelton and Yuzvinsky [43], for which we need another definition.

Definition 1.2. Suppose B is a nonnegatively graded $\mathbb{Q}$-algebra, finitely generated in degree 1, and not necessarily graded-commutative. Let $V = B_1$, a $\mathbb{Q}$-vector space. Then $B \cong \mathbb{T}(V)/R$, for some ideal of relations R. Let $W = R_2$, a subspace of $V \otimes V$. The quadratic dual of B is, by definition, the graded algebra

$$B^! = \mathbb{T}(V^*)/(W^\perp). \tag{1.8}$$

Clearly, quadratic duality is an involution: $(B^!)^! \cong B$. By reviewing the construction (1.6) carefully, one obtains the following.

Proposition 1.3 ([43]). *For any arrangement $\mathcal{A}$, we have $U(\mathfrak{h}(\mathcal{A})) \cong A^!$.*

On the other hand, the fundamental group $G = G(\mathcal{A})$ of an arrangement is 1-formal (in the sense of Sullivan [45]), from which it follows that $\operatorname{gr}_{\mathbb{Q}}(G) \cong \mathfrak{h}(\mathcal{A})$ as Lie algebras [26]. As an application, we could try to understand the LCS ranks, $\phi_i = \operatorname{rank}(G^{(i)}/G^{(i+1)})$, for $i \geq 1$. According to the Poincaré-Birkhoff-Witt Theorem, the associated graded algebra of an enveloping algebra $U(\mathfrak{g})$ (under the bracket-length filtration) is a polynomial algebra. In particular, the Hilbert series of $U(\mathfrak{g})$ is the same as that of the polynomial algebra $\mathbb{Q}[\mathfrak{g}]$.

Since $\mathfrak{h} \cong \operatorname{gr}_{\mathbb{Q}}(G)$ has ϕ_i generators of degree i, we obtain the formula

$$H(U(\mathfrak{h}), t) = \prod_{i \geq 1} (1 - t^i)^{-\phi_i}. \tag{1.9}$$

Understanding the LCS ranks $\{\phi_i\}$, then, is equivalent to knowing the Hilbert series of the quadratic dual $A^!$ of the Orlik-Solomon algebra. However, finding an explicit description is an open problem except for special classes of arrangements, as we see below.

Example 1 (continued). Continuing the example of n lines in $\mathbb{C}^2$, one can see that the element $z := \sum_{j=1}^n f_i$ is central in $\mathfrak{h}(\mathcal{A})$, and its quotient is just the free Lie algebra on $n-1$ generators, which we will call $\mathfrak{f}_{n-1}$. The quotient is (noncanonically) split, so $\mathfrak{h} \cong \mathfrak{f}_{n-1} \times \mathfrak{f}_1$.

On the level of enveloping algebras, $U(\mathfrak{h})$ is isomorphic to the tensor product of a tensor algebra with $n-1$ generators with a polynomial algebra $\mathbb{Q}[z]$. From (1.9), then

$$\prod_{i \geq 1} (1 - t^i)^{-\phi_i} = \frac{1}{(1-(n-1)t)(1-t)} \tag{1.10}$$

by multiplying the Hilbert series of the tensor algebra with that of the one-variable polynomial algebra.

2. Resolutions over the Orlik-Solomon algebra

We first outline some known results about free resolutions over the Orlik-Solomon algebra. We use [7] in particular as a reference for multiplicative structures in homological algebra. Consider the module $A_0 = \mathbb{Q}$, on which the positive degree part of A acts trivially. Recall that the Yoneda product makes the Ext groups $\operatorname{Ext}^\bullet_A(\mathbb{Q},\mathbb{Q})$ into a $\mathbb{Q}$-algebra, graded by cohomological degree. The grading from A induces a second grading on the algebra with $\operatorname{Ext}^p_A(\mathbb{Q},\mathbb{Q})_q = 0$ unless $q \leq -p$. For convenience, we let

$$\operatorname{Ext}^p_A(\mathbb{Q},\mathbb{Q})_{(r)} = \operatorname{Ext}^p_A(\mathbb{Q},\mathbb{Q})_{-p-r} \tag{2.1}$$

for $r \geq 0$, so that $\operatorname{Ext}^\bullet_A(\mathbb{Q},\mathbb{Q})$ is nonnegatively graded, and $\operatorname{Ext}^\bullet_A(\mathbb{Q},\mathbb{Q})_{(0)}$ is a subalgebra of $\operatorname{Ext}^\bullet_A(\mathbb{Q},\mathbb{Q})$. Note that $\dim_{\mathbb{Q}} \operatorname{Ext}^i_A(\mathbb{Q},\mathbb{Q})_{(0)} = b_{ii}$, the ith Betti number of the linear strand of a minimal free resolution of $\mathbb{Q}$ over A.

Löfwall [28] showed that $\operatorname{Ext}^\bullet_A(\mathbb{Q},\mathbb{Q})_{(0)} \cong A^!$ as algebras. Taking Hilbert series via (1.9), then, we get

$$\sum_{i\geq 0} b_{ii}t^i = \prod_{i\geq 1}(1-t^i)^{-\phi_i}, \tag{2.2}$$

which appeared in [36].

2.1. Koszul algebras

This leads to a definition.

Definition 2.1. A nonnegatively graded $\mathbb{Q}$-algebra B is *Koszul* if $\operatorname{Ext}^\bullet_B(\mathbb{Q},\mathbb{Q})_{(0)} = \operatorname{Ext}^\bullet_B(\mathbb{Q},\mathbb{Q})$.

There are numerous equivalent formulations: for detail and a more general treatment see [5, 23, 37]. From the discussion above, though, B being Koszul is equivalent to having the inclusion be an isomorphism $B^! \cong \operatorname{Ext}_B(\mathbb{Q},\mathbb{Q})$, as well as to the trivial module $\mathbb{Q}$ having a linear resolution. An important consequence (see (1.3)) is that

$$H(B,t)\cdot H(B^!,-t) = H(\mathbb{Q},t) = 1. \tag{2.3}$$

It follows that a Koszul algebra must be *quadratic*: that is, expressible as a quotient of a polynomial algebra by an ideal of relations generated in degree 2. This is a sufficient condition in the case that the ideal is generated by monomials [1], but in general it is difficult to decide if a given algebra is Koszul. A useful test is the following. An algebra B is Koszul if its ideal of relations possesses a quadratic Gröbner basis (also from [1]; see [23] for further discussion.)

The Koszul property has an interpretation in terms of rational homotopy theory, due to Papadima and Yuzvinsky [35]:

Theorem 2.2 ([35]). *If X is a connected, finite-type formal space, then $H^\bullet(X,\mathbb{Q})$ is Koszul if and only if the rational completion $X_{\mathbb{Q}}$ is an Eilenberg-Maclane space.*

In the case of Orlik-Solomon algebras, Björner and Ziegler [6] showed that $A(\mathcal{A})$ possesses a quadratic Gröbner basis if and only if $\mathcal{A}$ is a supersolvable arrangement. This leads to the following beautiful result:

Theorem 2.3 ([20]). *If $\mathcal{A}$ is a supersolvable arrangement of rank ℓ, then the lower central series ranks are given by the formula*

$$\prod_{i\geq 1}(1-t^i)^{-\phi_i} = \pi(\mathcal{A},-t)^{-1},$$

where $\pi(\mathcal{A},t)$ is the Poincaré polynomial of the arrangement.

Furthermore, the Poincaré polynomial of a supersolvable arrangement is known to factor as $\pi(\mathcal{A},t) = (1+m_1t)(1+m_2t)\cdots(1+m_\ell t)$ for certain combinatorially significant positive integers $\{m_i\}$, which makes the right-hand side of the identity above more attractive: see [31].

Proof. If $\mathcal{A}$ is supersolvable, the Orlik-Solomon algebra is Koszul, by the remark above. The LCS ranks are given by (2.2); using the Koszul property via (2.3), this generating function equals $H(A,-t)^{-1}$. The argument is completed by recalling that the Poincaré polynomial is the Hilbert series of the Orlik-Solomon algebra. □

Remark 2.4. Kohno [27] first established the LCS formula above for reflection arrangements of type A_ℓ. Falk and Randell [20] extended the formula to a class they called fiber-type arrangements, which Terao [46] found was the same as the supersolvable arrangements.

Problem 2.5. Examples of Koszul algebras defined by ideals that do not possess a quadratic Gröbner basis are known: see, for example, [40]. However, no such examples of Orlik-Solomon algebras are known. That is, if $\mathcal{A}$ is an arrangement for which $A(\mathcal{A})$ is Koszul, must $\mathcal{A}$ be supersolvable?

2.2. Decomposable arrangements

Papadima and Suciu [34] identified another class of arrangements for which the linear strand of a minimal free resolution has a nice structure, generalizing results of Schenck and Suciu [41]. The decomposable arrangements, defined below, are generally not Koszul, but they too have a LCS formula similar to that of Theorem 2.3.

For each subspace $X \in L_2(\mathcal{A})$, write $\mathfrak{h}_X = \mathfrak{h}(\mathcal{A}_X)$. This is the holonomy Lie algebra of a rank-2 arrangement, which we saw in Example 1. The natural projections $\pi_X\colon \mathfrak{h}(\mathcal{A}) \to \mathfrak{h}_X$ assemble to give a map

$$\pi\colon \mathfrak{h} \to \bigoplus_{X\in L_2(\mathcal{A})} \mathfrak{h}_X.$$

Holonomy Lie algebras are graded by bracket length, so $\mathfrak{h}' = \mathfrak{h}_{\geq 2}$. The restriction

$$\pi'\colon \mathfrak{h}' \to \bigoplus_{X\in L_2(\mathcal{A})} \mathfrak{h}'_X \tag{2.4}$$

is always surjective. The arrangement $\mathcal{A}$ is *decomposable* if the map (2.4) is an isomorphism. It is not hard to check that the degree-2 part, π_2, is always an isomorphism of vector spaces, but in general π' has a nonzero kernel in degrees 3 and higher.

A salient feature of this family of arrangements is an effective test for membership. Papadima and Suciu show that, remarkably, π' is an isomorphism if and only if π_3 is an isomorphism. [34, Theorem 2.4] Accordingly, one can decide if an arrangement is decomposable by counting dimensions in (2.4). Based on our calculations in Example 1, if X is a rank-2 flat, then $\mathfrak{h}'_X = \mathfrak{f}'_{m-1}$, where $m = |\mathcal{A}_X|$. By expanding (1.10), one can compute that $\dim_{\mathbb{Q}}(\mathfrak{h}_X)_3 = m(2 - 3m + m^2)/3$. Similarly, by expanding the generating function (1.9),

$$\dim_{\mathbb{Q}} \mathfrak{h}_3 = \frac{1}{3}(3b_{33} - 3b_{11}b_{22} + b_{11}^3 - b_{11}).$$

So $\mathcal{A}$ is decomposable if and only if

$$3b_{33} - 3b_{11}b_{22} + b_{11}^3 - b_{11} = \sum_{X \in L_2(\mathcal{A})} m_X(2 - 3m_X + m_X^2), \tag{2.5}$$

where $m_X = |\mathcal{A}_X|$.

Since for decomposable arrangements $\mathfrak{h}'$ is known explicitly, so is a generating function for the lower central series ranks of the fundamental group:

Theorem 2.6 ([34]). *If $\mathcal{A}$ is a decomposable arrangement, the lower central series ranks of $\pi_1(M(\mathcal{A}))$ are given by the formula*

$$\prod_{i \geq 1}(1 - t^i)^{-\phi_i} = (1 - t)^{m-n} \prod_{X \in L_2(\mathcal{A})} (1 - (m_X - 1)t)^{-1},$$

where n is the number of hyperplanes, $m_X = |\mathcal{A}_X|$, and $m = \sum_{X \in L_2(\mathcal{A})}(m_X - 1)$.

Example 2. The arrangement X_3 defined by $Q = xyz(x+y)(x+z)(y+z)$ has three triple points, so the right-hand side of (2.5) equals 18. By a computer calculation with Macaulay 2 [25], the first few Betti numbers of $\operatorname{Ext}_A(\mathbb{Q}, \mathbb{Q})$ are

```
0: 1 6 24 80 240
1: . .  1 12  84
2: . .  .  .   1
```

Here we use Macaulay 2 notation: that is, $b_{11} = 6$, $b_{22} = 24$, $b_{33} = 80$, and so on.

The left side of (2.5) equals 18 as well, which means X_3 is decomposable. By Theorem 2.6 and (2.2), the Betti numbers of the linear strand above are given by the generating function $(1 - 2t)^{-3}$.

2.3. The homotopy Lie algebra

Above, we saw that an isomorphism $\operatorname{Ext}_A(\mathbb{Q}, \mathbb{Q})_{(0)} \cong U(\mathfrak{h})$ relates the linear strand of a free resolution to the holonomy Lie algebra, and this was particularly satisfactory when A is Koszul. More generally, If B is any graded-commutative $\mathbb{Q}$-algebra, then $\operatorname{Ext}_B(\mathbb{Q}, \mathbb{Q})$ is a cocommutative Hopf alegebra, which means the primitive elements $\mathcal{P}\operatorname{Ext}_B(\mathbb{Q}, \mathbb{Q})$ form a graded Lie algebra, called the *homotopy Lie algebra* of B.

For an arrangement $\mathcal{A}$, let

$$\mathfrak{g} = \mathfrak{g}(\mathcal{A}) = \mathcal{P}\operatorname{Ext}_A(\mathbb{Q}, \mathbb{Q}),$$

which we will call the homotopy Lie algebra of the arrangement $\mathcal{A}$. Then $\mathfrak{h} = \mathfrak{g}_{(0)}$: the holonomy Lie algebra is the degree-0 subalgebra of $\mathfrak{g}$. The Lie algebra $\mathfrak{g}$ is bigraded, and it should be mentioned that $\mathfrak{g}$ is not a Lie algebra in the classical sense, but rather a "graded Lie algebra" or Lie superalgebra. This is to say that, for homogeneous elements $x \in \mathfrak{g}_{(p)}$ and $y \in \mathfrak{g}_{(q)}$, the bracket satisfies

$$[x, y] = -(-1)^{pq}[y, x],$$

and the Jacobi identity is replaced by

$$(-1)^{pr}[x, [y, z]] + (-1)^{pq}[y, [z, x]] + (-1)^{qr}[z, [x, y]] = 0$$

where, in addition, $z \in \mathfrak{g}_{(r)}$. Here, we are using the fact that the cohomology ring A is generated in degree 1: the signs in the general case are explained, for example, in [4].

If we let $\phi_{ij} = \dim_{\mathbb{Q}} \mathfrak{g}_{i,(j)}$ for $i, j \geq 0$, then $\phi_{i0} = \phi_i$, and the graded version of (1.9) reads

$$P_A(\mathbb{Q}, st^{-1}, t) = H(U(\mathfrak{g}), s, t) = \prod_{i,j \geq 0} \frac{(1 + s^i t^{2j+1})^{\phi_{i,(2j+1)}}}{(1 - s^i t^{2j})^{\phi_{i,(2j)}}}, \tag{2.6}$$

using (2.1).

Compared to the holonomy Lie algebra, little is known about the homotopy Lie algebra of an arrangement. On the positive side, the structure of $\mathfrak{g}(\mathcal{A})$ is described in [12] for arrangements obtained by intersecting supersolvable arrangements with certain linear subspaces. The resulting arrangements – a subclass of the hypersolvable arrangements – have a cohomology ring which is a Golod quotient of a Koszul algebra. We refer to [12] for details.

On the other hand, Roos [39] has shown that, for certain arrangements, $\mathfrak{g}(\mathcal{A})$ is badly behaved in two ways. First, $\mathfrak{g}(\mathcal{A})$ need not be finitely generated. In particular, $\mathfrak{g}(X_3)$ is not finitely generated. (Recall that we saw in Example 2 that the X_3 arrangement is decomposable, and so its holonomy Lie algebra is certainly finitely-generated.) Roos also has shown that the bigraded Hilbert series (2.6) need not be a rational function:

Example 3 ([39]). The arrangement of eight hyperplanes in $\mathbb{C}^3$ defined by $Q = xyz(x-y)(x-z)(y+z)(2x-y-z)(x-2y-z)$ has a transcendental Hilbert series $P_A(\mathbb{Q}, s, t)$. The Betti numbers of the linear strand are given by

$$H(U(\mathfrak{h}), t) = \left(\frac{1-t}{1-2t}\right)^7 \prod_{i \geq 3} (1 - t^i)^{-1}.$$

3. Part II: Resolutions over the exterior algebra

Some interesting, related free resolutions can be obtained by comparing a hyperplane complement with a torus, as follows. If $\mathcal{A}$ is an arrangement of n hyperplanes, its complement M can be regarded as the intersection of the torus $T = (\mathbb{C}^*)^n$ with a linear subspace of $\mathbb{C}^n$. Identify $H^*(T, \mathbb{Q})$ with the exterior algebra $E = \Lambda(V)$. Then the inclusion $i\colon M \hookrightarrow T$ induces a surjection in cohomology, $i^*\colon E \to A$, which is intrinsic to the combinatorics of the Orlik-Solomon algebra: see [14] in this volume.

This leads to some homological algebra over E. In this section, we will consider resolutions of $A = H^\bullet(M, \mathbb{Q})$ and its dual $A^* = H_\bullet(M, \mathbb{Q})$. By means of a beautiful theorem of Eisenbud, Popescu and Yuzvinsky [18], it turns out that these resolutions are closely related to the phenomenon of resonance discussed in Falk's survey [19].

3.1. The resolution of A over the exterior algebra

We begin with a fairly simple but illustrative example.

Example 4. For the X_3 arrangement, the first few Betti numbers of $\operatorname{Tor}^E(\mathbb{Q}, A)$, or equivalently of $\operatorname{Ext}_E(A, \mathbb{Q})$, are

```
0: 1 .  .  .  .   .   .
1: . 3  6  9 12  15  18
2: . 1  9 33 85 180 336
```

Certainly A does not possess a linear free resolution over E. However, there appear to be only two nonzero (interesting) rows: in other words, $b_{ij} = 0$ for $j > i + 2$. This reflects the fact that the Castelnuovo-Mumford regularity of A over E is at most $\ell - 1$; that is, for any arrangement of rank ℓ, we have $\dim_{\mathbb{Q}} \operatorname{Tor}_i^E(A, \mathbb{Q})_j = 0$ for $j \geq i + \ell$. (See [3, Lemma 2.5].) From the diagram, it also seems to be the case that $b_{i,i+1} = 3i$. We will see why in the continuation of this example. These numbers have a topological interpretation, given in Theorem 3.1, below.

Recall that the Lie algebra $\mathfrak{h}' = \mathfrak{h}_{\geq 2}$ inherits the bracket-length grading of $\mathfrak{h}$; therefore the quotient $\mathfrak{h}'/\mathfrak{h}''$ does as well. A result of Fröberg and Löfwall [24, Theorem 4.1(ii)] applies to show

$$\operatorname{Ext}_E(A, \mathbb{Q})_{(1)} \cong \mathfrak{h}'/\mathfrak{h}''$$

as a graded S-module, where $S = U(\mathfrak{h}/\mathfrak{h}') = \operatorname{Ext}_E(\mathbb{Q}, \mathbb{Q})$. Then $\dim_{\mathbb{Q}}(\mathfrak{h}'/\mathfrak{h}'')_i = \dim_{\mathbb{Q}}(\mathfrak{h}/\mathfrak{h}'')_{i+1}$ for $i \geq 1$, where $\mathfrak{h}/\mathfrak{h}''$ is the maximal metabelian quotient of the holonomy Lie algebra. That is, the first row of Betti numbers in the resolution of A can be regarded as the LCS ranks of a Lie algebra.

Once again, the 1-formality of arrangement groups makes it possible to interpret this topologically. Papadima and Suciu show in [32] that, if G is 1-formal, then the lower central series Lie algebra $\operatorname{gr}_{\mathbb{Q}}(G/G'')$ is isomorphic to $\mathfrak{h}/\mathfrak{h}''$. For any arrangement $\mathcal{A}$, then, let $G = \pi_1(M(\mathcal{A}))$, and for $i \geq 1$ let $\theta_i = \dim_{\mathbb{Q}}(\operatorname{gr}_i(G/G''))$, the *Chen ranks*. In [42], Schenck and Suciu prove the following:

Theorem 3.1 ([42]). *For all $i \geq 2$, $\theta_i = b_{i-1,i}$, where (b_{ij}) are the Betti numbers in the resolution of the Orlik-Solomon alegebra $A(\mathcal{A})$ over E.*

The proof in [42] is direct, using BGG duality to identify $\mathrm{Ext}_E(A, \mathbb{Q})_{(1)}$ as being a linearization of the Alexander invariant. This brings resonance into play, and we will return to this point in §3.4.

A combinatorial description even of the sublinear strand $b_{i-1,i}$ is unknown in general, but nice formulas exist for two special cases. The first is for decomposable arrangements, which were characterized by an isomorphism (2.4). In this case,

$$\mathfrak{h}'/\mathfrak{h}'' \cong \bigoplus_{X \in L_2(\mathcal{A})} \mathfrak{h}'_X/\mathfrak{h}''_X,$$

from [34, Theorem 6.2], so

$$\theta_k = \sum_{X \in L_2(\mathcal{A})} (k-1)\binom{|X|+k-3}{k} \tag{3.1}$$

for $k \geq 2$, by reducing to Chen's calculation [9] for free groups.

Example 4 (continued). We saw that the X_3 arrangement is decomposable. The only nonzero terms in (3.1) are contributed by the three triple points, so for $k \geq 1$, the sum (3.1) simplifies to

$$b_{k,k+1} = \theta_{k+1} = 3k,$$

as expected.

The second case is that of graphic arrangements. If Γ is a graph with m vertices, then $\mathcal{A}(\Gamma)$ is defined to be the arrangement in $\mathbb{C}^m$ with one hyperplane for each edge: $\{z_i - z_j = 0\colon \{i,j\} \in E(\Gamma)\}$. Then by [41, Lemma 6.9],

$$b_{k,k+1} = k(\kappa_2 + \kappa_3) \tag{3.2}$$

for all $k \geq 2$, where κ_s denotes the number of complete subgraphs in Γ with $s+1$ vertices. We refer to [41] for further explicit computations of the Betti numbers in this resolution and that of $\mathbb{Q}$ over A, which depend on some intricate combinatorics in the change of rings spectral sequence.

Problem 3.2. Give a direct, combinatorial interpretation of the integers

$$\dim_{\mathbb{Q}} \mathrm{Tor}_i^E(A, \mathbb{Q})_j$$

using the intersection lattice $L(\mathcal{A})$, for arbitrary i and j.

3.2. Homology of an arrangement complement

Let E^* denote the $\mathbb{Q}$-dual of E. By the universal coefficients theorem, $E^* \cong H_\bullet(T, \mathbb{Q})$. In homology, then, we have an inclusion (of coalgebras)

$$i_*\colon H_\bullet(M, \mathbb{Q}) \hookrightarrow H_\bullet(T, \mathbb{Q}) \cong E^*. \tag{3.3}$$

However, since the torus is an H-space, E^* is also an algebra: the exterior algebra on V^*. Let $\{e_i : H_i \in \mathcal{A}\}$ denote the standard basis of V, identifying it with $H^1(M, \mathbb{Q})$. Let $\{e_i^* : H_i \in \mathcal{A}\}$ be the dual basis.

Poincaré duality in the torus amounts to the following in the exterior algebra. Fix an ordered basis of V, or, equivalently, a choice of isomorphism $\det\colon E_n \cong \mathbb{Q}$. Then, for each p, there is a vector space isomorphism $\phi\colon E_p \to (E^*)_{n-p}$ given by

$$\phi(x)(y) = \det(xy). \tag{3.4}$$

Then, for all p, $0 \le -p \le n$,

$$\begin{aligned} A^*_p &= \operatorname{Hom}_{\mathbb{Q}}(A,\mathbb{Q})_p \\ &\cong \operatorname{Hom}_E(A,E)_{n+p} \qquad \text{by (3.4)}, \\ &\cong (\operatorname{ann} I)_{n+p}, \end{aligned} \tag{3.5}$$

where $\operatorname{ann} I$ denotes the annihilator ideal to the defining ideal I of the Orlik-Solomon algebra.

3.3. The resolution of A^* over the exterior algebra

Recall from [14, Section 2] that the ideal I is generated by elements $\partial(e_S)$, where the monomial e_S is indexed by a circuit: that is, a minimal dependent set of hyperplanes. With respect to the graded-lexicographic order, the leading monomial in $\partial(e_S)$ is $e_{S'}$, where, if i is the least index in S, then $S' = S - \{i\}$.

Additively, the initial ideal $\operatorname{in}(I) = \mathbb{Q}\{e_T\colon T \text{ contains a broken circuit}\}$.

The following construction captures the combinatorial nature of squarefree monomial ideals. As before let E be an exterior algebra with generators $\{e_1, e_2, \dots, e_n\}$. Let R be the polynomial algebra $\mathbb{Q}[x_1, x_2, \dots, x_n]$.

Definition 3.3. Let Δ be an abstract simplicial complex with vertices $[n] = \{1, 2, \dots, n\}$. The ideals I_Δ and J_Δ of R and E, respectively,

$$J_\Delta = (x_S \in R\colon S \notin \Delta) \text{ and } I_\Delta = (e_S \in E\colon S \notin \Delta)$$

are the *symmetric* and *exterior Stanley-Reisner ideals of* Δ.

Definition 3.4. Let Δ be a simplicial complex on vertices $[n]$ which is not the $n-1$ simplex. The combinatorial Alexander dual, $\Delta^\star$, is by definition the simplicial complex whose simplices are the complements of the nonsimplices of Δ:

$$\Delta^\star = \{\sigma \subseteq [n]\colon [n] - \sigma \notin \Delta\}.$$

It is straightforward to check that, for any Δ, we have

$$\operatorname{ann} I_\Delta = I_{\Delta^\star}. \tag{3.6}$$

For a fixed choice of arrangement and an order on its hyperplanes, the sets

$$\mathbf{nbc} = \{S \subset [n]\colon S \text{ does not contain a broken circuit}\}$$

form a simplicial complex called the *broken circuit complex* (see [6].) From the discussion above, we see $\operatorname{in}(I) = I_{\mathbf{nbc}}$.

Lemma 3.5. *For any homogeneous ideal I in E, we have* $\operatorname{in}(\operatorname{ann}(I)) = \operatorname{ann}(\operatorname{in}(I))$.

Proof. If $x \in \mathrm{ann}(I)$ is a homogeneous element, let e_S be its initial monomial. Replacing x by a nonzero scalar multiple, we get $x = e_S + x'$, where x' is supported on monomials $e_{S'}$ for which $e_S > e_{S'}$ in the term order. Then for any homogeneous $y \in I$, we may similarly write a multiple of y as $y = e_T + y'$, where e_T is the leading term of y. Since $e_S e_T$ is the leading monomial in xy, it must in fact be zero, since $xy = 0$. This shows $\mathrm{in}(\mathrm{ann}(I)) \subseteq \mathrm{ann}(\mathrm{in}(I))$.

Equality is established by checking that the Hilbert series of the two sides agree:

$$\begin{aligned} H(\mathrm{in}(\mathrm{ann}(I)), t) &= H(\mathrm{ann}(I), t) \\ &= (1+t)^n - t^n H(I, t^{-1}) \text{ by (3.5);} \\ &= (1+t)^n - t^n H(\mathrm{in}(I), t^{-1}) \\ &= H(\mathrm{ann}(\mathrm{in}(I)), t). \end{aligned}$$

□

Theorem 3.6 ([18]). *For any arrangement, A^* has a linear, minimal free resolution*

$$0 \longleftarrow A^* \longleftarrow E(\ell)^{b_{0,-\ell}} \longleftarrow E(\ell-1)^{b_{1,1-\ell}} \cdots \longleftarrow E(\ell-k)^{b_{k,k-\ell}} \longleftarrow \cdots$$

Sketch of proof: The first step is to reduce the problem to one of monomial ideals. Let us ignore degree shifts and replace A^* by $\mathrm{ann}\, I$ (using (3.5)). By Lemma 3.5 and (3.6), $\mathrm{in}(\mathrm{ann}\, I) = I_{\mathbf{nbc}^\star}$, since $\mathrm{in}(I) = I_{\mathbf{nbc}}$. If we could show that the monomial ideal $I_{\mathbf{nbc}^\star}$ has a linear resolution, then we would be done: the Gröbner deformation to the initial ideal is upper semicontinuous, so in particular if the initial ideal of $\mathrm{ann}(I)$ has a linear resolution, then so does $\mathrm{ann}(I)$.

It follows from a result of Aramova, Avramov, and Herzog [2] that a monomial ideal I_Δ in the exterior algebra has a linear resolution if and only if the corresponding squarefree monomial ideal J_Δ in the polynomial algebra R has a linear resolution. Such ideals have been studied extensively; in particular, Eagon and Reiner [15] show that J_Δ has a linear resolution if and only if the Stanley-Reisner ideal $J_{\Delta^\star}$ of the Alexander dual complex is Cohen-Macaulay.

Since Alexander duality is an involution, it is enough to know that the ideal $J_{\mathbf{nbc}}$ is Cohen-Macaulay. This amounts to a combinatorial condition on the simplicial complex; see [15] for details. The broken-circuit complex **nbc** is known to be *shellable* [38], a classical combinatorial property which implies $J_{\mathbf{nbc}}$ is Cohen-Macaulay. □

Then the Betti numbers of the resolution are given by (1.3):

$$\begin{aligned} \sum_{i \geq 0} b_{i,i-\ell} t^i &= (-t)^\ell H(A^*, -t)/H(E, -t) \\ &= (-t)^\ell \pi(\mathcal{A}, (-t)^{-1})/(1-t)^n \\ &= \chi(\mathcal{A}, t)/(1-t)^n, \end{aligned}$$

where $\chi(\mathcal{A}, t)$ denotes the characteristic polynomial of the arrangement.

3.4. Koszul modules

The definition of the quadratic dual of an algebra (1.8) admits a generalization to modules. Suppose B is a nonnegatively graded $\mathbb{Q}$-algebra and M is a finitely-generated left B-module, generated in degree 0. Once again, let $n = \dim_{\mathbb{Q}} B_1$, let $V = B_1$, and suppose further that M has a linear presentation $M = \operatorname{coker} f\colon B^k \to B^m$. Since f is given by a matrix with entries in B_1, it is determined by its degree-zero part, which is just a map of vector spaces $f_0\colon \mathbb{Q}^k \to V \otimes \mathbb{Q}^m$.

Then we define $M^!_B$ to be the left $B^!$-module given by the following linear presentation. Let $f_0^\perp\colon \mathbb{Q}^{mn-k} \to V^* \otimes \mathbb{Q}^m$ be a $\mathbb{Q}$-linear map onto the complement of the image of f_0. Define a map $f^!\colon (B^!)^{mn-k} \to (B^!)^m$ by letting it act in degree zero by $f_0^\perp$, and extending by the left action of $B^!$. Set

$$M^!_B = \operatorname{coker} f^!\colon (B^!)^{mn-k} \to (B^!)^m. \tag{3.7}$$

As in the case of algebras, the quadratic dual of a module can be understood in terms of resolutions. Let B be a Koszul algebra. If M is a left B-module, then the Yoneda product makes $\operatorname{Ext}_B(M,\mathbb{Q})$ a left $B^!$-module, since $B^! = \operatorname{Ext}_B(\mathbb{Q},\mathbb{Q})$. If $\{b_{ij}\}$ denote as usual the Betti numbers in the minimal free resolution of M, then

$$b_{ij} = \dim_{\mathbb{Q}} \operatorname{Ext}^i_B(M,\mathbb{Q})_{-j},$$

for all $i, j \geq 0$. The submodule of $\operatorname{Ext}_B(M,\mathbb{Q})$ corresponding to the linear strand turns out to be isomorphic to the quadratic dual of M: that is,

$$\operatorname{Ext}_B(M,\mathbb{Q})_{(0)} \cong M^!_B$$

as left $B^!$-modules. If the module M has a linear free resolution, we have:

$$\operatorname{Ext}_B(M,\mathbb{Q}) = \operatorname{Ext}_B(M,\mathbb{Q})_{(0)} \cong M^!_B.$$

In this case, M is called a *Koszul module.* It is not hard to check that quadratic duality for modules is an involution, which has the following good consequence.

Proposition 3.7 ([37]). *If B is a Koszul algebra, then M is a Koszul B-module if and only if $M^!_B$ is a Koszul $B^!$-module.*

That is, if M has a linear resolution, so does $M^!_B$. Additively, for $p \geq 0$, the pth term in the resolution of $M^!_B$ is $(M^*)_{-p} \otimes_{\mathbb{Q}} B^!(-p)$. In this language, Theorem 3.6 says that, for any arrangement, $A^*(-\ell)$ is a Koszul E-module. Let $F(\mathcal{A}) = \operatorname{Ext}_S(A^*(-\ell),\mathbb{Q})$, its quadratic dual R-module. By Proposition 3.7, $F(\mathcal{A})$ is a Koszul R-module. Since $F(\mathcal{A})^!_S \cong A^*(-\ell)$, the module $F(\mathcal{A})$ has a linear resolution of the form

$$0 \longleftarrow F(\mathcal{A}) \longleftarrow A_\ell \otimes S \longleftarrow A_{\ell-1} \otimes S(-1) \cdots \longleftarrow A_0 \otimes S(-\ell) \longleftarrow 0 \tag{3.8}$$

via the identification $A^*(-\ell)^*_{-p} \cong A_{\ell-p}$. It turns out that the differential is given by

$$e \otimes x \mapsto \sum_{i=1}^n ee_i \otimes x_i x,$$

where $\{x_i\}$ is the dual basis to the basis $\{e_i\}$ of A_1.

We note that the module $F(\mathcal{A})$ and its resolution (3.8) are originally introduced in [18] by means of Bernstein-Gelfand-Gelfand duality, as developed in Eisenbud-Fløystad-Schreyer [17] and explained in [16, Chapter 7]. BGG duality makes use of the special properties of the Koszul dual rings E and S; in particular, since $F(\mathcal{A})$ has a linear resolution, it implies the following:

Proposition 3.8. *For all j, there is an isomorphism of graded S-modules*

$$\mathrm{Ext}_S^{\ell-j}(F(\mathcal{A}),S) \cong \mathrm{Ext}_E(A,\mathbb{Q})_{(j)}.$$

In particular, for each fixed j, the Betti numbers $b_{i,i+j}$ in the resolution of A can be interpreted as the Hilbert series of an S-module, $\mathrm{Ext}_S^{\ell-j}(F(\mathcal{A}),S)$. With this in mind, Schenck and Suciu [42] proved Theorem 3.1 by understanding the S-module $\mathrm{Ext}_S^{\ell-1}(F(\mathcal{A}),S)$ as the linearized Alexander invariant (see [29], and more recently, [32, 33].)

3.5. Resonance

In Falk's lecture notes in this volume [19], he defines the resonance varieties of an arrangement, for $p \geq 1$, as

$$\mathcal{R}^p(\mathcal{A}) = \left\{a \in \overline{A}_1 - \{0\} : H^p(\overline{A},a) \neq 0\right\}, \tag{3.9}$$

where $(\overline{A},a)$ denotes the projective Orlik-Solomon algebra of $\mathcal{A}$ over $\mathbb{C}$, regarded as a cochain complex with a differential given by (right) multiplication by the element a. Since each $\mathcal{R}^p(\mathcal{A})$ is invariant under multiplication by $\mathbb{C}^*$, the orbits form a projective variety $\mathbb{P}\mathcal{R}^p(\mathcal{A})$ in $\mathbb{P}^{n-2}$.

We need to make a slight translation to fit the notation here. Recall from [19] that the derivation $\partial\colon E \to E$ defined by $\partial(e_i) = 1$ induces a well-defined differential on the Orlik-Solomon algebra; moreover, the chain complex $(A_\bullet,\partial)$ is exact, and one may identify $\overline{A}$ with $\ker\partial$ (or, equivalently, $\mathrm{im}(\partial)$.) A more general and geometric discussion may be found in [13].

If $a = \sum_{i=1}^n a_i e_i$, then $\partial(a) = \sum_{i=1}^n a_i$, so $\overline{A}_1 = \{a \in A_1 \colon \sum_{i=1}^n a_i = 0\}$. It is straightforward to check that the map $(\partial a + a\partial)\colon A \to A$ is multiplication by $\sum_{i=1}^n a_i$. Reading this as a chain homotopy on the cochain complex (A,a), multiplication by $\sum_{i=1}^n a_i$ is an isomorphism (over $\mathbb{C}$) unless this sum is zero, so (A,a) is exact unless $a \in \overline{A}_1$. On the other hand, it also says ∂ is a chain map provided $a \in \overline{A}_1$ which, together with exactness of ∂, gives a short exact sequence of chain complexes

$$0 \longrightarrow (\overline{A},a) \longrightarrow (A,a) \xrightarrow{\partial} (\overline{A},a)[-1] \longrightarrow 0.$$

Multiplication by any $a' \in A_1$ with $\partial(a') \neq 0$ gives a section, so the long exact sequence breaks up, and

$$H^p(A,a) = \begin{cases} H^p(\overline{A},a) \oplus H^{p-1}(\overline{A},a) & \text{for } a \in \overline{A}_1; \\ 0 & \text{otherwise} \end{cases}$$

for all p.

This means that we can look at resonance varieties using the resolution (3.8). The differential in the complex (3.8) is multiplication by the element $\omega = \sum_{i=1}^{n} e_i \otimes x_i \in A_1 \otimes S_1$, which we can specialize to the differential in any complex (A, a). Formally, for $a \in A_1$, define a homomorphism $S \to \mathbb{Q}$ by $x_i \mapsto a_i$ for $1 \leq i \leq n$. Let $\mathbb{Q}_a$ denote $\mathbb{Q}$, regarded as an S-module in this way. Then, as a consequence of Theorem 3.6,

Proposition 3.9. *For all* $0 \leq i \leq \ell$,

$$H^p(A, a) = \operatorname{Tor}^S_{\ell-p}(F(\mathcal{A}), \mathbb{Q}_a).$$

There is also a projective version of the module $F(\mathcal{A})$: we simply let $\overline{S} = S/(\sum_{i=1}^{n} x_i)$, and define

$$F(\overline{\mathcal{A}}) = F(\mathcal{A}) \otimes_S \overline{S}.$$

Then, by a modification of the same argument, $F(\overline{\mathcal{A}})$ has a linear resolution over $\overline{S}$:

$$0 \longleftarrow F(\overline{\mathcal{A}}) \longleftarrow \overline{A}_{\ell-1} \otimes \overline{S} \longleftarrow \overline{A}_{\ell-2} \otimes \overline{S}(-1) \cdots \longleftarrow \overline{A}_0 \otimes \overline{S}(-\ell+1) \longleftarrow 0 \tag{3.10}$$

with differential given by multiplying by the image of ω in $\overline{A} \otimes \overline{S}$. As long as $\partial(a) = 0$, the action of R on $\mathbb{Q}_a$ factors through $\overline{S}$, and we obtain

$$H^p(\overline{A}, a) = \operatorname{Tor}^{\overline{S}}_{\ell-1-p}(F(\overline{A}), \mathbb{Q}_a).$$

Since nonvanishing Tor groups appear with consecutive homological indices, we arrive at another corollary to Theorem 3.6, stated as [18, Theorem 4.1(ii)]:

Corollary 3.10. *For any arrangement* $\mathcal{A}$ *of rank* ℓ*, we have*

$$\emptyset = \mathcal{R}^0(\mathcal{A}) \subseteq \mathcal{R}^1(\mathcal{A}) \subseteq \mathcal{R}^2(\mathcal{A}) \subseteq \cdots \subseteq \mathcal{R}^{\ell-1}(\mathcal{A}) = \overline{A}_1.$$

As explained in [14], resonance is closely related to cohomology of local systems. This leads to a question:

Problem 3.11. If $\mathcal{L}_\lambda$ is a rank-one local system on the complement M of a rank ℓ hyperplane arrangement, is it necessarily the case that

$$H^i(M, \mathcal{L}_\lambda) \neq 0 \Rightarrow H^{i+1}(M, \mathcal{L}_\lambda) \neq 0$$

for all $i < \ell$? (From [14, Corollary 6.7], this follows from Corollary 3.10 for λ close to the trivial representation $\mathbb{1}$.)

3.6. Resonance and Betti numbers

One can obtain the first resonance variety directly from the module $F(\mathcal{A})$: Schenck and Suciu [42] show that

$$\mathcal{R}^1(\mathcal{A}) \cup \{0\} = V(\operatorname{ann} \operatorname{Ext}^{\ell-1}_S(F(\mathcal{A}), S)), \tag{3.11}$$

the subvariety of $\mathbb{C}^n$ defined by the vanishing of the annihilator ideal. A more complicated relationship occurs for the higher resonance varieties: see [11].

We already saw that the components of the resonance varieties are linear [19, Theorem 4.16], and the components of $\mathbb{P}\mathcal{R}^1(\mathcal{A})$ have empty intersections. Let h_r

denote the number of components of $\mathbb{P}\mathcal{R}^1(\mathcal{A})$ of (projective) dimension r, for $r \geq 0$. Using (3.11), these numbers bound the Hilbert series of $\mathrm{Ext}_S^{\ell-1}(F(\mathcal{A}), S)$, which is equivalent to the Betti numbers $\mathrm{Ext}_E^i(A, \mathbb{Q})_{(1)}$ (Proposition 3.8), and equivalent to the Chen ranks (Theorem 3.1). In the last formulation,

$$\theta_k \geq (k-1) \sum_{r \geq 1} h_r \binom{r+k-1}{k} \tag{3.12}$$

for sufficiently large k.

The Chen Ranks Conjecture ([44, 42]) states that (3.12) is in fact an equality for all sufficiently large k. It has been verified for graphic and decomposable arrangements, by comparing the explicit calculations of the Chen ranks in (3.2) and (3.1), respectively, with matching calculations of the resonance varieties. However, the general case remains open and makes a good problem with which to conclude these lecture notes:

Problem 3.12. Show that, for any arrangement $\mathcal{A}$,

$$\theta_k = (k-1) \sum_{r \geq 1} h_r \binom{r+k-1}{k} \qquad \text{for } k >> 0,$$

where h_r is the number of components of $\mathcal{P}\mathcal{R}^1(\mathcal{A})$ of dimension r, and θ_k is the rank of the kth lower central series quotient of G/G''.

References

[1] David J. Anick, *On the homology of associative algebras*, Trans. Amer. Math. Soc. **296** (1986), no. 2, 641–659. MR846601 (87i:16046)

[2] Annetta Aramova, Luchezar L. Avramov, and Jürgen Herzog, *Resolutions of monomial ideals and cohomology over exterior algebras*, Trans. Amer. Math. Soc. **352** (2000), no. 2, 579–594. MR1603874 (2000c:13021)

[3] Annetta Aramova, Jürgen Herzog, and Takayuki Hibi, *Gotzmann theorems for exterior algebras and combinatorics*, J. Algebra **191** (1997), no. 1, 174–211. MR1444495 (98c:13025)

[4] Luchezar L. Avramov, *Local algebra and rational homotopy*, Algebraic homotopy and local algebra (Luminy, 1982), Astérisque, vol. 113, Soc. Math. France, Paris, 1984, pp. 15–43. MR749041 (85j:55021)

[5] Alexander Beilinson, Victor Ginzburg, and Wolfgang Soergel, *Koszul duality patterns in representation theory*, J. Amer. Math. Soc. **9** (1996), no. 2, 473–527. MR1322847 (96k:17010)

[6] Anders Björner and Günter M. Ziegler, *Broken circuit complexes: factorizations and generalizations*, J. Combin. Theory Ser. B **51** (1991), no. 1, 96–126. MR1088629 (92b:52027)

[7] N. Bourbaki, *Éléments de mathématique. Algèbre. Chapitre 10. Algèbre homologique*, Springer Verlag, Berlin, 2007, Reprint of the 1980 original [Masson, Paris; MR0610795]. MR2327161

[8] Egbert Brieskorn, *Sur les groupes de tresses [d'après V. I. Arnol′d]*, Séminaire Bourbaki, 24ème année (1971/1972), Exp. No. 401, Springer, Berlin, 1973, pp. 21–44. Lecture Notes in Math., Vol. 317. MR0422674 (54 #10660)

[9] Kuo-Tsai Chen, *Integration in free groups*, Ann. of Math. (2) **54** (1951), 147–162. MR0042414 (13,105c)

[10] ———, *Extension of C^∞ function algebra by integrals and Malcev completion of π_1*, Advances in Math. **23** (1977), no. 2, 181–210. MR0458461 (56 #16664)

[11] Graham Denham and Henry K. Schenck, *A spectral sequence stratification of cohomology jump loci*, preprint.

[12] Graham Denham and Alexander I. Suciu, *On the homotopy Lie algebra of an arrangement*, Michigan Math. J. **54** (2006), no. 2, 319–340. MR2252762 (2007f:17039)

[13] Alexandru Dimca, *Singularities and topology of hypersurfaces*, Universitext, Springer Verlag, New York, 1992. MR1194180 (94b:32058)

[14] Alexandru Dimca and Sergey Yuzvinsky, *Lectures on Orlik-Solomon algebras*, this volume, 83–110.

[15] John A. Eagon and Victor Reiner, *Resolutions of Stanley-Reisner rings and Alexander duality*, J. Pure Appl. Algebra **130** (1998), no. 3, 265–275. MR1633767 (99h:13017)

[16] David Eisenbud, *The geometry of syzygies*, Graduate Texts in Mathematics, vol. 229, Springer Verlag, New York, 2005, A second course in commutative algebra and algebraic geometry. MR2103875 (2005h:13021)

[17] David Eisenbud, Gunnar Fløystad, and Frank-Olaf Schreyer, *Sheaf cohomology and free resolutions over exterior algebras*, Trans. Amer. Math. Soc. **355** (2003), no. 11, 4397–4426 (electronic). MR1990756 (2004f:14031)

[18] David Eisenbud, Sorin Popescu, and Sergey Yuzvinsky, *Hyperplane arrangement cohomology and monomials in the exterior algebra*, Trans. Amer. Math. Soc. **355** (2003), no. 11, 4365–4383 (electronic). MR1986506 (2004g:52036)

[19] Michael Falk, *Geometry and combinatorics of resonant weights*, this volume, 155–176.

[20] Michael Falk and Richard Randell, *The lower central series of a fiber-type arrangement*, Invent. Math. **82** (1985), no. 1, 77–88. MR808110 (87c:32015b)

[21] ———, *On the homotopy theory of arrangements. II*, Arrangements—Tokyo 1998, Adv. Stud. Pure Math., vol. 27, Kinokuniya, Tokyo, 2000, pp. 93–125. MR1796895 (2002b:32044)

[22] R. Fox and L. Neuwirth, *The braid groups*, Math. Scand. **10** (1962), 119–126. MR0150755 (27 #742)

[23] R. Fröberg, *Koszul algebras*, Advances in commutative ring theory (Fez, 1997), Lecture Notes in Pure and Appl. Math., vol. 205, Dekker, New York, 1999, pp. 337–350. MR1767430 (2001i:16046)

[24] R. Fröberg and C. Löfwall, *Koszul homology and Lie algebras with application to generic forms and points*, Homology Homotopy Appl. **4** (2002), no. 2, part 2, 227–258 (electronic), The Roos Festschrift volume, 2. MR1918511 (2003g:13018)

[25] Daniel R. Grayson and Michael E. Stillman, *Macaulay 2, a software system for research in algebraic geometry*, Available at http://www.math.uiuc.edu/Macaulay2/.

[26] Toshitake Kohno, *On the holonomy Lie algebra and the nilpotent completion of the fundamental group of the complement of hypersurfaces*, Nagoya Math. J. **92** (1983), 21–37. MR726138 (85d:14032)

[27] ———, *Série de Poincaré-Koszul associée aux groupes de tresses pures*, Invent. Math. **82** (1985), no. 1, 57–75. MR808109 (87c:32015a)

[28] Clas Löfwall, *On the subalgebra generated by the one-dimensional elements in the Yoneda Ext-algebra*, Algebra, algebraic topology and their interactions (Stockholm, 1983), Lecture Notes in Math., vol. 1183, Springer, Berlin, 1986, pp. 291–338. MR846457 (88f:16030)

[29] W. S. Massey, *Completion of link modules*, Duke Math. J. **47** (1980), no. 2, 399–420. MR575904 (81g:57004)

[30] John W. Milnor and John C. Moore, *On the structure of Hopf algebras*, Ann. of Math. (2) **81** (1965), 211–264. MR0174052 (30 #4259)

[31] P. Orlik and H. Terao, *Arrangements of hyperplanes*, Grundlehren der Mathematischen Wissenschaften, no. 300, Springer Verlag, 1992.

[32] Stefan Papadima and Alexander I. Suciu, *Chen Lie algebras*, Int. Math. Res. Not. (2004), no. 21, 1057–1086. MR2037049 (2004m:17043)

[33] ———, *Algebraic invariants for right-angled Artin groups*, Math. Ann. **334** (2006), no. 3, 533–555. MR2207874 (2006k:20078)

[34] ———, *When does the associated graded Lie algebra of an arrangement group decompose?*, Comment. Math. Helv. **81** (2006), no. 4, 859–875. MR2271225 (2007h:52028)

[35] Stefan Papadima and Sergey Yuzvinsky, *On rational $K[\pi, 1]$ spaces and Koszul algebras*, J. Pure Appl. Algebra **144** (1999), no. 2, 157–167. MR1731434 (2000k:55017)

[36] Irena Peeva, *Hyperplane arrangements and linear strands in resolutions*, Trans. Amer. Math. Soc. **355** (2003), no. 2, 609–618 (electronic). MR1932716 (2004c:13020)

[37] Alexander Polishchuk and Leonid Positselski, *Quadratic algebras*, University Lecture Series, vol. 37, American Mathematical Society, Providence, RI, 2005. MR2177131 (2006f:16043)

[38] J. Scott Provan and Louis J. Billera, *Decompositions of simplicial complexes related to diameters of convex polyhedra*, Math. Oper. Res. **5** (1980), no. 4, 576–594. MR593648 (82c:52010)

[39] Jan-Erik Roos, *The homotopy Lie algebra of a complex hyperplane arrangement is not necessarily finitely presented*, Experimental Mathematics **17** (2008), no. 2, 129–143.

[40] Jan-Erik Roos and Bernd Sturmfels, *A toric ring with irrational Poincaré-Betti series*, C. R. Acad. Sci. Paris Sér. I Math. **326** (1998), no. 2, 141–146. MR1646972 (2000e:14042)

[41] Henry K. Schenck and Alexander I. Suciu, *Lower central series and free resolutions of hyperplane arrangements*, Trans. Amer. Math. Soc. **354** (2002), no. 9, 3409–3433 (electronic). MR1911506 (2003k:52022)

[42] ———, *Resonance, linear syzygies, Chen groups, and the Bernstein-Gelfand-Gelfand correspondence*, Trans. Amer. Math. Soc. **358** (2006), no. 5, 2269–2289 (electronic). MR2197444 (2007a:52026)

[43] Brad Shelton and Sergey Yuzvinsky, *Koszul algebras from graphs and hyperplane arrangements*, J. London Math. Soc. (2) **56** (1997), no. 3, 477–490. MR1610447 (99c:16044)

[44] Alexander I. Suciu, *Fundamental groups of line arrangements: enumerative aspects*, Advances in algebraic geometry motivated by physics (Lowell, MA, 2000), Contemp. Math., vol. 276, Amer. Math. Soc., Providence, RI, 2001, pp. 43–79. MR1837109 (2002k:14029)

[45] Dennis Sullivan, *Infinitesimal computations in topology*, Inst. Hautes Études Sci. Publ. Math. (1977), no. 47, 269–331 (1978). MR0646078 (58 #31119)

[46] Hiroaki Terao, *Modular elements of lattices and topological fibration*, Adv. in Math. **62** (1986), no. 2, 135–154. MR865835 (88b:32032)

[47] S. Yuzvinsky, *Orlik-Solomon algebras in algebra and topology*, Uspekhi Mat. Nauk **56** (2001), no. 2(338), 87–166. MR1859708 (2002i:14047)

Graham Denham
Department of Mathematics
University of Western Ontario
London, Ontario N6A 5B7
Canada
e-mail: gdenham@uwo.ca

Progress in Mathematics, Vol. 283, 59–82

Pencils of Plane Curves and Characteristic Varieties

Alexandru Dimca

Abstract. We explore the relation between the positive dimensional irreducible components of the characteristic varieties of rank 1 local systems on a plane curve arrangement complement and the associated pencils of plane curves. We discuss from this point the minimal arrangements and the fibered complements introduced by Falk and Yuzvinsky in their recent paper [17].

Mathematics Subject Classification (2000). Primary 14C21, 14F99, 32S22; Secondary 14E05, 14H50.

Keywords. Local system, constructible sheaf, twisted cohomology, characteristic variety, pencil of plane curves.

1. Introduction

Let C be a curve arrangement in the complex projective plane $\mathbb{P}^2$ and let $M = M(C) = \mathbb{P}^2 \setminus C$ be the corresponding complement. The characteristic varieties $\mathcal{V}_m(M)$ (resp. the resonance varieties $\mathcal{R}_m(M)$) describe the jumping loci for the dimension of the twisted cohomology group $H^1(M, \mathcal{L})$, with $\mathcal{L}$ a rank 1 local system on the complement M, (resp. a rank 1 local system $\mathcal{L}$ close to the trivial local system), see for details Section 3 below. Since we are interested only in the first cohomology groups, the more general case when M is a *hypersurface complement* in some $\mathbb{P}^n$ for $n \geq 2$ can be reduced to the situation at hand, i.e., $n = 2$, by using a generic 2-dimensional linear section and applying the Zariski Theorem on fundamental groups, see for instance [11], p. 25.

The *main aim* of this paper is to study the translated components of the characteristic variety $\mathcal{V}_1(M)$. According to Arapura's results, such a component $W = W(f, \rho)$ is described by a pair (f, ρ) where:

(a) f is a surjective morphism $M \to S = \mathbb{P}^1 \setminus B$, which is induced by a pencil $\mathcal{C}$ of plane curves, see Proposition 2.2.

(b) ρ is a torsion character such that W is the translate by ρ of a subtorus constructed via f.

Our results in this paper can be described as follows.

The set of mappings f arising in (a) above are parametrized by the (rationally defined) maximal isotropic linear subspaces $E \subset H^1(M,\mathbb{C})$. (In fact, in the case $\dim E = 1$, not all rationally defined maximal isotropic linear subspaces yield components in $\mathcal{V}_1(M)$, see Remark 6.3 (ii)). When $\dim E \geq 2$, then this maximal isotropy condition is equivalent to asking E to be an irreducible component of the resonance variety $\mathcal{R}_1(M)$, see Corollary 3.15, Corollary 3.17, Proposition 6.1 and Corollary 6.2. Moreover in this case the rationality condition is automatically fulfilled, see Remark 3.16.

If the arrangement C is given (e.g. the equations $f_j = 0$ for the components C_j of C are known), then the map f associated to a (rationally defined) maximal isotropic linear subspace $E \subset H^1(M,\mathbb{C})$ can be constructed explicitly, see Propositions 3.19 and 3.20. These two results can be regarded as a non-proper Castelnuovo-De Franchis Lemma, see [4], [6] and especially [14] for an in-depth discussion.

However, it is not clear whether this construction is *combinatorial* in the case of a line arrangement.

The 1-dimensional case is the most mysterious, and Suciu's example of such a component for the deleted B_3-arrangement given in [25], [26] played a key role in our understanding of this question. We consider this component in detail in Examples 3.11 and 6.7, together with its generalization given by the $\mathcal{A}_m$-arrangements discussed in [7] and [9], as a good test for our results, see Examples 6.8 and 6.9.

In Section 4, Theorem 4.1 is our analog of the main results of Falk and Yuzvinsky in [17], where they study the relation between the irreducible components of the resonance varieties for a *line arrangement* $\mathcal{A}$ and the associated pencils of plane curves via a combinatorial approach, based on the description of the irreducible components of the resonant variety $\mathcal{R}_1(M(\mathcal{A}))$ in terms of generalized Cartan matrices, obtained by Libgober and Yuzvinsky [21]. We obtain a necessary numerical condition involving self-intersection numbers for the existence of an essential positive dimensional irreducible component of a characteristic variety. This is practically the same condition as that in Theorem 4.1.1 in Libgober [20], which was established via the use of adjunction ideals.

In Section 5 we discuss the complements M which are in an obvious way $K(\pi,1)$-spaces, i.e., for which the mapping $f : M \to S$ considered above is a fibration, and we conclude with Example 5.2 where several of the above features are clearly illustrated. In particular we point out several differences with the case of line arrangements.

In [13] we emphasize the key role played in this setting by the constructible sheaves obtained as direct images of local systems on M under the mapping $f : M \to S$. These results are briefly reviewed in Section 6. Next we associate to a map $f : M \to S$ as above a finite abelian group $T(f)$, such that the torsion character ρ

is determined by a character $\tilde{\rho}$ of $T(f)$, see also [13]. Then we compute this group $T(f)$ in terms of the pencil associated to f, see Theorem 6.6. This result explains why usually $T(f) = 1$ and hence there are no translated components associated to f. Several examples are discussed, including a recent one discovered by Z. Raza, see [23].

We would like to thank Alexander Suciu for interesting and stimulating discussions on the subject of this paper, and for suggesting several improvements of the presentation.

2. On rational maps from $\mathbb{P}^2$ to $\mathbb{P}^1$

Let $f : \mathbb{P}^2 \to \mathbb{P}^1$ be a rational map. Then there is a minimal non-empty finite set $A \subset \mathbb{P}^2$ such that f is defined on $U = \mathbb{P}^2 \setminus A$. We recall the following basic fact, see also Proposition 2.4 in [13].

Proposition 2.1. *Any morphism $f : U \to \mathbb{P}^1$ is given by a pencil $\mathcal{C} : \alpha_1 P_1 + \alpha_2 P_2$ of plane curves having the base locus $V(P_1, P_2)$, the minimal finite set A. This pencil is unique up-to an automorphism of $\mathbb{P}^1$.*

Let $C \subset \mathbb{P}^2$ be a reduced curve such that $C = \cup_{j=1,r} C_j$, with C_j an irreducible curve of degree d_j. We set $M = \mathbb{P}^2 \setminus C$. For the proof of the following result we refer to Proposition 2.6 in [13].

Proposition 2.2. *Let $B \subset \mathbb{P}^1$ be a finite set and denote by S the complement $\mathbb{P}^1 \setminus B$. For any surjective morphism $f : M \to S$ there is a pencil $\mathcal{C} : \alpha_1 P_1 + \alpha_2 P_2$ of curves in $\mathbb{P}^2$ such that any irreducible component C_j of C appears in one of the three following cases.*

(1) *C_j is contained in a curve $\mathcal{C}_b$ in the pencil $\mathcal{C}$, corresponding to a point $b \in B$;*
(2) *C_j is strictly contained in a curve $\mathcal{C}_s$ in the pencil $\mathcal{C}$, corresponding to a point $s \in S$;*
(3) *C_j is a horizontal component, i.e., C_j intersects the generic fiber $\mathcal{C}_t$ of the pencil $\mathcal{C}$ outside the base locus.*

Moreover, C_j appears in the first case above if and only if the homology class γ_j of a small loop around C_j satisfies $H_1(f)(\gamma_j) \neq 0$ in $H_1(S, \mathbb{Z})$.

Definition 2.3. In the setting of Proposition 2.2, we say that the curve arrangement C is *minimal with respect to the mapping* $f : M \to S$ if any component C_j of C is of type (1), i.e., C_j is contained in a curve $\mathcal{C}_b$ in the pencil $\mathcal{C}$, corresponding to a point $b \in B$. We say that the curve arrangement C is *special with respect to the mapping* $f : M \to S$ if some component C_j of C is of type (2), i.e., C_j is strictly contained in a curve $\mathcal{C}_s$ in the pencil $\mathcal{C}$, corresponding to a point $s \in S$.

Remark 2.4. If $|B| > 1$, then the base locus $\mathcal{X}$ of the pencil $\mathcal{C}$ is just the intersection of any two distinct fibers $\mathcal{C}_b \cap \mathcal{C}_{b'}$ for b, b' distinct points in B. Note also that the second case (2) above cannot occur if all the fibers $\mathcal{C}_s$ for $s \in S$ are irreducible.

Note that the fibers $\mathcal{C}_s$ may be non-reduced, i.e., we consider them usually as divisors. Saying that C_j is contained in $\mathcal{C}_s$ means that $\mathcal{C}_s = m_jC_j + \cdots$, with $m_j > 0$. On the other hand, C is always regarded as a reduced curve.

3. Characteristic varieties and resonance varieties

3.1. Local systems on S

Here we return to the notation $S = \mathbb{P}^1 \setminus B$, with $B = \{b_1, \dots, b_k\}$ a finite set of cardinals $|B| = k > 0$. If δ_i denotes an elementary loop based at some base point $b \in B$ and turning once around the point b_i, then using the usual choices, the fundamental group of S is given by

$$\pi_1(S) = < \delta_1, \dots, \delta_k \mid \delta_1 \cdots \delta_k = 1 > . \tag{3.1}$$

It follows that the first integral homology group is given by

$$H_1(S) = \mathbb{Z} < \delta_1, \dots, \delta_k > / < \delta_1 + \cdots + \delta_k = 0 > . \tag{3.2}$$

Hence, the rank 1 local systems on S are parametrized by the $(k-1)$-dimensional algebraic torus

$$\mathbb{T}(S) = \mathrm{Hom}(H_1(S), \mathbb{C}^*) = \{\lambda = (\lambda_1, \dots, \lambda_k) \in (\mathbb{C}^*)^k \mid \lambda_1 \cdots \lambda_k = 1\}. \tag{3.3}$$

Here $\lambda_j \in \mathbb{C}^*$ is the monodromy about the point $b_j \in B$. For $\lambda \in \mathbb{T}(S)$, we denote by $\mathcal{L}_\lambda$ the corresponding rank 1 local system on S.

3.2. Local systems on M

Let γ_j be an elementary loop around the irreducible component C_j, for $j = 1, \dots, r$. Then it is known, see for instance [11], p. 102, that

$$H_1(M) = \mathbb{Z} < \gamma_1, \dots, \gamma_r > / < d_1\gamma_1 + \cdots + d_r\gamma_r = 0 > \tag{3.4}$$

where d_j is the degree of the component C_j. It follows that the rank 1 local systems on M are parametrized by the algebraic group

$$\mathbb{T}(M) = \mathrm{Hom}(H_1(M), \mathbb{C}^*) = \{\rho = (\rho_1, \dots, \rho_r) \in (\mathbb{C}^*)^r \mid \rho_1^{d_1} \cdots \rho_r^{d_r} = 1\}. \tag{3.5}$$

The connected component $\mathbb{T}^0(M)$ of the unit element $1 \in \mathbb{T}(M)$ is the $(r-1)$-dimensional torus given by

$$\mathbb{T}^0(M) = \{\rho = (\rho_1, \dots, \rho_r) \in (\mathbb{C}^*)^r \mid \rho_1^{e_1} \cdots \rho_r^{e_r} = 1\} \tag{3.6}$$

with $D = G.C.D.(d_1, \dots, d_r)$ and $e_j = d_j/D$ for $j = 1, \dots, r$.

Remark 3.3. If $d_1 = 1$, then $\{\gamma_2, \dots, \gamma_r\}$ is a basis for $H_1(M)$ and the torus $\mathbb{T}(M)$ can be identified with $(\mathbb{C}^*)^{r-1}$ under the projection $\rho \mapsto (\rho_2, \dots, \rho_r)$.

Now the computation of the twisted cohomology groups $H^m(M, \mathcal{L}_\rho)$ is one of the major problems. The case when $\mathcal{L}_\rho = \mathbb{C}_M$ is easy, and the result depends on the local singularities of the plane curve C. In fact $\dim H^0(M,\mathbb{C}) = 1$, $\dim H^1(M,\mathbb{C}) = r-1$ and $H^m(M,\mathbb{C}) = 0$ for $m \geq 3$. To determine the remaining Betti number $b_2(M) = \dim H^2(M,\mathbb{C})$ is the same as determining the Euler characteristic $\chi(M) = 3 - \chi(C)$ and this can be done, e.g. by using the formula for $\chi(C)$ given in [11], p. 162.

In the sequel we concentrate on the case $\mathcal{L}_\rho \neq \mathbb{C}_M$ and assume $\chi(M)$ known. Then we have $H^m(M,\mathcal{L}_\rho) = 0$ for $m = 0$ and $m \geq 2$, and $\dim H^2(M,\mathcal{L}_\rho) - \dim H^1(M,\mathcal{L}_\rho) = \chi(M)$, see for instance [12], p. 49. To study these cohomology groups, one idea is to study the characteristic varieties

$$\mathcal{V}_m(M) = \{\rho = (\rho_1,\ldots,\rho_r) \in \mathbb{T}(M) \mid \dim H^1(M,\mathcal{L}_\rho) \geq m\}. \tag{3.7}$$

Definition 3.4. An irreducible component W of such an m-th characteristic variety $\mathcal{V}_m(M)$ is called a *coordinate component* (resp. a *translated coordinate component*) if W is contained in a subgroup $\mathbb{T}_j$ of $\mathbb{T}(M)$ defined by an equality $\rho_j = 1$ for some j (resp. there is a torsion character $\rho \in \mathbb{T}(M)$ such that $W \subset \rho\mathbb{T}_j$ for some j). An irreducible component W which is not a translated coordinate component is called a *global component.*

Note that if $1 \in W$, then W is a coordinate component if and only if W is a translated coordinate component.

Let $C(j)$ be the plane curve obtained from C by discarding the j-th component C_j. Let $M(j) = \mathbb{P}^2 \setminus C(j)$ be the corresponding complement. Then the inclusion $\iota_j : M \to M(j)$ induces an epimorphism $H_1(M) \to H_1(M(j))$ and hence an embedding $\iota_j^* : \mathbb{T}(M(j)) \to \mathbb{T}(M)$.

Definition 3.5. An irreducible component W of the m-th characteristic variety $\mathcal{V}_m(M)$ is called a *non-essential component, or a pull-back component* if $W = \iota_j^*(W_j)$ for some j and some irreducible component W_j of the m-th characteristic variety $\mathcal{V}_m(M(j))$, see [2], [17], [20]. An irreducible component W which is not non-essential is called an *essential component.*

Remark 3.6. The notions of coordinate and (non-)essential component depend on the curve arrangement $C = \cup C_i$, i.e., on the chosen embedding of M into $\mathbb{P}^2$. So they are not invariants of the surface M. For more on this see [2].

Assume given a surjective morphism $f : M \to S$ such that $f_\sharp : \pi_1(M) \to \pi_1(S)$ is surjective. This gives rise to an embedding $f^* : \mathbb{T}(S) \to \mathbb{T}^0(M)$, which implies in particular $k \leq r$. More precisely, if we start with $\mathcal{L}_\lambda \in \mathbb{T}(S)$, then the monodromy ρ_j of the pull-back local system $f^*\mathcal{L}_\lambda = \mathcal{L}_\rho$ is given by:

(i) $\rho_j = 1$ if the component C_j is not in the first case of Proposition 2.2, and by
(ii) $\rho_j = \lambda_i^{m_j}$ if the component C_j is in the first case of Proposition 2.2, i.e., $g_1(C_j) = b_i$ in the notation from the proof of Proposition 2.2. Recall that m_j is the multiplicity of C_j in $f^{-1}(b_i)$.

Corollary 3.7. *With the above notation, the pull-back local system* $f^*\mathcal{L}_\lambda = \mathcal{L}_\rho$ *satisfies* $\rho_j \neq 1$ *for all* $j = 1, \dots, r$ *if and only if:*

(i) *The curve* C *consists exactly of the fibers of the associated pencil* $\mathcal{C}$ *corresponding to the points in* B.

(ii) *For all* $j = 1, \dots, r$, *if we set* $g_1(C_j) = b_{i(j)}$, *then* $\lambda_{i(j)}^{m_j} \neq 1$.

3.8. Arapura's results

We recall here some of the main results from [1], applied to the rank 1 local systems on M, with some additions from [20], [16] and [13].

Theorem 3.9. *Let* W *be an irreducible component of* $\mathcal{V}_1(M)$ *and assume that* $d_W := \dim W \geq 1$. *Then there is a surjective morphism* $f_W : M \to S_W$ *with connected generic fiber* $F(f_W)$, *and a torsion character* $\rho_W \in \mathbb{T}(M)$ *such that*

$$W = \rho_W \otimes f_W^*(\mathbb{T}(S_W)).$$

More precisely, the following hold.

(i) $S_W = \mathbb{P}^1 \setminus B_W$, *with* B_W *a finite set satisfying* $k_W := |B_W| = d_W + 1$.

(ii) *For any local system* $\mathcal{L} \in W$, *the restriction* $\mathcal{L}|F(f_W)$ *of* $\mathcal{L}$ *to the generic fiber of* f_W *is trivial, i.e.,* $\mathcal{L}|F(f_W) = \mathbb{C}_{F(f_W)}$.

(iii) *If* $1 \in W$ *and* $\mathcal{L} \in W$, *then* $\dim H^1(M, \mathcal{L}) \geq -\chi(S_W) = d_W - 1$ *and equality holds with finitely many exceptions.*

(iv) *If* $1 \in W$, *then* $d_W \geq 2$.

(v) *If* $1 \notin W$ *and* $d_W \geq 2$, *then the subtorus* $W' = f_W^*(\mathbb{T}(S_W))$ *is another irreducible component of* $\mathcal{V}_1(M)$. *In this situation,* W' *is a coordinate component if and only if* W *is a translated coordinate component.*

For a detailed discussion of the proof of the following result we refer to the proof of Theorem 3.6 in [13].

Remark 3.10. Conversely, if $f : M \to S$ is a morphism with a generic connected fiber and with $\chi(S) < 0$, then $W_f = f^*(\mathbb{T}(S))$ is an irreducible component in $\mathcal{V}_1(M)$ such that $1 \in W_f$ and $\dim W_f \geq 2$, see [1], Section V, Prop. 1.7. Some basic situations of this general construction are the following.

(i) *The local components*, see for instance [25], subsection (2.3) in the case of line arrangements. The case of curve arrangements runs as follows. Let $p \in \mathbb{P}^2$ be a point such that there is a degree d_p and an integer $k_p > 2$ such that:

(1) the set $A_p = \{j \mid p \in C_j \text{ and } \deg C_j = d_p\}$ has cardinality k_p;

(2) $\dim < f_j \mid j \in A_p >= 2$, with $f_j = 0$ being an equation for C_j.

If $\{P, Q\}$ is a basis of this 2-dimensional vector space, then the associated pencil induces a map

$$f_p : M \to S_p$$

where S_p is obtained from $\mathbb{P}^1$ by deleting the k_p points corresponding to the curves C_j, for $j \in A_p$. In this way, the point p produces an irreducible component in $\mathcal{V}_1(M)$, namely

$$W_p = f_p^*(\mathbb{T}(S_p))$$

of dimension $k_p - 1$, which is called local because it depends only on the chosen point p. Note that in the case of line arrangements p can be chosen to be any point of multiplicity at least 3.

(ii) *The components associated to neighborly partitions*, see [21], corresponds exactly to pencils associated to the line arrangement, as remarked in [17], see the proof of Theorem 2.4. It is known that each non-local component (passing through the origin) has dimension at most 4, see [22, Theorem 7.2].

All these points are illustrated by the following.

Example 3.11. This is a key example discovered by A. Suciu, see Example 4.1 in [25] and Example 10.6 in [26]. Consider the line arrangement in $\mathbb{P}^2$ given by the equation

$$xyz(x-y)(x-z)(y-z)(x-y-z)(x-y+z)=0.$$

We number the lines of the associated affine arrangement in $\mathbb{C}^2$ (obtained by setting $z=1$) as follows: $L_1 : x=0$, $L_2 : x-1=0$, $L_3 : y=0$, $L_4 : y-1=0$, $L_5 : x-y-1=0$, $L_6 : x-y=0$ and $L_7 : x-y+1=0$; see the pictures in Example 4.1 in [25] and Example 10.6 in [26]. We consider also the line at infinity $L_8 : z=0$. As stated in Example 4.1 in [25], there are

(i) Seven local components: six of dimension 2, corresponding to the triple points, and one of dimension 3, for the quadruple point.

(ii) Five components of dimension 2, passing through 1, coming from the following neighborly partitions (of braid subarrangements): (15|26|38), (28|36|45), (14|23|68), (16|27|48) and (18|37|46). For instance, the pencil corresponding to the first partition is given by $P = L_1L_5 = x(x-y-z)$ and $Q = L_2L_6 = (x-z)(x-y)$. Note that $L_3L_8 = yz = Q-P$, is a decomposable fiber in this pencil.

(iii) Finally, there is a 1-dimensional component W in $\mathcal{V}_1(M)$ with

$$\rho_W = (1,-1,-1,1,1,-1,1,-1) \in \mathbb{T}(M) \subset (\mathbb{C}^*)^8$$

and $f_W : M \to \mathbb{C}^*$ given by

$$f_W(x:y:z) = \frac{x(y-z)(x-y-z)^2}{(x-z)y(x-y+z)^2}$$

or, in affine coordinates

$$f_W(x,y) = \frac{x(y-1)(x-y-1)^2}{(x-1)y(x-y+1)^2}.$$

Then $W \subset \mathcal{V}_1(M)$ and $W \cap \mathcal{V}_2(M)$ consists of two characters, ρ_W above and

$$\rho'_W = (-1,1,1,-1,1,-1,1,-1).$$

Note that this component W is a translated coordinate component. This is related to the fact that the associated pencil is special. For more on this arrangement see Example 6.7.

It is clear that any non-essential component is a coordinate component. The following converse result on positive dimensional coordinate components W of $\mathcal{V}_m(M)$ was obtained by Libgober [20]. For reader's convenience we include a neater proof.

Proposition 3.12. *Any positive dimensional translated coordinate component of $\mathcal{V}_m(M)$ is non-essential.*

Proof. Let $W = \rho_W \otimes f_W^*(\mathbb{T}(S_W))$ be a positive dimensional irreducible component of $\mathcal{V}_m(M)$. Assume W is contained in the subtorus of $\mathbb{T}^r$ given by $\rho_j = 1$. It follows that the corresponding component $\rho_{W,j}$ of the character ρ is 1, and that the torus $\mathbb{T}_W = f_W^*(\mathbb{T}(S_W))$ is also contained in the same subtorus. The discussion before Corollary 3.7 implies that the corresponding component C_j of C is not in the first case of Proposition 2.2. This in turn implies the existence of an extension $f(j) : M(j) \to S_W$, whose generic fibers are still connected (being obtained from those of f by adding at most finitely many points). It follows that $W = \iota_j^*(W_j)$, with $\iota_j : M \to M(j)$ the inclusion and $W_j = \rho_j \otimes f(j)^*(\mathbb{T}(S_W))$, where the character ρ_j is obtained from ρ_W by discarding the j-th component. To show that W_j is an irreducible component in $\mathcal{V}_m(M(j))$, we can use Proposition 6.1 in the case $\chi(S_W) < 0$ and Corollary 6.2 in the case $S_W = \mathbb{C}^*$. In both cases we have to use in addition the equality

$$\begin{aligned}\dim H^1(M, \mathcal{L}_1 \otimes f_W^*\mathcal{L}_2) &= \dim H^1(S_W, R^0 f_{W*}\mathcal{L}_1 \otimes \mathcal{L}_2)\\ &= \dim H^1(S_W, R^0 f(j)_*\mathcal{L}_1' \otimes \mathcal{L}_2) = \dim H^1(M(j), \mathcal{L}_1' \otimes f(j)^*\mathcal{L}_2).\end{aligned}$$

Here $\mathcal{L}_1 \in \mathbb{T}(M)$ (resp. $\mathcal{L}_1' \in \mathbb{T}(M(j))$ is the local system corresponding to ρ_W (resp. ρ_j), the first and the third equalities hold for a generic $\mathcal{L}_2$, while the middle equality comes from $\mathcal{L}_1' = \iota_{j,*}\mathcal{L}_1$ and $f(j) \circ \iota_j = f_W$. □

In view of this result, it is natural to study first the non-coordinate positive dimensional components. Indeed, the other components come from simpler arrangements, involving fewer components C_j's. The situation of translated components is different, e.g. the component W studied in Example 3.11 is NOT coming from a simpler arrangement. The case of 0-dimensional components is very interesting as well, see [2], [25], [26].

3.13. Resonance varieties

Let $H^*(M, \mathbb{C})$ be the cohomology algebra of the surface M with $\mathbb{C}$-coefficients. Right multiplication by an element $z \in H^1(M, \mathbb{C})$ yields a cochain complex $(H^*(M, \mathbb{C}), \mu_z)$. The *resonance varieties* of M are the jumping loci for the degree 1 cohomology of this complex:

$$\mathcal{R}_m(M) = \{z \in H^1(M, \mathbb{C}) \mid \dim H^1(H^*(M, \mathbb{C}), \mu_z) \geq m\}. \tag{3.8}$$

One of the main results in [16] gives the following. For the case of hyperplane arrangements see [8].

Theorem 3.14. *The exponential map* $\exp : H^1(M,\mathbb{C}) \to \mathbb{T}^0(M)$ *induces for any* $m \geq 1$ *an isomorphism of analytic germs*

$$(\mathcal{R}_m(M), 0) \simeq (\mathcal{V}_m(M), 1).$$

The following easy consequence will play a key role.

Corollary 3.15. *The irreducible components of* $\mathcal{R}_1(M)$ *are precisely the maximal linear subspaces* $E \subset H^1(M,\mathbb{C})$, *isotropic with respect to the cup product on* M,

$$\cup : H^1(M,\mathbb{C}) \times H^1(M,\mathbb{C}) \to H^2(M,\mathbb{C}),$$

and such that $\dim E \geq 2$.

Proof. Let E be a component of $\mathcal{R}_1(M)$. By the above theorem there is a component W in $\mathcal{V}_1(M)$ such that $1 \in W$ and $T_1W = E$. By Theorem 3.9 we can write $W = f^*(\mathbb{T}(S))$, and hence $T_1W = f^*(H^1(S,\mathbb{C}))$ is isotropic with respect to the cup product, since the cup product on $H^1(S,\mathbb{C})$ is trivial. Maximality of E comes from the fact that E is a component of $\mathcal{R}_1(M)$. The restriction $\dim E \geq 2$ comes from Theorem 3.9, (v). □

Remark 3.16. It follows from the proof of Corollary 3.15 that any maximal isotropic linear subspaces $E \subset H^1(M,\mathbb{C})$ is *rationally defined*, i.e., there is a linear subspace $E_{\mathbb{Q}} \subset H^1(M,\mathbb{Q})$ such that $E = E_{\mathbb{Q}} \otimes_{\mathbb{Q}} \mathbb{C}$ under the identification $H^1(M,\mathbb{C}) = H^1(M,\mathbb{Q}) \otimes_{\mathbb{Q}} \mathbb{C}$. Indeed, one can take $E_{\mathbb{Q}} = f^*(H^1(S,\mathbb{Q}))$.

We can restate the above corollary as follows.

Corollary 3.17. *If* $f : M \to S$ *is a surjective morphism with connected generic fiber* F *and* $\rho \in \mathbb{T}(M)$ *is a torsion character such that* $W = \rho \otimes f^*(\mathbb{T}(S))$ *is an irreducible component of* $\mathcal{V}_1(M)$ *with* $\dim W \geq 2$, *then* $K_f = f^*(H^1(S,\mathbb{C}))$ *is a (rationally defined) maximal isotropic subspace in* $H^1(M,\mathbb{C})$ *with respect to the cup-product.*

Proof. Using Theorem 3.9, (vi), we can take $\rho = 1$. Then K_f is exactly the tangent space at $1 \in W$, and, by Theorem 3.14, an irreducible component of $\mathcal{R}_1(M)$. In addition, K_f is obviously an isotropic subspace in $H^1(M,\mathbb{C})$ with respect to the cup-product. It should be maximal, since any strictly larger isotropic subspace would contradict the fact that K_f is an irreducible component of $\mathcal{R}_1(M)$. □

We say that an irreducible component E of some $\mathcal{R}_m(M)$ (which is a vector subspace in $H^1(M)$) is a *coordinate component*, resp. a *non-essential component*, if it corresponds under the above isomorphism to a coordinate (resp. non-essential) component of $\mathcal{V}_m(M)$. Proposition 3.12 can be reformulated as follows.

Proposition 3.18. *An irreducible component of* $\mathcal{R}_m(M)$ *is non-essential if and only if it is a coordinate component.*

An irreducible component E of some $\mathcal{R}_m(M)$ which is not a coordinate component is called a *global component* in [17]. This is compatible with our Definition 3.4 above.

Let us consider the component W_E introduced in Remark 6.3, in the case $\dim E \geq 2$. Then W_E corresponds to a mapping $f_E : M \to S_E$ as in Theorem 3.9. We have the following result.

Proposition 3.19. *If the arrangement C is given, then the mapping f_E is determined by the vector subspace $E \subset H^1(M, \mathbb{C})$.*

Proof. First we know that S_E is obtained from $\mathbb{P}^1$ by deleting a subset B, with $|B| - 1 = \dim E$. Let $k := \dim E$ and assume that the points in B are $(0 : 1)$ and $(1 : b_j)$, for some $b_j \in \mathbb{C}$, $j = 1, \ldots, k$. If $(u : v)$ are the homogeneous coordinates on $\mathbb{P}^1$, then the cohomology group $H^1(S_E, \mathbb{C})$ has a basis given by

$$\omega_j = \frac{d(v - b_j u)}{v - b_j u} - \frac{du}{u}$$

where $j = 1, \ldots, k$. As explained in Proposition 2.2, the mapping f_E corresponds to a pencil (P, Q), where P and Q are homogeneous polynomials in $\mathbb{C}[X, Y, Z]$, of the same degree and without common factors. In terms of this pencil, one has

$$f_E^*(\omega_j) = \frac{d(Q - b_j P)}{Q - b_j P} - \frac{dP}{P}.$$

So, in down-to-earth terms, the question is: how to determine the pencil (P, Q) from the vector space of 1-forms with logarithmic poles

$$E =< f_E^*(\omega_j) \mid j = 1, \ldots, k >?$$

Using only logarithmic poles allows us to work with rational differential forms rather than cohomology classes and this is essential for this proof. Start with the curve C_1 in the curve arrangement C and consider the subset

$$E_1 = \{\omega \in E \mid \int_{\gamma_1} \omega = 0\}.$$

Two cases may occur.

Case 1. $(E_1 = E)$ This case occurs exactly when C_1 is not a connected component in any of the $(k+1)$ special fibers of the pencil (P, Q) corresponding to the set B. If this happens, we discard the curve C_1 and test the next curve C_2 and so on.

Case 2. $(E_1 \neq E)$ Then C_1 is a component of a special fiber of the pencil, say of the fiber $\mathcal{C}_{b_1} : Q - b_1 P = 0$. Note that any form $\omega \in E$ can be written as a sum

$$\omega = \sum_j a_j f_E^*(\omega_j)$$

and, with this notation, one has

$$\int_{\gamma_1} \omega = 2\pi i a_1 m(C_1)$$

where $i^2 = -1$ and $m(C_1)$ is the multiplicity of the irreducible curve C_1 in the divisor $\mathcal{C}_{b_1}$. It follows that $\omega \notin E_1$ if and only if $a_1 \neq 0$. This shows that $Q - b_1 P$ is the G.C.D. of the denominators of the forms $\omega \in E \setminus E_1$. After we have determined the polynomial $Q - b_1P$ as above, we discard all the curves C_j in the arrangement which are contained in the support of the divisor $\mathcal{C}_{b_1}$. From the remaining curves in C, we can find a new curve, say C_2, such that

$$E_2 := \{\omega \in E \mid \int_{\gamma_2} \omega = 0\} \neq E.$$

(Such a curve exists since $|B| \geq 3$.) Then C_2 is a component in a new fiber of the pencil (P, Q), say of the fiber $\mathcal{C}_{b_2} : Q - b_2P = 0$ and hence $Q - b_2P$ is the G.C.D. of the denominators of the forms $\omega \in E \setminus E_2$. The two homogeneous polynomials $Q - b_1P$ and $Q - b_2P$ span the same vector space as the polynomials P and Q, i.e., they determine the same pencil up to an automorphism of $\mathbb{P}^1$. □

Now we treat the special case of rationally defined maximal isotropic subspace in $H^1(M, \mathbb{C})$ of dimension 1.

Proposition 3.20. *Let E be a rationally defined maximal isotropic subspace in $H^1(M, \mathbb{C})$ of dimension 1. Then there is a surjective mapping $f_E : M \to \mathbb{C}^*$ with connected generic fiber such that $E = f_E^*(H^1(\mathbb{C}^*, \mathbb{C}))$.*

Proof. Let $f_j = 0$ be a homogeneous reduced equation for the component C_j in the curve arrangement C, for $j = 1, \ldots, r$. Assume that $\deg f_j = d_j$. A basis of the $(r-1)$-dimensional vector space $H^1(M, \mathbb{Q})$ is given by the 1-forms

$$\eta_k = \frac{1}{d_k}\frac{df_k}{f_k} - \frac{1}{d_{k+1}}\frac{df_{k+1}}{f_{k+1}}$$

for $k = 1, \ldots, r-1$. Using this, we see that any 1-dimensional subspace in $H^1(M, \mathbb{Q})$ has a unique generator (up to a $\pm$-sign) of the form

$$\eta = \sum_{j=1,r} m_j \frac{df_j}{f_j}$$

where m_j are relatively prime integers, i.e., $G.C.D.(m_1, \ldots, m_r) = 1$, such that

$$\sum_{j=1,r} d_j m_j = 0.$$

It follows that the rational fraction

$$f = \prod_{j=1,r} f_j^{m_j}$$

is homogeneous of degree 0 and hence induces a morphism $f : M \to \mathbb{C}^*$. The fact that f is surjective is obvious, while the connectivity of the generic fiber follows from Bertini's Theorem, see [24], p. 79, using the condition $G.C.D.(m_1, \ldots, m_r) = 1$. The equality $E = f^*(H^1(\mathbb{C}^*, \mathbb{C}))$ is obvious by taking $\frac{dt}{t}$ as a basis of $H^1(\mathbb{C}^*, \mathbb{C})$. □

4. Minimal arrangements

In this section we prove the following result which applies to an arbitrary plane curve arrangement. For a different approach in the case of line arrangements see [17].

Theorem 4.1. *Let $C = \cup_{i=1,r} C_i$ be a plane curve arrangement in $\mathbb{P}^2$, having r irreducible components C_i, for $i = 1, r$. Let $M = \mathbb{P}^2 \setminus C$ be the corresponding complement. Then, for $d \geq 2$, the following are equivalent.*

(i) *there is a global d-dimensional irreducible component E in the resonance variety $\mathcal{R}_1(M)$;*
(ii) *there is a global d-dimensional irreducible component W in the characteristic variety $\mathcal{V}_1(M)$;*
(iii) *there is a pencil $\mathcal{C}$ of plane curves on $\mathbb{P}^2$ with an irreducible generic member and having $d+1$ fibers $\mathcal{C}_t$ whose reduced supports form a partition of the set of irreducible components C_i, for $i = 1, r$.*

Moreover, for $d = 1$, (i) always fails and (ii) implies (iii).

Proof. For $d \geq 2$, the equivalence between (i) and (ii) follows from Theorem 3.14 and Theorem 3.9, claim (vi). And the equivalence between (ii) and (iii) follows from Arapura's results recalled in subsection 3.8 combined with Proposition 2.2 and Corollary 3.7. The case $d = 1$ follows from Theorem 3.9, claim (i). □

Note that the condition (ii) above can be reformulated as the existence of a global irreducible component W in the characteristic variety $\mathcal{V}_1(M)$ such that $1 \in W$ and $\dim H^1(M, \mathcal{L}) = d - 1 \geq 1$ for a generic local system $\mathcal{L} \in W$.

The following result is similar to Theorem 4.1.1 in Libgober [20] and closely related to the discussion in [21], just before Proposition 7.2.

Corollary 4.2. *Assume the equivalent statements in Theorem 4.1 above hold. Let $f : \mathbb{P}^2 \to \mathbb{P}^1$ be the rational morphism associated to the pencil $\mathcal{C}$. Let $\pi : X \to \mathbb{P}^2$ be a sequence of blowing-ups such that $g = f \circ \pi$ is a regular morphism on X. If $\tilde{C}$ denotes the proper transform of the (reduced) curve C under π, then the self intersection number of $\tilde{C}$ is non-positive, i.e.,*

$$\tilde{C} \cdot \tilde{C} \leq 0.$$

Proof. There is a partition of $\tilde{C} = \cup_{i=1,d+1} \tilde{C}_i$ of $\tilde{C}$ as a union of disjoint curves $\tilde{C}_i$, such that $g(\tilde{C}_i) = b_i$ for $i = 1, \ldots, d+1$. It follows that

$$\tilde{C} \cdot \tilde{C} = \sum_{i=1,d+1} \tilde{C}_i \cdot \tilde{C}_i.$$

Each curve is contained in the support of the positive divisor $g^{-1}(b_i)$, and hence by Zariski's Lemma, see [3], p. 90, we get

$$\tilde{C}_i \cdot \tilde{C}_i \leq 0.$$

□

Example 4.3. Assume that C is a line arrangement, i.e., $d_j = 1$ for all $j = 1, \dots, r$. Let $\mathcal{X}$ be the base locus of a pencil as in Corollary 4.2, $k = d + 1$ the number of sets in the associated partition $(\tilde{C}_i)_i$ of the set of lines in C. Then it is shown in [17] that the following hold.

(i) For each base point $p \in \mathcal{X}$, the multiplicity

$$n_p = \text{mult}_p(\mathcal{C}_b)$$

is independent of $b \in B$.

(ii) $\sum_{p \in \mathcal{X}} n_p = D^2$, where D is the degree of the pencil.

(iii) $\sum_{j=1,r} m(C_j) = kD$, where $m(C_j) \geq 1$ is the multiplicity with which C_j occurs in the corresponding fiber of the pencil.

To resolve the indeterminacy points of the associated pencil (i.e., to determine the map $\pi : X \to \mathbb{P}^2$), one has in this case just to blow-up once the points in the base locus $\mathcal{X}$. This is a direct consequence of the property (i) above.

Assume moreover that $m(C_j) = 1$ for all $j = 1, r$, i.e., all $\mathcal{C}_b$ for $b \in B$ are reduced.

Then, again by (i) above, it follows that $\text{mult}_p(C) = kn_p$ for any $p \in \mathcal{X}$. On the other hand, by (iii) we get $\deg C = kD$. Finally, in this very special case we get

$$\tilde{C} \cdot \tilde{C} = C \cdot C - \sum_{p \in \mathcal{X}} \text{mult}_p(C)^2 = k^2D^2 - k^2D^2 = 0$$

since $C \cdot C = (\deg C)^2$. This happens for instance in Example 3.4 in [17]: the Ceva arrangement given by the pencil $ax^d + by^d + cz^d = 0$ with $(a : b : c) \in \mathbb{P}^2$ satisfying $a + b + c = 0$. There are three special fibers, corresponding to $x^d - y^d$, $y^d - z^d$, $z^d - x^d$.

In the case of a general line arrangement, the condition $\tilde{C} \cdot \tilde{C} \leq 0$ may bring new non-trivial information on the arrangement. In particular it can be used as a test for candidates to the base locus $\mathcal{X}$ of a pencil associated to a given arrangement. In Example 3.6 in [17] the B_3-arrangement consists of nine lines, and the base locus $\mathcal{X}$ consists of three points of multiplicity 4 and four other points of multiplicity 3. As a result we have

$$\tilde{C} \cdot \tilde{C} = C \cdot C - \sum_{p \in \mathcal{X}} \text{mult}_p(C)^2 = 81 - 3 \times 16 - 4 \times 9 = -3.$$

This latter arrangement is associated to the pencil $((x^2 - y^2)z^2 : (y^2 - z^2)x^2)$ which has again three special fibers (this time non-reduced!), corresponding to $(x^2 - y^2)z^2$, $(y^2 - z^2)x^2$, $(x^2 - z^2)y^2$. Up to a linear change of coordinates, this is the same B_3-arrangement as in our Examples 3.11 and 6.7.

5. Fibered complements and $K(\pi, 1)$-spaces

Let $f : M \to S$ be a morphism associated to a plane curve pencil $\mathcal{C}$ with base locus $\mathcal{X}$, as in Theorem 4.1 and Corollary 4.2 above. Consider a fiber $\mathcal{C}_s$ of the pencil

$\mathcal{C}$ corresponding to $s \in S = \mathbb{P}^1 \setminus B$. We say that $\mathcal{C}_s$ is a *special fiber* of $\mathcal{C}$ if either $\mathcal{C}_s \setminus \mathcal{X}$ is singular , or if $\mathcal{C}_s \setminus \mathcal{X}$ is smooth and there exists a point $p \in \mathcal{X}$ such that

$$\mu(\mathcal{C}_s, p) > \min_{t \in S} \mu(\mathcal{C}_t, p),$$

where μ denotes the Milnor number of an isolated singularity.

Let $\mathcal{C}_{spec}$ be the union of all the special fibers in the pencil $\mathcal{C}$. (There is a finite number of such fibers, and they are easy to identify, see [19], [10]). Let B' be the union of B and the set of all $s \in S$ such that $\mathcal{C}_s$ is a special fiber. We call $C' = C \cup \mathcal{C}_{spec}$ the *extended plane curve arrangement* associated to the plane curve arrangement C and denote by M' the corresponding complement. We set $S' = \mathbb{P}^1 \setminus B'$. With this notation, we have the following result.

Proposition 5.1. *The restriction $f' : M' \to S'$ is a locally topologically trivial fibration. The plane curve arrangement complement M' is a $K(\pi, 1)$-space.*

Proof. Note that for any $p \in \mathcal{X}$, the family of plane curve isolated singularities $(\mathcal{C}_s, p)$ for $s \in S'$ is a μ-constant family. Using the relation between μ-constant families and Whitney regular stratifications, as well as Thom's First Isotopy Lemma (see for instance [11], pp. 11–16 and especially the proof of Proposition (1.4.1) on p. 20), we get the first claim above. The second claim is an obvious consequence, as explained already in [17]. □

Example 5.2. We discuss now the following curve arrangement, considered already in Example 4.8 in [15]. The curve C consists of the following: three lines $C_1 : x = 0$, $C_2 : y = 0$ and $C_3 : z = 0$ and a conic $C_4 : x^2 - yz = 0$. The corresponding pencil can be chosen to be $f = (x^2 : yz)$, and the set $B = \{b_1, b_2, b_3\}$ is given by $b_1 = (0 : 1)$, $\mathcal{C}_1 = 2C_1$, $b_2 = (1 : 0)$, $\mathcal{C}_2 = C_2 \cup C_3$, and $b_3 = (1 : -1)$, $\mathcal{C}_3 = C_4$.

The base locus of this pencil is $\mathcal{X} = \{p_1 = (0 : 0 : 1), p_2 = (0 : 1 : 0)\}$ and it is easy to check that there no special fibers. It follows that $f : M \to S$ is a locally topologically trivial fibration with fiber $\mathbb{C}^*$ (a smooth conic minus two points). Hence the complement M of this curve arrangement is a $K(\pi, 1)$-space, where the group π fits into an exact sequence

$$1 \to \mathbb{Z} \to \pi \to \mathbb{F}_2 \to 1$$

with $\mathbb{F}_2$ denoting the free group on two generators.

Note also that

$$\text{mult}_{p_1} \mathcal{C}_1 = 2 > 1 = \text{mult}_{p_1} \mathcal{C}_2,$$

hence the property (i) in Example 4.3 does not hold for arbitrary curve arrangements.

A local system $\mathcal{L}$ on S is given by a triple $(\lambda_1, \lambda_2, \lambda_3)$ with $\lambda_1 \cdot \lambda_2 \cdot \lambda_3 = 1$. With this notation, the pull-back local system $f^*\mathcal{L}$ is given by $(\rho_1, \rho_2, \rho_3, \rho_4)$, where $\rho_1 = \lambda_1^2$, $\rho_2 = \rho_3 = \lambda_2$ and $\rho_4 = \lambda_3$ (recall the discussion before Corollary 3.7). With this, it is easy to check that the irreducible component given by $f^*\mathbb{T}(S)$ coincide with the irreducible component obtained in Example 4.8 in [15] by computation using the associated integrable connections.

Indeed, the ρ_i's satisfy the equation $\rho_1 \cdot \rho_2 \cdot \rho_3 \cdot \rho_4^2 = 1$, see Equation (3.5), and hence we can use (ρ_1, ρ_2, ρ_4) to parametrize the torus $\mathbb{T}(M)$. It follows that the irreducible component given by $W = f^*\mathbb{T}(S)$ is parametrized by $\rho_1 = \lambda_1^2$, $\rho_2 = \lambda_2$ and $\rho_4 = \lambda_3$, i.e., W is given by the equation

$$\rho_1 \cdot \rho_2^2 \cdot \rho_4^2 = 1$$

which appears in Example 4.8 in [15] with slightly different notation.

Let $\widehat{C}$ be the proper transform on C under the blowing-up $\widehat{\pi} : \widehat{X} \to \mathbb{P}^2$ of the two points in $\mathcal{X}$. Then

$$\widehat{C} \cdot \widehat{C} = C \cdot C - 3^2 - 3^2 = 7 > 0.$$

This is not in contradiction with Corollary 4.2, since in order to resolve the indeterminacy points of f in this case, we have to blow the points $\widehat{p}_1, \widehat{p}_2 \in \widehat{X}$ corresponding to the tangents of the conic C_4 at p_1 and p_2. The new multiplicities are

$$\operatorname{mult}_{\widehat{p}_1} \widehat{C} = \operatorname{mult}_{\widehat{p}_2} \widehat{C} = 3$$

and hence for the new proper transform $\tilde{C}$ we get

$$\tilde{C} \cdot \tilde{C} = \widehat{C} \cdot \widehat{C} - 3^2 - 3^2 = -11.$$

6. Translated components

We start by the following key result, which can be regarded as a strengthening of Remark 3.10.

For $\mathcal{L}_\rho \in \mathbb{T}(S)$, note that $\mathcal{F} = R^0 f_*(\mathcal{L}_\rho)$ is in general no longer a local system on S, but a *constructible sheaf*. By definition, there exists a minimal finite set $\Sigma = \Sigma(\mathcal{F}) \subset S$, called the *singular support* of $\mathcal{F}$, such that $\mathcal{F}|(S \setminus \Sigma)$ is a local system, see [12], p. 87. These objects enter in the following result.

Proposition 6.1. *If $\mathcal{L}_\rho$ is a rank 1 local system on M such that $\mathcal{L}_\rho|F$ is trivial, then*

$$\dim H^1(M, \mathcal{L}_\rho \otimes f^{-1}\mathcal{L}) \geq -\chi(S) + |\Sigma(R^0 f_*(\mathcal{L}_\rho))|$$

with equality for all but finitely many local systems $\mathcal{L} \in \mathbb{T}(S)$. In particular, if $W_{f,\rho} = \rho \otimes f^*(\mathbb{T}(S))$ is a positive dimensional irreducible component of $\mathcal{V}_1(M)$, then $\dim W_{f,\rho} = -\chi(S) + 1 > 0$ and $W_{f,\rho}$ is an irreducible component of $\mathcal{V}_q(M)$, for any $1 \leq q \leq q(f,\rho) := -\chi(S) + |\Sigma(R^0 f_*(\mathcal{L}_\rho))|$. Conversely, any positive dimensional irreducible component of $\mathcal{V}_q(M)$ for $q \geq 1$ is of this type.

For the proof of this result we refer to Corollary 4.7 in [13].

Corollary 6.2. *If $f : M \to \mathbb{C}^*$ is a surjective morphism with connected generic fiber F and $\rho \in \mathbb{T}(M)$ is a torsion character such that $W = \rho \otimes f^*(\mathbb{T}(\mathbb{C}^*))$ is an irreducible component of $\mathcal{V}_1(M)$ with $\dim W = 1$, then $K_f = f^*(H^1(\mathbb{C}^*, \mathbb{C}))$ is a rationally defined maximal isotropic subspace in $H^1(M, \mathbb{C})$ with respect to the cup-product.*

Conversely, let $f : M \to \mathbb{C}^*$ be a surjective morphism with a connected generic fiber. Assume that $K_f = f^*(H^1(\mathbb{C}^*, \mathbb{C}))$ is a rationally defined maximal isotropic subspace in $H^1(M, \mathbb{C})$ with respect to the cup-product. Then for any character $\rho \in \mathbb{T}(M)$ with $\mathcal{L}_\rho|F = \mathbb{C}_F$ for a generic fiber F of f and $\rho \notin f^*(\mathbb{T}(\mathbb{C}^*))$, the translated subtorus

$$W_{f,\rho} = \rho \otimes f^*(\mathbb{T}(\mathbb{C}^*))$$

is either an irreducible component in $\mathcal{V}_1(M)$ such that $\dim W_{f,\rho}=1$, or $H^1(M,\mathcal{L})= 0$ for $\mathcal{L} \in W_{f,\rho}$ with finitely many exceptions. This latter case occurs if and only if $\Sigma(R^0 f_*(\mathcal{L}_\rho)) = \emptyset$.

Proof. Assume that K_f is not maximal, and let $K \supset K_f$ be a maximal isotropic subspace in $H^1(M, \mathbb{C})$ with respect to the cup-product. Then $\dim K \geq 2$ and there is a morphism surjective $f_1 : M \to S_1$, with connected generic fiber such that $K = K_{f_1}$. But then W is strictly contained in $W_1 = \rho \otimes f_1^*(\mathbb{T}(S_1))$, which is a component of $\mathcal{V}_1(M)$ by Proposition 6.1, a contradiction.

For the converse part, just note that, using the same argument as above, $W_{f,\rho}$ cannot be strictly contained in a component of $\mathcal{V}_1(M)$. □

Remark 6.3. (i) Unlike the case of maximal isotropic subspaces in $H^1(M, \mathbb{C})$ of dimension at least 2, which are automatically rationally defined, see Remark 3.16, there are a lot of non-rationally defined maximal isotropic subspaces in $H^1(M, \mathbb{C})$ of dimension 1 as soon as a rationally defined one exists. To see this, use the semi-continuity of the dimension of $H^1(H^*(M, \mathbb{C}), \mu_z)$ with respect to $z \in H^1(M, \mathbb{C})$.

(ii) If E is a maximal isotropic subspace in $H^1(M, \mathbb{C})$ of dimension at least 2, then it has at least one associated component W_E in $\mathcal{V}_1(M)$ corresponding to E under the bijection in Theorem 3.14. On the other hand, if E is a rationally defined maximal isotropic subspace in $H^1(M, \mathbb{C})$ of dimension 1, it is quite possible that there is no associated component in $\mathcal{V}_1(M)$. As an explicit example, consider the case of the line arrangement $xyz = 0$ in $\mathbb{P}^2$. Then $M = (\mathbb{C}^*)^2$ and any 1-dimensional subspace $E_\mathbb{Q} \subset H^1(M, \mathbb{Q})$ gives rise to a rationally defined maximal isotropic subspace in $H^1(M, \mathbb{C})$ of dimension 1. However, it is well known that $\mathcal{V}_1(M) = \{1\}$ in this case.

Let W be a translated irreducible component of $\mathcal{V}_1(M)$, i.e., $1 \notin W$. Then, as in Theorem 3.9, there is a torsion character $\rho \in \mathbb{T}(M)$ and a surjective morphism $f : M \to S$ with connected generic fiber F such that

$$W = \rho f^*(\mathbb{T}(S)). \tag{6.1}$$

We say in this situation that the component W is associated to the mapping f. In this section we give detailed information on the torsion character $\rho \in \mathbb{T}(M)$ in terms of the geometry of the associated mapping $f : M \to S$.

6.4. The general setting

Let F be the generic fiber of the mapping $f : M \to S$, i.e., F is the fiber of the topologically locally trivial fibration $f' : M' \to S'$ associated to f as in the previous section. Then, we have an exact sequence

$$H_1(F) \xrightarrow{i'_*} H_1(M') \xrightarrow{f'_*} H_1(S') \to 0 \tag{6.2}$$

as well as a sequence

$$H_1(F) \xrightarrow{i_*} H_1(M) \xrightarrow{f_*} H_1(S) \to 0 \tag{6.3}$$

which is not necessarily exact in the middle, i.e., the group

$$T(f) = \frac{\ker f_*}{\operatorname{im} i_*} \tag{6.4}$$

is in general non-trivial. Here $i : F \to M$ and $i' : F \to M'$ denote the inclusions, and homology is taken with $\mathbb{Z}$-coefficients if not stated otherwise. The group $\mathrm{Hom}(T(f), \mathbb{C}^*)$ parametrizes the possible choices for ρ for a given morphism f. For more on this group $T(f)$, in particular for its precise relation to the multiple fibers of f, we refer to the fifth section in [13]. Here we describe an alternative approach for the computation of this group, which yields a geometric way to construct a set of generators for $T(f)$.

6.5. The computation of the group $T(f)$

In order to simplify the presentation, we assume in this subsection that at least one of the curves C_j in the curve arrangement C is a line. This covers the case of line arrangements and of curve arrangements in the affine plane $\mathbb{C}^2$. More specifically, we assume that C_1 is the line at infinity in $\mathbb{P}^2$ and hence $M = \mathbb{C}^2 \setminus \cup_{j>1}(\mathbb{C}^2 \cap C_j)$. We assume that $\infty \in B$ and set $B_1 = B \setminus \{\infty\} \subset \mathbb{C}$.

For each $b \in B$, consider the following divisor on $\mathbb{C}^2$,

$$D_b = \overline{g_1^{-1}(b)} \cap \mathbb{C}^2 = \sum_a m_{ba} C_{ba}, \tag{6.5}$$

where g_1 is the extension of f from the proof of Proposition 2.2, $m_{ba} \geq 1$ are integers and C_{ba} are irreducible curves in the arrangement C. Note that $D_b = 0$ if and only if $\overline{g_1^{-1}(b)} = C_1$.

Recall the larger set $B' \supset B$ obtained from B by adding the bifurcation points of $f : M \to S$. Set $C(f) = B' \setminus B$ and assume from now on that $C(f)$ is non-empty. (Otherwise f is a fibration and hence $T(f) = 1$). For each $c \in C(f)$, consider the following divisor on $\mathbb{C}^2$,

$$D_c = \overline{g_1^{-1}(c)} \cap \mathbb{C}^2 = \sum_{a'} m'_{ca'} C'_{ca'} + \sum_{a''} m''_{ca''} C''_{ca''}, \tag{6.6}$$

where g_1 is the extension of f as above, $m'_{ca'} \geq 1$ and $m''_{ca''} \geq 1$ are integers, the irreducible curves $C'_{ca'}$ are curves in our arrangement (corresponding to case (2) in Proposition 2.2), and $C''_{ca''}$ are the new curves to be deleted from M in order to obtain M'. Since $f : M \to S$ is a surjection, it follows that there is at least one term in the second sum in the equality (6.6). At least some of the first type sums are non-trivial if and only if the arrangement C is special.

Note that M is obtained from $\mathbb{C}^2$ by deleting the irreducible curves C_{ba}, $C'_{ca'}$ and possibly some horizontal components C_h. For each irreducible curve Z in $\mathbb{C}^2$ we denote by $\gamma(Z)$ the elementary oriented loop associated to Z. It follows that $H_1(M)$ is a free $\mathbb{Z}$-module with a basis given by

$$\gamma(C_{ba}), \gamma(C'_{ca'}), \gamma(C_h) \tag{6.7}$$

see for instance [11], p. 103.

Similarly, $H_1(S)$ is a free $\mathbb{Z}$-module with a basis given by δ_b for $b \in B_1$, where δ_b is an elementary loop based at b as in subsection (3.1). In terms of these bases, the morphism $f_* : H_1(M) \to H_1(S)$ is described as follows. Let $\alpha \in H_1(M)$ be given by

$$\alpha = \sum_{b \in B} \alpha_{ba}\gamma(C_{ba}) + \sum \alpha_{ca'}\gamma(C'_{ca'}) + \sum \alpha_h \gamma(C_h). \tag{6.8}$$

Here $a \in A_b$, an index set depending on b. This is not written explicitly, in order to keep the notation simpler. Similar remarks apply to the indices a' and a'' below. Then the relation $\sum_{b \in B} \delta_b = 0$ yields

$$f_*(\alpha) = \sum_{b \in B_1} (\sum_a m_{ba}\alpha_{ba} - \sum_a m_{\infty a}\alpha_{\infty a})\delta_b. \tag{6.9}$$

In particular $\alpha \in \ker f_*$ if and only if

$$\sum_a m_{ba}\alpha_{ba} = \sum_a m_{\infty a}\alpha_{\infty a} \tag{6.10}$$

for all $b \in B_1$. Next, M' is obtained from M by deleting the curves $C''_{ca''}$. Hence, it follows that $H_1(M')$ is a free $\mathbb{Z}$-module with a basis given by

$$\gamma(C_{ba}), \gamma(C'_{ca'}), \gamma(C_h), \gamma(C''_{ca''}). \tag{6.11}$$

The inclusion $j : M' \to M$ induces a morphism $j_* : H_1(M') \to H_1(M)$ which, at coordinate level with respect to the chosen bases, is just the obvious projection. Similarly, $H_1(S')$ is a free $\mathbb{Z}$-module with a basis given by δ_b for $b \in B_1$, and δ_c for $c \in C(f)$.

In terms of these bases, the morphism $f'_* : H_1(M') \to H_1(S')$ is described as follows. Let $\beta \in H_1(M')$ be given by

$$\beta = \sum \beta_{ba}\gamma(C_{ba}) + \sum \beta_{ca'}\gamma(C'_{ca'}) + \sum \beta_h\gamma(C_h) + \sum \beta_{ca''}\gamma(C''_{ca''}). \tag{6.12}$$

Then $f'_*(\beta) = E_1 + E_2$, where

$$E_1 = \sum_{b \in B_1} (\sum_a m_{ba}\beta_{ba} - \sum_a m_{\infty a}\beta_{\infty a})\delta_b \tag{6.13}$$

and

$$E_2 = \sum_{c \in C(f)} (\sum_{a'} m'_{ca'}\beta_{ca'} + \sum_{a''} m''_{ca''}\beta_{ca''} - \sum_a m_{\infty a}\beta_{\infty a})\delta_c. \tag{6.14}$$

To see this, use the relation

$$\sum_{b\in B}\delta_b+\sum_{c\in C(f)}\delta_b=0.$$

The exact sequence (6.2) yields $\operatorname{im} i'_* = \ker f'_*$. On the other hand $\operatorname{im} i_* = j_*(\operatorname{im} i'_*)$. It follows that $\alpha \in H_1(M)$ as above is in $\operatorname{im} i_*$ if and only if there is a $\beta \in H_1(M')$ such that:

(i) β is a lifting of α, i.e., $j_*(\beta)=\alpha$;

(ii) $f'_*(\beta)=0$.

In terms of our bases this means that

(i′) $\beta_{ba}=\alpha_{ba}$, $\beta_{ca'}=\alpha_{ca'}$, $\beta_h=\alpha_h$;

(ii′) The coordinates of α satisfy the equation (6.10) (equivalently, $E_1=0$), and there is a choice of coordinates $\beta_{ca''}$ such that one has

$$\sum_{a''} m''_{ca''}\beta_{ca''}=\sum_a m_{\infty a}\alpha_{\infty a}-\sum_{a'} m'_{ca'}\alpha_{ca'} \tag{6.15}$$

for each $c\in C(f)$.

Let $m''(c)=G.C.D.\{m''_{ca''}\}$ where a'' takes all the possible values (this set of indices is non-empty). Equation (6.15) has a solution if and only if the right-hand side is divisible by $m''(c)$. This explains the following construction. Consider the finite abelian group

$$G(f)=\bigoplus_{c\in C(f)}\mathbb{Z}/m''(c)\mathbb{Z} \tag{6.16}$$

and let

$$\theta:\ker f_*\to G(f)$$

be the morphism sending $\alpha\in\ker f_*$ to the element in $G(f)$ having as its c-coordinate the class of

$$\sum_a m_{\infty a}\alpha_{\infty a}-\sum_{a'} m'_{ca'}\alpha_{ca'}$$

modulo $m''(c)$.

Let $m'(c)=G.C.D.\{m'_{ca'}\}$ where a' takes all the possible values (this set can be empty and in this case we set $m'(c)=0$) and set $m(c)=G.C.D.(m'(c),m''(c))$. For $b\in B$, set $m(b)=G.C.D.\{m_{ba}\}$ and then $m(f)=L.C.M.\{m(b)\mid b\in B\}$.

The above discussion is summarized in the following.

Theorem 6.6. *With the above notation, there is an isomorphism $T(f)\simeq\operatorname{im}\theta$. The group $T(f)\simeq\operatorname{im}\theta$ is the sum of the cyclic subgroup $T_0(f)$ spanned by*

$$(m(f),m(f),\dots,m(f))$$

in the group $G(f)$ defined in (6.16) and of the subgroup

$$T'(f)=\bigoplus_{c\in C(f)} m(c)\mathbb{Z}/m''(c)\mathbb{Z}\subset G(f).$$

In particular we get the following:

(i) *If none of the fibers D_c is multiple, i.e., $m(c) = 1$ for all $c \in C(f)$, then $T(f) = T'(f) \simeq G(f)$. More generally, if $m(c)|m(f)$ for all $c \in C(f)$, then $T(f) = T'(f)$.*
(ii) *If C is a minimal arrangement with respect to $f : M \to S$, i.e., there are no curves of type $C'_{ca'}$, then $T(f) = T_0(f)$ is a cyclic group. If in addition $m''(c)|m(f)$ for all $c \in C(f)$, then $T(f) = 1$.*
(iii) *If $f : M \to S$ has no multiple fibers i.e., if $m''(c) = 1$ for all $c \in C(f)$, then there are no translated components in $\mathcal{V}_1(M)$ associated to f.*

The case (ii) above does not rule out the possibility of a *translated global component*, but explains in a sense why such a component should be quite exceptional. Recall that by Theorem 4.1, a translated global component corresponds to a minimal arrangement. Conversely, a translated coordinate component is more likely to occur, and then it is related to a special arrangement, as in the deleted B_3-arrangement revisited below in Example 6.7.

Example 6.7 (The deleted B_3-arrangement). We return to Example 3.11 and apply the above discussion to this test case. The corresponding mapping $f : M \to \mathbb{C}^*$ has $B = \{0, \infty\}$ and $C(f) = \{1\}$. Indeed, with obvious notation, we get the following divisors: $D_0 = L_1 + L_4 + 2L_5$, $D_\infty = L_2 + L_3 + 2L_7$ and $D_1 = L_6 + 2L$ where $L : x + y - 1 = 0$ is exactly the line from the B_3-arrangement that was deleted in order to get Suciu's arrangement. Moreover, the associated fibration $f' : M' \to S'$ in this case is just the fibration of the B_3-arrangement discussed in [17], Example 4.6.

The line L is the only new component that has to be deleted, therefore $m''(1) = 2$. Since none of the fibers D_c (in our case there is just one, for $c = 1$) is multiple, Theorem 6.6 implies that

$$T(f) = \mathbb{Z}/2\mathbb{Z}.$$

Let $\gamma_i = \gamma(L_i)$. We know that $\rho(\gamma_i) = \pm 1$ and to get the exact values we proceed as follows. First note that we can choose $\rho(\gamma_1) = 1$, since the associated torus is

$$f^*(\mathbb{T}(\mathbb{C}^*)) = \{(t, t^{-1}, t^{-1}, t, t^2, 1, t^{-2}, 1) \mid t \in \mathbb{C}^*\}.$$

(In fact the choice $\rho(\gamma_1) = -1$ produces the character ρ'_W introduced in Example 3.11.) Next let $\alpha = \sum_{i=1,7} \alpha_i \gamma_i \in H_1(M)$. Then $\alpha \in \ker f_*$ if and only if

$$\alpha_1 + \alpha_4 + 2\alpha_5 = \alpha_2 + \alpha_3 + 2\alpha_7. \tag{6.17}$$

In our case, the morphism $\theta : \ker f_* \to \mathbb{Z}/2\mathbb{Z}$ is given by $\alpha \mapsto \alpha_2 + \alpha_3 - \alpha_6$. It follows that $\gamma_6 \in \ker f_*$ and $\theta(\gamma_6) = 1 \in \mathbb{Z}/2\mathbb{Z}$ from which it follows that $\rho(\gamma_6) = -1$.

Next $\gamma_1 + \gamma_2 \in \ker f_*$ and $\theta(\gamma_1 + \gamma_2) = 1 \in \mathbb{Z}/2\mathbb{Z}$. It follows that $\rho(\gamma_1)\rho(\gamma_2) = -1$, i.e., $\rho(\gamma_2) = -1$. The reader can continue in this way and get the value of $\rho = \rho_W$ given above in Example 3.11.

Example 6.8 (A more general example: the $\mathcal{A}_m$-arrangement). Let $\mathcal{A}_m$ be the line arrangement in $\mathbb{P}^2$ defined by the equation

$$x_1 x_2 (x_1^m - x_2^m)(x_1^m - x_3^m)(x_2^m - x_3^m) = 0.$$

This arrangement is obtained by deleting the line $x_3 = 0$ from the complex reflection arrangement associated to the full monomial group $G(3,1,m)$ and was studied in [7] and in [9]. The deleted B_3-arrangement studied above is obtained by taking $m = 2$.

Consider the associated pencil

$$(P,Q) = (x_1^m(x_2^m - x_3^m), x_2^m(x_1^m - x_3^m)).$$

Then the set B consists of two points, namely $(0:1)$ and $(1:0)$, and the set C is the singleton $(1:1)$, see for instance [17], Example 4.6. It follows that $m'(c) = 1$, $m''(c) = m$ and hence via Theorem 6.6 we get

$$T(f) = \frac{\mathbb{Z}}{m\mathbb{Z}}.$$

We expect $(m-1)$ 1-dimensional components in $\mathcal{V}_1(M)$, and this is precisely what has been proved in [7], or in Thm. 5.7 in [9]. There are $r = 2 + 3m$ lines in the arrangement, and to describe these components we use the coordinates

$$(z_1, z_2, z_{12:1}, \ldots, z_{12:m}, z_{13:1}, \ldots, z_{13:m}, z_{23:1}, \ldots, z_{23:m})$$

on the torus $(\mathbb{C}^*)^r$ containing $\mathbb{T}(M)$. Here z_j is associated to the line $x_j = 0$, for $j = 1, 2$, and $z_{ij:k}$ is associated to the line $x_i - w^k x_j$, where $i, j = 1, 3$, $k = 1, \ldots, m$, and $w = \exp(2\pi\sqrt{-1}/m)$. All the above 1-dimensional components have the same associated 1-dimensional subtorus

$$\mathbb{T} = f^*(\mathbb{T}(\mathbb{C}^*)) = \{(u^m, u^{-m}, 1, \ldots, 1, u^{-1}, \ldots, u^{-1}, u, \ldots, u) \mid u \in \mathbb{C}^*\}$$

where $f : M \to \mathbb{C}^*$ is the morphism associated to the pencil (P,Q), and each element 1, u^{-1} and u is repeated m times. The associated maximal isotropic subspace E in $H^1(M,\mathbb{C})$ is spanned by the 1-form

$$\omega = m\frac{dx_1}{x_1} - m\frac{dx_2}{x_2} - \sum_{k=1,m} \frac{dx_1 - w^k dx_3}{x_1 - w^k x_3} + \sum_{k=1,m} \frac{dx_2 - w^k dx_3}{x_2 - w^k x_3}.$$

The patient reader may check that for any $\alpha \in H^1(M,\mathbb{C})$, the vanishing $\alpha \wedge \omega = 0$ in $H^2(M,\mathbb{C})$ implies that α is a multiple of ω (this is the maximality condition in this 1-dimensional case). Let γ_c be an elementary loop about one line L in the fiber $\mathcal{C}_c$, with multiplicity 1, e.g., $L : x_1 - x_2 = 0$. Similarly, let γ_b be an elementary loop about one line L' in the fiber $\mathcal{C}_b$, with multiplicity 1, where $b = \infty = (0:1)$, e.g., $L' : x_2 - x_3 = 0$. And let γ_0 be an elementary loop about one line L_0 in the fiber $\mathcal{C}_0$, with multiplicity 1, where $0 = (1:0)$, e.g., $L_0 : x_1 - x_3 = 0$. One can show easily that:

(i) the classes $[\gamma_c]$ and $[\gamma_b + \gamma_0]$ in the group $T(f)$ are independent of the choices made;

(ii) $[\gamma_c] = -[\gamma_b + \gamma_0]$ is a generator of $T(f)$.

It follows that a torsion character $\rho \in \mathbb{T}(M)$ such that $\mathcal{L}_\rho | F = \mathbb{C}_F$ and inducing a non-trivial character $\tilde{\rho} : T(f) \to \mathbb{C}^*$ is given by

$$\rho = (1, 1, w^k, \ldots, w^k, w^{-k}, \ldots, w^{-k}, 1, \ldots, 1)$$

for $k = 1, \dots, m-1$. Here $\tilde{\rho}([\gamma_c]) = w^k$ and ρ is normalized by setting the last m components equal to 1.

Example 6.9. In a celebrated paper [5], Broughton considered the polynomial $f(x,y) = x(xy-1)$. The associated mapping $f : \mathbb{C}^2 \to \mathbb{C}$ has no singular points, but the fibers $F_t = f^{-1}(t)$ are not all diffeomorphic.

Z. Raza introduced in [23] what can be called *generalized Broughton polynomials*

$$f_{p,q}(x,y) = x^p[xy(x+2)\cdots(x+q)-1]$$

where the integers p, q satisfy $p \geq 1$, $q \geq 1$, with the convention

$$f_{p,1}(x,y) = x^p(xy-1).$$

He considers the following two irreducible rational affine curves in $\mathbb{C}^2$:

$$C_0 : g_q(x,y) = 0$$

where $g_q(x,y) = xy(x+2)\cdots(x+q) - 1$ for $q > 1$ and $g_1(x,y) = xy - 1$, and

$$C_1 : f_{p,q}(x,y) = 1.$$

In other words, C_0 is a component of the 0-fiber of $f_{p,q}$ (the other component being the line $L : x = 0$ with multiplicity p), while C_1 is the 1-fiber of $f_{p,q}$ (which is clearly the generic fiber of $f_{p,q}$). Consider the complement $M = \mathbb{C}^2 \setminus (C_0 \cup C_1)$.

By computing the cohomology ring of M, Raza shows that the resonance varieties of M are trivial, i.e., $\mathbb{R}_k(M) = 0$ for any $k > 0$.

Note next that $f = f_{p,q}$ induces a map $f_1 = f - 1 : M \to \mathbb{C}^*$, having one multiple fiber $x = 0$ of multiplicity p when $p > 1$. It follows from the description of the group $T(f)$, that $T(f) = \mathbb{Z}/p\mathbb{Z}$ and hence, we see that there are exactly $(p-1)$ associated 1-dimensional translated components. If we identify $\mathbb{T}(M) = (\mathbb{C}^*)^2$ by associating to a local system $\mathcal{L} \in \mathbb{T}(M)$ the two monodromies (λ_0, λ_1) about the curves C_0 and C_1, and in a similar way $\mathbb{T}(\mathbb{C}^*) = \mathbb{C}^*$, then the induced morphism

$$f_1^* : \mathbb{T}(\mathbb{C}^*) \to \mathbb{T}(M)$$

is just $\lambda \mapsto (1, \lambda)$.
With these identifications, the above $(p-1)$ associated 1-dimensional translated components of $\mathcal{V}_1(M)$ are given by $W_j = \epsilon_j \times \mathbb{C}^* \subset (\mathbb{C}^*)^2$, where $\epsilon_j = \exp(2\pi i j/p)$ for $j = 1, 2, \dots, p-1$.

One of the difficult points is to show that there are no other positive dimensional components.

It was thought that the number of translated components in $\mathcal{V}_k(M)$ is small in comparison to the number of lines or, more generally, the number of irreducible components of a plane curve arrangement. This example shows that this is not true if we go beyond the class of line arrangements.

References

[1] D. Arapura: Geometry of cohomology support loci for local systems. I, *J. Algebraic Geom.* **6** (1997), 563–597.

[2] E. Artal Bartolo, J. Carmona Ruber, J.I. Cogolludo Agustin: Essential coordinate components of characteristic varieties, *Math. Proc. Cambridge Philos. Soc.* **136** (2004), no. 2, 287–299.

[3] W. Barth, C. Peters, A. Van de Ven: *Compact Complex Surfaces*, Springer Verlag, 1984.

[4] A. Beauville, Annulation du H^1 pour les fibrés en droites plats, in: *Complex algebraic varieties* (Bayreuth, 1990), 1–15, Lecture Notes in Math., vol. 1507, Springer, Berlin, 1992.

[5] S.A. Broughton: Milnor numbers and the topology of polynomial hypersurfaces, *Invent. Math.* **92** (1988), 217–241.

[6] F. Catanese, Moduli and classification of irregular Kaehler manifolds (and algebraic varieties) with Albanese general type fibrations, *Invent. Math.* **104** (1991), no. 2, 263–289.

[7] D.C. Cohen: Triples of arrangements and local systems. *Proc. Amer. Math. Soc.* **130** (2002), no. 10, 3025–3031.

[8] D.C. Cohen, A.I. Suciu: Characteristic varieties of arrangements, *Math. Proc. Cambridge Philos. Soc.* **127** (1999), 33–54.

[9] D.C. Cohen, G. Denham, A.I. Suciu: Torsion in Milnor fiber homology, *Algebraic and Geometric Topology* **3** (2003),511–535.

[10] F. Delgado, H. Maugendre: Special fibers and critical locus for a pencil of plane curve singularities, *Compositio Math.* **136** (2003), 69–87.

[11] A. Dimca: *Singularities and Topology of Hypersurfaces*, Universitext, Springer Verlag, 1992.

[12] A. Dimca: *Sheaves in Topology*, Universitext, Springer Verlag, 2004.

[13] A. Dimca: Characteristic varieties and constructible sheaves, *Rend. Lincei Mat. Appl.* **18** (2007), 365–389.

[14] A. Dimca: On the isotropic subspace theorems, *Bull. Math. Soc. Sci. Math. Roumanie* **51** (2008), no. 4, 307–324.

[15] A. Dimca, L. Maxim: Multivariable Alexander invariants of hypersurface complements, *Transactions Amer. Math. Soc.* **359** (2007), no. 7, 3505–3528.

[16] A. Dimca, S. Papadima and A. Suciu: Formality, Alexander invariants, and a question of Serre, math.AT/0512480.

[17] M. Falk, S. Yuzvinsky: Multinets, resonance varieties, and pencils of plane curves, *Compositio Math.* **143** (2007), no. 4, 1069–1088.

[18] R. Hartshorne: *Algebraic Geometry*, GTM 52, Springer, 1977.

[19] D. T. Lê, C. Weber: Équisingularité dans les pinceaux de germes de courbes planes et C^0-suffisance, *Enseign. Math. (2)* **43** (1997), no. 3–4, 355–380.

[20] A. Libgober: Characteristic varieties of algebraic curves, in: C.Ciliberto et al.(eds), *Applications of Algebraic Geometry to Coding Theory, Physics and Computation*, pp. 215–254, Kluwer, 2001.

[21] A. Libgober, S. Yuzvinsky: Cohomology of the Orlik-Solomon algebras and local systems, *Compositio Math.* **121** (2000), 337–361.

[22] J. Pereira, S. Yuzvinsky: Completely reducible hypersurfaces in a pencil, math.AG/0701312.

[23] Z. Raza: Broughton polynomials and characteristic varieties, to appear in *Studia Scient. Math. Hungarica.*

[24] A. Schinzel: *Selected topics on polynomials.* University of Michigan Press, Ann Arbor, Mich., 1982.

[25] A. Suciu: Translated tori in the characteristic varieties of complex hyperplane arrangements. Arrangements in Boston: a Conference on Hyperplane Arrangements (1999). *Topology Appl.* **118** (2002), no. 1–2, 209–223.

[26] A. Suciu: Fundamental groups of line arrangements: enumerative aspects. Advances in algebraic geometry motivated by physics (Lowell, MA, 2000), 43–79, *Contemp. Math.* **276**, Amer. Math. Soc., Providence, RI, 2001.

Alexandru Dimca
Laboratoire J.A. Dieudonné
UMR du CNRS 6621
Université de Nice Sophia-Antipolis
Parc Valrose
06108 Nice Cedex 02
France
e-mail: `dimca@math.unice.fr`

Progress in Mathematics, Vol. 283, 83–110

Lectures on Orlik-Solomon Algebras

Alexandru Dimca and Sergey Yuzvinsky

Abstract. The first part of this survey is an elementary introduction into Orlik-Solomon algebras. The remaining part is devoted to more recent results, such as the description of the geometric meaning of the Orlik-Solomon complex in terms of the computation of rank 1 local system cohomology via logarithmic connections.

Mathematics Subject Classification (2000). Primary: 52C35, 05B35; Secondary: 16E05, 15A75.

1. Introduction

These notes are based on the set of lectures given by the second author during the CIMPA School at the Galatasarai University of Istanbul in June of 2007. The lectures were devoted to the Orlik-Solomon (OS) algebras of hyperplane arrangements. These algebras appeared first from theorems by Arnold, Brieskorn, and Orlik-Solomon as the cohomology algebras of the complements to complex hyperplane arrangements. Then they were shown to play an important part in theory of multivariable hypergeometric functions, conformal field theory, cohomology of the Milnor fibers of nonisolated singularities, and Alexander invariants of projective curves. Results about OS algebras usually intertwine several areas of mathematics such as pure algebra, combinatorics, topology, differential and algebraic geometries.

The first part of this survey is an elementary introduction into the topic that follows closely chapters 2–4 of the survey [46]. All definitions are given but for proofs we refer the reader to [46].

Section 5 is devoted to more recent results. Here we view an OS algebra A as a cochain complex under multiplication by an element a of degree 1 and study its cohomology $H^*(A,a)$. When a is in sufficiently general position, $H^*(A,a)$ can be described completely – see Theorem 5.2 ([45, 39]). To get more information about $H^*(A,a)$, we consider the resonance varieties introduced in [25]. The most results have been gotten about R^1_m, that is the set of a such that $\dim H^1(A, a) \geq m$. Some basic properties of irreducible components of R^1_m were obtained in [33] and [12]. They continue to hold for M a smooth quasi-projective variety which is 1-formal, see Theorem 6.12 below, but fail if the extra condition of formality is not present, see for details [20].

In [33], using the results of Arapura [4], there appeared a connection of the irreducible components of R^1_1 with pencils of plane curves and interesting sets of lines in the projective plane called nets. Here we follow the paper [26] where nets were generalized to multinets and the equivalence between multinets, pencils of curves, and irreducible components of R^1_1 was proved directly. We also mention the very recent results about the number of completely reducible fibers of pencils of hypersurfaces or, equivalently, about the dimensions of the components of R^1_1.

In the final section we describe the geometric meaning of the OS complex in terms of the computation of rank 1 local system cohomology via the logarithmic connections, see Proposition 6.5. This result depends on some Hodge theoretic properties recall in Proposition 6.1 and on the notion of an admissible local system, see Definition 6.3. Arapura's results mentioned above concern the characteristic varieties $\mathcal{V}_k(M)$ of a smooth, quasi-projective complex variety M. They are recalled in Theorem 6.9 and Theorem 6.10.

The resonance varieties can also be defined in this more general setting, and their relation to the characteristic varieties is described in Theorem 6.12. The irreducible components of resonance varieties satisfy a number of very specific properties, called the resonance obstructions. These obstructions, and some of their applications, are discussed at the end of the paper.

We would like to thank the local organizers and Professor Muhammed Uludag in particular for making the school happen and being a pleasant experience.

2. Orlik-Solomon ideals of exterior algebras

In this section, the pure algebraic definition of the OS ideal (and OS algebra) of an arrangement is given. We state some simple properties of the OS ideals.

2.1. Combinatorics of an arrangement

The notation we introduce here will be used throughout the notes. We denote by $\mathcal{A}$ an arrangement in an affine space $V \approx K^\ell$ for a fixed arbitrary field K and $\ell \geq 1$. This means that $\mathcal{A}$ is a finite set of affine hyperplanes of V. The cardinality of $\mathcal{A}$ will be usually denoted by n. Very often we will fix an arbitrary linear order on $\mathcal{A}$, i.e., put $\mathcal{A} = (H_1, \ldots, H_n)$.

Any time when it is convenient, we fix a linear basis $(x_1, \ldots, x_\ell)$ of V^* and identify V with K^ℓ using the dual basis in V. Then in order to define a hyperplane H of K^ℓ it suffices to fix a degree 1 polynomial $\alpha_H \in S = K[x_1, \ldots, x_\ell]$ such that H is the zero locus of α_H. This polynomial is uniquely defined up to multiplication by a nonzero element from K. We will abbreviate α_{H_i} by α_i. If the hyperplanes are linear subspaces, i.e., the defining polynomials are homogeneous, $\mathcal{A}$ is called *central.*

In order to define the Orlik-Solomon (abbreviated as OS) algebra of $\mathcal{A}$ we do not need to know the hyperplanes; it suffices to know the combinatorics of $\mathcal{A}$, i.e., its *intersection poset* $L(\mathcal{A})$. Elements of this poset $L = L(\mathcal{A})$ are all subspaces of V that are intersections of some hyperplanes from $\mathcal{A}$ (including V itself as the intersection of the empty set of hyperplanes). The partial order on L is the reverse inclusion of subspaces. In particular L always has the unique minimal element V but has the unique maximal element $\bigcap_{i=1}^{n} H_i$ if and only if this intersection is nonempty. In the latter case we can use a translation and always assume that this intersection contains 0 (equivalently, all α_H are homogeneous). In this case $\mathcal{A}$ is called *central.* If the above intersection is 0, $\mathcal{A}$ is called *essential.*

The poset L is far from arbitrary. If $\mathcal{A}$ is central, then L is a lattice whose atoms (i.e., the elements succeeding the minimal element) are the hyperplanes themselves. It also has the following extra properties:

(i) it is *atomic*, i.e., its every element is the join (the least upper bound) of some atoms;

(ii) it is *ranked*, i.e., every nonrefinable chain ($V < X_1 < \cdots < X_r = X$) from V to a fixed $X \in L$ has the same number of elements, namely the codimension of X (following lattice theory we call this number the rank of X and denote it by $\mathrm{rk}X$);

(iii) for every $X, Y \in L$ the following *semimodular inequality* holds,

$$\mathrm{rk}X + \mathrm{rk}Y \geq \mathrm{rk}(X \vee Y) + \mathrm{rk}(X \wedge Y),$$

where the symbols $\vee$ and $\wedge$ are used respectively for the operations of join and meet (i.e., the greatest lower bound).

Lattices satisfying the above properties are called *geometric.* The rank $\mathrm{rk}L$ of a geometric lattice L is the maximal rank of its elements. Clearly $\mathrm{rk}L \leq \ell$ and $\mathrm{rk}L = \ell$ if and only if the arrangement is essential. Notice that any nonempty closed interval of a geometric lattice, i.e., the subposet of elements lying inclusively between two fixed ones, is again a geometric lattice.

Among many properties of geometric lattices we need to record a property of its homology. For an arbitrary poset, all its flags (i.e., linearly ordered subsets) form a simplicial complex on the set of all its elements. The homotopy invariants of this complex are attributed to the poset itself. If the poset is a lattice, then by its homotopy invariants one usually means those of its subposet with the maximal and minimal elements deleted. In this sense the following result is a part of the celebrated Folkman's theorem [27]: the reduced homology (with integer coefficients) of a geometric lattice vanishes except perhaps in the highest dimension (equal to $\mathrm{rk}L - 2$).

In order to understand the structure of the intersection poset $L(\mathcal{A})$, if $\mathcal{A}$ is not central we apply the simple *coning* construction. If $\mathcal{A}$ is an arbitrary (not necessarily central) arrangement of n hyperplanes in K^ℓ, then by embedding K^ℓ

into the projective space $\mathbb{P}K^\ell$ and adjoining to $\mathcal{A}$ the hyperplane $H_0 = \mathbb{P}K^\ell \setminus K^\ell$ we obtain an arrangement of $n+1$ projective hyperplanes in $\mathbb{P}K^\ell$. The cone $c\mathcal{A}$ over this arrangement in $K^{\ell+1}$ is called the *cone over* $\mathcal{A}$. It is a central arrangement of $n+1$ hyperplanes. Notice also that every nonempty central arrangement $\mathcal{B}$ in K^ℓ can be viewed as the cone over an arrangement $\mathcal{A}$. Indeed fix an $H \in \mathcal{A}$ and a linear functional α_H such that $H = \ker \alpha_H$. Then we can define the affine hyperplane $\bar{H}$ of K^ℓ by the equation $\alpha_H = 1$ and the arrangement $d\mathcal{A} = \{H' \cap \bar{H} | H' \in \mathcal{A} \setminus \{H\}\}$ in $\bar{H}$. It is easy to see that $c(d\mathcal{A})$ is linearly isomorphic to $\mathcal{A}$. The arrangement $d\mathcal{A}$ is called *a deconing* of $\mathcal{A}$.

Now the poset $L(\mathcal{A})$ is the subposet of the (geometric) lattice $L(c\mathcal{A})$. It can be defined as containing all the atoms but H_0 and being closed with respect to the join. If we substitute here an arbitrary geometric lattice for $L(c\mathcal{A})$ and its arbitrary atom for H_0, we obtain a definition of a *geometric semilattice.* An axiomatic definition can be found in [43]. The most important for us property of a geometric semilattice is that each of its closed intervals is a geometric lattice whence Folkman's theorem holds for it.

2.2. Definition and simple properties of OS algebras

Let k be a commutative ring and $\mathcal{A} = (H_1, \dots, H_n)$ be an arrangement. Let E be the exterior algebra over k with generators $e_1, \dots, e_n$ of degree 1. Notice that the idexes define a bijection from $\mathcal{A}$ to the generating set. Sometimes we will denote the generator corresponding to $H \in \mathcal{A}$ by e_H. The algebra E is graded via $E = \oplus_{p=1}^n E_p$ where $E_1 = \oplus_{j=1}^n k e_j$ and $E_p = \bigwedge^p E_1$. For every k, the k-module E_j is free and has the distinguished basis consisting of monomials $e_S = e_{i_1} \cdots e_{i_p}$ where $S = \{i_1, \dots, i_p\}$ runs through all the subsets of $[n] = \{1, 2, \dots, n\}$ of cardinality p and $i_1 < i_2 < \cdots < i_p$. The graded algebra E is a (commutative) DGA with respect to the differential ∂ of degree -1 uniquely defined by the conditions $\partial e_i = 1$ for every $i = 1, \dots, n$ and the graded Leibniz formula. Then for every $S \subset [n]$ of cardinality p,

$$\partial e_S = \sum_{j=1}^{p} (-1)^{j-1} e_{S_j}$$

where S_j is the complement in S to its jth element.

It is important to notice that ∂e_S is a pure (or decomposable) element of E for every S. Indeed if again $S = \{i_1, \dots, i_p\} \subset [n]$, then a straightforward check shows that

$$\partial e_S = (e_{i_2} - e_{i_1})(e_{i_3} - e_{i_1}) \cdots (e_{i_p} - e_{i_1}). \tag{1}$$

For every $S \subset [n]$, put $\cap S = \bigcap_{i \in S} H_i$ and call S *dependent* if $\cap S \neq \emptyset$ and the set of linear polynomials $\{\alpha_i | i \in S\}$ is linearly dependent.

Definition 2.1. The OS ideal of $\mathcal{A}$ is the ideal $I = I(\mathcal{A})$ of E generated by all e_S with $\cap S = \emptyset$ and all ∂e_S with S dependent. The algebra $A = A(\mathcal{A}) = E/I(\mathcal{A})$ is called the OS algebra of $\mathcal{A}$.

We denote the canonical projection $E \to A$ by π. Since π induces a k-linear isomorphism of E_1 and A_1 we will identify these k-modules.

Notice that for any nonempty $S \subset [n]$ and $i \in S$ one has $e_i \partial e_S = \pm e_S$ whence $I(\mathcal{A})$ contains e_S for every dependent set S. This implies that $A(\mathcal{A})$ is generated as a k-module by the images of e_S such that $\cap S \neq \emptyset$ and S is independent.

In Definition 2.1, the set of generators can be made much smaller. The minimal by inclusion dependent sets of $[n]$ are called *circuits*. It is easy to prove that an OS ideal is generated by all e_S with $\cap S = \emptyset$ and by all ∂e_T with T a circuit.

Clearly I is a homogeneous ideal of E whence A is a graded algebra. Write $A = \oplus A_p$ where $A_p = E_p/I \cap E_p$. Since every set of more than rk$L$ elements is dependent, the maximal degree of elements of A is rk$L$. Notice that factoring V by $\bigcap_i H_i$ makes $\mathcal{A}$ essential and does not change A. Thus, unless we are dealing with subarrangements, we will always assume that $\mathcal{A}$ is essential, i.e., rk$L = \ell$.

2.3. Grading and filtration by the intersection lattice

In this subsection we discuss a finer grading on A. We say here that an algebra U is *graded* by a lattice P if $U = \oplus_{X \in P} U_X$ where U_X are subspaces of U and $U_X + U_Y = U_{X \vee Y}$ for every $X, Y \in P$. We say also that U is *filtered* by P if there is a monotone embedding of P into the lattice of subalgebras of U ordered opposite to inclusion.

The finest grading on E is the grading by the Boolean lattice of all subsets of $[n]$. In general, the OS ideal is not homogeneous in this grading. Consider the coarser grading on E assigning to each e_S ($S \subset [n]$) the set $\cap S$. This is a grading by the lattice $\bar{L}(\mathcal{A})$ where $\bar{L}(\mathcal{A}) = L(\mathcal{A})$ if $\mathcal{A}$ is central and otherwise it is $L(\mathcal{A})$ completed by the unique maximal element $\emptyset$. We write $E = \oplus_{X \in \bar{L}(\mathcal{A})} E_X$ where E_X is spanned by e_S with $\cap S = X$. Notice that this grading is in general incomparable with the standard grading by $|S|$.

Proposition 2.2. *The OS algebra $A = A(\mathcal{A})$ is graded by the poset $L(\mathcal{A})$ and this grading is finer than the standard grading $A = \oplus_{p=0}^n A_p$. More precisely*

$$A_p = \oplus_{X \in L(\mathcal{A}), \mathrm{rk} X = p} A_X.$$

The above grading of A by $L(\mathcal{A})$ induces a filtration of A that can be defined independently. Among subarangements of an arrangement $\mathcal{A}$ there are the ones corresponding to elements of $L(\mathcal{A})$. For every $X \in L(\mathcal{A})$ we put $\mathcal{A}_X = \{H \in \mathcal{A} | X \subset H\}$. We also put $[n]_X = \{i \in [n] | H_i \in \mathcal{A}_X\}$. Notice that all the subarrangements $\mathcal{A}_X$ are central. They can be completely characterized also by the property of being *closed.* This means that with several hyperplanes they contain all hyperplanes dependent on them.

Proposition 2.3. *The natural homomorphism $i : A(\mathcal{A}_X) \to A = A(\mathcal{A})$ is injective and the mapping $X \mapsto A(\mathcal{A}_X)$ is the filtration of A by $L = L(\mathcal{A})$.*

This filtration is intimately related to the grading of A by L.

Proposition 2.4. *The filtration $\{A(\mathcal{A}_X)\}$ is induced by the grading*

$$A = \oplus_{Y \in L(\mathcal{A})} A_Y(\mathcal{A}).$$

More precisely, $A_Y(\mathcal{A}_X) = A_Y(\mathcal{A})$ for every $X, Y \in L(\mathcal{A})$ such that $X \subset Y$, whence $A(\mathcal{A}_X) = \oplus_{Y \leq X} A_Y(\mathcal{A})$.

If an arrangement $\mathcal{A}$ is central, then A receives another piece of structure from E. In this case I is generated over k by $\{e_T \partial e_S\}$ where T is arbitrary and S is dependent. Using the Leibniz rule we have

$$\partial(e_T \partial e_S) = \partial e_T \partial e_S$$

whence $I(\mathcal{A})$ is invariant with respect to ∂. Thus it induces a differential

$$\partial_A : A \to A$$

that defines the structure of commutative DGA on A. The relations of ∂_A with the gradings and filtration of A are as follows:

1. $\partial_A(A_X) \subset A_{r-1}$ for every $X \in L(\mathcal{A})$ of rank r;
2. $\partial_A(A(\mathcal{A}_X) \subset A(\mathcal{A}_X)$ for every $X \in L(\mathcal{A})$.

Finally if the arrangement is not empty, the restriction of ∂_A to any $A(\mathcal{A}_X)$ is acyclic. In particular it is true for $A = A(\mathcal{A}) = A(\mathcal{A}_U)$ where U is the unique maximal element of $L(\mathcal{A})$. Indeed the multiplication by any e_i with $H_i \in \mathcal{A}_X$ is clearly a homotopy between the identity and 0 maps of the complex (A, ∂) to itself.

2.4. Gröbner basis

Recall that we fix an arbitrary linear order on an arrangement $\mathcal{A}$. This order induces the *deg-lex linear order* on the set of all standard monomials e_S of E: if $S = (i_1, \dots, i_p)$ and $T = (j_1, \dots, j_q)$ (where $i_1 < \dots < i_p$ and $j_1 < \dots < j_q$), then $e_S < e_T$ if either $p < q$ or $p = q$ and there exists m, $1 \leq m \leq p$, such that $i_r = j_r$ for $r < m$ and $i_m < j_m$. The basis of E consisting of standard monomials is multiplicative up to $\pm$ (i.e., the product of two standard monomials is either 0 or a standard monomial perhaps with minus) and the deg-lex order is multiplicative (i.e., invariant under multiplication by monomials) if one ignores zero monomials and minuses. Thus we can apply the theory of Gröbner bases to the ideal $I = I(\mathcal{A})$.

Theorem 2.5. *Let $B = \{\partial e_S | S$ is a circuit $\} \cup \{e_T | T$ is minimal with $\cap T = \emptyset\} \subset E$. Then B is a Gröbner basis of I.*

Let us reformulate the theorem. First the initial monomials of elements from B, i.e., their largest monomials in the deg-lex order, can be easily defined combinatorially. Since e_T is a monomial itself its initial monomial coincides with it. The initial monomial of ∂e_S is e_{S_1} (recall that S_1 is obtained from S by deleting its smallest element). Thus these initial monomials correspond to the *broken circuits*, i.e., independent subsets S of $[n]$ such that $S \cup \{i\}$ is a circuit for some $i < j$ for all $j \in S$.

Denote by $In(B)$ and $In(I)$ the sets of the initial monomials of elements of B and I respectively. Then the statement of Theorem 2.5 means that

$$\langle In(B)\rangle = In(I)$$

where the left-hand side is the set of all monomials divisible by some monomials from In(B). Now denote by $C = C(\mathcal{A})$ (resp. $\bar{C}$) the natural linear complement to $\langle In(B)\rangle$ (resp. $In(I)$) namely the free k-module spanned by all the monomials of E not divisible by any element of $In(B)$ (resp. not in $In(I)$). The monomials from C are called **nbc**-monomials and the respective subsets of $[n]$ (i.e., subsets not containing any broken circuits) are called **nbc**-sets. Theorem 2.5 implies that the projection to A is injective on C and the image form is a (homogeneous) basis of A.

Corollary 2.6. *All the k-modules A and A_X $(X \in L(\mathcal{A}))$ are free.*

2.5. Deletion-restriction exact sequence

In this subsection, we construct a short exact sequence for OS algebras that is very useful for proving properties of these algebras by induction on n.

Let $\mathcal{A}$ be a nonempty arrangement and fix a hyperplane H' from it. Then we can consider two more arrangements: $\mathcal{A}' = \mathcal{A} \setminus \{H'\}$ (obtained by the *deletion* of H' from $\mathcal{A}$) and $\mathcal{A}'' = \{H \cap H' | H \in \mathcal{A}'\}$ (obtained by the *restriction* of $\mathcal{A}$ to H'). Let us emphasize that $\mathcal{A}''$ is an arrangement of hyperplanes in the space H'. Also while the cardinality of $\mathcal{A}'$ is always $n-1$, the cardinality of $\mathcal{A}''$ is at most $n-1$ and may be smaller. For any object O attached to an arrangement we will put $O = O(\mathcal{A})$, $O' = O(\mathcal{A}')$, and $O'' = O(\mathcal{A}'')$.

Since the results we are after do not depend on the choice of order on the arrangements, we make that choice in the most convenient way for our goal. First of all, H' will be the last element of $\mathcal{A}$, i.e., $H' = H_n$. Second, let $\lambda : \mathcal{A}' \to \mathcal{A}''$ be defined by $\lambda(H) = H \cap H_n$. Then we choose linear orders on $\mathcal{A}$, $\mathcal{A}'$, and $\mathcal{A}''$ so that the order on $\mathcal{A}'$ is induced by the order on $\mathcal{A}$ (i.e., $\mathcal{A}' = (H_1, \dots, H_{n-1})$) and $\lambda(H)$ precedes $\lambda(\bar{H})$ only if H precedes $\bar{H}$ for every $H, \bar{H} \in \mathcal{A}'$. Notice that the second condition can be satisfied for every linear order on $\mathcal{A}''$ if one orders preimages of λ according to this order and takes any linear orders inside those preimages.

Now extend λ to the ordered subsets of $\mathcal{A}'$ via

$$\lambda(H_{i_1}, \dots, H_{i_p}) = (\lambda(H_{i_1}), \dots, \lambda(H_{i_p})).$$

Notice that the image is a sequence of elements from $\mathcal{A}''$ perhaps with repetitions but always in nondecreasing order. The extended λ can be also considered as the map $S \mapsto \lambda(S)$ from the set of subsets of $[n-1]$ to the nondecreasing subsequences of $[n'']$ where $n'' = |\mathcal{A}''|$.

Now we can consider the natural sequence of k-linear maps:

$$0 \longrightarrow E' \xrightarrow[i]{} E \xrightarrow[j]{} E'' \longrightarrow 0 \tag{2}$$

where i is the natural embedding and j is defined via $j(e_S) = e_{\lambda(S\setminus\{n\})}$ if $n \in S$ and $j(e_S) = 0$ otherwise. Here for every subsequence T of n'' we denote by e''_T the product of generators e''_k of E'' taken in the order on T. Notice that although the sequence (2) is clearly exact in the terms E' and E'' and $ji = 0$, it does not have to be exact in E.

Theorem 2.7. *The sequence* (2) *generates the sequence of* k*-maps*

$$0 \longrightarrow A' \xrightarrow[\bar{i}]{} A \xrightarrow[\bar{j}]{} A'' \longrightarrow 0 \tag{3}$$

that is exact.

2.6. Central-affine exact sequence

In this subsection we construct another exact sequence relating the OS algebras of an arrangement $\mathcal{A}$ and its cone $c\mathcal{A}$.

Let $\mathcal{A} = (H_1, \dots, H_n)$ and recall that $c\mathcal{A}$ consists of cones cH_i over H_i ($i = 1, \dots, n$) and the extra hyperplane H_0. In order to consider the OS algebras of these two arrangements at the same time, we need to deal with two exterior algebras E generated by $e_1, \dots, e_n$ and cE generated by $e_0, e_1, \dots, e_n$. Clearly $E = cE/(e_0)$. Denote by E' the subalgebra of cE generated by all the differences $e_i - e_j$ and by I' the ideal of E' generated by ∂e_S for all dependent subsets S of $\{0, 1, \dots, n\}$. It follows from (1) that $\partial e_S \in E'$ whence the definition of I' indeed makes sense. Notice that E is isomorphic to E' under the isomorphism defined by $e_i \mapsto e_i - e_0$ ($i = 1, 2, \dots, n$).

If B is a factor of an exterior algebra over a graded ideal we put $B\langle e_0\rangle = B \otimes_k E(e_0)$ where $E(e_0)$ is the exterior algebra with the single generator e_0 of degree 1 and the tensor product is taken in the category of skew-commutative algebras. The OS ideal of $c\mathcal{A}$ is $cI = I'(cE)$ and the OS algebra $cA = cE/cI$ of $c\mathcal{A}$ is naturally isomorphic to $(E'/I')\langle e_0\rangle$. Furthermore the OS algebra A of $\mathcal{A}$ is $cE/(cI + (e_0)) \cong E'/I'$. In particular $cA = A\langle e_0\rangle$.

Proposition 2.8. *For each* p *we have the decomposition*

$$cA_p = A_p \oplus A_{p-1}$$

where the embedding of the first summand is induced by the identity map on e_S ($S \subset [n]$) *and that of the second summand is induced by* $e_S \mapsto e_0 e_S$.

3. Homological interpretation of OS algebras

In this section we give the interpretation of the OS-algebra as the local (Whitney) homology of the intersection poset. We introduce a product on the relative atomic complex making it a differential graded algebra (DGA) which lifts the interpretation from cohomology to the chain complex level and makes it more transparent.

3.1. Relative atomic complex of a lattice

In this subsection, we review certain facts about homology of lattices. We want to accommodate both central and noncentral arrangements. Because of that we use the letter L in this subsection for an arbitrary upper semilattice. This class of posets consists of complete lattices and lattices stripped of their maximal elements. The unique minimal and maximal (if it exists) elements of L are denoted by $\check{0}$ and $\check{1}$ respectively. The set of atoms of L is denoted by $\mathcal{A} = \mathcal{A}(L)$ and we put $n = |\mathcal{A}|$. An arbitrary linear order is fixed on $\mathcal{A}$ which allows us to identify $\mathcal{A}$ with $[n]$. A set $\sigma \subset [n]$ is called *bounded* if it has an upper bound in L whence its least upper bound $\bigvee \sigma$ exists in L. Otherwise it is called unbounded.

All homologies are taken with coefficients in a fixed commutative ring k. As above $H_*(L)$ denotes the homology of the flag complex of $L \setminus \{\check{0}, \check{1}\}$. There exist a few smaller complexes homotopic to the flag complex. We recall one of them, namely the simplicial atomic complex $\Delta = \Delta(L)$. It is the complex on $[n]$ whose simplexes are all subsets σ that are bounded in $L \setminus \{\check{1}\}$.

Even more important for us is the *relative atomic complex* $D = D(L)$. This is the chain complex that is the free module over a commutative ring k with a basis consisting of all bounded in L sets $\sigma \subset [n]$. The grading on D assigns degree $|\sigma|$ to σ. (Notice the unusual grading.) The differential d on D is defined as follows. If $\sigma = \{i_1, \dots, i_p\}$ and $\sigma_j = \sigma \setminus \{i_j\}$, then

$$d(\sigma) = \sum_{\{j \in \sigma \mid \bigvee \sigma_j = \bigvee \sigma\}} (-1)^{j-1} \sigma_j.$$

We will use the following simple properties of D.

1. D is the direct sum of the complexes $D(X)$ $(X \in L)$ where $D(X)$ is spanned by subsets $\sigma \subset [n]$ such that $\bigvee \sigma = X$.
2. For every p and $X \in L$ the group $H_p(D(X))$ is naturally isomorphic to $\tilde{H}_{p-2}(L_X)$ where $L_X = \{Y \in L | Y \leq X\}$. Notice that L_X is always a lattice.
3. If L_X is a geometric lattice for every $X \in L$, then $H_p(D(X)) = 0$ for all $p \neq \mathrm{rk} X$.

This follows immediately from (2) and Folkman's theorem.

The last property allows us, in the case where L_X is a geometric lattice, to exhibit nice generators of $H_*(D(X))$ as a k-module. We call a set $\sigma \subset [n]$ *independent* if it is bounded and $\bigvee \sigma_i < \bigvee \sigma$ for every $i \in \sigma$. Clearly if $\mathcal{A}$ is a hyperplane arrangement and $L = L(\mathcal{A})$ this definition coincides with the definition in Section 1.

In the general case, each independent set σ of atoms is a cycle in D whence it defines a homology class ζ_σ in $H_{|\sigma|}(D(X))$ where $X = \bigvee \sigma$. Besides if L is a geometric lattice, then $\mathrm{rk}(\bigvee \sigma) \leq |\sigma|$. Thus in this case, each set σ of atoms such that $\bigvee \sigma = X$ and $|\sigma| = \mathrm{rk} X$ is independent. This implies that $D_{\mathrm{rk} X}(X)$ consists of cycles only whence $H_*(D)$ is generated by $\{\zeta_\sigma | \sigma \text{ is independent}\}$.

3.2. Multiplicative structure on the relative atomic complex

Here we consider only the case where each L_X is geometric. We introduce a bilinear (over k) multiplicative structure on D that converts it into a differential graded algebra (DGA). This structure is defined via the product of generators. Let σ and

τ be bounded sets of atoms. Then

$$\sigma \cdot \tau = \begin{cases} 0, & \text{if } \sigma \cup \tau \text{ is unbounded,} \\ 0, & \text{otherwise and } \operatorname{rk}(\bigvee(\sigma \cup \tau)) \neq \operatorname{rk}(\bigvee(\sigma)) + \operatorname{rk}(\bigvee(\tau)), \\ \epsilon(\sigma, \tau)\sigma \cup \tau, & \text{otherwise} \end{cases} \tag{4}$$

where $\epsilon(\sigma, \tau)$ is the sign of the *shuffle* of σ and τ, i.e., the permutation of $\sigma \cup \tau$ putting all elements of τ after elements of σ and preserving fixed orders inside these sets.

Proposition 3.1. *The multiplication defined in* (4) *converts D to a (commutative) DGA.*

The proof consists of case-by-case checking the Leibniz formula.

From the definition of the multiplication we obtain the following.

Proposition 3.2. *Every independent set σ of atoms is the product of all its one-element subsets (taken in the natural order). In particular the algebra $H_*(D)$ is generated by $\zeta_i = \zeta_{\{i\}}$ $(i \in [n])$.*

It is easy to see that if a set σ of atoms is unbounded or "dependent" (i.e., bounded and not independent), then the product (in any order) of its all one-element subsets is 0.

3.3. An interpretation of OS algebras

For the arrangement case, that is if $\mathcal{A}$ is an arrangement and $L = L(\mathcal{A})$ is its intersection poset, we obtain an interpretation of the OS algebra. Recall that although L is not a lattice if $\mathcal{A}$ is not central, every L_X is a geometric lattice.

Theorem 3.3. *The correspondence $e_i \mapsto \zeta_i$ $(i = 1, 2, \ldots, n)$ generates an isomorphism of algebras $A(\mathcal{A}) \to H_*(D)$.*

Theorem 3.3 allows us to give a combinatorial formula for the Hilbert series $H(A, t) = \sum_{p \geq 0} \dim_R A_p t^p$ of A.

Corollary 3.4.

$$H(A, t) = \sum_{p \geq 0} \sum_{X \in L, \operatorname{rk} X = p} |\mu(X)| t^p$$

where $\mu(X) = \mu(V, X)$ is the value of the Möbius function of L on X.

The right-hand side of the above equality is called the *Poincarè polynomial* of L and is denoted by $P(L, t)$.

Example 3.5. Let $\mathcal{A}$ be the arrangement in K^3 given by the linear functionals $x, y, z, x - y, x - z, y - z$. If $K = \mathbb{R}$, then $\mathcal{A}$ can be obtained from the arrangement of reflection hyperplanes of the Coxeter group of type A_3 by factoring out the mutual intersection of these hyperplanes. The arrangement $\mathcal{A}$ is the first non-trivial one in the series of essential arrangements whose complements for $K = \mathbb{C}$

are classifying spaces of pure braid groups. For this reason, all these arrangements are called *braid* arrangements.

The intersection lattice L of $\mathcal{A}$ has six atoms 1, 2, 3, 4, 5, 6 that correspond to the hyperplanes in the order they are written. The rank-2 elements can be identified with the sets of atoms below them. In this sense they are $\{1,2,4\}$, $\{1,3,5\}$, $\{2,3,6\}$, $\{4,5,6\}$, $\{1,6\}$, $\{2,5\}$, $\{3,4\}$. The Möbius function is easy to compute. In particular we obtain

$$H(A,t) = P(L,t) = 1 + 6t + 11t^2 + 6t^3.$$

An interesting observation is that $H(A,t) = (1+t)(1+2t)(1+3t)$. This property has deep generalizations which can be found, for example in [37], section 4.6.

4. Cohomology algebra of hyperplane complement

Here we give another homological interpretation of $A(\mathcal{A})$ when $\mathcal{A}$ is a complex arrangement. This is a combination of the celebrated Arnold-Brieskorn and Orlik-Solomon theorems ([7, 36]). A weak form of the result claims that $A(\mathcal{A})$ is isomorphic to the cohomology ring of the complement M to $\mathcal{A}$. To give a natural isomorphism we need to involve more tools.

Throughout this section $K = \mathbb{C}$ and $k = \mathbb{Z}$. All cohomologies are also taken with coefficients in $\mathbb{Z}$.

4.1. Topological induction

Let $(\mathcal{A}', \mathcal{A}, \mathcal{A}'')$ be the deletion-restriction triple (see subsection 2.5) defined by $H_n \in \mathcal{A}$. Denote the respective complements by M', M, and M''. Notice that while $\mathcal{A}$ and $\mathcal{A}'$ are arrangements in V, $\mathcal{A}''$ is an arrangement in H_n whence $M'' = H_n \setminus \bigcup_{i \neq n}(H_i \cap H_n)$. The goal of this subsection is to exhibit an exact sequence connecting cohomology of M', M, and M''.

Since $M'' = M' \cap H_n$, it makes sense to consider a tubular neighborhood N of M'' in M' and the fiber bundle $N \to M''$ with the fiber homeomorphic to $\mathbb{C}$. Then the restriction of the fibration to $N_0 = N \setminus M''$ is the fiber bundle with the fiber $\mathbb{C}^*$. Since both of these bundles are restrictions of trivial bundles over H_n they are trivial, i.e., up to a homeomorphism $(N, N_0) = M'' \times (\mathbb{C}, \mathbb{C}^*)$. Thus $H^*(N, N_0) = H^*(M'') \otimes H^*(\mathbb{C}, \mathbb{C}^*)$. Denote by t the generator of $H^2(\mathbb{C}, \mathbb{C}^*) \approx \mathbb{Z}$ corresponding to the natural orientation of $\mathbb{C}$. Then multiplication by t (identified with $1 \otimes t$) defines a group isomorphism

$$\tau : H^*(M'') \to H^*(N, N_0)[2] \tag{5}$$

increasing degrees by 2. (It is just a specialization of the Thom isomorphism to our simple situation.)

Now consider the embeddings $\mu : N \to M'$ and $\nu : (N, N_0) \to (M', M)$ and put $\alpha = \mu^*$ and $\alpha_1 = \nu^*$. Since ν is the excision of $U = M' \setminus N$ from (M', M), α_1

is the excision isomorphism $\alpha_1 : H^*(M', M) \to H^*(N, N_0)$. Then we obtain from (5) the group isomorphism

$$\alpha_1^{-1}\tau : H^*(M'') \to H^*(M', M)[2]. \tag{6}$$

Substituting the isomorphism (6) into the exact sequence of the pair (M, M') we have the exact sequence

$$\cdots \longrightarrow H^p(M') \xrightarrow{i^*} H^p(M) \xrightarrow{\tau^{-1}\alpha_1\delta} H^{p-1}(M'') \xrightarrow{j^*} H^{p+1}(M')\cdots \tag{7}$$

where $i : M \to M'$ and $j : (M', \emptyset) \to (M', M)$ are natural embeddings and $\delta : H^p(M) \to H^{p+1}(M', M)$ is the connecting homomorphism. In the rest of this section we put $\delta' = \tau^{-1}\alpha_1\delta$.

4.2. Basic cohomology classes

If H is a hyperplane of V, then $V \setminus H$ is homotopy equivalent to $\mathbb{C}^*$ (via projection along H) whence $H^1(V\setminus H)$ is generated by a canonical element β_H (corresponding to the natural orientation of $\mathbb{C}^*$). If $H = H_m \in \mathcal{A}$ and $i_{H_m} : M \to V \setminus H_m$ is the natural inclusion, then we put $\beta_m = i^*_{H_m}(\beta_{H_m}) \in H^1(M)$. We call elements β_m the *basic cohomology classes* of $H^*(M)$. Substituting $\mathcal{A}'$ and $\mathcal{A}''$ for $\mathcal{A}$ we obtain the basic cohomology classes $\beta'_m \in H^1(M')$ and $\beta''_m \in H^1(M'')$ $(m = 1, \ldots, n-1)$.

The naturality of the basic cohomology classes immediately implies the following properties for every p and $m = 1, \ldots, n-1$.

(i) $i^*(\beta'_m) = \beta_m$.
(ii) $\delta(\beta_m) = 0$.
(iii) $\delta(a\beta_m) = \delta(a)\beta'_m$ for every $a \in H^*(M)$.
(iv) $\tau^{-1}(\alpha(\beta'_m)t) = \beta''_m$.
(v) $\alpha_1(\delta(\beta_n)) = t$.

(We always denote by the juxtaposition both the multiplication in the absolute cohomology and the action of the absolute cohomology on the relative cohomology.)

The properties (i)–(v) immediately imply the following.

Proposition 4.1. *In the sequence* (7) *we have (up to sign)*

$$i^*(\beta'_{m_1}\cdots\beta'_{m_p}) = \beta_{m_1}\cdots\beta_{m_p},$$
$$\delta'(\beta_{m_1}\cdots\beta_{m_p}\beta_n) = \beta''_{m_1}\cdots\beta''_{m_p},$$

and

$$\delta'(\beta_{m_1}\cdots\beta_{m_p}) = 0$$

for every $1 \le m_1, \ldots, m_p \le n-1$.

In order to find other properties of products of β_i it is convenient to represent these classes by differential forms (using the de Rham theorem). The canonical generator of $H^1(\mathbb{C}^*)$ can be represented by the form $\frac{1}{2\pi\sqrt{-1}}\frac{dz}{z}$. Pulling this form back to M we represent β_m by the form

$$\omega_m = \frac{1}{2\pi\sqrt{-1}}\frac{d\alpha_m}{\alpha_m}$$

where α_m is (as above) a linear polynomial on V whose zero locus is H_m.

Now for every $S = \{i_1, \dots, i_p\} \subset [n]$ $(i_1 < \dots < i_p)$ put $\omega_S = \omega_{i_1} \wedge \dots \wedge \omega_{i_p}$ and $\partial\omega_S = \sum_{j=1}^{p}(-1)^{j-1}\omega_{S_j}$ where $S_j = S \setminus \{i_j\}$. Then we have the following.

Proposition 4.2. (i) *If* $\cap S = \emptyset$, *then* $\omega_S = 0$.
(ii) *If* S *is dependent, then* $\partial\omega_S = 0$.

Applying the de Rham map to the relations of forms in Proposition 4.2 we obtain similar relations for the cohomology classes. For a set S as above put $\beta_S = \beta_{i_1} \cdots \beta_{i_p}$ and define $\partial\beta_S$ similarly to $\partial\omega_S$.

Corollary 4.3. (i) *If* $\cap S = \emptyset$, *then* $\beta_S = 0$.
(ii) *If* S *is dependent, then* $\partial\beta_S = 0$.

4.3. Cohomology of M

Since the ring $H^*(M)$ is graded commutative and $\beta_m \in H^1(M)$ $(m = 1, \dots, n)$, the assignment $e_m \mapsto \beta_m$ defines a ring homomorphism $\phi : E \to H^*(M)$. Using Corollary 4.3 one can prove the main result of this section.

Theorem 4.4. *The homomorphism* ϕ *induces an isomorphism* $\psi : A \to H^*(M)$.

The following corollaries are immediate.

Corollary 4.5. *The de Rham homomorphism defines an isomorphism between the ring of differential forms generated by* ω_i $(i = 1, \dots, n)$ *and* $H^*(M)$.

Corollary 4.6. *The Poincarè polynomials of* M *and* L *coincide* (*cf. Corollary* 3.4).

5. Cohomology of OS complexes

5.1. Definition of OS complexes

In this section we study the OS algebra A of an arbitrary arrangement as a cochain complex under multiplication by an element $a \in A_1$. This complex first appeared in work on hypergeometric functions, mostly by Aomoto in [2, 3] and in the related paper on cohomology of local systems [23]. The algebraic study of this complex (mostly for generic a) was started in [45] where Theorem 5.2 was proved. A part of this theorem can be deduced also from the determinantal formula from [39]. The generic case results give lower bounds on dimensions of cohomology for arbitrary a. This was discovered in [24] for a more general situation.

A general approach to the nongeneric case was started in [25] where the resonance varieties R_1^p were introduced. The conjecture by Falk that the irreducible components of R_1^1 are linear was first proved in [12] and [31] independently. This property was generalized to arbitrary R^p_m in [32] and [11]. The other properties of R^1_m are from [33].

For some results of this section, the basic ring k has to be a field of zero characteristic; so we assume from the beginning that it is so. We define the cochain complex (A, a) for every $a \in E_1 = A_1$ as

$$0 \to A_0 \to A_1 \to \cdots \to A_\ell \to 0$$

where the differential is multiplication by a.

The main question studied here is how $\dim H^*(A, a)$ depends on a. First it is easy to show that for a central arrangement $H^*(A, a) = 0$, unless $\sum_{H \in \mathcal{A}} a_H = 0$ where $a = \sum_{H \in \mathcal{A}} a_H e_H$. Indeed the differential

$$(1/\sum_{H \in \mathcal{A}} a_H)\partial$$

gives a homotopy of the identity and zero maps of the complex (A, a) to itself. To study elements a in sufficiently general position with non-vanishing cohomology we need to introduce certain techniques which is a fact of interest by itself.

5.2. A complex of sheaves on L^{op}

In this subsection we represent the complex (A, a) as the complex of global sections of a complex of sheaves. The topological space on which these sheaves are defined is the poset $L^{\mathrm{op}} = L^{\mathrm{op}}(\mathcal{A})$, i.e., the set of all the nonempty intersections of hyperplanes from $\mathcal{A}$ as before but ordered *by inclusion*. This poset has the unique maximal element, V, but it has a unique minimal element if and only if $\mathcal{A}$ is central. A subset of L^{op} is called *increasing* if with each element Y it contains all elements of L^{op} larger than Y. Each $X \in L^{\mathrm{op}}$ defines the increasing set $L^X = \{Y \in L^{\mathrm{op}} | Y \geq X\}$ and the subarrangement $\mathcal{A}_X = \{H \in \mathcal{A} | X \subset H\}$ of $\mathcal{A}$. The standard (*order*) topology on L^{op} is given by all the increasing subsets of L^{op}. This topology has the minimal base consisting of sets L^X for all $X \in L^{\mathrm{op}}$.

Sheaves on L^{op} can be identified with functors on L^{op} viewed as the small category whose objects are elements and morphisms are pairs $(X \leq Y)$. If $\mathcal{F}$ is a contravariant functor from L to an additive category, then the respective presheaf is defined on the basic open set L^X as $\mathcal{F}(X)$ with the restriction $\mathcal{F}(X) \to \mathcal{F}(Y)$ $(X \leq Y)$ given by the morphism $\mathcal{F}(X \leq Y)$. Then the sheafification gives for every open subset $U \subset L^{\mathrm{op}}$ the object of sections on U as

$$\Gamma(U, \mathcal{F}) = \{(a_X) \in \oplus_{X \in U} \mathcal{F}(X) | \mathcal{F}(X \leq Y)(a_X) = a_Y, \ X, Y \in U\}.$$

Fix p, $0 \leq p \leq \ell$, and for each $X \in L^{\mathrm{op}}$ put $\mathcal{F}^p(X) = A_p(\mathcal{A}_X)$. Recall from section 2.3 that $A = \oplus_{Z \in L^{\mathrm{op}}} A_Z$ where if $\dim Z = \ell - p$, then $A_Z = A_p(\mathcal{A}_Z)$ (cf. Proposition 2.4). The linear spaces A_Z serve as building blocks, namely $A_p(\mathcal{A}_X) = \oplus A_Z$ where the summation is taken over all $Z \in L$ of dimension $\ell - p$ such that $Z \geq X$ (see Proposition 2.4). Thus if $Y \geq X$, then $\mathcal{F}^p(Y)$ is a direct summand of $\mathcal{F}^p(X)$ and there is a projection $\mathcal{F}^p(X) \to \mathcal{F}^p(Y)$. These projections define a contravariant functor on L whence a sheaf $\mathcal{F}^p$ on L^{op}. Notice that $\mathcal{F}^p$ is supported on the subposet of L^{op} consisting of all elements of dimension less than or equal to $\ell - p$.

Proposition 5.1. (i) *For every p we have*

$$\Gamma(L^{\mathrm{op}}, \mathcal{F}^p) = A_p.$$

(ii) *For every $X \in L^{\mathrm{op}}$ and $p < \ell - \dim X$ we have*

$$\Gamma(L^X, \mathcal{F}^p) = A_p(\mathcal{A}_X).$$

(iii) *For every p, the sheaf $\mathcal{F}^p$ is flasque.*

Now we fix an $a \in A_1$ and combine the sheaves $\mathcal{F}^p$ in a complex of sheaves. In order to do that, for every $X \in L^{\mathrm{op}}$ put $a(X) = \sum a_i e_i$ where summation is taken over all indexes $i \in [n]_X$. Denote by $d_a(X)$ the left multiplication by $a(X)$. Clearly for every p, the set $(d_a(X))_{X \in L^{\mathrm{op}}}$ is a homomorphism $d_a(p) : \mathcal{F}^p \to \mathcal{F}^{p+1}$ and these homomorphisms form the complex of sheaves

$$\mathcal{F}^0 \to \mathcal{F}^1 \to \cdots \to \mathcal{F}^\ell \to 0. \tag{8}$$

Moreover $\mathcal{E} = \ker(d_a) \subset \mathcal{F}^0$ is the sheaf whose only nonzero stalk is at V and it is equal to k. Thus we can augment the complex (8) on the left and obtain the complex

$$0 \to \mathcal{E} \to \mathcal{F}^0 \to \mathcal{F}^1 \to \cdots \mathcal{F}^\ell \to 0. \tag{9}$$

It is clear that the complex (9) is exact (and thus a flasque resolution of $\mathcal{E}$) if and only if the complex $(A(\mathcal{A}_X), d_a(X))$ is acyclic for every $X \in L^{\mathrm{op}} \setminus \{V\}$.

5.3. Cohomology for a general position a

In order to state the main theorem we need to introduce the notion of irreducible arrangement. Instead we introduce the opposite notion of a reducible one. An arrangement $\mathcal{A}$ is *reducible* if its ambient space can be represented non-trivially as $V = V_1 \oplus V_2$ and the arrangement can be non-trivially partitioned $\mathcal{A} = \mathcal{A}_1 \cup \mathcal{A}_2$ such that for $H \in \mathcal{A}_1$ we have $H = H' \oplus V_2$ with H' being a hyperplane in V_1 and for $H \in \mathcal{A}_2$ we have $H = V_1 \oplus H''$ with H'' being a hyperplane in V_2. In other words there exists a partition of a basis in V in two classes B_1 and B_2 such that the equation for α_H for $H \in \mathcal{A}_i$ involves only coordinates from B_i $(i = 1, 2)$.

Theorem 5.2. (i) *Let $\mathcal{A} = \{H_1, \ldots, H_n\}$ be an arbitrary arrangement with the intersection poset L of rank ℓ. Suppose $a = \sum_{i=1}^n a_i e_i \in A_1$ satisfies the condition*

$$\sum_{X \subset H_i} a_i \neq 0 \tag{10}$$

for all $X \in L^{\mathrm{op}}$ such that the arrangement $\mathcal{A}_X$ is irreducible. Then one has $H^p(A, a) = 0$ for every $p < \ell$.

(ii) *Let $\mathcal{A}$ and L be as in* (i) *but $\mathcal{A}$ is central. Let a satisfy* (10) *for $X \in L^{\mathrm{op}}$ not equal to the minimal element but $\sum_{i=1}^n a_i = 0$. Then $H^p(A, a) = 0$ for every $p < \ell - 1$ and $\dim H^{\ell-1}(A, a) = \dim H^\ell(A, a) = \bar{\chi}(\mathcal{A})$ (recall that $\bar{\chi}(\mathcal{A})$ is the beta-invariant of $\mathcal{A}$ equal to $(H(A,t)/(1+t))_{|t=-1}$).*

Remark 5.3. The conditions of Theorem 5.2 are not necessary for the vanishing of $H^p = H^p(A,a)$ for $p < \ell$. If the complex (9) is not exact, then it defines cohomology sheaves $\mathcal{H}^p$ (not all vanishing) and the spectral sequence with $E_2^{p,q} = H^p(L^{\mathrm{op}}, \mathcal{H}^q)$ that converges to H^{p+q}. In principle, one can use this spectral sequence to compute $H^*(A,a)$ in particular cases.

The particular case that is important for applications is given by the class of general position arrangements, i.e., such that all $\mathcal{A}_X$ for $\mathrm{rk}X > 1$ are reducible. For these arrangements the condition of Theorem 5.2 gives $a_i \neq 0$ for all i. This condition can be significantly relaxed.

Proposition 5.4. *Let $\mathcal{A}$ be a general position arrangement and $a \neq 0$. Then $H^p(A,a) = 0$ for every $p < \ell$.*

5.4. Nets and multinets

From now on to the end of the section we will consider only the arrangements of lines in an affine or projective plane. Because of that we change the notation and denote an arrangement by $\mathcal{L}$ instead of $\mathcal{A}$. In this subsection we introduce an extra combinatorial structure that may exist on an arrangement of lines. This structure allows one to construct an irreducible component of R_1^1 and vice versa - any irreducible component defines such a structure.

Let $\mathcal{L} = \{\ell_1, \ldots, \ell_n\}$ be an arrangement of lines in the complex projective plane. A point contained in three or more lines will be called a *multiple point.*

Let $m : \mathcal{L} \to \mathbb{Z}_{>0}$ be a function which assigns to each $\ell \in \mathcal{L}$ a positive integer multiplicity $m(\ell)$. The pair $(\mathcal{L}, m)$ is called a *multi-arrangement.*

Definition 5.5. Let k, d be integers such that $k \geq 3$ and $d > 0$. A *weak (k,d)-multinet* supported on a multi-arrangement $(\mathcal{L}, m)$ is a partition $\mathcal{N}$ of $\mathcal{L}$ into k classes $\mathcal{L}_1, \ldots \mathcal{L}_k$, such that

(1) $\sum_{\ell \in \mathcal{L}_i} m(\ell) = d$, independenalyt of i;
(2) Let $\mathcal{X}$ be the set of all intersections of lines from different classesa; then for each $p \in \mathcal{X}$, $\sum_{\ell \in \mathcal{L}_i, p \in \ell} m(\ell)$ is constant, independent of i.

A *multinet* is a weak multinet satisfying the additional property

(3) For $1 \leq i \leq k$, for any $\ell, \ell' \in \mathcal{L}_i$, there is a sequence of $\ell = \ell_0, \ell_1, \ldots, \ell_r = \ell'$ such that $\ell_{j-1} \cap \ell_j \notin \mathcal{X}$ for $1 \leq j \leq r$.

We will call $\mathcal{X}$ the *base locus* – it does not need to include all multiple points of $\mathcal{L}$. Note that if $\mathcal{N}$ is a multinet, then $\mathcal{X}$ determines $\mathcal{N}$: construct a graph Γ with vertex set $\mathcal{L}$ and an edge from ℓ to ℓ' when $\ell \cap \ell' \notin \mathcal{X}$. Then the classes $\mathcal{L}_i$ are the components of Γ. We will see below (Remark 5.8) that every weak multinet can be refined to a multinet with the same base locus.

The second condition says that the number of lines from $\mathcal{L}_i$ passing through $p \in \mathcal{X}$, counting multiplicities, is the same for every i. This number is denoted n_p. If $n_p = 1$ for every $p \in \mathcal{X}$, then $(\mathcal{N}, \mathcal{X})$ is called a *net.* In this case condition (3) follows from (2), and $m(\ell) = 1$ for every $\ell \in \mathcal{L}$. The converse of the last statement is false – there are multinets with $m(\ell) = 1$ for every $\ell \in \mathcal{L}$ that are not nets.

Simple examples of multinets are given by the sets of (the projectivizations of) all reflection hyperplanes of the groups of Coxeter types A_3 and B_3 given respectively by equations $xyz(x-y)(x-z)(y-z)=0$ and $x^2y^2z^2(x^2-y^2)\cdot(x^2-z^2)(y^2-z^2)=0$ where the exponents of linear factors denote the multiplicities of the respective projective lines. It is easy to see that the multinet given by A_3 is really a net.

Combinatorially a k-net corresponds to a set of $k-2$ mutually orthogonal latin squares; in particular 3-nets correspond to quasi-groups.

In studies of components of resonance varieties, nets in $\mathbb{CP}^2$ first appeared implicitly in [33] and explicitly with a partial classification in [44]. See also [30].

The following lemma is very elementary.

Lemma 5.6. *Suppose $\mathcal{N}$ is a weak (k,d)-multinet. Then*

(1) $\sum_{\ell\in\mathcal{L}} m(\ell)=dk$.
(2) $\sum_{p\in\mathcal{X}} n_p^2=d^2$.
(3) *For each* $\ell\in\mathcal{L}$, $\sum_{p\in\mathcal{X}\cap\ell} n_p=d$.

5.5. Multinets and resonance variety

Recall that we denote the generator of A corresponding to $\ell\in\mathcal{L}$ by e_ℓ. A subspace R of A^1 is *isotropic* if $ab=0$ for every $a,b\in R$. Then, perforce, an isotropic subspace of dimension at least two is contained in R^1_1.

Theorem 5.7. *Suppose a multi-arrangement $(\mathcal{L},m)$ supports a weak (k,d)-multinet $\mathcal{N}$. Consider the linear space of vectors $u=\sum u_\ell e_\ell$ from A_1 $(u_\ell\in K,\ell\in\mathcal{L})$ satisfying the following conditions such that u_ℓ is constant on every class $\mathcal{L}_j$ and $\sum_{\ell\in L}u_\ell=0$. Then this space is a $(k-1)$-dimensional isotropic subspace of A^1. The other way around, if $\mathcal{L}$ supports a global resonance component of dimension $k-1$, then $\mathcal{L}$ supports a (k,d)-multinet for some d.*

Remark 5.8. Theorem 5.7 has as a consequence that any weak multinet can be refined to a multinet. For example, consider an arrangement of five concurrent lines, one having multiplicity 2. Partition the lines into three classes, of two lines each, counting multiplicities. This is a weak $(3,2)$-multinet. Its multinet refinement is the partition into five singletons, with each line having multiplicity 1, i.e., a $(5,1)$-net.

For an alternative proof of Theorem 5.7 see [34], and for an extension of it to curve arrangements see [10].

5.6. Multinets and pencils of plane curves

It is interesting that multinets can be related to pencils of plane algebraic curves. In light of the connection with resonance established in the preceding section, this construction may be viewed as a concrete realization of the mapping of the complement of $\mathcal{L}$ to a curve predicted by the theorem of Arapura [4]. The existence of such a pencil was proved in [33], where it was used to derive restrictions on parameters of nets. The constructions in [26], while less general, are much more elementary than Arapura's

We will identify a homogeneous polynomial in three variables with the projective plane curve it defines, and often refer to either as a "curve." A one-dimensional linear system of curves is called a *pencil.* One can think of a pencil as a line in the projective space of homogeneous polynomials of some fixed degree. Thus any two distinct curves generate a pencil, and conversely a pencil is determined by any two of its curves C_1, C_2. An arbitrary curve in the pencil is then $aC_1 + bC_2$, $[a : b] \in \mathbb{CP}^1$. Every two curves in a pencil intersect in the same set of points $\mathcal{X} = C_1 \cap C_2$, called the base of the pencil. We will always assume that our pencils have no fixed components, that is, that the base locus is a finite set of points.

The two curves C_1, C_2 determine a rational map $\pi : \mathbb{CP}^2 \rightarrowtail \mathbb{CP}^1$ via $p \mapsto [C_2(p) : -C_1(p)]$ whose indeterminacy locus is the base of the pencil. The (closure of the) fiber of π over $[a : b]$ is the curve $aC_1 + bC_2$, and each point outside the base locus lies in a unique such curve. The map π is uniquely determined by the pencil, up to linear change of coordinates in $\mathbb{CP}^1$. For convenience we will often call π a "pencil," when no confusion will result. For more about pencils the reader may consult [29].

A curve of the form $\prod_{i=1}^{q} \alpha_i^{m_i}$, where α_i is a linear form and m_i is a positive integer, for $1 \leq i \leq q$, will be called *completely reducible.* Usually we restrict our consideration to pencils with at least two completely reducible fibers.

Lemma 5.9. *Suppose π is a pencil with no fixed components and at least two completely reducible fibers C_1 and C_2. Let $p \in C_1 \cap C_2$ be a base point, let n_i be the multiplicity of C_i at p, and $n_p = \min\{n_1, n_2\}$. Then*

(1) *no two fibers of π are tangent at p,*
(2) *if $n_p = 1$, then the fibers of π are nonsingular at p,*
(3) *if $n_p > 1$, then the fibers of π have nodal singularities of multiplicity n_p at p, and*
(4) *if π has another completely reducible fiber, then $n_1 = n_2$, i.e., the multiplicities of all of these fibers at p are the same.*

We say the pencil π is *connected* if for every fiber its proper transform under the blow-up of $\mathbb{CP}^2$ at the points of $\mathcal{X}$ is connected. Equivalently, by Lemma 5.9(ii) and (iii), π is connected if and only if its generic fiber is irreducible.

Definition 5.10. A *pencil of Ceva type* (or "Ceva pencil") is a connected pencil of plane curves (with a finite base), in which three or more fibers are completely reducible.

The condition of Definition 5.10 means that, as a line in the projective space $\mathbb{CP}^N$ of degree d curves ($N = \binom{d+2}{2} - 1$), a Ceva pencil is a trisecant to the subvariety of completely reducible curves, which is the symmetric product $S^d(\mathbb{CP}^2)$, realized as a projection of the Segre variety $(\mathbb{CP}^2)^d$ in $\mathbb{CP}^M$, $M = 3^d - 1$. The terminology comes from the first example below.

Example 5.11. Consider the Fermat pencil $ax^d + by^d + cz^d = 0$, $[a : b : c] \in \mathbb{CP}^2$, $a + b + c = 0$. There are three singular values: $[1 : -1 : 0]$, $[0 : 1 : -1]$, and $[-1 : 0 : 1]$. The corresponding fibers $x^d - y^d$, $y^d - z^d$, and $z^d - x^d$, are completely reducible. The components of each of the singular fibers meet in a single point outside the base locus, so the pencil is connected. The resulting arrangement of $3d$ lines, called the Ceva arrangement (so-named in [5]), supports a $(3, d)$-net; in the case $d = 2$ the arrangement of all the linear factors of the completely reducible fibers is the A_3 type arrangement considered in the previous subsection. Moreover the partition into fibers coincides with the partition into classes of the net.

Example 5.12. The other "classical" example of a pencil of Ceva type is the Hesse pencil of cubics $a(x^3 + y^3 + z^3) + 3bxyz$, $[a : b] \in \mathbb{CP}^1$ which share the same nine inflection points. In this case there are four completely reducible fibers, each of which is a product of three distinct lines, which meet in pairs outside the base locus, so again the pencil is connected. The resulting arrangement of twelve lines in $\mathbb{CP}^2$ is called the Hessian arrangement [37]. It supports a $(4, 3)$-net, and in fact is the only known example of a line arrangement supporting a $(4, d)$-net for any d [44].

Example 5.13. The connected pencil $[(x^2 - y^2)z^2 : (y^2 - z^2)x^2]$ has three singular fibers $(x^2 - y^2)z^2$, $(y^2 - z^2)x^2$, and $(z^2 - x^2)y^2$, which are completely reducible but not reduced. The resulting arrangement of nine lines is the B_3 arrangement considered in the previous subsection and again the partitions into fibers and classes coincides.

Now we state the main result that generalizes the examples.

Proposition 5.14. *A pencil of Ceva type of degree d with k completely reducible fibers induces a (k, d)-multinet on the line arrangement $\mathcal{L}$ consisting of the components of its completely reducible fibers.*

Conversely let a multiarrangement $(\mathcal{L}, m)$ support a (k, d)-multinet and $C_i = \prod_{\ell \in \mathcal{L}_i} \alpha_\ell^{m(\ell)}$. Then all curves C_i, $i = 1, \dots, k$ are fibers of the Ceva pencil generated by two of them.

Here is an example of a pencil which satisfies the second, but not the first condition of Definition 5.10, and yields a weak multinet that is not a multinet.

Example 5.15. Let $C_1 = x^d$ and $C_2 = y^d$. Then every curve $aC_1 + bC_2$ except C_1 and C_2 is the product of d distinct lines. All these fibers are completely reduced, with components meeting only inside the base locus. Thus the pencil is not connected. The associated weak multinet is a multinet only in case $d = 1$. (Compare with Remark 5.8.)

Conversely, if $d = 1$, then any (k, d)-multinet is a (k, d)-net. For every k a $(k, 1)$-net is a pencil of k lines partitioned into singletons. Such an arrangement supports a weak (k/m,m)-multinet for every m dividing k. From the point of view of resonance varieties, $(k, 1)$ nets correspond to local components [25].

This rank-two example is special. We conjecture that for rank-three pencils the connectedness condition is unnecessary, that is, the connectedness follows from the other condition of Definition 5.10.

Not every weak multinet arises from a pencil, connected or not. In rank two, consider the multi-arrangement with defining polynomial $x^2y^2(ax-by)(cx+dy)$. The induced weak (3,2)-multinet corresponds to a pencil if and only if $ad-bc=0$. Here is a rank-three example.

Example 5.16. Consider the Hesse pencil $a(x^3+y^3+z^3)-3bxyz$ of Example 5.12. Let C_0, C_1, C_2, C_3 be the four completely reducible fibers, whose components together form the Hessian arrangement $\mathcal{L}$ of twelve lines. The base $\mathcal{X}$ of the pencil consists of the nine inflection points of any of the smooth fibers of the pencil. Now define a weak multinet structure on $\mathcal{L}$ with the base locus $\mathcal{X}$ and three classes formed by the irreducible components of respectively C_0C_1, C_2^2, C_3^2. This is a weak $(3,6)$-multinet that is not a multinet. It is easy to check directly that the pencil $[C_2^2 : C_3^2]$ does not contain C_0C_1 as a fiber. It is interesting to note that the reason is that the four points

$$[a:b]=[0:1],[1:1],[1:\xi],[1:\xi^2],$$

where $\xi^3=1$, corresponding to the special fibers of Hesse pencil, cannot be ordered to form a harmonic set.

In conclusion of this section we state a very recent result about the upper bound on the number k of completely reducible fibers in a Ceva pencil.

Theorem 5.17. *A Ceva pencil of degree $d>1$ cannot have more than four completely reducible fibers.*

This result can be also formulated in at least two different equivalent ways.

(i) For a (k,d) multinet in $\mathbb{CP}^2$ if $d>1$, then $k<5$.
(ii) Every non-local irreducible resonance component has dimension either 2 or 3.

A weaker inequality $k\le 5$ was proved for nets in [33] and for multinets with all $m(\ell)=1$ in [26]. Then this inequality was proved in the general case in [38]. The stronger inequality $k\le 4$ was proved in [42] for nets and recently generalized [47] to the general case.

6. Relation to local system cohomology

In this section M denotes a smooth quasi-projective complex variety and $H^*(M)$ denotes the cohomology with complex coefficients. Let (Z,D) be a good compactification of M, i.e.

(i) D is a normal crossing divisor with smooth irreducible components in a smooth projective variety Z;
(ii) there is an isomorphism $\pi : Z\setminus D \to M$.

In this setting there is a Hodge-Deligne spectral sequence

$$E_1^{p,q} = H^q(Z, \Omega_Z^p(logD)) \Rightarrow H^{p+q}(M) \tag{11}$$

degenerating at E_1 and inducing the Hodge filtration F of the Deligne mixed Hodge structure on $H^{p+q}(M)$, see [14].

Proposition 6.1. *If the Deligne mixed Hodge structure on some cohomology space $H^m(M)$ is pure of type (m, m), then*

(i) $H^0(Z, \Omega_Z^m(logD)) = H^m(M)$ *and*
(ii) $H^q(Z, \Omega_Z^p(logD)) = 0$ *for* $p + q = m$ *and* $q > 0$.

We list below several cases when this property holds.

Example 6.2. (a) When M is a hyperplane arrangement complement, the cohomology space $H^m(M)$ is pure of type (m, m) for all $m \geq 0$, see [18]. This purity result implies that $j^* = 0$ in the exact sequence (7). This splitting plays a key role in proving Theorem 4.4 which relates the topology and the combinatorics of M.

(b) When M is a smooth rational curve arrangement complement in the projective plane (i.e. any irreducible component is either a line or a smooth conic), the cohomology space $H^m(\mathcal{U})$ is pure of type (m, m) for all $m \geq 0$.

(c) $H^m(M)$ is pure of type (m, m) for all $m \leq 1$ when M is an affine hypersurface complement. This follows from the fact that $g = (g_1, \ldots, g_s) : M \to (\mathbb{C}^*)^s$ induces an isomorphism at the H^m-level all $m \leq 1$. Here we look at M as the complement in $\mathbb{C}^n$ of a set of hypersurfaces $H_j : g_j(x_1, \ldots, x_n) = 0$ for $j = 1, \ldots, s$. The case when M is a projective hypersurface complement is similar.

Let $\mathbb{T}(M) = \mathrm{Hom}(\pi_1(M), \mathbb{C}^*)$ be the character variety of M. This is an algebraic group whose identity irreducible component is an algebraic torus $\mathbb{T}(M)_1 \simeq (\mathbb{C}^*)^{b_1(M)}$. Consider the exponential mapping

$$\exp : H^1(M, \mathbb{C}) \to H^1(M, \mathbb{C}^*) = \mathbb{T}(M) \tag{12}$$

induced by the exponential function $\exp : \mathbb{C} \to \mathbb{C}^*$, $t \mapsto \exp(-2\pi i t)$.

One case when all this can be made explicit is when M is an affine hypersurface complement as in Example 6.2, case (c). For $\lambda = (\lambda_1, \ldots, \lambda_s) \in \mathbb{T}(M) = (\mathbb{C}^*)^s$, let $\mathcal{L}_\lambda$ be the corresponding rank-1 local system on M. Let $\alpha_j \in \mathbb{C}$ be such that $\exp(-2\pi i \alpha_j) = \lambda_j$ for $j = 1, \ldots, s$. Then $\mathcal{L}_\lambda$ is the local system of horizontal sections of the connection

$$\nabla_\alpha : \mathcal{O}_\mathcal{U} \to \Omega^1_\mathcal{U}$$

given by $\nabla_\alpha(u) = du + u \cdot \omega_\alpha$ where

$$\omega_\alpha = \sum_{j=1,s} \alpha_j \frac{dg_j}{g_j}.$$

In other words, this gives $\exp(\omega_\alpha) = \mathcal{L}_\lambda$. Alternatively, if we look at M as a subset of $\mathbb{CP}^n$, then we can use the formula

$$\omega_\alpha = \sum_{j=0,s} \alpha_j \frac{df_j}{f_j} \tag{13}$$

where we set $\alpha_0 = -\sum_{j=1,s} d_j \cdot \alpha_j$, $f_0 = x_0$ and $H_0 : f_0 = 0$. Here $d_j = \deg(H_j)$.

The best way to compute $H^m(M, \mathcal{L}_\lambda)$ is using the *logarithmic connections*. More precisely, the pull-back of the connection ∇_α under an embedded resolution $\sigma : (Z, D) \to (\mathbb{CP}^n, H)$, where $H = \cup_{i=0,s} H_s$, is a logarithmic connection $\tilde{\nabla}_\alpha$ on the trivial line bundle on Z with poles along D. Let ρ_i be the residue of the connection $\tilde{\nabla}_\alpha$ along the irreducible component D_i of D. When D_i is the proper transform of some component H_j of H one has $\rho_i = \alpha_j$.

Definition 6.3. A choice of residues $\alpha = (\alpha_0, \alpha_1, \dots, \alpha_s)$ for $\mathcal{L}_\lambda$ as above is an *admissible choice of residues* for $\mathcal{L}_\lambda$ if $\rho_i \notin \mathbb{N}_{>0}$ for all irreducible components D_i of D. A rank-1 local system $\mathcal{L}_\lambda$ is *admissible* if there is some admissible choice of residues for it.

Remark 6.4. It is easy to see using Hironaka's embedded resolution of singularities by blowing-up smooth subvarieties in the singular locus of H, that for any i there is a relation

$$\rho_i = \sum_{j=1,s} n_{ij} \alpha_j$$

with $n_{ij} \in \mathbb{Z}$ (see [23] for similar formulas and note that negative coefficients occur due to the presence of the hyperplane at infinity). The above condition $\rho_i \notin \mathbb{N}_{>0}$ is clearly satisfied if all α_j are sufficiently small. In other words, there is a neighborhood $U(\mathbf{1})$ of the trivial local system $\mathbf{1} \in \mathbb{T}(M)$ formed entirely by admissible local systems.

By Deligne's results in [13], for an admissible choice of residues one has an E_1-spectral sequence

$$E_1^{p,q} = H^q(Z, \Omega_Z^p(logD)) \Rightarrow H^{p+q}(M, \mathcal{L}_\lambda) \tag{14}$$

whose differential d_1 is induced by $\tilde{\nabla}_\alpha$, see also [16], (Thm. 3.4.11 (i)). The above discussion proves the following result, see Proposition 4.5 in [19].

Proposition 6.5. *Assume that $\alpha = (\alpha_0, \alpha_1, \dots, \alpha_s)$ is an admissible choice of residues for $\mathcal{L}_\lambda$ and that the cohomology groups $H^m(M)$ are pure of type (m, m) for all $m \leq k$. Then*

$$H^m(M, \mathcal{L}_\lambda) = H^m(H^*(M), \omega_\alpha \wedge)$$

for all $m \leq k$ and $H^{k+1}(H^*(M), \omega_\alpha \wedge)$ is a subspace in $H^{k+1}(M, \mathcal{L}_\lambda)$.

When M is a hyperplane arrangement complement, this is exactly the argument used in [23] and [40]. This explains geometrically why the OS complexes (alias the Aomoto complexes) introduced in section 5.1 above play such a central role in this theory.

Proposition 6.5, Remark 6.4 and Example 6.2 yield the following.

Corollary 6.6. *If M is an affine hypersurface arrangement complement, then there is a neighborhood $U(\mathbf{1})$ of the trivial local system $\mathbf{1} \in \mathbb{T}(M)$ such that*

$$H^1(M, \mathcal{L}_\lambda) = H^1(H^*(M), \omega_\alpha \wedge)$$

for any local system $\mathcal{L}_\lambda \in U(\mathbf{1})$, α *being an arbitrary choice of admissible residues for* $\mathcal{L}_\lambda$.

Corollary 6.7. *If* $M = M(\mathcal{A})$ *is a hyperplane arrangement complement, then there is a neighborhood* $U(\mathbf{1})$ *of the trivial local system* $\mathbf{1} \in \mathbb{T}(M)$ *such that*

$$H^m(M, \mathcal{L}_\lambda) = H^m(H^*(M), \omega_\alpha \wedge) \tag{15}$$

for any $m \in \mathbb{N}$, *and any local system* $\mathcal{L}_\lambda \in U(\mathbf{1})$, α *being an arbitrary choice of admissible residues for* $\mathcal{L}_\lambda$.

Remark 6.8. When $M = M(\mathcal{A})$ is a hyperplane arrangement complement, there is a special embedded resolution σ as above, obtained by blowing up the *dense* edges of $\mathcal{A}$. An edge X is called dense if the corresponding hyperplane arrangement $\mathcal{A}_X$ is irreducible as defined in section 5.3 above. One may compare Theorem 5.2 to Theorem 6.4.18 in [16]. Note also that the relation between the claims (i) and (ii) in Theorem 5.2 can be perhaps better understood in view of Proposition 6.4.1 in [16].

In a quite general setting, namely for a finite type CW-complex M, one introduces the following definitions.

The *characteristic varieties* of M are the jumping loci for the cohomology of M, with coefficients in rank-1 local systems:

$$\mathcal{V}^i_k(M) = \{\rho \in \mathbb{T}(M) \mid \dim H^i(M, \mathcal{L}_\rho) \geq k\}. \tag{16}$$

When $i = 1$, we use the simpler notation $\mathcal{V}_k(M) = \mathcal{V}^1_k(M)$.

The *resonance varieties* of M are the jumping loci for the cohomology of the complex $H^*(H^*(M, \mathbb{C}), \alpha\wedge)$, namely:

$$R^i_k(M) = \{\alpha \in H^1(M, \mathbb{C}) \mid \dim H^i(H^*(M, \mathbb{C}), \alpha\wedge) \geq k\}. \tag{17}$$

When $i = 1$, we use the simpler notation $R_k(M) = R^1_k(M)$.

Foundational results on the structure of the cohomology support loci for local systems on quasi-projective algebraic varieties or, more generally, on quasi-Kähler manifolds, were obtained by Beauville [6], Green and Lazarsfeld [28], Simpson [41], Campana [9] (for the proper case), and Arapura [4] (for the non-compact case and first characteristic varieties $\mathcal{V}_1(M)$).

Theorem 6.9. *Let* X *be a compact Kähler manifold,* D *a normal crossing divisor and set* $M = X \setminus D$, *i.e.,* M *is a quasi-Kähler manifold. Then the strictly positive dimensional irreducible components of the first characteristic variety* $\mathcal{V}_1(M)$ *are translated subtori in* $\mathbb{T}(M)$ *by elements of finite order. When* M *is proper, then all the components of* $\mathcal{V}^i_k(M)$ *are translated subtori in* $\mathbb{T}(M)$ *by elements of finite order.*

In the non-proper case, we get unitary characters $\rho_j \in \mathbb{T}(M)$ as isolated points of the first characteristic varieties $\mathcal{V}_1(M)$), see Arapura [4]. Recent results by N. Budur give a precise description in terms of complex tori and convex rational polytopes of some related characteristic varieties, see [8], and imply in particular that these unitary characters ρ_j's are actually torsion points in $\mathbb{T}(M)$.

In the sequel we assume that M is quasi-projective and concentrate ourselves on the strictly positive dimensional irreducible components of the first characteristic variety $\mathcal{V}_1(M)$. They have the following rather explicit description, given by Arapura [4].

Theorem 6.10. *Let W be a d-dimensional irreducible component of the first characteristic variety $\mathcal{V}_1(M)$, with $d > 0$. Then there is a regular morphism $f : M \to S$ onto a smooth curve S with $b_1(S) = d$ such that the generic fiber F of f is connected, and a torsion character $\rho \in \mathbb{T}(M)$ such that the composition*

$$\pi_1(F) \xrightarrow{i_\sharp} \pi_1(M) \xrightarrow{\rho} \mathbb{C}^*,$$

where $i : F \to M$ is the inclusion, is trivial and

$$W = \rho \cdot f^*(\mathbb{T}(S)).$$

Moreover, the following hold.

(i) *If $H^1(M)$ is pure of weight 2, e.g. when M is a hypersurface complement in an affine or projective space, then S is obtained from $\mathbb{CP}^1$ by deleting a finite set of points.*
(ii) *If $W = \rho \cdot f^*(\mathbb{T}(S))$ is a translated component in $\mathcal{V}_1(M)$ with $\dim W > 2$, then $W' = f^*(\mathbb{T}(S))$ is also a component in $\mathcal{V}_1(M)$ and $\mathbf{1} \in W'$. If $H^1(M)$ is pure of weight 2, then the same holds for $\dim W > 1$.*

The claim (i) explains the important role played by pencils of curves, nets and multinets in the sections 5.4-4.6 above. See also [17]. The claim (ii) allows us to restate Theorem 5.17 as follows.

Corollary 6.11. *If $M = M(\mathcal{A})$ is a hyperplane arrangement complement, then any non-local irreducible component W of the the first characteristic variety $\mathcal{V}_1(M)$ satisfies $\dim W \leq 3$.*

Corollaries 6.6 and 6.7 suggest that there is a close relation between the resonance varieties $R^i_k(M)$ and the germs of the characteristic varieties $\mathcal{V}^i_k(M)$ at the trivial representation $\mathbf{1} \in \mathbb{T}(M)$. A very general relation of this type is the following, see [20].

Theorem 6.12. *Assume that M is 1-formal. Then the irreducible components E of the resonance variety $R_1(M)$ are linear subspaces in $H^1(M, \mathbb{C})$ and the exponential mapping (12) sends these irreducible components E onto the irreducible components W of $\mathcal{V}_1(M)$ with $1 \in W$. Moreover, if $H^1(M, \mathbb{C})$ is pure of weight 2, then the irreducible components E of the resonance variety $R_1(M)$ are precisely the maximal isotropic subspaces in $H^1(M, \mathbb{C})$ of dimension at least 2.*

Remark 6.13. Recall that a group G is 1-formal if its Malcev Lie algebra, as constructed by Quillen, is quadratically presented; see [1], Chapter 3 for details.

A path-connected topological space M is 1-formal if the fundamental group $G = \pi_1(M)$ has this property. The class of 1-formal varieties is large enough, as it contains all the projective smooth varieties and any hypersurface complement in $\mathbb{P}^n$, see [20] for references and further discussion. In fact, if the Deligne mixed Hodge structure on $H^1(M)$ is pure of weight 2, then the smooth quasi-projective variety M is 1-formal, see [35]. Note also that there are varieties with $H^1(M)$ pure of weight 1, but not 1-formal.

Using Theorem 6.12, one can show that the resonance varieties $R_k(M)$ for a 1-formal smooth, quasi-projective variety M satisfy the following special properties, called *resonance obsructions* in [20].

Let $R_1(M) = \bigcup_\alpha R^\alpha$ be the decomposition of $R_1(M)$ into irreducible components.

1. **Linearity.** Each component R^α is a linear subspace of $H^1(M)$.
2. **Isotropicity.** If $R^\alpha \neq \{0\}$, then R^α is a p-isotropic subspace of dimension at least $2p+2$, for some $p = p(\alpha) \in \{0,1\}$.
3. **Genericity.** If $\alpha \neq \beta$, then $R^\alpha \bigcap R^\beta = \{0\}$.
4. **Filtration by dimension.** For $1 \leq k \leq \dim H^1(M)$,
$$R_k(M) = \bigcup_\alpha R^\alpha,$$
where the union is taken over all components R^α such that $\dim R^\alpha > k+p(\alpha)$. By convention, the union equals $\{0\}$ if the set of such components is empty.

One says that R^α is 0-isotropic (resp. 1-isotropic) if the restriction of the cup-product to $R^\alpha \times R^\alpha$ has a 0-dimensional (resp. a 1-dimensional) image.

Using these obstructions, it is often possible to show that some finitely presented groups cannot be quasi-projective, i.e. isomorphic to fundamental groups of smooth quasi-projective varieties. In the case of specific classes of groups, e.g. the Artin right angle groups or the Bestvina-Brady groups, it is possible to give the precise lists of such groups which are in addition quasi-projective, see [20], [21]. Similar techniques lead to new insights into the multivariable Alexander polynomials associated to smooth quasi-projective varieties, see [22].

References

[1] J. Amorós, M. Burger, K. Corlette, D. Kotschick, D. Toledo, *Fundamental groups of compact Kähler manifolds*, Math. Surveys Monogr., vol. 44, Amer. Math. Soc., Providence, RI, 1996.

[2] K. Aomoto, Un théorème du type de Matsushima-Murakami concernant l'intégrale des fonctions multiformes, J. Math. Pures Appl. **52** (1973), 1–11.

[3] K. Aomoto, Les équations aux différences linéaires des fonctions multiformes, J. Fac. Sci. Univ. Tokyo, sec. IA **22** (1975), 271–297.

[4] D. Arapura, Geometry of cohomology support loci for local systems I, Journal of Algebraic Geometry, **6** (1997), 563–597.

[5] G. Barthel, F. Hirzebruch, and T. Höfer, *Geradenkonfigurationen und Algebraische Fächen*, Aspects of Mathematics, D4. Friedr. Vieweg & Sohn, Braunschweig, 1987.

[6] A. Beauville: Annulation du H^1 pour les fibrés en droites plats, in: Complex algebraic varieties (Bayreuth, 1990), 1–15, Lecture Notes in Math., vol. 1507, Springer, Berlin, 1992.

[7] E. Brieskorn, Sur les groupes de tresses, Séminaire Bourbaki 1971/72, LNM **317**, Springer Verlag, 1973, 21–44.

[8] N. Budur: Unitary local systems, multiplier ideals, and polynomial periodicity of Hodge numbers, Advances in Math. **221** (2009), 217–250.

[9] F. Campana: Ensembles de Green-Lazarsfeld et quotients resolubles des groupes de Kähler, J. Alg. Geometry 10(2001), 599–622.

[10] J.I. Cogolludo Agustin, D. Matei, Cohomology algebra of plane curves, weak combinatorial type and formality, arxiv:0711.1951.

[11] D. Cohen and P. Orlik, Arrangements and local systems, Math. Res. Lett. **7** (2000), 299–316.

[12] D. Cohen and A. Suciu, Characteristic varieties of arrangements, Math. Proc. Cambridge Phil. Soc **127** (1999), 33–53.

[13] P. Deligne, *Equations différentielles à points singuliers réguliers,* Lecture Notes in Math., **163**, Springer, Berlin (1970).

[14] P. Deligne, Théorie de Hodge II, Publ. Math. IHES, **40**, 5–57 (1972).

[15] P. Deligne, P. Griffiths, J. Morgan, D. Sullivan, *Real homotopy theory of Kähler manifolds*, Invent. Math. **29** (1975), no. 3, 245–274.

[16] A. Dimca, *Sheaves in Topology*, Universitext, Springer Verlag, 2004.

[17] A. Dimca, Pencils of Plane Curves and Characteristic Varieties, this volume, 59–82.

[18] A. Dimca and G.I. Lehrer, Purity and equivariant weight polynomials, in: Algebraic Groups and Lie Groups, editor G.I.Lehrer, Cambridge University Press, 1997.

[19] A. Dimca, L. Maxim: Multivariable Alexander invariants of hypersurface complements, Trans. Amer. Math. Soc. **359** (2007), no. 7, 3505–3528.

[20] A. Dimca, S. Papadima, A. Suciu, Topology and geometry of cohomology jump loci, Duke Math. J. **148** (2009), no. 3, 405–457.

[21] A. Dimca, S. Papadima, A. Suciu, Quasi-Kähler Bestvina–Brady groups, J. Algebraic Geom. **17** (2008), no. 1, 185–197.

[22] A. Dimca, S. Papadima, A. Suciu, Alexander polynomials: Essential variables and multiplicities, Int. Math. Research Notices vol. **2008** (2008), no. 3, Art. ID rnm119, 36 pp.

[23] H. Esnault, V. Schechtman, and E. Viehweg, Cohomology of local systems of the complement of hyperplanes, Invent. math. **109** (1992), 557–561; Erratum, ibid. **112** (1993) 447.

[24] D. Eisenbud, S. Popoescu, and S. Yuzvinsky, Hyperplane arrangement cohomology and monomials in the exterior algebra, Transactions of AMS **355** (2003), 4365–4383.

[25] M. Falk, Arrangements and cohomology, Ann. Combin. **1** (1997), 135–157.

[26] M. Falk and S. Yuzvinsky, Multinets, resonance varieties, and pencils of plane curves, Compositio Math. **143** (2007), no. 4, 1069–1088.

[27] J. Folkman, The homology groups of a lattice, J. Math. and Mech. **15** (1966), 631–636.

[28] M. Green, R. Lazarsfeld, Higher obstructions to deforming cohomology groups of line bundles, J. Amer. Math. Soc. 4(1991), no. 1, 87–103.

[29] P. Griffiths and J. Harris, Principles of Algebraic Geometry, Wiley Interscience, New York, 1978.

[30] Y. Kawahara. The non-vanishing cohomology of Orlik-Solomon algebras, preprint, 2005.

[31] A. Libgober, Characteristic varieties of algebraic curves, in: *Applications of algebraic geometry to coding theory, physics and computation* (Eilat, 2001), NATO Sci. Ser. II Math. Phys. Chem., vol. 36, Kluwer Acad. Publ., Dordrecht, 2001, pp. 215–254.

[32] A. Libgober, First order deformations of local systems with non vanishing cohomology, Topology Appl. **118** (2002), 159–168.

[33] A. Libgober and S. Yuzvinsky. Cohomology of the Orlik-Solomon algebras and local systems, Compositio mathematica, **121** (2000), 337–361.

[34] M. A. Marco Buzunariz, Resonance varieties, admissible line combinatorics, and combinatorial pencils, arxiv:math.CO/0505435.

[35] J. W. Morgan: The algebraic topology of smooth algebraic varieties, Inst. Hautes Études Sci. Publ. Math. 48 (1978), 137–204.

[36] P. Orlik and L. Solomon, Combinatorics and topology of complements of hyperplanes, Invent. math. **56** (1980), 167–189.

[37] P. Orlik and H. Terao, Arrangements of Hyperplanes, Springer Verlag, Berlin–Heidelberg–New York, 1992.

[38] J. Pereira and S. Yuzvinsky, Completely reducible hypersurfaces in a pencil, Advances in Math. **219** (2008), 672–688.

[39] V. Schechtman and A. Varchenko, Arrangements of hyperplanes and Lie algebra homology, Invent. Math. **106** (1991), 139–194.

[40] V. Schechtman, H. Terao, and A. Varchenko, Local systems over complements of hyperplanes and the Kac-Kazhdan condition for singular vectors, J. Pure Appl. Algebra, **100**, 93–102 (1995)

[41] C. Simpson, Subspaces of moduli spaces of rank 1 local systems, Ann. Sci. École Norm. Sup. 26(1993), no. 3, 361–401.

[42] J. Stipins, On finite k-nets in the complex projective plane, Ph. D. thesis, The University of Michigan, 2007.

[43] M. Wachs and J. Walker, On geometric semilattices, Order **2** (1986), 367–385.

[44] S. Yuzvinsky, Realization of finite abelian groups by nets in $\mathbb{P}^2$, Compos. Math., **140(6)** (2004),1614–1624.

[45] S. Yuzvinsky, Cohomology of the Brieskorn-Orlik-Solomon algebras, Comm. in Algebra **23** (1995), 5339–5354.

[46] S. Yuzvinsky, Orlik-Solomon algebras in algebra, topology, and geometry, Russian Math. Surveys **56** (2001), 294–364.

[47] S. Yuzvinsky, A new bound on the number of special fibers in a pencil of curves, Proc. Amer. Math. Soc. **137** (2009), 1641–1648.

Alexandru Dimca
Laboratoire J.A. Dieudonné
UMR du CNRS 6621
Université de Nice Sophia Antipolis
Parc Valrose
06108 Nice Cedex 02
France
e-mail: dimca@math.unice.fr

Sergey Yuzvinsky
University of Oregon
Eugene, OR 97403
USA
e-mail: yuz@math.uoregon.edu

Progress in Mathematics, Vol. 283, 111–153

Local Systems and Constructible Sheaves

Fouad El Zein and Jawad Snoussi

Abstract. The article describes local systems, integrable connections, the equivalence of both categories and their relations to linear differential equations. We report in details on regular singularities of connections and on singularities of local systems which leads to the theory of intermediate extensions and the decomposition theorem.

Mathematics Subject Classification (2000). Primary 32S60, 32S40; Secondary 14F40.

Keywords. Algebraic geometry, analytic geometry, local systems, linear differential equations, connections, constructible sheaves, perverse sheaves, Hard Lefschetz theorem.

Introduction

The purpose of these notes is to indicate a path for students that starts from a basic theory in undergraduate studies, namely the structure of solutions of *Linear Differential Equations* which is a classical subject in mathematics (see Ince [9]) that has been constantly enriched with developments of various theories and ends in a particular subject of research in contemporary mathematics, namely perverse sheaves.

We report in these notes on the developments that occurred with the introduction of sheaf theory and vector bundles in the works of Deligne [4] and Malgrange [3,2)].

Instead of continuing with differential modules as developed by Kashiwara and explained in [12], a subject already studied in a Cimpa school, we shift our attention to the geometrical aspect represented by the notion of *Local Systems* which describe on one side the structure of solutions of linear differential equations and on the other side the cohomological higher direct image of a constant sheaf by a proper smooth differentiable morphism.

Then we introduce the theory of *Connections* on vector bundles generalizing to analytic varieties the theory of linear differential equations on a complex disc.

The definition of local systems is easily extended to varieties of dimension n while it is more elaborate to extend the notion of differential equations into the concept of connections. We note that Deligne has established an equivalence of categories between local systems and integrable connections.

In particular this point of view explains how the connections named after Gauss and Manin are defined by the cohomology of families of algebraic or analytic varieties (precisely by a smooth proper morphism). A background to this result is the classical construction of solutions of differential equations as integrals along cycles of relative differential forms on algebraic families of varieties defined by a smooth proper morphism. DeRham resolutions of local systems are obtained via the associated integrable connections.

Singularities in the fields of algebraic and analytic geometry appear in the study of linear differential equations with meromorphic coefficients on the punctured complex disc. In particular a basic result of Fuchs on equations with *Regular Singularity* is at the origin of the theory and leads to the notion of meromorphic connections with regular singularity.

The work of P. Deligne in 1970 [4] pointed out the developments of this theory to higher dimensional varieties in algebraic and analytic geometry.

Constructible sheaves. Singularity theory in mathematics, which arises for example with the vanishing of the differential of a morphism, has had important developments in algebraic geometry; in particular Whitney's and Thom's stratification theory [10] contributed to a further generalization of local systems, namely the concept of constructible sheaves which appears in the study of cohomology theory of the fibers of any algebraic morphism. This concept is used in séminaire de géométrie algébrique [6] by Grothendieck's school and in an important article [11] on Chern classes for singular algebraic varieties by MacPherson.

Among the complexes of sheaves with constructible cohomology, the perverse sheaves have important special properties, since they are related to the theory of differential modules in the sense that the DeRham complex defined by a holonomic differential module is a complex with constructible cohomology sheaves which is in fact a perverse sheaf.

Complexes with constructible cohomology sheaves are preserved through derived direct images by a proper algebraic morphism (and in general by the six classical operations). The concept and the proofs are based on Thom-Whitney stratification of varieties and morphisms and a result, proved by Mather, known as the Thom-Mather isotopy lemma describing local topological triviality along strata.

Decomposition theorem. This theorem is stated here to illustrate how it is possible to develop a basic classical result such as Lefschetz's theorem via the above tools. The proof is beyond the scope of this exposition. The reader will not see here the use of regularity necessary in the proof, nor can we present Hodge theory which is hidden in the hypothesis of a geometric local system. We do however mention further references where it is possible to find more results on the subject.

Family of Elliptic Curves. The appendix gives an explicit computation of the monodromy of a local system and the Gauss-Manin connection defined by the family of elliptic curves, a mathematical subject that should serve as a test example for every mathematician.

The reader can expect to learn from this expository paper various points of view of the subject in topology, geometry and analysis in the direction of the decomposition theorem. To cover recent developments in the theory, the reader is adviced to consult the various books listed after the references. Finally, the theory of Local Systems and Constructible Sheaves plays an important role in the theory of Arrangement of Hyperplanes and we refer the reader to expository and research articles by experts on this subject presented in this summer school.

1. Local Systems

We study here sheaves of groups with topological interest known as local systems or locally constant sheaves. They arise in mathematics as solutions of linear differential equations, as higher direct image of a constant sheaf by a proper differentiable submersive morphism of manifolds and as representations of the fundamental group of a topological space. Local systems can be enriched with structures reflecting geometry such as the notion of Hodge structures.

1.1. Background in undergraduate studies

The affine differential equation $zu'(z) = 1$ with complex variable z, well known by students, is singular at the origin, since we can apply Cauchy's theorem on the existence of a unique solution with given initial condition only for $z \neq 0$. We put $u'(z) = \frac{1}{z}$, then for any $a \neq 0$ there exists an analytic solution near a for $|z-a| < |a|$, $u(z) = \sum_{n\geq 0} \frac{(-1)^n}{(n+1)a^{n+1}}(z-a)^{n+1}$. In particular, for $a = 1$ we define in this way the function $u(z) = \log z$ solution of the equation satisfying the condition $u(1) = 0$. Then we can extend the above local solution into the global function $\log z = r + i\theta, \theta \in]-\pi, \pi[$ for $z = re^{i\theta}$. The main point of interest in our study, due to the singularity of the equation at zero, comes down in this case to the fact that $\log z$ cannot be extended in a continuous function beyond the above domain in the complex plane, since its limit near a negative real number $-r$ along a path in the upper half plane is $\log r + i\pi$ and differs by $2i\pi$ with its limit $\log r - i\pi$ along a path in the opposite half plane.

Such a function is an inverse to the exponential map e^z, but other inverse maps can be written as $\log z + 2ki\pi$ and are always defined on $\mathbb{C}$-{ ray }. They are called various determinations of the logarithm. The exponential map $e^z : \mathbb{C} \to \mathbb{C}^*$ is said to be a covering and a determination of $\log z$ is a section of such a covering.

However our interest is in linear differential equations, for example $zu'(z) - \alpha u(z) = 0$, for $\alpha \in \mathbf{R}$, with solutions $z^\alpha = r^\alpha e^{i\alpha\theta}$. When we cross the negative reals the solution is multiplied by $e^{i2\pi\alpha}$. We express this property by introducing the one-dimensional vector space $\mathbb{C}z^\alpha$ of all the solutions defined on a simply connected open subset of $\mathbb{C}^*$ and the linear endomorphism $T : \mathbb{C}z^\alpha \to \mathbb{C}z^\alpha$ called a monodromy, acting as $e^{i2\pi\alpha} Id$ on this linear space.

From another point of view, the monodromy extends to a morphism from $\mathbb{Z}$ to the group of linear automorphisms of the one-dimensional vector space $\mathbb{C}z^\alpha$ defined by $n \mapsto T^n$. We obtain in this special case the representation of the fundamental group $\pi_1(\mathbb{C}^*)$ identified with $\mathbb{Z}$, defined by the differential equation.

1.2. Definition and properties

To define local systems we use the language of sheaf theory for which basic references are Godement [5] and Warner [14], then we describe here the relation with the topology of the base space, precisely the fundamental group.

The constant sheaf. On a topological space M, an abelian group G defines a constant sheaf denoted by G_M (or also by G), whose sections on a connected open subset is the group itself with the identity as a restriction morphism to smaller connected open subsets.

Definition 1.1. Let A be a ring; a local system $\mathcal{L}$ on a connected topological space M with fiber an $A-$module L, is a sheaf locally isomorphic to the constant sheaf defined by L, i.e., at each point v in M there exists an open neighbourhood U of v in M and an isomorphism of $A-$modules on the restriction of $\mathcal{L}$ to the constant sheaf L_U on U: $\mathcal{L}_{|U} \simeq L_U$.

There exists a covering U_i of M and isomorphisms of modules $\varphi_{i,j} : L_{U_{i,j}} \to L_{U_{i,j}}$ constant on each connected component of $U_{i,j} = U_i \cap U_j$, called transition transformations, whose restrictions to triple intersections $U_{i,j,k} = U_i \cap U_j \cap U_k$ satisfy $\varphi_{i,j|U_{i,j,k}} \circ \varphi_{j,k|U_{i,j,k}} = \varphi_{i,k|U_{i,j,k}}$.

We will be mainly interested in $\mathbb{Q}$-local systems $\mathcal{L}$ with finite dimensional $\mathbb{Q}$-vector spaces as fiber (said to be of finite rank). The transition morphisms $\varphi_{i,j}$ are defined by matrices in $GL(n, \mathbb{Q})$ constant on each connected component of $U_{i,j} = U_i \cap U_j$.

We will be concerned with local systems arising in two natural subjects. The first will consist of the higher direct image cohomology sheaves by a proper submersive morphism and the second is defined by the solutions of linear differential systems.

Properties. The inverse image of a local system $\mathcal{L}$ by a continuous map $f : N \to M$ is defined as the locally constant sheaf $f^{-1}(\mathcal{L})$ on N.

Lemma 1.2. *A local system $\mathcal{L}$ on the interval* $[0, 1]$ *is constant.*

Proof. There exists a finite number of intervals $[t_i, t_{i+1}]$ such that the restriction of $\mathcal{L}$ is constant on each interval. Each element a_i in the fiber L at a point $t \in [t_i, t_{i+1}]$ defines a unique section on $[t_i, t_{i+1}]$ which extends to sections on $[t_{i-1}, t_i]$ and $[t_{i+1}, t_{i+2}]$ and successively to a section on $[0, 1]$. The extension operation has an inverse defined by the restriction of global sections to the point $t \in [0, 1]$, hence it is an isomorphism. □

From now on we suppose the topological space M *locally path connected and locally simply connected* (each point has a basis of connected neighbourhoods $(U_i)_{i\in I}$ with trivial fundamental groups, i.e., $\pi_0(U_i) = e$ and $\pi_1(U_i) = e$).

Remark 1.3. On complex algebraic varieties, we refer to the transcendental topology and not the Zariski topology to define local systems.

1.2.1. Monodromy. Let $\gamma : [0,1] \to M$ be a loop in M with origin a point v and let $\mathcal{L}$ be a $\mathbb{Q}$-local system on M with fiber L at v. The inverse image $\gamma^{-1}(\mathcal{L})$ of the local system is isomorphic to the constant sheaf defined by L on $[0,1]$: $\gamma^{-1}\mathcal{L} \simeq L_{[0,1]}$.

Definition 1.4 (Monodromy). The composition of the linear isomorphisms

$$L \simeq \mathcal{L}_v = \mathcal{L}_{\gamma(0)} \simeq \Gamma([0,1],\mathcal{L}) \simeq \mathcal{L}_{\gamma(1)} = \mathcal{L}_v \simeq L$$

is denoted by T and called the monodromy along γ. It depends only on the homotopy class of γ.

Proof. Given a homotopy H defined on $[0,1]^2$ between two loops γ and γ' we lift $\mathcal{L}$ by H to $[0,1]^2$ where we apply an argument similar to the interval case, by covering the square with products $[x_i, x_{i+1}] \times [y_j, y_{j+1}]$ such that the restrictions of the inverse image of $\mathcal{L}$ are constant (proofs using homotopy are standard and must be worked once in detail, see for instance the invariance of the primitive of an analytic function constructed along two homotopic loops in Cartan [1] p. 59). □

Proposition 1.5. *Let M be a topological space connected, simply connected, locally path connected and locally simply connected, then a local system $\mathcal{L}$ on M is isomorphic to a constant sheaf L_M.*

Proof. Let a be a fixed point in M and let $L = \mathcal{L}_a$ denote the stalk of the sheaf $\mathcal{L}$ at a. For any point $x \in M$, two paths γ and γ' from a to x define equal isomorphisms $\gamma_* = \gamma'_* : \mathcal{L}_a \to \mathcal{L}_x$ since $(\gamma'.\gamma^{-1})$ is homotopic to the identity, hence $(\gamma'.\gamma^{-1})_* = Id$. We define an isomorphism of sheaves $\varphi : L_M \to \mathcal{L}$ such that for all points $x \in M$, $\varphi_x : (L_M)_x \simeq \mathcal{L}_a \to \mathcal{L}_x$ is equal to γ_*, then φ is well defined since φ_x is independent of the choice of the path. □

1.3. Local systems and Representations of the fundamental group

The notion of local system can be introduced as the theory of representations of the fundamental group of a topological space. This fact will be presented in terms of equivalence of categories that we recall now.

1.3.1. An equivalence of two categories $\mathcal{C}$ and $\mathcal{D}$ consists of a functor $F : \mathcal{C} \to \mathcal{D}$, a functor $G : \mathcal{D} \to \mathcal{C}$, and two natural isomorphisms $a : F \circ G \to Id_{\mathcal{D}}$ and $b : Id_{\mathcal{C}} \to G \circ F$.

An interesting criteria states that a functor $F : \mathcal{C} \to \mathcal{D}$ defines an equivalence of categories if and only if it is:

1) full, i.e. for any two objects A_1 and A_2 of $\mathcal{C}$, the map $Hom_{\mathcal{C}}(A_1, A_2) \to Hom_{\mathcal{D}}(F(A_1), F(A_2))$ induced by F is surjective

2) faithful, i.e. for any two objects A_1 and A_2 of $\mathcal{C}$, the map $Hom_{\mathcal{C}}(A_1, A_2) \to Hom_{\mathcal{D}}(F(A_1), F(A_2))$ induced by F is injective and
3) essentially surjective, i.e. each object B in $\mathcal{D}$ is isomorphic to an object of the form $F(A)$, for A in $\mathcal{C}$.

Definition 1.6. Let L be a $\mathbb{Q}$−vector space. A representation of a group G is a homomorphism of groups

$$G \xrightarrow{\rho} Aut_{\mathbb{Q}}(L)$$

from G to the group of $\mathbb{Q}$−linear automorphisms of L or equivalently a linear action of G on L.

The monodromy of a local system $\mathcal{L}$ defines a representation of the fundamental group $\pi_1(M, v)$ of a topological space M on the stalk at v, $\mathcal{L}_v = L$,

$$\pi_1(M, v) \xrightarrow{\rho} Aut_{\mathbb{Q}}(\mathcal{L}_v)$$

which characterizes local systems on connected spaces in the following sense.

Proposition 1.7. *Let M be a connected topological space. The above correspondence is an equivalence between the following categories:*

i) *$\mathbb{Q}$-local systems with fiber a vector space L on M,*
ii) *Representations of the fundamental group $\pi_1(M, v)$ by linear automorphisms of a $\mathbb{Q}$-vector space L.*

Proof. The representation associated in (i) to the local system is defined by the monodromy along a path as we have seen above, and this correspondence is functorial.

ii) To simplify the proof we use the existence of a universal covering $P : \tilde{M} \to M$ of M. The group $\pi_1(M, v)$ acts on $\tilde{M}$ and can be identified with the group of covering transformations. Given a representation $\rho : \pi_1(M, v) \to Aut\, L$, we define its associated local system $\mathcal{L}$ via the introduction of the constant sheaf $L_{\tilde{M}}$ on $\tilde{M}$; we put:

$$\Gamma(U, \mathcal{L}) := \{s \in \Gamma(P^{-1}(U), L_{\tilde{M}}) : \forall u \in P^{-1}(U), \forall \gamma \in \pi_1(M, v), s(\gamma.u) = \rho(\gamma).s(u)\},$$

i.e., the sections of $\mathcal{L}$ on U are the equivariant sections of $L_{\tilde{M}}$ on $P^{-1}(U)$ under the action of $\pi_1(M, v)$. Then the functoriality of the construction can be checked. □

Given a linear automorphism $T \in Aut_{\mathbb{Q}}L$, we define an action of T on a representation ρ by conjugation as

$$T.\rho : \pi_1(M, v) \to Aut_{\mathbb{Q}}(\mathcal{L}_v) : \gamma \in \pi_1(M, v) \mapsto T \circ \rho(\gamma) \circ T^{-1};$$

then we deduce from the proposition a correspondence between:

i) *Isomorphism classes of $\mathbb{Q}$-local systems on M with fiber L at a fixed point v and*
ii) *Classes of representations of the fundamental group $\pi_1(M, v)$ in finite dimensional $\mathbb{Q}$-vector spaces L modulo the above action of the group of linear automorphisms $Aut_{\mathbb{Q}}L$ on the representations.*

Let $L = \mathcal{L}_v$; then for each isomorphism $T : L \to L'$ we deduce a local system $\mathcal{L}'$ isomorphic to $\mathcal{L}$ as shown in the following diagram where $\gamma \in \pi_1(M, v)$:

$$\begin{array}{ccc} L & \xrightarrow{\rho(\gamma)} & L \\ T\downarrow & & T\downarrow \\ L' & \xrightarrow{T\circ\rho(\gamma)\circ T^{-1}} & L' \end{array}$$

Reciprocally a linear automorphism $\mathcal{T}$ of $\mathcal{L}$ is defined by the action of $T \in Aut_{\mathbb{Q}} L$ where T is induced by $\mathcal{T}$ at v.

Remark 1.8. In the above equivalence the vector space L is viewed as the fiber of the local system at the reference point v for the fundamental group. In another point of view the vector space L is identified with the space of global sections of the constant inverse image of the local system $\mathcal{L}$ on the universal covering of M, in which case L is called the space of multivalued sections of $\mathcal{L}$ [4].

Corollary 1.9. *The group of global sections of the local system $\mathcal{L}$ is isomorphic to the invariant subspace of the fiber L at the reference point v under the action of the representation ρ,*

$$H^0(M, \mathcal{L}) \simeq L^\rho : = \{a \in L \mid \rho(\alpha)(a) = a, \forall \alpha \in \pi_1(M, v)\}.$$

Proof. The above proposition applied to the constant sheaf $\mathbb{Z}_M$ states that the space of morphisms $\varphi \in Hom(\mathbb{Z}, \mathcal{L})$ is isomorphic to the space of morphisms of the trivial representation $\mathbb{Z}$ into ρ. On one side $Hom(\mathbb{Z}, \mathcal{L})$ is isomorphic to the space of global sections $H^0(M, \mathcal{L})$ via $\varphi \mapsto s = \varphi(1)$ and on the other side the morphisms from the trivial representation $\mathbb{Z}$ to $\mathcal{L}$ are defined by elements $a \in L$ satisfying the above formula where $\rho(\alpha) = Id$ since it is the image by φ of the trivial action on $\mathbb{Z}$. □

Example. i) Let D_r be an open complex disc centered at 0 of radius r; then a local system on $D_r^* = D_r - \{0\}$ is defined by a vector space L and a linear automorphism T on L.

ii) More generally for $M = D_r - \{x_1, \ldots, x_n\}$ a disc with n points deleted, $\pi_1(M)$ is the free group on n generators corresponding to a loop around each point; hence the representations of $\pi_1(M)$ are defined by the choice of n linear automorphisms T_i on L.

1.3.2. Cohomology. Let $\mathcal{L}$ be a local system on D^*, with fiber L at some point and monodromy $T : L \to L$; we prove that its cohomology is

$$H^0(D^*, \mathcal{L}) \simeq ker\,(T - Id), \quad H^1(D^*, \mathcal{L}) \simeq coker\,(T - Id)$$
$$\text{and } H^i(D^*, \mathcal{L}) \simeq 0 \text{ for } i > 1,$$

hence it is defined as the cohomology of the complex

$$L \xrightarrow{\partial} L : \partial(b) = (T - Id)(b).$$

The cohomology is computed via Cěch definition; we consider the covering of D^* by the two open sets, north $U_n = \{D - D \cap i\mathbf{R}^-\}$ (complement of negative imaginary numbers) and south $U_s = \{D - D \cap i\mathbf{R}^+\}$ (complement of positive imaginary numbers) and their intersections $U_n \cap U_s = U^+ \cup U^-$ where $U^+ = \{a + ib \in D : a > 0\}, U^- = \{a + ib \in D : a < 0\}$. The associated Čech complex C_1 is defined as

$$(C_1) \qquad \begin{array}{c} H^0(U_n, \mathcal{L}) \oplus H^0(U_s, \mathcal{L}) \xrightarrow{\partial_1} H^0(U^+, \mathcal{L}) \oplus H^0(U^-, \mathcal{L}) : \\ \partial_1(a_1 + a_{-1}) = (a_{-1} - a_1)_{|U_n \cap U_s}. \end{array}$$

The fundamental group of D^* is generated by the loop γ defined by $e^{2i\pi t}$ for $t \in [0, 1]$, hence according to the definition as a representation, the local system is determined by its stalk $L = \mathcal{L}_1$ at 1 and the monodromy T image of γ. Since all the open subsets are simply connected we have isomorphisms: $\varphi_1 : H^0(U_n, \mathcal{L}) \simeq \mathcal{L}_1 = L$, $\psi_1 : H^0(U^+, \mathcal{L}) \simeq \mathcal{L}_1 = L$, $\varphi_{-1} : H^0(U_s, \mathcal{L}) \simeq \mathcal{L}_{-1}$, $\psi_{-1} : H^0(U^{-1}, \mathcal{L}) \simeq \mathcal{L}_{-1}$ with the fibers L at $1 \in U_n$ and $\mathcal{L}_{-1}$ at $-1 \in U_s$. Moreover we have isomorphisms $\alpha : L = \mathcal{L}_1 \simeq \mathcal{L}_{-1}$ defined by a path from $\{1\}$ to $\{-1\}$ and $\delta : \mathcal{L}_{-1} \simeq \mathcal{L}_1 = L$ defined by a path from $\{-1\}$ to $\{1\}$ such that $\delta \circ \alpha = T$. We introduce the complex

$$(C_2) \qquad L \oplus \mathcal{L}_{-1} \xrightarrow{\partial_2} \mathcal{L}_{-1} \oplus L : \partial_2(b_1, b_{-1}) = (b_{-1} - \alpha(b_1), \delta(b_{-1}) - b_1).$$

The morphisms $\varphi_1, \psi_1, \psi_{-1}$ and φ_{-1} above can be combined to define a quasi-isomorphism of complexes $\varphi : C_1 \to C_2$. Then we introduce the complex

$$(C_3) \qquad L \xrightarrow{\partial_3} L : \partial_3(b) = (T - Id)(b)$$

and the morphism $D : C_3 \to C_2$ defined by $D^0(b) = (b, \alpha(b))$ in degree 0 and $D^1(a) = (0, a)$ in degree 1. Finally we can check that D is a quasi-isomorphism, since for example in C_2, $\ker \partial_2 \simeq \{(b_1, b_{-1}) | \alpha(b_1) = b_{-1} \text{ and } \delta(b_{-1}) = b_1\}$, hence $\ker \partial_2 \simeq \{b \in L | T(b) = b\}$.

1.4. System of n linear first-order differential equations and local systems

We consider here holomorphic equations, however the theory can be developed for differentiable equations.

Definition 1.10. A first-order holomorphic system of n linear differential equations is written as

$$\frac{du}{dz} = A(z)u$$

where $u \in \mathbb{C}^n$, z is a coordinate on an open subset U of $\mathbb{C}$ and $A : U \to End(\mathbb{C}^n)$ is a holomorphic map in the vector space of endomorphisms of $\mathbb{C}^n$. When A is independent of z, the system is said to have constant coefficients.

Classically the system is referred to as homogeneous in n unknowns $u_i(z) \in \mathbb{C}$, $i \in [1, n]$, with holomorphic coefficients as entries of A, $a_{ij} : U \to \mathbb{C}$, $i, j \in [1, n]$, and is written in the form

$$\frac{du_i}{dz} = \sum_{j=1}^{n} a_{ij}(z)u_j(z), \ i = 1, 2, \ldots, n.$$

This is one equation and it is not true that we have n distinct equations with independent variables.

If we consider a differentiable map $A : I \to End(\mathbf{R}^n)$ on an interval I of $\mathbf{R}$, then Cauchy's theorem confirms the existence of global solutions, defined on the whole interval, which form a real vector space of dimension n, the isomorphism with $\mathbf{R}^n$ being determined by the initial condition given at a fixed time $t \in I$ with varying position $u \in \mathbf{R}^n$.

The extension of Cauchy's theorem to the holomorphic case can be found in the book of Cartan [1]. This proves the existence of unique local solutions with fixed initial conditions. A subtle point to study is the existence and behavior of a global solution on $U \subset \mathbb{C}$, namely to decide whether a local solution can be extended to all U. This behavior is a central point in our subject here.

Let $\gamma : I \to U$ denote a path in U defined on a real interval. Identifying $\mathbb{C}^n$ with $\mathbf{R}^{2n}$ we deduce by composition a map $A \circ \gamma : I \to \mathbf{R}^{2n}$ defining a real differential system. Applying the result on the existence of global solutions on I, it is not difficult to check that the local solutions can be extended along each path.

The problem arises when we extend a solution along a non-trivial loop. Since a solution defined near the origin and extended along a path does not necessarily coincide, upon first return to the origin, with itself; that is, we don't obtain necessarily the original solution. In conclusion, holomorphic solutions cannot be extended necessarily to the whole open set. However the extensions along two paths with the same origin and the same end point coincide at the same end point if the two paths are homotopic.

Corollary 1.11. *Global solutions are defined on a simply connected open subset V in U and form a complex vector space of dimension n.*

The notion of local system is the abstract concept which describes this behavior and the basic properties of the space of solutions.

Proposition 1.12. *Given a homogeneous system of n linear first-order differential equations with holomorphic coefficients on an open subset U of $\mathbb{C}$, the sheaf defined by holomorphic global solutions on each open subset $V \subset U$ forms a local system $\mathcal{L}$ on U.*

Proof. The restriction of $\mathcal{L}$ to a simply connected open subset V is isomorphic to the constant sheaf $\mathbb{C}^n_V$. □

1.5. Connections and Local Systems

We introduce the concept of connections directly on analytic manifolds. The generalization of the concept of a system of n linear first-order differential equations is in two directions. First, since the coordinate space may be of dimension higher than 1, we are concerned with partial differential equations in many variables and second, the definition is compatible with transition transformations on the manifold.

Definition 1.13. Let F be a locally free holomorphic $\mathcal{O}_X$-module on a complex analytic manifold X. A connection on F is a $\mathbb{C}_X$-linear map

$$\nabla : F \to \Omega^1_X \otimes_{\mathcal{O}_X} F$$

satisfying the following condition for all sections f of F and φ of $\mathcal{O}_X$:

$$\nabla(\varphi f) = d\varphi \otimes f + \varphi \nabla f,$$

known as the Leibnitz condition.

1.5.1. Properties. We define a morphism of connections as a morphism of $\mathcal{O}_X$-modules which commutes with ∇. The definition of ∇ extends to differential forms in degree p as a $\mathbb{C}$−linear map

$$\nabla^p : \Omega^p_X \otimes_{\mathcal{O}_X} F \to \Omega^{p+1}_X \otimes_{\mathcal{O}_X} F \text{ such that } \nabla^p(\omega \otimes f) = d\omega \otimes f + (-1)^p \omega \wedge \nabla f.$$

The connection is said to be integrable if its curvature $\nabla^1 \circ \nabla : F \to \Omega^2_X \otimes_{\mathcal{O}_X} F$ *vanishes* (the curvature is a linear morphism).

Then it follows that the composition of maps $\nabla^{i+1} \circ \nabla^i = 0$ vanishes for all $i \in \mathbb{N}$ for an integrable connection. In this case a DeRham complex is associated to ∇,

$$(\Omega^*_X \otimes_{\mathcal{O}_X} F, \nabla) : = F \to \Omega^1_X \otimes_{\mathcal{O}_X} F \cdots \Omega^p_X \otimes_{\mathcal{O}_X} F \stackrel{\nabla^p}{\to} \cdots \Omega^n_X \otimes_{\mathcal{O}_X} F.$$

The contraction of ∇ with a vector field X is denoted by ∇_X. For two vector fields X, Y, let $[X, Y]$ denote the vector field defined as the bracket of X and Y, then the connection is integrable if and only if $\nabla_{[X,Y]} = \nabla_X \nabla_Y - \nabla_Y \nabla_X$ for all X, Y.

Proposition 1.14. *The horizontal sections* F^∇ *of a connection* ∇ *on a module* F *on an analytic smooth variety* X*, are defined as the solutions of the differential equation*

$$F^\nabla = \{f : \nabla(f) = 0\}.$$

When the connection is integrable, F^∇ *is a local system of dimension equal to* $\dim F$.

Proof. Based on the relation between differential equations and connections, locally we can find a small open subset $U \subset X$ isomorphic to an open set of $\mathbb{C}^n$ such that $F_{|U}$ is isomorphic to $\mathcal{O}^m_U$. This isomorphism corresponds to the choice of a frame $\{e_i\}_{i \in [1,m]}$ of F on U and extends to the tensor product of F with the module of differential forms: $\Omega^1_U \otimes F \simeq (\Omega^1_U)^m$. The connection matrix Ω_U is a matrix of differential forms $\{\omega_{ij}\}_{i,j \in [1,m]}$, sections of Ω^1_U, defined as follows: its i-th column is the transpose of the line image of $\nabla(e_i)$ in $(\Omega^1_U)^m$. Then the restriction of ∇ to U corresponds to a connection on $\mathcal{O}^m_U$ denoted ∇_U and defined on sections $y = (y_1, \ldots, y_m)$ of $\mathcal{O}^m_U$ on U; written in column form $\nabla_U {}^t y = d({}^t y) + \Omega_U {}^t y$ or

$$\nabla_U \begin{pmatrix} y_1 \\ \vdots \\ y_m \end{pmatrix} = \begin{pmatrix} dy_1 \\ \vdots \\ dy_m \end{pmatrix} + \Omega_U \begin{pmatrix} y_1 \\ \vdots \\ y_m \end{pmatrix},$$

the equation is in $End(T, F)_{|U} \simeq (\Omega^1 \otimes F)_{|U}$ where T is the tangent bundle to X.

Let $(x_1, \ldots, x_n)$ denote the coordinates of $\mathbb{C}^n$, then ω_{ij} decomposes as

$$\omega_{ij} = \sum_{k \in [1,n]} \Gamma^k_{ij}(x) dx_k$$

so that the equation of the coordinates of horizontal sections is given by linear partial differential equations, for $i \in [1,m]$ and $k \in [1,n]$,

$$\frac{\partial y_i}{\partial x_k} + \sum_{j\in[1,m]} \Gamma_{ij}^k(x) y_j = 0.$$

The solutions form a local system of dimension m, since the Frobenius condition is satisfied by the integrability hypothesis on ∇. □

Remark 1.15. In terms of the basis $e = (e_1, \dots, e_m)$ of $F_{|U}$, a section s is written as

$$s = \sum_{i\in[1,m]} y_i e_i \text{ and } \nabla s = \sum_{i\in[1,m]} dy_i \otimes e_i + \sum_{i\in[1,m]} y_i \nabla e_i \text{ where}$$

$$\nabla e_i = \sum_{j\in[1,m]} \omega_{ij} \otimes e_j.$$

The connection appears as a global version of linear differential equations, independent of the choice of local coordinates on X.

Remark 1.16. The natural morphism $\mathcal{L} \to (\Omega_X^* \otimes_{\mathbb{C}} \mathcal{L}, \nabla)$ defines a resolution of $\mathcal{L}$ by coherent modules, hence induces isomorphisms on cohomology

$$H^i(X, \mathcal{L}) \simeq H^i(R\Gamma(X, (\Omega_X^* \otimes_{\mathbb{C}} \mathcal{L}, \nabla)))$$

where we take hypercohomology on the right. On a smooth differentiable manifold X, the natural morphism $\mathcal{L} \to (\Omega_X^* \otimes_{\mathbb{C}} \mathcal{L}, \nabla)$ defines a soft resolution of $\mathcal{L}$ and induces isomorphisms on cohomology

$$H^i(X, \mathcal{L}) \simeq H^i(\Gamma(X, (\Omega_X^* \otimes_{\mathbb{C}} \mathcal{L}, \nabla)).$$

1.5.2. Connections defined by local systems. We associate to a local system $\mathcal{L}$ on X, a vector bundle $\mathcal{L}_X := \mathcal{O}_X \otimes_{\mathbb{C}} \mathcal{L}$ on X. The transition transformations are deduced from the corresponding transformations of $\mathcal{L}$. Then a connection is defined on $\mathcal{L}_X$ as

$$\forall g \in \Gamma(U, \mathcal{O}), \forall s \in \Gamma(U, \mathcal{L}), \quad \nabla(g \otimes s) = dg \otimes s.$$

The connection is well defined since the transition transformations are defined by matrices with locally constant coefficients.

Example. i) Let D_r be an open complex disc centered at 0 of radius r; then the connection on the trivial line bundle $\mathcal{O}_{D_r^*}$, defined as $\nabla u = du - \frac{\alpha}{z} u dz$, admits for flat sections the vector space of solutions of the equation $du = \frac{\alpha}{z} u dz$ generated by all determinations z^α on open subsets of D_r^*. Since the extension of z^α along a loop around the origin produces the new determination $e^{\alpha(\log z + 2i\pi)} = e^{2i\pi\alpha} z^\alpha$, the flat sections define a local system on D_r^* with fiber $\mathbb{C}$ and monodromy $T = e^{2i\pi\alpha} Id$ on $\mathbb{C}$.

ii) More generally for $M = D_r - \{b_1, \dots, b_n\}$ a disc with n points deleted, the connection on the trivial line bundle $\mathcal{O}_M$ defined as $\nabla u = du - (\Sigma_{i\in[1,n]} \frac{\alpha_i}{z - b_i}) u dz$

admits for flat sections the vector space of solutions of the equation $du = (\Sigma_{i\in[1,n]}\frac{\alpha_i}{z-b_i})udz$ generated by all determinations $\Pi_{i\in[1,n]}(z-b_i)^{\alpha_i}$ on open subsets of M. They define a local system with fiber $\mathbb{C}$ and monodromy $T_k = e^{2i\pi\alpha_k}Id$ around each point b_k. The bundle associated to the local system is isomorphic to $\mathcal{O}_M$, hence trivial.

Theorem 1.17 (Deligne). *The functor* $(F,\nabla)\mapsto F^\nabla$ *is an equivalence between the category of integrable connections on* X *and the category of complex local systems on* X *with quasi-inverse defined by* $\mathcal{L}\mapsto\mathcal{L}_X$.

Proof. i) The correspondence giving horizontal sections is functorial. The canonical morphism, compatible with the connections, $\mathcal{O}_X\otimes_{\mathbb{C}}F^\nabla\to F : g\otimes s\mapsto gs$, is an isomorphism. In fact, since the connection is integrable, there exists locally a basis consisting of horizontal sections, hence locally every section s of (F,∇) is written as a sum $s=\sum_{i\in[1,n]}g_ih_i$ where h_i is horizontal and g_i is an analytic function, then $\nabla(s)=\sum_{i\in[1,n]}g_ih_i$.

ii) Let $\mathcal{L}$ denote a local system on X; then the canonical morphism $\mathcal{L}\to\mathcal{L}_X$: $s\mapsto 1\otimes s$ is an isomorphism onto $\mathcal{L}_X^\nabla$. □

1.6. Fibrations and local systems (Gauss-Manin connection)

We prove that the i-th group of a rational cohomology of the fibers of a proper submersion of manifolds $f : M\to N$ form a local system $R^if_*\mathbb{Q}$ on N for all integers i. First we recall notions on the higher direct image of a sheaf.

1.6.1. Cohomology via sheaf theory techniques. In this section we recall notions on cohomology constructed via sheaf theory. Basic references are Godement [5] and Warner [14]. The cohomology theory attaches to a topological space M a group $H^i(M,\mathbb{Q})$ in each degree i, known as the i-th cohomology group, and to a continuous map $f : M\to N$ of topological spaces a linear function on the groups (for this reason the cohomology is an invariant said to be linear depending on the topology underlying possibly another structure on M, differentiable, for example). The first technical constructions were simplicial, based on a triangulation of the space, but later a cohomology with various coefficients constructed via sheaf theory proved to be of more flexible use in various domains of mathematics.

Real coefficients. The real field $\mathbf{R}$ defines, on a topological space M, a constant sheaf denoted by $\mathbf{R}_M$ or simply $\mathbf{R}$.

On a differentiable manifold M, the DeRham complex of differential forms $\mathcal{E}_M^*$ is a resolution of the constant sheaf $\mathbf{R}_M$ (Poincaré's lemma).

The DeRham theorem asserts that *the* i*-th cohomology of the complex of global sections* $\Gamma(M,\mathcal{E}_M^*)$ *is isomorphic to the* i*-th cohomology vector space:*

$$H^i(M,\mathbf{R})\simeq H^i(\Gamma(M,\mathcal{E}_M^*)).$$

This construction of cohomology is compatible with the structure of analysis. It is so important that it can be considered as a definition for a first approach to cohomology.

The pushforward functor f_* and its derived functors. We may think of a map of topological spaces $f : M \to N$ as a family of spaces consisting of the fibers $M_v = f^{-1}(v)$ at various points v of N, whose cohomologies can be viewed as a family of groups. However the fact that the union M of these fibers has itself the structure of a topological space, as well as the presence of a topology on the parameter space N and the continuity of the morphism f, can be used to obtain a richer structure on the family of cohomologies $\{H^i(M_v, \mathbb{Z})\}_v$, $v \in N$, namely the structure of a sheaf $R^i f_* \mathbb{Z}$ defined by the "presheaf" associated to an open subset U in N the group $H^i(f^{-1}(U), \mathbb{Z})$.

Definition 1.18. Let $f : M \to N$ be a continuous map and $\mathcal{F}$ be a sheaf of abelian groups on M.

i) The direct image sheaf $f_*\mathcal{F}$ is associated to the presheaf on N defined by the global sections on inverses of open sets: $U \to \Gamma(f^{-1}(U, \mathcal{F})$.
ii) For any f_*-acyclic resolution $\mathcal{K}^*$ of $\mathcal{F}$ on M, the complex of sheaves $Rf_*\mathcal{F}$ on N is defined as $Rf_*\mathcal{F}: = f_*\mathcal{K}^*$ on N and called the higher direct image of the sheaf $\mathcal{F}$. Its i-th cohomology sheaves is defined as $R^i f_*: = \mathcal{H}^i(f_*\mathcal{K}^*)$ and called the i-th derivative of the direct image functor (flabby or fine resolutions are examples of acyclic resolutions).

If we view f as giving rise to the family of fibers $f^{-1}(v)$ and if f is proper, the sheaf $R^i f_* \mathbb{Z}_M$ gives rise to the family of cohomology of the fibers $H^i(f^{-1}(v), \mathbb{Z}) \simeq (R^i f_* \mathbb{Z}_M)_v$. The sheaf structure contains more information than merely the family of cohomology of the fibers (for example the monodromy invariant recalled below).

Even if f is not proper the direct image is still interesting, for example in the case of the embedding j of the punctured disk in $\mathbb{C}$ the fiber at 0 of $R^1 j_* \mathbb{Z}$ is isomorphic to $\mathbb{Z}$ generated by the Poincaré dual of the homology class of a loop around 0.

Example. Let X be an analytic variety, Y a normal crossing divisor (NCD) in X and $j : X - Y \to X$ the open embedding. The local information on the topology near Y as a product of discs punctured or not is reflected in the higher direct image $Rj_*\mathbb{C}$ of the constant sheaf $\mathbb{C}$ on $X - Y$. Let $y \in Y$ and U_y a neighbourhood of y isomorphic to a product of complex discs D^n such that $D^n - D^n \cap Y \simeq D^{*p} \times D^q$, then $(R^i j_* \mathbb{C})_y \simeq H^i(D^{*p}, \mathbb{C}) \simeq \wedge^i(\mathbb{C}^p)$.

The result follows from a general relation known as Künneth's formula for a product of spaces and a general statement: $(R^i j_* \mathbb{C})_y$ *is isomorphic to the inductive limit of* $H^i(U_y, \mathbb{C})$ *for small open sets* U_y.

We see in this example that in practice we don't need to go back to the definition and construct a flabby resolution of $\mathbb{C}$ to compute the cohomology groups.

However, it happens that one needs to work directly on the complex level, as in Hodge theory where Deligne needed filtered complexes to define the weight and the Hodge filtrations on the cohomology of algebraic varieties. In this case, although the analytic DeRham complex Ω^*_{X-Y} is not fine, the direct image complex $j_*\Omega^*_{X-Y}$ is quasi-isomorphic to $Rj_*\mathbb{C}$ since we can find enough small Stein open sets

U_y with acyclic cohomology for coherent analytic coefficients sheaves. Moreover it was necessary to introduce a subcomplex of forms having logarithmic singularities along Y in order to define the correct Hodge filtration (see (2.2.1.)).

Theorem 1.19 (Differentiable fibrations). *Let $f : M \to N$ be a proper differentiable submersive morphism of manifolds. For each point $v \in N$ there exists an open neighbourhood U_v of v such that the differentiable structure of the inverse image $M_{U_v} = f^{-1}(U_v)$ decomposes as a product of a fibre at v with U_v:*

$$f^{-1}(U_v) \xrightarrow[\simeq]{\varphi} U_v \times M_v \quad \textit{such that} \quad pr_1 \circ \varphi = f_{|U_v}.$$

The proof follows from the existence of a tubular neighbourhood of the submanifold M_v. Let $V \subset M$ containing M_v be isomorphic to an open neighbourhood of the zero section in the normal bundle $N_{M_v/M}$ and endowed with a retraction $P : V \to M_v$. Since the differential of the map $P \times f_{|V} : V \to M_v \times N$ is invertible on the compact manifold M_v and the restriction of $P \times f_{|V}$ is injective on M_v, there exists an open neighbourhood $V' \subset V$ such that the restriction of $P \times f_{|V}$ to V' is an open embedding. Since f is proper we find U_v such that $f^{-1}(U_v)$ is included in V' and satisfies the statement of the theorem.

We will retain the restriction that if M, N are smooth complex algebraic varieties and f is a smooth algebraic proper morphism, the theorem will apply only to the underlying differentiable structure. Neither the algebraic structure nor the underlying analytic structure decompose into a product, since two smooth nearby fibers are not necessarily isomorphic as analytic or algebraic varieties but only as differentiable varieties.

Remark. i) The morphism obtained by composition $P = pr_2 \circ \varphi : M_{U_v} \to M_v$ induces for each point $w \in U_v$ a diffeomorphism of the fibers $M_w \simeq M_v$, equal to the identity on M_v for $w = v$. It defines a retraction by deformation from M_{U_v} onto M_v.

ii) Let $f : (M, \partial M) \to N$ be a differentiable morphism of manifolds with boundary. Suppose f proper and submersive, as well as its restriction to the boundary ∂M of M. Then f is locally differentially trivial on N, that is at each point v in N there exists a commutative diagram

$$\begin{array}{ccc} U \times (f^{-1}(v),\ \partial f^{-1}(v)) & \simeq & (f^{-1}(U),\ \partial f^{-1}(U)) \\ pr_U \searrow & & \swarrow f_{|U} \\ & U & \end{array}$$

where U is an open neighbourhood of v in N and the isomorphism is a differentiable morphism of manifolds with boundary.

1.6.2. Locally constant cohomology. The differentiable result above for a proper submersive morphism $f : M \to N$ has a linear version on the cohomology of the fibers.

Proposition 1.20 (Locally constant cohomology). *In each degree i, the cohomology sheaf of the fibers $R^i f_* \mathbb{Z}$ is constant on a small neighbourhood U_v of any point v of fiber $H^i(M_v, \mathbb{Z})$, i.e., there exists an isomorphism between the restriction $(R^i f_* \mathbb{Z})_{|U_v}$ with the constant sheaf $H^i_{U_v}$ defined on U_v by the vector space $H^i = H^i(M_v, \mathbb{Z})$.*

Proof. Let U_v be isomorphic to a ball in $\mathbf{R}^n$ over which f is trivial, then for any small ball B_ρ included in U_v, the restriction $H^i(M_{U_v}, \mathbb{Z}) \to H^i(M_{B_\rho}, \mathbb{Z})$ is an isomorphism since M_{B_ρ} is a deformation retract of M_{U_v}. □

1.6.3. Complex algebraic case. Let $f : X \to V$ be an algebraic, proper and smooth morphism of complex algebraic varieties (analytically f^a is a submersion); then f defines a differentiable locally trivial fiber bundle on V (that is the trivialisations are differentiable but not necessarily analytic). The problem of discovering properties to distinguish such classes of local systems (called geometric) is a fundamental problem in geometry.

We still denote by f the differentiable morphism $X^{dif} \to V^{dif}$ associated to f; then the complex of real differential forms $\mathcal{E}^*_X$ is a fine resolution of the constant sheaf $\mathbf{R}$ and $R^i f_* \mathbf{R} \simeq \mathcal{H}^i(f_* \mathcal{E}^*_X)$.

Example. Let $f : X \to S^1$ be a locally trivial fibration with typical fiber F at some point t. The direct image sheaves $R^i f_* \mathbb{Q}$ are local systems on S^1. In this case the fibration is defined by a monodromy homeomorphism $T : F \to F$ which induces on cohomology isomorphisms $T_i : H^i(F, \mathbb{Q}) \to H^i(F, \mathbb{Q})$, the monodromy of $R^i f_* \mathbb{Q}$ on S^1 where the fiber $(R^i f_* \mathbb{Q})_t$ at t is identified with $H^i(F, \mathbb{Q})$. It follows that: $H^1(S^1, R^i f_* \mathbb{Q}) \simeq Coker(T_i - Id)$ and $H^0(S^1, R^i f_* \mathbb{Q}) \simeq Ker(T_i - Id)$.

Example (Geometric local system). Let $f \colon X \to V$ be a smooth and proper morphism of smooth analytic varieties. It follows that f is a differentiable bundle on V (since f is a submersion), i.e., every point y in V has a neighbourhood U_y such that $f^{-1}(U_y)$ is diffeomorphic to a product $U_y \times X_y$ of U_y with the fiber of X at y. Namely let $\gamma \colon [0,1] \to V$ be a differentiable path in V between two points y_0 and y_1; then it defines a diffeomorphism $\gamma_* \colon X_{y_0} \to X_{y_1}$ inducing an isomorphism γ^* on cohomology.

This isomorphism on a cohomology depends on the path up to homotopy and hence defines a representation of the fundamental group $\pi_1(V, y_0)$ on the cohomology $H^i(X_{y_0}, \mathbb{Z})$ or equivalently, the family $H^i(X_y, \mathbb{Z})$ forms a local system on V. In this example the structure of the sheaf on the higher direct cohomology is defined by the cohomology of the fibers and the monodromy. The monodromy in this case is induced by the diffeomorphism defined on the fiber $f^{-1}(v)$ by a trivialization of $f_{|\gamma}$; in particular it is compatible with the cup-product on cohomology.

Suppose now that V is a punctured disc D^*, then $\pi_1(D^*, t)$ is isomorphic to $\mathbb{Z}$. The action of a generator (a circle through t) is the monodromy operator on $H^i(X_t, \mathbb{Z})$ and denoted by T.

Suppose again V is a disc but f is smooth only over D^*. Then the monodromy is related to the singularities of the fiber X_0 at the origin of D. For example, it is a necessary condition that the monodromy defined by f over D^* is trivial for f to be smooth over D. A morphism f over D^* giving rise to a non-trivial monodromy cannot be extended to a smooth morphism over D. The local system is called geometric to remind us that it is constructed as a direct image of a cohomology of algebraic varieties. Such local systems reflect special topological properties of algebraic varieties. The concept of Variation of Hodge structures is introduced in order to accommodate such additional properties. At the expense of hard technical constructions, such structures lead to results subsequent to the relevant geometry.

1.6.4. Relative DeRham complex. Let $f : X \to V$ be a smooth morphism of analytic manifolds. The bundle of relative differential forms is defined as $\Omega^1_{X/V} := \Omega^1_X / f^*(\Omega^1_V)$ and $\Omega^p_{X/V} := \wedge^p \Omega^1_{X/V}$ so that the differential d on Ω^p_X induces a differential on the relative forms, and a relative DeRham complex $(\Omega^*_{X/V}, d)$ is defined and can be extended for any local system $\mathcal{L}$ to a complex $\Omega^*_{X/V}(\mathcal{L}) : = (\Omega^*_{X/V} \otimes_{\mathbb{C}} \mathcal{L}, \nabla)$. At a point $v \in V$, there exists an isomorphism $\Omega^*_{X(v)} \simeq \Omega^*_{X/V} \otimes \mathbb{C}(v)$ where $\mathbb{C}(v) = \mathcal{O}_{V,v}/\mathcal{M}_{V,v} \simeq \mathbb{C}$. The complex of holomorphic forms is not adequate to compute a cohomology, but Grothendieck showed the interest in the notion of hypercohomology (see later 3.6) which is used in the next result, the best that we can hope for and which is indeed proved in [4].

Theorem 1.21 (Deligne). *There exist natural isomorphisms of holomorphic bundles on V,*

$$R^p f_*(\mathcal{L}) \otimes \mathcal{O}_V \simeq R^p f_* \Omega^*_{X/V}(\mathcal{L}).$$

This result generalizes the classical DeRham theorem in the case where V is reduced to a point but with coefficients in a local system in Remark 1.16. It leads to a description of the Gauss-Manin connection on $R^p f_* \Omega^*_{X/V}(\mathcal{L}$ with $R^p f_*(\mathcal{L})$ isomorphic to the horizontal sections.

Definition 1.22. Suppose f is a locally trivial topological fibration; then the connection defined by the local system $R^p f_* \mathcal{L}$ on the bundle $R^p f_* \Omega^*_{X/V}(\mathcal{L})$ is the Gauss-Manin connection.

2. Singularities of Local Systems and Systems of differential equations with meromorphic coefficients: Regularity

In the previous paragraph we obtained a general result on the equivalence of the two categories defined by Local Systems on one side and flat Connections on a manifold on the other side; moreover Gauss-Manin connections are associated to the cohomology of the fibers of a smooth morphism of smooth varieties.

A morphism, in general acquires singular fibers at some critical values in the space of parameters, thus the Gauss-Manin connection is defined on the complement of the set of critical values (known also as the discriminant of the morphism). Hironaka's result on desingularisation suggests that the case of normal crossing divisor as discriminant is the most important to study.

Historically, connections which are meromorphic on analytic varieties and holomorphic on the complement of a divisor were studied first. As usual it is natural to start the study on a small disc, where an asymptotic property near a singularity known as a regular singularity has been described first.

The corresponding singularities of local systems appears in the DeRham complex defined by the connection. The later discovery of perverse sheaves will be presented in the third section. The references to this section are [3] and [4].

2.1. System with meromorphic coefficients on the complex disc

Let $K = \mathcal{M}_0$ denote the field of germs of meromorphic functions at 0. We consider the above homogeneous system of m linear first-order differential equations with coefficients $a_{ij}(z)$ meromorphic at 0,

$$\frac{du}{dz} = A(z)u, \quad A(z) \text{ matrix with entries } a_{ij}(z) \text{ meromorphic at } 0.$$

Taking the coefficients of the system in K is a convenient way to make sure that 0 is the only singular point of the coefficients. Equivalently the system is meromorphic on a disc D_r where we shrink the radius enough to have a unique singular point at the origin (we may suppose $r = 1$ for convenience). The solutions are vectors $u(z) = (u_1(z), \ldots, u_m(z))$ of holomorphic functions on any sector in D^* (simply connected regions of the disc defined by the rays of angle θ satisfying $\theta_1 < \theta < \theta_2$); there exist always solutions since the restriction of the system to such sectors is holomorphic. The main point of study here is that the solutions cannot be extended in general to univalent solutions on the whole disc.

Definition 2.1. A fundamental matrix of solutions consists of a basis of solutions of m vectors where each vector is written as a column of m holomorphic functions defined on a sector.

In fact it is convenient to write the equation in matrix form as

$$\frac{dU}{dz}(z) = A(z)U(z), \tag{2.1}$$

where $A(z)$ is a matrix with entries $a_{ij}(z) \in K$ and $U(z)$ represents a matrix of m independent solutions. This equation can be treated as a first-order equation.

Example. Let Γ be a constant (m, m) complex matrix. It is easy to check that a solution of the matrix equation: $z\frac{d}{dz}U(z) - \Gamma\, U(z) = 0$ on the complex disc D^* is given on any sector by $U(z) = \exp((\log z)\,\Gamma)$ where $\log z$ is a determination of the logarithm. In particular the equation defined by the matrix

$$A(z) = \frac{1}{z}\Gamma, \tag{2.2}$$

where Γ is a Jordan matrix with eigenvalue α, admits the solutions $u_i(z) = (z^\alpha(\log z)^i, z^\alpha(\log z)^{i-1}, \ldots, z^\alpha, 0, \ldots)$ for $0 \le i < m$ forming the columns of the exponential matrix $\exp(\log z\,\Gamma)$.

2.1.1. Multivalued solutions. For certain canonical constructions it is convenient to avoid the choice of a sector, then there is an advantage to introducing the space of multivalued solutions [4].

By definition, *multivalued functions* on D_r^* are holomorphic functions on the universal covering $\widetilde{D}_r^*$ of D_r^* (called the Poincaré half plane H for $r = 1$),

$$\widetilde{D}_r = \{t \in \mathbb{C}, |z = e^{2i\pi t}| < r\}, \ \pi : \widetilde{D}_r \to D_r^* : t \mapsto z = e^{2i\pi t}. \tag{2.3}$$

Definition 2.2. Let $\mathcal{F}$ be a sheaf of complex vector spaces on D_r^*. A multivalued section of $\mathcal{F}$ is a section of $\pi^{-1}\mathcal{F}$.

The covering π admits sections s on sectors of D_r^*. Given a multivalued section $\tilde{f}$ of $\mathcal{F}$, a section s of π defines a section $\tilde{f} \circ s$ of $\mathcal{F}$ on a sector of D_r^*.

Definition 2.3. The multivalued solutions of a homogeneous system of m linear first-order differential equations on D_r^* form a vector space of finite dimension, solutions of the differential operator obtained on $\widetilde{D}_r$ by the change of variable $z = e^{2i\pi t}$.

Example. The general theory below is similar to the above example (2.2) that we discuss again.

i) The matrix $\tilde{U}(t) = \exp((2i\pi t)\Gamma)$ consists of multivalued solutions on the Poincaré half plane H of the equation: $\frac{d}{dt}\tilde{U}(t) = 2i\pi\Gamma\,\tilde{U}(t)$.

ii) *Monodromy.* If we view the solutions in z near a point $z_0 \neq 0$ as sections of a *local system* $\mathcal{L}$ and we follow a solution $U(z)$ along a circle, we obtain upon the first return to z_0 a new basis of sections of $\mathcal{L}$: $\exp(2i\pi\Gamma)U(z) = TU(z)$, where T is the monodromy and $\log T = 2i\pi\Gamma$.

On H, $\tilde{U}(t)$ consists of sections of $\pi^{-1}\mathcal{L}$ and satisfies:

$$\tilde{U}(t+1) = \exp(2i\pi\Gamma)\tilde{U}(t), \text{ that is, } \exp(2i\pi(t+1)\Gamma) = \exp(2i\pi\Gamma)\exp(2i\pi t\Gamma).$$

iii) If we introduce the fibre bundle $\mathcal{L}_{D^*} = \mathcal{O}_{D^*} \otimes \mathcal{L}$, the *holomorphic sections* defined by $\tilde{U}'(t) = \exp(-2i\pi t\Gamma)\tilde{U}(t)$ have period 1 on H, that is,

$$\begin{aligned}\tilde{U}'(t+1) &= \exp((-2i\pi t - 2i\pi)\Gamma)\tilde{U}(t+1)\\ &= \exp(-2i\pi t\Gamma)(\exp -2i\pi\Gamma)\exp(2i\pi\Gamma)\tilde{U}(t) = \tilde{U}'(t).\end{aligned}$$

Hence if $(u_1'(t), \ldots, u_m'(t))$ is a multivalued solution, then the product of the matrix $\exp(-t\log T)$ with the vector $(u_1'(t), \ldots, u_m'(t))$ is the inverse image of a global holomorphic section of $\mathcal{L}_{D^*}$.

2.1.2. Canonical form of the solutions of the meromorphic system. We consider again the above general system (2.1) defined by $A(z)$ lifted to H. Let $\widetilde{S}(t) = (\widetilde{U}_1(t), \ldots, \widetilde{U}_m(t))$ be a set of m independent multivalued solutions; each vector $\widetilde{U}_k(t)$ has m holomorphic functions on $H = \widetilde{D}^*$ as components.

A *Monodromy* is induced on the solutions by the action of the translation: $t \mapsto t+1$ on $\widetilde{D}^*$. As the coefficients of the system defined by $A(2i\pi t)$ are of period

1, the substitution of t by $t+1$ transforms the basis $\widetilde{S}(t)$ of the vector space of solutions into another basis, hence there is a matrix $C \in GL(m, \mathbb{C})$ such that

$$\widetilde{S}(t+1) = \widetilde{S}(t)C.$$

The matrix C defines a linear transformation of the space of solutions, that is, the monodromy transformation.

The logarithm of the monodromy. Let Γ be a matrix such that $e^{2i\pi\Gamma} = C$ with eigenvalues λ satisfying the condition $0 \leq Re(\lambda) < 1$; then the the matrix $\widetilde{S}(t)e^{-2i\pi t\Gamma}$ has period 1 in t. Considering the change of variable $z = e^{2i\pi t}$, we get a *matrix* $\Sigma(z)$ *with coefficients holomorphic on* D_r^* such that

$$\Sigma(z) \in Gl(m, \mathcal{O}_{D_r^*}) : \Sigma(e^{2i\pi t}) = \widetilde{S}(t)e^{-2i\pi t\Gamma};$$

then for each determination of $log z$, the columns of $\Sigma(z)e^{\Gamma log z}$ form a basis of the vector space of solutions (called also a fundamental system of solutions) since if we put formally $z = e^{2i\pi t}$ and $2i\pi t = \log z$, we recover $\widetilde{S}(t)$. For example, for $m = 1$ and $\Gamma = \alpha$, the solution $z^\alpha = e^{2i\pi\alpha t}$ satisfies $e^{2i\pi\alpha(t+1)} = e^{2i\pi\alpha t}e^{2i\pi\alpha}$, hence the matrix C has one entry $e^{2i\pi\alpha}$.

Remark 2.4. The condition $0 \leq Re(\lambda) < 1$ on the eigenvalues is arbitrary and we could add to λ an integer.

In summary we have:

Proposition 2.5. *Let* $A(z)$ *be a* (m, m) *matrix with coefficients holomorphic on* D_r^*, *meromorphic at* 0, *and consider the equation*

$$\frac{dU}{dz} = A(z)U.$$

There exists a matrix Γ *with constant complex coefficients and an* (m, m) *matrix* $\sum(z)$ *with coefficients holomorphic on* D_r^* *such that a fundamental system of multivalued solutions of the equation is of the form* $\widetilde{S}(t) = \Sigma(e^{2i\pi} t)e^{2i\pi t\Gamma}$, *or equivalently* $S(z) = \Sigma(z)e^{log z\Gamma}$ *is a solution on any sector of* D_r^* *with a fixed determination of* $\log z$. *The monodromy matrix is then defined as* $C = e^{2i\pi\Gamma}$.

Remark 2.6. i) To construct the matrix C we need to choose a basis of the solutions and then study the action of T, that is the transformation of the solutions by the change of variable $\theta + 2i\pi$ (one turn around zero). Then C and Γ are defined up to conjugation by the matrix of change of the basis. Hence we can reduce Γ to the Jordan canonical form in the example (2.2).

ii) If the matrix $A(z)$ is of the form $\frac{B(z)}{z}$ where $B(z)$ is holomorphic such that the difference of two eigenvalues of $B(0)$ is never a non-zero integer, the matrix C is conjugate to $\exp(2i\pi B(0))$ [3,1), p 137].

2.1.3. Equivalent Systems. Consider a matrix $M(z) \in GL(n, K)$ and let

$$G(z) = M^{-1}(AM - \frac{dM}{dz}),$$

then $V = M^{-1}U$ is a solution of the equation $\frac{dV}{dz} = G(z)V$ if and only if U is a solution of the equation $\frac{dU}{dz} = A(z)U$.

Definition 2.7. Let $A(z)$ and $G(z)$ be two matrices with meromorphic coefficients at 0; the two systems $\frac{dU}{dz} = A(z)U$ and $\frac{dV}{dz} = G(z)V$ are said to be equivalent if there exists an invertible matrix $M(z)$ such that $G(z)$ is related to $A(z)$ by the above formula.

2.1.4. Regular singularity.

Theorem 2.8 (Regular singular point). *Given a system $\frac{dU}{dz} = A(z)U$, the following conditions are equivalent:*

i) *the system is equivalent to a system of the form $\frac{dV(z)}{dz} = \frac{B(z)}{z}V(z)$ where $B(z)$ is a matrix with holomorphic coefficients;*
ii) *the system is equivalent to a system of the form $\frac{dV(z)}{dz} = \frac{\Gamma}{z}V(z)$ where Γ is a matrix with constant coefficients;*
iii) *there exists a matrix $\sum(z)$ with meromorphic coefficients at 0 such that a fundamental system of solutions is given as $S(z) = \Sigma(z)e^{\log z\Gamma}$.*

Example of a non-regular point. The singularity of the equation $z^2\frac{df}{dz} + f = 0$ is not regular since it has $e^{\frac{1}{z}}$ as solution.

Definition 2.9. A system $\frac{dU}{dz} = A(z)U$ has a regular singular point at 0 if the equivalent conditions of the theorem are satisfied.

2.1.5. Linear differential equations on a punctured disc. Let $D_\varepsilon = \{z \in \mathbb{C} : |z| < \varepsilon\}$ denote the complex open disc of radius ε centered at 0. A set of $n+1$ meromorphic functions $a_i(z)$ for $i \in [0, n]$ on the complex disc D_r with a unique isolated singular point at 0 defines a differential operator

$$P = \sum_{i\in[0,n]} a_i(z)(\frac{d}{dz})^{n-i}$$

of degree n if $a_0(z) \not\equiv 0$, acting on the holomorphic functions on D_ε for $\varepsilon < r$. The study of the solutions u on D_ε^* of the equation

$$(\frac{d}{dz})^n u(z) = -\frac{1}{a_0(z)}(\sum_{i\in[1,n]} a_i(z)(\frac{d}{dz})^{n-i}u(z))$$

can be reduced to the case of a linear system if we introduce the new variables

$$u_0 = u, \ \ u_1 = (\frac{d}{dz})u, \dots, u_i = (\frac{d}{dz})^i u, \dots, u_{n-1} = (\frac{d}{dz})^{n-1}u,$$

then the system is written as

$$\frac{d}{dz}u_0 = u_1, \ \frac{d}{dz}u_i = u_{i+1}, \dots, \frac{d}{dz}u_{n-1} = -\frac{1}{a_0(z)}\Big(\sum_{i\in[1,n]} a_i(z)u_{n-i}\Big).$$

Corollary 2.10. i) *The holomorphic solutions of the differential equation $P(u) = 0$ on a simply connected open subset $U \subset D_r^* = D_r - \{0\}$ form a complex vector space of dimension n.*

ii) *The sheaf of solutions on D_r^* is a local system.*

The corollary follows from the existence and uniqueness of local holomorphic solutions of the equation. The holomorphic version of Cauchy conditions for the existence of solutions of the equation $P(u) = 0$ apply near any point $z_0 \neq 0 \in D_r^*$ and show the existence of a unique holomorphic solution u_1 with given initial values for its derivatives to the order $n-1$ at z_0. The solutions near z_0, being in correspondence with these initial values in $\mathbb{C}^n$, form a complex vector space of dimension n on a neighbourhood of z_0 and extend on a simply connected open subset.

Remark 2.11. i) The regularity condition has been introduced by Fuchs as follows: the order of the pole of $\frac{a_i}{a_0}$ at 0 is at most i. Later we will extend the notion of regularity to meromorphic connections always associated to a differential equation [3,1), p 143].

ii) Although the solutions are sections of $\mathcal{O}_{D_r^*}$ which is a trivial fibre bundle, the local system $\mathcal{L}$ is not necessarily trivial ($\mathcal{L}$ defines an analytic bundle denoted by $\mathcal{L} \otimes \mathcal{O}_{D_r^*}$ which is trivial).

Example. A determination of the function $z^\alpha = e^{\alpha \log z}$ is the solution of the equation $z\frac{du}{dz} - \alpha u = 0$ on any sector. The value of one determination on the complement of a ray in D_r is multiplied by $e^{2i\pi\alpha}$ after extension across the ray in the positive orientation.

2.1.6. Monodromy. Let P be a holomorphic differential operator on a punctured disc D_r^* and consider the vector space E of solutions near a point v in D_r^*. The extension of a solution along a circle S^1 through v defines the *invertible linear monodromy operator* $T : E \to E$.

2.2. Connections with Logarithmic singularities

The object of regularity is to study the behavior of a connection near singularities along a divisor Y on X. Since resolution of singularities leads naturally to a normal crossing divisor (NCD), special techniques have been developed in this case by Deligne, based on the logarithmic complex.

2.2.1. Logarithmic Complex. i) Let X be a smooth complex algebraic variety. A normal crossing divisor Y in X is defined by a system of local parameters of the regular local ring $\mathcal{O}_{X,y}$; then Y is the union of its irreducible components Y_i and is written as $Y = \cup_{i\in I} Y_i$. For all subsets M of I, let $Y_M = \cap_{i\in M} Y_i$,

$Y_M^* = Y_M - \cup_{i\notin M} Y_i$; then given a general point $y \in Y_M^*$ there exist analytic local coordinates $z_j, j \in [1,n]$ at y such that Y_M is defined by $z_i = 0$ for $i \leq p$. A neighbourhood $U(y)$ of y is isomorphic to D^{p+k} and $U(y)^* \simeq (D^*)^p \times D^k$ where D is a complex disc, denoted with a star when the origin is deleted, with fundamental group $\Pi_1(U(y)^*)$ a free abelian group generated by p elements representing classes of closed paths around the origin, one for each D^* in the various one-dimensional axes with coordinate z_i.

ii) Let $j : X - Y \to X$ denote the open embedding; the sheaf $\Omega_X^m(\operatorname{Log} Y)$ of differential forms of degree m with logarithmic poles on Y is the subsheaf of $j_*\Omega_{X-Y}^m$ defined locally near y by

$$\Omega_X^m(\operatorname{Log} Y)_y := \{\omega \in (j_*\Omega_X^m)_y : f\omega \in \Omega_{X,y}^m \text{ and } fd(\omega) \in \Omega_{X,y}^{m+1}\}$$

where f is a reduced equation of Y near y.

2.2.2. Properties. The above definition is independent of the choice of coordinate equations of the components Y_i, $i \leq p$ of Y near y.

From now on we will suppose the component Y_i of Y smooth.

i) In terms of the equations z_i, the forms $\frac{dz_i}{z_i}$ for $i \leq p$ and dz_i for $p < i < n$ form an $\mathcal{O}_{X,y}$-basis of $\Omega_X^1(\operatorname{Log} Y)_y$.
ii) $\Omega_X^m(\operatorname{Log} Y) \simeq \wedge^m \Omega_X^1(\operatorname{Log} Y)$.
iii) There exists a global residue morphism $Res_i : \Omega_X^1(\operatorname{Log} Y) \to \mathcal{O}_{Y_i}$ with value, the restriction to Y_i of the locally defined coefficient of $\frac{dz_i}{z_i}$.

Definition 2.12. Let F be a vector bundle on X. A connection with logarithmic poles along Y, $\nabla : F \to \Omega_X^1(\operatorname{Log} Y) \otimes_{\mathcal{O}_X} F$ has a matrix with logarithmic poles.

It extends to $\nabla^i : \Omega_X^i(\operatorname{Log} Y) \otimes_{\mathcal{O}_X} F \to \Omega_X^{i+1}(\operatorname{Log} Y) \otimes_{\mathcal{O}_X} F$; it is integrable if $\nabla^1 \circ \nabla = 0$, so that a logarithmic complex $\Omega_X^*(\operatorname{Log} Y)(F) \colon = (\Omega_X^*(\operatorname{Log} Y) \otimes_{\mathcal{O}_X} F, \nabla)$ is defined in this case.

The composition map: $R_i \otimes Id \circ \nabla : F \to \Omega_X^1(\operatorname{Log} Y) \otimes_{\mathcal{O}_X} F \to \mathcal{O}_{Y_i} \otimes F$ vanishes on the product $\mathcal{I}_{Y_i} F$ of F with the ideal $\mathcal{I}_{Y_i}$ defining Y_i. It induces a linear map called the residue endomorphism of the connection,

$$Res_i(\nabla) : F \otimes_{\mathcal{O}_X} \mathcal{O}_{Y_i} \to F \otimes_{\mathcal{O}_X} \mathcal{O}_{Y_i}.$$

At the point $y \in Y$, the residue Res_i induces a linear endomorphism $Res_i(y)$ on the fibre $F(y)$ of the vector bundle defined by F.

Let $U(y)$ denote a polydisc D^n with center y. The universal covering $\widetilde{U}(y)$ of $D^{*,p} \times D^{n-p}$ is defined by

$$\{t = (t_1, \ldots, t_n) \in \mathbb{C}^n : \forall i \leq p, \operatorname{Im} t_i > 0 \text{ and } \forall i > p, \mid t_i \mid < \varepsilon\},$$

with $t \to (e^{2i\pi t_1}, \ldots, e^{2i\pi t_p}, t_{p+1}, \ldots, t_n) \in D^{*,n}$ as covering map. We denote by $\mathcal{L}$ the local system defined by the horizontal sections, by L its fibre and by $\widetilde{L}$ the global sections of the inverse image of $\mathcal{L}$ on $\widetilde{U}(y)$. The monodromy action T_j for $j \leq p$ is defined on $\widetilde{U}(y)$ by the formula: $T_j v(t) = v(t_1, \ldots, t_j + 1, \ldots, t_n)$ for any section v. Moreover there is an isomorphism between the fibre $F(y)$ and the vector space $\widetilde{L}$ (see the proof of the next theorem); via this isomorphism, we have the following relation proved in ([4], II,5).

Lemma 2.13. $T_i = \exp(-2i\pi Res_i(\nabla))$.

In what follows we need to choose a section τ of the projection $\mathbb{C}$ on $\mathbb{C}/\mathbb{Z}$, given for example by a region of $\mathbb{C}$ defined by the conditions on real z, $\Re z \in [0,1[$; such a section will appear, when we fix a determination of the logarithm, in the proof in [4] (see also [3,2)]) of the following result due to Manin in dimension 1.

Theorem 2.14 (Logarithmic extension). *Let Y be a NCD in X, F_{X^*} a holomorphic vector bundle on $X - Y$ and ∇ a connection on F_{X^*}. There exists a locally free module F_X on X which extends F_{X^*}, moreover the extension is unique if the following conditions are satisfied:*

i) *∇ has logarithmic poles with respect to F_X,*

$$\nabla\colon F_X \to \Omega^1_X(\log Y) \otimes F_X$$

ii) *The eigenvalues of the residues of ∇ with respect to F_X belong to the image of τ.*

Proof. a) The local system $\mathcal{L}$ is said to be *locally unipotent along Y if at any point $y \in Y$ all T_j are unipotent*, in which case the extension we describe is called *canonical.*

First we work locally on a neighbourhood of a point y of Y, which amounts to supposing from now on X^* *isomorphic to a product of discs, punctured or not.* Let $\mathcal{L} = F^{\nabla}_{X^*}$ and let $\widetilde{L}$ denote the vector space of multivalued sections of $\mathcal{L}$, that is the subspace of horizontal sections of the analytic sheaf $\widetilde{F}_{X^*}$ on $\widetilde{X^*}$. The bundle F_{X^*} is isomorphic to $\mathcal{O}_{X^*}\otimes_{\mathbb{C}}\widetilde{L}$. In such cases of local unipotent monodromy actions, the endomorphisms $N_i = Res_i\nabla$ are nilpotent and related to the logarithm of the monodromy by the formula: $N_i = -\frac{1}{2i\pi}\operatorname{Log} T_i = \frac{1}{2i\pi}\sum_{k>0}(I - T_i)^k/k$. The connection on $\mathcal{O}_X\otimes_{\mathbb{C}}\widetilde{L}$ defined by the matrix

$$\nabla = \Sigma_{i\leq p} N_i\, dz_i/z_i$$

is isomorphic to the local extension we are looking for. For $v \in \widetilde{L}$, we define a section $\tilde{v} \in \widetilde{F}_{X^*}$ via the action of the monodromy, explicitly described by the formula

$$\tilde{v} = (\exp(2i\pi\Sigma_{i\leq p} t_i N_i)).v.$$

Notice that the exponential is a linear sum of multiples of $Id - T_j$ with analytic coefficients, hence its action defines an analytic section.

We have, for all $t \in \widetilde{X^*}$, $\tilde{v}(t + e_j) = \tilde{v}(t)$, since for $e_j = (0, \dots, 1_j, \dots, 0)$,

$$\begin{aligned}\tilde{v}(t + e_j) &= [\exp(2i\pi N_j)\exp(2i\pi\Sigma_{i\leq p} t_i N_i)].v(t + e_j)\\ &= [\exp(2i\pi\Sigma_{i\leq p} t_i N_i]\exp(2i\pi N_j).v(t + e_j)\end{aligned}$$

where $\exp(2i\pi N_j).v(t+e_j) = T_j^{-1}.v(t+e_j) = T_j^{-1}T_j.v(t) = v(t)$; hence $\tilde{v}$ is the inverse image of a section of F_{X^*} denoted by

$$\tilde{\mathrm{v}} = (\exp(\Sigma_{i\leq p}(\log z_i)N_i)).v$$

where $2i\pi t_i = \log z_i$, moreover $\nabla\tilde{v} = \Sigma_{i\leq p}\widetilde{N_i.v} \otimes \frac{dz_i}{z_i}$.

Let $j: X^* \to X$ be the inclusion, then we can describe F_X as a subsheaf of $j_*F_{X^*}$ by the condition that a basis of $\widetilde{L}$ is sent onto a basis of $F_{X,y}$.

b) In general the local system is defined by a representation of $\Pi_1(X^*)$ on the vector space L, i.e., the action of commuting automorphisms T_i for $i \in [1,p]$ indexed by the local components Y_i of Y at y. The automorphisms T_i decompose as a product of commuting automorphisms, semi-simple and unipotent $T_i = T_i^s T_i^u$. Since L is a $\mathbb{C}$ - vector space, T_i^s can be represented by the diagonal matrix of its eigenvalues. If we consider families of eigenvalues λ_i for each T_i, we have the spectral decomposition of L:

$$L = \oplus_{\lambda.} L^{\lambda.}, \qquad L^{\lambda.} = \cap_{i\in[1,n]}(\cup_{j>0} \ker\,(T_i - \lambda_i I)^j)$$

where the direct sum is over all families $(\lambda.) \in \mathbb{C}^p$. The logarithm of T_i is defined as the sum

$$\operatorname{Log} T_i = \operatorname{Log} T_i^s + \operatorname{Log} T_i^u.$$

$\operatorname{Log} T_i^s$ is the diagonal matrix formed by $\log \lambda_i$ for all eigenvalues λ_i of T_i^s and for a fixed determination of log, while $\operatorname{Log} T_i^u = -\Sigma_{k\geq 1}(1/k)(I - T_i^u)^k$ is defined as a finite sum of nilpotent endomorphisms. Let $N_i = -\frac{1}{2i\pi}\operatorname{Log} T_i^u$ and $D_i = -\frac{1}{2i\pi}\operatorname{Log} T_i^s$, then the i-th residue is $D_i + N_i$ with eigenvalues $\alpha_i \in [0,1[$ such that $\lambda_i = e^{-2i\pi\alpha_i}$.

Let $\mathcal{U}_X^{\lambda.}$ denote the vector bundle defined by $\mathcal{O}_X$ with the connection $\nabla^{\lambda.}$ defined by the matrix

$$\Sigma_{i\leq p}\alpha_i dz_i/z_i \quad , \quad \alpha_i = -\frac{1}{2i\pi}\log\lambda_i$$

where the determination of $\log\lambda_i$ is such that α_i is in the image of the section τ. Let $\mathcal{U}^{\lambda.}$ denote the local system of horizontal sections. The local system $\mathcal{L}$ on X^* decomposes into

$$\mathcal{L} = \oplus_{\lambda.}(\mathcal{U}^{\lambda.}) \otimes \mathcal{L}^{\lambda.}$$

where $\mathcal{L}^{\lambda.}$ is unipotent, then we put $F_X = \oplus_{\lambda.}\mathcal{U}_X^{\lambda.} \otimes \mathcal{L}_X^{\lambda.}$ where $\mathcal{L}_X^{\lambda.}$ is the extension of $\mathcal{O}_{X^*} \otimes \mathcal{L}^{\lambda.}$ defined above in (a).

c) The crucial step is in the uniqueness since the patching process of the local system extends uniquely to a patching process of the bundle F_X. This result is explained with details in [3,2)].

A basic ingredient in the proof is that the eigenvalues of the residue are constants along Y_i and F_X is unique up to isomorphism if we suppose the eigenvalues of the residues to be in the image of the section τ.

Local description of F_X as a subsheaf of $j_*F_{X^*}$.

Let t_i be the coordinates of the product of p upper half planes and $n-p$ discs, then the universal covering map of a neighbourhood of y is given by $z_j = \exp(2i\pi t_j), j = 1,\dots,p$ and $z_j = t_j$ for $j > p$. We associate to an element v of $\widetilde{L}^{\lambda.}, \lambda. = e^{-2i\pi\alpha.}$, the section $\tilde{v}$ of F_X defined by the formula

$$\tilde{\mathrm{v}} = \exp(2i\pi\Sigma_{j\leq p}t_j(\alpha_j I + N_j)).v = \Pi_{j\leq p}z_j^{\alpha_j}\exp(\Sigma_{j\leq p}\log z_j N_j).v$$

It can be checked that this section descends to a section near $y \in Y_M$ in X^*. A basis of L is sent on a basis of $(F_X)_y$ and we have

$$\nabla\tilde{v} = \Sigma_{j\leq p}[\widetilde{\alpha_j v} + \widetilde{N_j.v}] \otimes \frac{dz_j}{z_j}.$$

Notice that a different choice for the section τ would add an integer k to α_j, hence would multiply the basis $\tilde{\mathrm{v}}$ by z_j^k and modify the extension, but a different choice of the determination of $\log z_j$ would add an integer $2i\pi k$ and hence change v by $T^{-k}v$ in the expression of $\tilde{\mathrm{v}}$ which is only a linear transformation of the basis and does not modify the extension. □

The main application of the above construction is proved in ([4], II,6).

Theorem 2.15 (Logarithmic DeRham cohomology). *The integrable connection ∇ defines a DeRham complex with coefficients in the canonical extension F_X of a flat bundle on X^*, quasi-isomorphic to $Rj_*F^{\nabla}_{X^*}$,*

$$Rj_*F^{\nabla}_X \cong \Omega^*_X(\operatorname{Log} Y) \otimes F_X.$$

In particular $H^i(X - Y, F^{\nabla}_{X^*}) \simeq H^i(R\Gamma(X, \Omega^*_X(\operatorname{Log} Y) \otimes F_X))$.

2.3. Meromorphic Connections on the disc

To study the behavior of a connection near singularities along any divisor Y (not necessarily a NCD) on X, it is useful to introduce meromorphic connections. We start with the disc case.

Let $\mathcal{O}_D[0]$ denote the sheaf of holomorphic functions on D^*, meromorphic at 0. We define now a meromorphic connection on $\mathcal{O}_D[0]$-modules

Definition 2.16. Let D be a complex disc and M a locally free $\mathcal{O}_D[0]$-module of finite rank. A meromorphic connection on M is a $\mathbb{C}$-linear operator $\nabla : M \to M$ satisfying

$$\forall h \in \mathcal{O}_D[0], \forall u \in M, \quad \nabla_z(hu) = \frac{dh}{dz}u + h\nabla_z u.$$

The definition is for modules on a disc and will be extended to varieties. In this definition we did contract the differential form dz with ∂_z.

Let K be the germ of $\mathcal{O}_D[0]$ at 0, then the above definition applies for the K-vector space E, germ of M at 0.

If $\mathbf{e} = (e_1,\dots,e_n)$ is a basis of E over K, we can write $\nabla_z e_i = \sum_j a_{ji}(z)e_j$, then ∇_z is defined by the matrix

$$A = (a_{ji}(z)) \in End(n, K)$$

with respect to this basis; then we have in matrix form $\nabla_z \mathbf{e} = \mathbf{e}A$ and for $u = \sum u_i(z)e_i$, $\nabla_z u = \sum_i (\frac{du_i}{dz} + \sum_j a_{ji}(z)u_j(z))e_i$.

If $\mathbf{f} = (f_1, \ldots, f_n)$ is a new basis defined on $\mathbf{e}$ by a matrix B ($\mathbf{f} = \mathbf{e}B$), then $\nabla_z \mathbf{f} = \mathbf{e}(\frac{dB}{dz} + AB) = \mathbf{f}B^{-1}(\frac{dB}{dz} + AB)$.

Example. Let v be a vector in E with matrix Q on $\mathbf{f}$, $v = \mathbf{f}Q = \mathbf{e}BQ$ and let $P = BQ$ such that $v = \mathbf{e}P$; then $dQ/dz = -(B^{-1}(\frac{dB}{dz} + AB))Q$ is equivalent to $dP/dz = -AP$ (the horizontality of v is independent of the basis).

2.3.1. Connections and Systems of linear differential equations. The equation $\nabla_z u = 0$ is equivalent for $u = \mathbf{e}U$ to the system $\frac{dU(z)}{dz} = -AU$. Hence a system defines a connection with respect to the canonical basis of K^n.

We deduce from the expression of the matrix of a connection with respect to a basis that equivalent classes of systems of linear differential equations give isomorphic meromorphic connections.

Corollary 2.17. *There is a correspondence between equivalent classes of systems of linear differential equations and isomorphism classes of meromorphic connections on the disc.*

Example. i) The differential operator $P = z(d/dz) - \alpha$ defines on the sheaf $\mathcal{O}_D[0]$ the connection $\nabla_z(f) = P(f)$.

ii) Let $M = Ke^{\frac{1}{z}}$; this notation means that we consider M as a K subspace of the inductive limit of $\Gamma(D_\varepsilon^*, \mathcal{O})$ so as to induce the natural connection. Hence we consider $\nabla_z(fe^{\frac{1}{z}})\colon = \frac{df}{dz}e^{\frac{1}{z}} - f\frac{1}{z^2}e^{\frac{1}{z}}$, then (M, ∇_z) is isomorphic to (K, ∇'_z) such that $\nabla'_z(f) = \frac{df}{dz} - f\frac{1}{z^2}$.

2.4. Regular meromorphic connections

For any divisor Y on X we denote by $\mathcal{O}_X[Y]$ the sheaf of rings of holomorphic functions on $X^* = X - Y$, meromorphic along Y. It is a coherent sheaf of rings since $\mathcal{O}_X[Y]$ is locally isomorphic to $\mathcal{O}_X[h^{-1}]$ where h is a local equation of Y.

Definition 2.18. i) Let Y be a divisor on X and F a vector bundle on $X - Y$. An $\mathcal{O}_X[Y]$-coherent module $\widetilde{F}$ with an isomorphism $\widetilde{F}_{|X-Y} \simeq F$ is called a meromorphic extension of F.

ii) A connection on $\widetilde{F}$ is defined as a $\mathbb{C}-$linear map $\widetilde{F} \to \Omega^1_X \otimes_{\mathcal{O}_X[Y]} \widetilde{F}$ satisfying the usual (Leibnitz) condition.

iii) A connection ∇ on F is said to be meromorphic with respect to $\widetilde{F}$ if it extends to a connection on $\widetilde{F}$.

iv) A coherent module $\widetilde{F}$ is effective if there exists an $\mathcal{O}_X$-coherent module G such that $\widetilde{F} \simeq \mathcal{O}_X[Y] \otimes_{\mathcal{O}_X} G$.

We recover the definition on the disc D since Ω^1_D is free of rank 1 generated by dz ($\Omega^1_D \otimes_{\mathcal{O}_D} F \simeq F$).

Let $u : Z \to X$ be a morphism on a connected analytic manifold Z such that $u^{-1}(Y)$ is a divisor on Z. The inverse image $u^*\nabla$ of the meromorphic connection is defined on the vector bundle u^*F and its meromorphic extension $u^*\widetilde{F} =$

$\mathcal{O}_D \otimes_{u^{-1}\mathcal{O}_X} u^{-1}\widetilde{F}$ as follows. Near a point $a \in Z$ with local coordinates $(z_1, \dots, z_r)$ and $b = u(a)$ with local coordinates $(x_1, \dots, x_n)$, u is defined by $x_i = u_i(z_1, \dots, z_p)$. For a section $f \in \widetilde{F}_b$ such that $\nabla f = \sum dx_i \otimes f_i$ where $f_i \in \widetilde{F}_b$, we define $(u^*\nabla)(u^* f) = \sum du_i \otimes u^* f_i$. It can be checked that the definition of $u^*\nabla$ is independent of all choices and that the inverse image of the local system of flat sections of ∇ consists of the flat sections of $u^*\nabla$ (write u as a composition of a projection and an immersion).

Definition 2.19 (Regularity in dimension 1). A connection (F, ∇) meromorphic with respect to $\widetilde{F}$ on a disc D is said to be regular at 0 if the system defining the flat sections is regular, that is if we can choose a basis $e = (e_1, \dots, e_m)$ near 0 of $\widetilde{F}$ over K such that the matrix of the connection has a simple pole

$$z\nabla_z e_i = -\sum_j b_{ij}(z) e_j, \quad b_{ij}(z) \in \mathcal{O}_{D,0}.$$

Example. We deduce in the one-dimensional case, by 2.8.ii) and Remark 2.6.i) on the reduction of the logarithm of the monodromy via Jordan form, that a regular connection on a disc, meromorphic at 0, is isomorphic to a direct sum of meromorphic connections of the form $M_{\alpha,\ell}$, where $M_{\alpha,\ell}$ is the meromorphic connection with a basis $e_1, \dots, e_\ell$ such that

$$z\nabla e_i = \alpha\, e_i + e_{i+1} \text{ for } i < l \quad \text{and } z\nabla e_\ell = \alpha\, e_\ell.$$

$M_{\alpha,\ell}$ can be realized as the vector subspace of $\widetilde{\mathcal{O}}$ over K with basis

$$e_i, \quad i \in [1, l] : e_i = e_{l,l+1-i} \text{ where } e_{\ell,j} = z^\alpha \frac{(\log z)^{j-1}}{(j-1)!}.$$

Definition 2.20 (Regularity). i) A connection (F, ∇) meromorphic with respect to $\widetilde{F}$ is said to be regular if for any morphism $u : D \to X$ such that $u^{-1}(Y) = \{0\}$ its inverse $u^*(F, \nabla)$ is regular on D with respect to $u^*\widetilde{F}$.

ii) A connection (F, ∇) meromorphic with respect to $\widetilde{F}$ has logarithmic poles along a normal crossing divisor Y in X if there exists a bundle G on X such that $\widetilde{F} \simeq G \otimes_{\mathcal{O}_X} \mathcal{O}_X[Y]$ and ∇ restricts to $G \to \Omega^1_X(\operatorname{Log} Y) \otimes_{\mathcal{O}_X} G$.

2.4.1. Riemann-Hilbert correspondence.

Theorem 2.21. *The functor $(\widetilde{F}, \nabla) \to (\widetilde{F}, \nabla)_{|X-Y}$ on a complex analytic manifold X with a divisor Y induces an equivalence of the following categories:*

i) *the category of flat meromorphic and regular connections along Y,*
ii) *the category of analytic flat connections on $X - Y$,*
iii) *the category of finite rank complex local systems on $X - Y$.*

The proof in [3,2), prop. 5.1] is based on the canonical logarithmic extension across a NCD obtained by the repeated blowing up process until we transform Y into a NCD.

2.4.2. Algebraic results. Let X be a smooth complex algebraic variety and F an algebraic vector bundle on X with its Zariski topology. A connection $\nabla : F \rightarrow \Omega^1_X \otimes F$ is said to be algebraic if its image is in the algebraic tensor product with *algebraic differential forms* on X; then an analytic connection denoted also by $\nabla : F^{an} \rightarrow \Omega^{1,an}_X \otimes F^{an}$ extends ∇. The algebraic connection ∇ is integrable if and only if the associated analytic ∇ is, and in this case the analytic flat sections $F^{an,\nabla}$ form a local system $\mathcal{L}$ for the transcendental topology on $X - Y$.

Regularity. The main difference between algebraic and analytic varieties consists in the fact that an algebraic variety X can always be embedded into a proper algebraic variety $\overline{X}$ such that $Y = \overline{X} - X$ is a divisor in $\overline{X}$. We can moreover suppose $\overline{X}$ smooth if X is. Let $j : X \rightarrow \overline{X}$ be the inclusion, then we consider the algebraic extension j_*F , its analytic extension $(j_*F)^{an}$ and the extension $j_*^{an}(F^{an})$ (for example in the case of $\mathbb{C}^*$ and $F = \mathcal{O}_{\mathbb{C}^*}$ we get the meromorphic functions at 0 and in the second case the essential singularities of functions at 0). The sheaf $(j_*F)^{an}$ is a vector bundle on $\overline{X} - Y$ meromorphic on $\overline{X}$ along Y with a connection ∇.

Definition 2.22 (Regularity at ∞). The algebraic connection (F, ∇) is regular at ∞ if $((j_*F)^{an}, \nabla)$ is regular as an analytic meromorphic connection along Y.

This definition is independent from the choice of $\overline{X}$.

The following version of the Riemann-Hilbert correspondence is proved in [3,2)].

Theorem 2.23. *The functor $(F, \nabla) \rightarrow (F^{an}, \nabla)$ is an equivalence of the following categories:*

i) *the category of algebraic flat connections on X regular at ∞,*
ii) *the category of analytic flat connections on X^{an}.*

Hence with the category of finite rank complex local systems on X^{an}.

3. Singularities of local systems: Constructible Sheaves

In the previous sections local systems were attached to proper differentiable fibrations. We explore here the structure of higher direct images of constant sheaves by algebraic morphisms, subsequent to the special properties of algebraic morphisms (for example Bertini's theorem on the general fiber of a morphism of algebraic varieties, the counterpart of Sard's theorem in differential geometry).

More precisely, Thom-Whitney stratifications for proper morphisms are introduced to describe the relative behavior of singular fibers varying on a general base space of dimension higher than 1. As a consequence, we are lead to introduce constructible sheaves to describe the derived image by an algebraic morphism.

In parallel, local systems were attached to differential equations with holomorphic coefficients. The corresponding subject, which is not treated in this section, would be the structure differential modules on analytic varieties, the correspondence being via the DeRham complex attached to such modules.

The references to these subsections are the papers of Lê-Teissier [10],[2] and the book of Goresky-MacPherson [7].

3.1. Stratification theory

3.1.1. Bertini's result. Let $f\colon X \to V$ be a morphism on a smooth variety X of dimension m into a variety V of dimension n; *then there exists a Zariski open subset S_n in V such that the restriction of f to S_n is smooth.*

The fibers of f on S_n are smooth.

An algebraic morphism f can be completed, that is f can be factorized as an open immersion followed by a proper morphism. This important property explains the fact that we don't need f to be proper.

One can try again to describe the restriction $f_n : f^{-1}(V - S_n) \to (V - S_n)$ of f to $V - S_n$. However $f^{-1}(V - S_n)$ may be a singular variety now, nevertheless it follows from Thom's work:

For a proper morphism f, there exists an open set S_{n-1} in $V - S_n$ such that any point $y \in S_{n-1}$ has an open neighbourhood U_y of y in S_{n-1} satisfying the following property: the topological structure of $f^{-1}(U_y)$ decomposes into a product: $f^{-1}(U_y) \simeq U_y \times f^{-1}(y)$.

Hence a sequence of subspaces S_j, locally closed in V and adapted to f, can be constructed in this way (and will be called a stratification of V adapted to f when it satisfies additional topological properties). This is the subject of the next result.

Example. 1) Let $P\colon \mathbb{C}^n \to \mathbb{C}$ be a morphism defined by a polynomial. Bertini's result shows that there exists a subset $A \subset \mathbb{C}$ containing a finite number of critical values such that the restriction of P to $\mathbb{C}^n - P^{-1}(A)$ is smooth.

However P may not define a locally trivial fibration on $\mathbb{C} - A$ if it is not proper.

The morphism $P\colon \mathbb{C}^2 \to \mathbb{C}$ defined by the polynomial $X^2Y + X$ has no critical value. Let $X^* = \mathbb{C}^2 - P^{-1}(0)$; for any $a \neq 0$ the morphism $X_a^* \times \mathbb{C}^* \to X^* : ((x,y),b) \mapsto (\frac{xb}{a}, \frac{ya}{b})$ is an algebraic trivialization over $\mathbb{C}^*$. In particular the fiber over $\mathbb{C}^*$ is smooth and irreducible while the fiber at 0 is the union of $X = 0$ and $XY + 1 = 0$ hence reducible. Hence, the morphism is not topologically locally trivial on $\mathbb{C}$ [S.A. Broughton, On the topology of polynomial hypersurfaces, AMS Proceedings of Symposia in Pure Math. volume 40 (1983) Part 1].

2) In the previous example, the morphism is not proper but can be compactified by considering the family of polynomials $P_t^h = X^2Y + XZ^2 - tZ^3$ homogeneous in X, Y, Z defining a variety $V(P_t^h) \subset \mathbb{P}^2 \times \mathbb{C}$. The points $((x,y,z),t) = ((0,1,0),t)$ are singular on the fibers of the projection $V(P_t^h) \to \mathbb{C}$ to the coordinate t such that $(0,1,0)$ is contained and singular in all compactified fibers of P.

In general algebraic varieties have the property that they can be compactified; and the morphisms can be factorized by an embedding followed by a proper projection.

For example, a polynomial $P(x_1,\dots,x_n)$ of degree m in n variables defines a quasi-projective variety $X \subset \mathbb{C}^n \subset \mathbb{P}^n(\mathbb{C})$ such that its closure is defined by $P_h(x_0,x_1,\dots,x_n) = x_0^m P(\frac{x_1}{x_0},\dots,\frac{x_n}{x_0})$. To compactify the morphism we introduce the hypersurface $\tilde{X} \subset \mathbb{P}^n \times \mathbb{C}$ defined as $\{x,t) \in \mathbb{P}^n \times \mathbb{C} : P_h(x_0,x_1,\dots,x_n) - tx_0^m = 0\}$.

The open subset $\tilde{X} \cap (\mathbb{C}^n \times \mathbb{C})$, complement of the hyperplane $H_\infty : x_0 = 0$, is the graph of the morphism defined by $t = P(x_1,\dots,x_n)$. The study of the fibration on X is related to the fibration on $\tilde{X}$.

3) Let $p : \widetilde{\mathbb{C}}^2 \to \mathbb{C}^2$ denote the blowing -up of 0 in $\mathbb{C}^2$; the fiber over $\mathbb{C}^2 - \{0\}$ is reduced to a point and the restriction of p is an isomorphism over $\mathbb{C}^2 - \{0\}$; at the origin the fiber is a projective line.

Let D be a line in $\mathbb{C}^2$ and $\bar{D}$ its strict transform, then the image of $p' : (\widetilde{\mathbb{C}}^2 - \bar{D}) \to \mathbb{C}^2$ is $(\mathbb{C}^2 - D) \cup \{0\}$.

3.1.2. Stratification. Let V be a complex analytic space (resp. algebraic variety) endowed with a decreasing sequence of sub-analytic spaces (resp. algebraic subvarieties) $V = V_d \supseteq V_{d-1} \cdots \supseteq V_0 \supseteq V_{-1} = \emptyset$. The various subspaces $S_l := V_l \setminus V_{l-1}$ (called strata) form a partition of V by locally closed subspaces ($V = \cup_l S_l$). A partition of V is called a (Whitney) stratification when it is subject to the following properties.

(1) *Smoothness.* $S_l := V_l \setminus V_{l-1}$ is either empty or a locally closed analytic (resp. algebraic) subset of pure dimension l and *the connected components of S_l are a finite number of non-singular varieties.*

(2) *(Local normal topological triviality).* Given a point v in a strata S_l in V (S_l is smooth but V is not necessarily smooth along S_l), we consider a local embedding in a complex space $\mathbb{C}^n$ of a neighbourhood U_v of v in V and a transversal section through v, that is an analytic (resp. algebraic) smooth subspace $\mathcal{N}_{n-l}$ of $\mathbb{C}^n$ intersecting transversally in $\mathbb{C}^n$ each strata $S_j \cap U_y, j \geq l$ adjacent to S_l such that $S_l \cap \mathcal{N}_{n-l}$ is a zero-dimensional subspace containing v.

The intersection with a small ball of center v and small radius r: $N(v) = \mathcal{N}_{n-l} \cap V \cap B_r^{2n}$ is called a normal slice. The boundary $L(v) = \partial N(v)$ of $N(v)$ is called the link at v.

The normal slice and the link are canonically partitioned (or stratified) as transverse intersections of partitioned spaces. For r small enough, the homeomorphic type of the pair $(N(v), L(v))$ is a topological local invariant (independent of the embedding, the choice of $\mathcal{N}_{n-l}$ and the point v varying in the (connected) strata ([7] p 41). The partitioned normal slice is homeomorphic to a cone on the link with its canonical partition with respect to the vertex (identified with v) and the product partition on $L(v) \times]0,1]$. Moreover there exist standard (transcendental) neighbourhoods W_v of v in V satisfying:

$$W_v \simeq (N \times (W_v \cap S_l) \simeq (N \times (\mathbb{C}^l)$$

this being a homeomorphism respecting the partitions.

Historically, Whitney introduced two conditions on the stratifications that were enough to obtain local topological triviality as it has been shown by the following *Thom-Mather isotopy lemma.*

Lemma 3.1. *Every stratum Y of a Whitney stratified algebraic set has a neighbourhood which is the total space of a locally trivial topological fibre bundle with base space the stratum.*

Hence we may consider this lemma as an existence theorem for the above stratifications.

Example (Whitney umbrella). Consider the surface $W : x^2 - zy = 0$ in : $\mathbb{C}^3$. Let A_z be the z-axis defined by $x = 0$, $y = 0$. The singular subset is A_z, but the link at a point $z \neq 0$ has a different topological type than the link at 0.

3.1.3. Topological structure of algebraic morphisms.

Theorem 3.2 (Thom-Whitney). *Let $f : X \to V$ be a proper algebraic map of algebraic varieties. There exist finite algebraic Whitney stratifications $\mathcal{X}$ of X and $\mathcal{S}$ of V such that, given any connected component S of a stratum S_l of $\mathcal{S}$ on V:*

1) *$f^{-1}(S)$ is a union of connected components of strata of $\mathcal{X}$ each of which is mapped submersively to S; in particular, every fiber $f^{-1}(y)$ is stratified by its intersection with the strata of $\mathcal{X}$.*
2) *For all points $y \in S$ there exists a transcendental open neighbourhood U of y in S and a stratum-preserving homeomorphism $h : U \times f^{-1}(y) \simeq f^{-1}(U)$ such that $f \circ h$ is the projection to U.*

Definition 3.3 (Stratification of f). A pair of stratifications $\mathcal{X}$ and $\mathcal{S}$ as above is called a (Thom-Whitney) stratification of f.

The proof is based on Thom Isotopy Lemmas, adapted to the algebraic setting [10].

Definition 3.4 (Constructible sheaves). Let X be an analytic (resp. algebraic) variety and A be a ring. A sheaf $\mathcal{F}$ in the category of A_X-modules is constructible if there exists a stratification (resp. with algebraic strata) $\mathcal{X}$ of X such that its restriction to each stratum is a local system (for the transcendental topology).

A linear version of the result of Thom is:

Proposition 3.5. *The i-th higher direct image sheaf $R^i f_* \mathbb{Z}_X$ by an algebraic morphism is constructible.*

Precisely, the restriction of $R^i f_* \mathbb{Z}_X$ in degree i, the cohomology of the fibers, is locally constant on each stratum of a Thom-Whitney stratification.

3.1.4. Basic properties. i) *The category of constructible sheaves is abelian.* Let $f : \mathcal{F} \to \mathcal{G}$ be a morphism of constructible sheaves, then *ker* f and *coker* f are constructible sheaves.

The proof is based on the following result. If $\mathcal{F}$ (resp. $\mathcal{G}$) is constructible with respect to a stratification $\mathcal{S}_1$ (resp. $\mathcal{S}_2$), then there exists a finer stratification for which $\mathcal{F}$ and $\mathcal{G}$ are constructible.

ii) Let $0 \to \mathcal{F}_1 \to \mathcal{F} \to \mathcal{F}_2 \to 0$ be a short exact sequence of sheaves, *if* $\mathcal{F}_1$ *and* $\mathcal{F}_2$ *are constructible, then* $\mathcal{F}$ *is also constructible.*

iii) Let $f : X \to V$ be an algebraic morphism, then *the inverse image of a constructible sheaf on* V *is constructible on* X *and the direct image of a constructible sheaf on* X *is constructible on* V.

3.2. Cohomologically Constructible sheaves

In general not only the cohomology of complexes is interesting but also the complexes themselves; however since there is no preference between resolutions of complexes, there is a need to define a category which identifies all resolutions in some sense. Verdier did find the correct definition of the category by considering morphisms up to homotopy and inverting quasi-isomorphisms (morphisms inducing isomorphisms on cohomology), constructing in this way the derived category of abelian sheaves. Inside this category of abelian sheaves on a variety, we are interested in the subcategory of complexes whose cohomology are constructible sheaves. The correspondence between differential modules and their associated DeRham complexes is a good example where we need derived categories.

3.2.1. The derived category of abelian sheaves $D^+(M, \mathbb{Z})$. In the previous cohomological constructions we needed to choose an acyclic resolution to define the direct image functors. The existence of various acyclic resolutions gives the necessary flexibility for computation; however we need to justify such construction, that is to prove that the various resolutions give isomorphic objects.

At first sight the resolution itself may appear to be of no interest, while only its cohomology is of interest. However, DeRham resolution has already established its own interest in analysis, and the increasing use of cohomology in mathematics has shown that one might need to work with the complex itself. For example, considering a diagram of continuous maps $N \xrightarrow{f} V \xrightarrow{g} M$, the higher direct images $R^k(g \circ f)_*\mathcal{F}$ of a sheaf $\mathcal{F}$ on N by $g \circ f$ is linked to the higher direct images $R^i g_*(R^j f_*\mathcal{F})$ only via a spectral sequence, hence they can never be recovered completely unless we keep some knowledge of the complex $Rf_*\mathcal{F}$ itself instead of its cohomology.

How we should formulate this knowledge without losing flexibility in the choice of acyclic resolutions is the problem solved by Verdier [13]. The basic idea is to consider a category where the complex remains the object but to modify the morphisms of complexes to signal that our interest is in fact in its cohomology. In the first step of the construction the morphisms of complexes are considered up to homotopy. This step already transforms significantly the category. For example, the inverse of a morphism of complexes $f : K \to K'$ is a morphism $g : K' \to K$ such that $g \circ f$ (resp. $f \circ g$) is only homotopic to the identity. This is already an important modification of the category, since Deligne,for example, notices that any morphism of complexes is isomorphic to an injective morphism [2] in such a category.

In the second step a quasi-isomorphism, that is a morphism of complexes inducing an isomorphism on cohomology, is set to be an isomorphism by declaring invertible all quasi-isomorphisms.

Rigorous proofs and elaborate constructions are needed to develop this concept of Grothendieck-Verdier derived category (see Illusie's article which gives motivations behind these constructions in [8]; see also the Springer book of Iversen in the series Universitext (1986) and earlier work in a book by Cartan and Eilenberg published by Princeton University Press (1956)).

The category obtained from the category of complexes of abelian sheaves on a topological space M by the above two-step construction is called the derived category of abelian sheaves $D^+(M, \mathbb{Z})$.

3.2.2. Hypercohomology. That is *cohomology with coefficients in a complex of sheaves.* Let

$$\mathcal{L}^* = (\cdots \to \mathcal{L}_j \to \mathcal{L}_{j+1} \to \cdots)$$

be a complex of abelian sheaves on M. The hypercohomology $R^i\Gamma(M, \mathcal{L}^*)$, is the cohomology of the derived functor defined by global sections Γ on M with value in the complex $\mathcal{L}^*$. The definition is in two steps. In the first step one constructs a quasi-isomorphism of $\mathcal{L}^*$ with a complex of Γ-acyclic sheaves $\mathcal{A}^*$ (for example fine or flabby or injective sheaves) that is a morphism of complexes $g : \mathcal{L}^* \to \mathcal{A}^*$ inducing isomorphisms on cohomology. In the second step one takes the cohomology of global sections as a definition of hypercohomology

$$R^i\Gamma(M, \mathcal{L}^*) := H^i(\Gamma(M, \mathcal{A}^*)).$$

In general, for each left exact (resp. right exact) functor F from the category of complexes of sheaves to an abelian category, *derived functors denoted RF (resp. LF) are defined in a similar way.*

Remark 3.6. The classical cohomology theory with coefficients in the group $\mathbb{Z}$ can be viewed as a special case of the cohomology with coefficients in the constant sheaf $\mathbb{Z}$.

However the original topological construction of homology (dual to cohomology, see Spanier, Massey, and Dieudonné for historical remarks) remains basic for the intuition in topological problems that motivates the interest in cohomology and still gives powerful methods of computation as in the case of the first construction of intersection cohomology by MacPherson with Goresky.

3.2.3. The derived category of c-constructible sheaves $D_c^b(X,\mathbb{Q})$. *The main result states that the higher direct image sheaves of constructible sheaves by algebraic morphisms are constructible.*

This is the main reason to study cohomology with coefficients in constructible sheaves and for more flexibility in deriving functors. We need to work in the subcategory of the Grothendieck-Verdier derived category of sheaves of $\mathbb{Q}$-modules consisting of complexes of sheaves whose cohomology sheaves are constructible.

Definition 3.7. i) A complex of sheaves of $\mathbb{Q}_X-$modules is c-constructible (or cohomologically constructible) if its cohomology sheaves are constructible.

ii) Let $D(X)$ denote the derived category of $\mathbb{Q}_X$-modules. The full sub-category of $D(X)$ whose objects are c-constructible sheaves (resp. bounded, bounded at left, bounded at right) is denoted by $D_c(X,\mathbb{Q})$ (resp. $D_c^b(X,\mathbb{Q})$, $D_c^{+\infty}(X,\mathbb{Q})$, $D_c^{-\infty}(X,\mathbb{Q})$).

Proposition 3.8. *The higher direct image of a c-constructible complex by a proper algebraic morphism $f : X \to V$ is c-constructible, hence the derived functor Rf_* : $D_c^b(X,\mathbb{Q}) \to D_c^b(Y,\mathbb{Q})$ is well defined.*

The result follows from Thom-Whitney stratification theory for an algebraic morphism.

4. From Lefschetz theorems to the decomposition theorem

To illustrate the power of the various objects introduced in the last three sections, we give a statement of the decomposition theorem in [2]. However it is not possible to give a proof, since either we deduce the result from the proof in positive characteristic, or we need to develop Hodge theory and in both cases there is still a long way left to the interested reader.

The classical Hard Lefschetz theorem on a non-singular complex projective variety $X \hookrightarrow \mathbb{P}^m$ of dimension n with a class $\eta \in H^2(X,\mathbb{Q})$ of a hyperplane section $X_t = H_t \cap X$, states that the iterated cup-product

$$H^{n-i}(X,\mathbb{Q}) \xrightarrow{\eta^i} H^{n+i}(X,\mathbb{Q}) \tag{4.1}$$

is an isomorphism for $i \in [0,n]$.

The relative case. For a smooth projective morphism $f : X \to V$ where $\dim X = n$ and $\dim V = s$, the class of a hyperplane section that defines a section $\eta \in R^2 f_*\mathbb{Q}_X$ induces by cup-product a map

$$R^{n-s-i} f_*\mathbb{Q}_X \xrightarrow{\eta^i} R^{n-s+i} f_*\mathbb{Q}_X \tag{4.2}$$

which is an isomorphism.

If we study the cohomology of X via the fibration on V, we see that we need to consider the cohomology of V with coefficients in the geometric local systems $R^i f_*\mathbb{Q}_X$. Such local systems underlie various geometric invariants that have been abstracted into the theory of polarized variations of Hodge structures, and many of the important theorems on *geometric local systems* follow uniquely from this underlying Hodge theory.

For example, in the abelian category of finite local systems which is noetherian and artinian, *the sheaves $R^i f_* \mathbb{Q}_X$ are semisimple*, which means that they split into a finite direct sum of irreducible local subsystems (with no non-trivial local subsystems).

The study of such geometric local systems in Hodge theory has become a central object in the study of cohomology of algebraic varieties.

The isomorphisms η^i in the relative case are compatible with the Hodge structure on the cohomology of the fibres X_t of f. Deligne, using Hodge theory, deduced the degeneration of Leray's spectral sequence, from the relative version of Lefschetz's result. By definition, such a spectral sequence is associated to the canonical filtration τ on $Rf_*\mathbb{Q}_X$ defined by truncation, that is to say that if $\mathcal{I}_X$ is an injective resolution of $\mathbb{Q}_X$, then the filtered complex $(f_*\mathcal{I}_X, \tau)$ is defined up to a filtered quasi-isomorphism in the derived category of filtered complexes on the base V; then the associated spectral sequence corresponding to the filtration τ is defined up to isomorphism. Such filtration τ defines a filtration L on the cohomology of X,

$$H^j(V, R^i f_* \mathbb{Q}_X) \Longrightarrow Gr^i_L H^{i+j}(X, \mathbb{Q})$$

and the degeneration statement asserts that there exists natural isomorphisms:

$$Gr^i_L H^{i+j}(X, \mathbb{Q}) \simeq H^j(V, R^i f_* \mathbb{Q}_X).$$

In particular there exist non-canonical isomorphisms of rational vector spaces

$$H^n(X, \mathbb{Q}) \simeq \oplus_{i+j=n} H^j(V, R^i f_* \mathbb{Q}_X).$$

The degeneration translates in the derived category of sheaves on V, into a non-canonical decomposition of the derived direct image complex as a direct sum of its cohomology

$$Rf_* \mathbb{Q}_X \simeq \oplus R^i f_* \mathbb{Q}_X[-i]. \tag{4.3}$$

4.1. The decomposition theorem for projective morphisms

A natural question is to find how far we can relax the hypothesis and at the same time preserve the result. In fact the theorems as stated are false for a non-necessarily smooth projective morphism. In order to formulate similar results in the presence of singularities of spaces as well of the morphism, various objects and tools introduced in the last two decades have proved to be fundamental objects in the study of the topology as well of the geometry of varieties and have led to spectacular extensions of the results [2]. In particular the following are basic subjects in the theory:

- Thom-Whitney stratification,
- Perverse sheaves and Intermediate extension of a local system.
- General Intersection theory on cohomology.

Since the restriction to a strata f/S is a locally trivial topological bundle, the higher direct cohomology sheaf $(R^i f_* \mathbb{Q}_X)/S$ is locally constant on S. Then we say that $R^i f_* \mathbb{Q}_X$ is constructible on V and $Rf_* \mathbb{Q}_X$ is cohomologically constructible on V.

The category of perverse sheaves. A subcategory of the derived category $D^+(V, \mathbb{Q})$ of $\mathbb{Q}$–sheaves on a variety V, which is abelian, called the category of perverse sheaves, has been introduced in [2] following earlier work in ([7], 1). It appeared to be a fundamental object in the study of topological and geometrical properties of the morphism f.

A complex of sheaves K in $D^+(V, \mathbb{Q})$ is defined to be perverse if the following property is satisfied: there exists a stratification $\mathcal{S}$ of V such that for each strata S, $i_S : S \to V$, the restriction $H^n(i_S^* K) = 0$ for $n > -\dim S$ and $H^n(Ri_S^! K) = 0$ for $n < -\dim S$.

When K is constructible, these conditions show that the restriction of K to the open strata, is reduced to a local system $\mathcal{L}$ in degree $-\dim X$.

The perverse truncation. The main interest in the subcategory of perverse sheaves follows from the construction of a cohomological functor defined on the derived category $D^+(V, \mathbb{Q})$ with value in the category of perverse sheaves, constructed inductively with respect to a stratification of V. Namely, the notion of perverse truncation ${}^p\tau^i$ of a complex K is constructed in [2] and then the notion of i-th perverse cohomology ${}^p\mathcal{H}^i(K)$ is defined as the cone of the morphism ${}^p\tau^{i-1}(K) \to {}^p\tau^i(K)$ so as to fit in a triangle ${}^p\tau^{i-1}(K) \to {}^p\tau^i(K) \to {}^p\mathcal{H}^i(K)$. Perverse cohomology sheaves in various degrees fit together in a long exact sequence in the abelian category and in fact such an exact sequence is the best way to compute these objects, as in any cohomology theory.

The Intermediate extension. Research in the above field has been motivated first by the discovery by Goresky and MacPherson ([7], 1) of special objects called Intersection complexes. They are uniquely defined by local systems on locally closed subsets of V. Their construction uses the above Whitney stratification on a singular variety V and in an essential way the local topological triviality of the various strata S_l.

In the abelian category of perverse sheaves which is noetherian and artinian, the irreducible sheaves are Intersection complexes defined by irreducible local systems.The following is Deligne's construction of Intersection complexes.

Let $\mathcal{S} = \{S_l\}_{l \leq d}$ of V (dimension $S_l = l$) and let $j_0 : V - \overline{S}_0 \to V$, $j_l : V - \overline{S}_l \to V - \overline{S}_{l-1}$ for $0 < l < d$ denote the embedding. The Intermediate extension compatible with $\mathcal{S}$ of a local system $\mathcal{L}$ on the big open strata S_d is defined as:

$$j_{!*}\mathcal{L}[d] = \tau_{\leq -1} Rj_{0*} \cdots \tau_{\leq -l-1} Rj_{l*} \cdots \tau_{\leq -d} Rj_{d-1*}\mathcal{L}[d]$$

where for all sheaves $\mathcal{F}$ constructible with respect to $\mathcal{S}$, we have

$$(Rj_{l*}\mathcal{F})_v \simeq R\Gamma(L_{S_{l,v}}, \mathcal{F})$$

where $L_{S_{l,v}}$ *is the link of* S_l *at* v*; then* $\tau_{\leq -l-1}$ *truncates the cohomology up to degree* $\leq -l-1$.

The Decomposition theorem. Now we are in a position where we can state the version of the decomposition for a projective morphism and a geometric local system ([2], 6.2.4) so that we don't need to invoke explicitly the Hodge theoretical properties which underly such local system and can refer to ([2], 6.2.10).

Let $f : X \to V$ *be a projective morphism and* $\mathcal{L}$ *be a geometric local system on a smooth open subset* U *of* X, $j : U \to X$, $\dim X = n$. *The map defined by iterated cup-product with the class of a hyperplane section* $\eta \in R^2 f_* \mathbb{Q}_X$,

$$ {}^p\mathcal{H}^{-i}(Rf_*(j_{!*}\mathcal{L}[n])) \xrightarrow{\eta^i} {}^p\mathcal{H}^i(Rf_*(j_{!*}\mathcal{L}[n])), \tag{4.4}$$

is an isomorphism.

Then, the degeneration of the perverse Leray's spectral sequence defined by the perverse filtration on $Rf_*(j_{!*}\mathcal{L}[n])$ follows and leads to the decomposition:

There exists a non-canonical isomorphism in the derived category

$$ Rf_*(j_{!*}\mathcal{L}[n])) \xrightarrow{\cong} \oplus_i {}^p\mathcal{H}^i(Rf_*(j_{!*}\mathcal{L}[n]))[-i]. \tag{4.5}$$

Remark 4.1. The proof given in [2] is deduced from the theory in positive characteristic. The interested reader can find a proof via differential modules in the paper by Saito M.: Modules de Hodge polarisables. *Publ. RIMS, Kyoto univ.*, 24 (1988), 849–995.

An interesting paper, for constant coefficients, by De Cataldo M.A.A., Migliorini L.: The Hodge Theory of algebraic maps. *Ann. scient. Ec. Norm. Sup.* (2005), is easier to read. Finally, a paper by El Zein F.: Topology of Algebraic Morphisms, Contemporary Mathematics 474;
http://arxiv.org/abs/math/0702083, states the theorem in the form of a geometrical decomposition formula and treats the isolated singularity case. Such treatment applies also in the general case (to appear later).

References

[1] Cartan H.: Théorie élémentaire des fonctions analytiques d'une ou plusieurs variables complexes *Hermann, Paris*, 1961.

[2] Beilinson A.A., Bernstein J., Deligne P.: Analyse et Topologie sur les espaces singuliers Vol. I, *Astérisque* 100, Soc. Math. France, Paris, 1982.

[3] Borel A. et al.: Algebraic D-modules, *Perspectives in Math. 2*, Academic Press, Boston, 1987.
1) Haefliger A.: Local theory of meromorphic connections. 2) Malgrange B.: Regular Connections, after Deligne.

[4] Deligne P.: Equations différentielles à points singuliers réguliers. *Lecture Notes in Math.* 163, Springer, 1970.

[5] Godement R.: Topologie algébrique et théorie des faisceaux, *Hermann, Paris*, 1971.

[6] [SGA]. Séminaire de Géométrie algébrique. par Deligne P. SGA $4\frac{1}{2}$ *Lecture Notes in Math.* 569, Springer, Berlin, 1977.

[7] Goresky M., Macpherson R. : 1) Intersection homology II. *Inv.Math.* 72 (1983), 77–129. 2) Stratified Morse theory, *Ergebnisse der Mathematik*, 3. Folge, Band 14, Springer Verlag, Berlin–Heidelberg, 1988.

[8] Illusie L.: Catégories dérivées et dualité, travaux de J.L. Verdier, *Enseign. Maths.* 36 (1990), 369–391.

[9] Ince E.L.: Ordinary differential equations, 1926. Dover, New York, 1956.

[10] Lê D. T.-Teissier B.: Cycles évanescents, sections planes et conditions de Whitney II. *Proceedings of Symp. in pure math.* 40, 1983.

[11] MacPherson R. : Chern classes for singular varieties, *Annals of Math.* 100 (1974), 423–432.

[12] Pham F.: Singularités des systèmes différentiels de Gauss-Manin, *Progress in Math.*, Birkhäuser, Basel, 1979.

[13] Verdier J.L.: Des Catégories dérivées des catégories abéliennes, *Astérisque*, 239 Soc. Math. France, 1996.

[14] Warner, F.W.: Foundations of Differentiable Manifolds and Lie Groups. *Graduate Texts in Math.* 94, Springer, 1983.

Literature

In spite of the title, our article does not cover recent developments in the theory. In fact we were heading directly to the variation of Hodge structures underlying geometric local systems, but even this aim has not been attained. Instead an appendix with a basic example with Hodge theoretical oriented references has been added. For a wide view open to other applications in mathematics we suggest the following books covering developments in the theory with an extensive list of research articles in the field.

- Dimca A.: Sheaves in topology, *Universitext*, Springer Verlag, 2004.
- Kashiwara M., Schapira P.: Sheaves on manifolds, *Grundlehren math. Wissensch.* 292, Springer Verlag, 1990, 2003.
- Schürman J.: Topology of Singular Spaces and Constructible Sheaves, *Monografie Mathematyczne, New series*, Polish Academy of Sciences, Birkhäuser, Basel, 2003.

Appendix A. Example: Family of Elliptic Curves

To illustrate the various features of the subject we go back to its origin and compute the local system and Gauss-Manin connection defined by the family of elliptic curves.

A.1. Riemann Surfaces

The classical theory of Riemann surfaces originated in the study of the 'algebraic functions' $w = w(z)$ satisfying the equation with analytic coefficients

$$a_0(z)w^n + a_1(z)w^{n-1} + \cdots + a_n(z) = 0, \quad a_0(z) \not\equiv 0.$$

The main problem consists in the fact that there is no such continuous function $w(z)$ since there exist in general n values of the function for each z. However it is possible to define on each simply connected open subset U in $\mathbb{C}$, a holomorphic function $w(z)$ solution of the equation, called a branch of the function. The behavior of such branches is useful to understand the integrals of rational functions R of z and w,

$$F(z) = \int_{z_0}^{z} R(z, w(z))dz.$$

The beautiful idea of Riemann, to interpret such a branch as a section of a covering space of $\mathbb{C}$, is at the origin of the introduction of the notion of manifolds in modern geometry. This rich subject is treated here as an example, but it is also an historical subject in the field basic in mathematics.

A.1.1. Elliptic Curves. We consider the equation parameterized by a variable t,

$$w^2 = z(z-1)(z-t). \tag{A.1}$$

The point of view of manifolds consists in the introduction of the complex curve S_t defined by the equation $P_t(z,w) = w^2 - z(z-1)(z-t) = 0$ in the variables z, w in $\mathbb{C}^2$. The implicit function theorem shows that for $t \neq 0$ and $t \neq 1$, the curve is smooth. Moreover since the degree in z is 3, this curve which is not compact, is homeomorphic to a torus minus one point(the proof uses Weierstrass periodic meromorphic function $\mathbb{P}$ and its derivative).

This strongly suggests that we study the torus itself, that is compactify the curve. This operation, known as the projective completion, appeared to be so rich in mathematics that it became common to view the non-compact varieties as compact varieties minus a locus "at infinity" (or open varieties).

Such a Riemann surface S_t can be represented as a cover of the Riemann sphere $\mathbb{P}^1$ via the projection onto the variable z that extends to the projective curve onto the projective space. The historical technique to understand such a curve via the cuts from 0 to 1 and from t to ∞ in the $z-$plane may be confusing, unless it is coupled with this covering point of view (see [2] for a complete description of the theory).

A.1.2. The local system. The theory of the Weierstrass function $\mathcal{P}(z)$ constructed as the sum of the series with indices in $\mathbb{Z}^2$ and its relation to its derivative $\mathcal{P}'(z)$ define an isomorphism between the torus with its analytic structure as a quotient of $\mathbb{C}$ by a lattice isomorphic to $\mathbb{Z}^2$ and the elliptic curve (see the book of Cartan on analytic functions). It follows that the cohomology space $H^1(S_t, \mathbb{Z})$ of the smooth elliptic curve, for $t \neq 0$, $t \neq 1$ is generated by two cycles γ and δ with intersection matrix $J = \begin{pmatrix} 0 & 1 \\ -1 & 0 \end{pmatrix}$.

Now if the parameter t varies in $\mathbb{C} - \{0,1\} = P^1 - \{0,1,\infty\}$, we consider the family $S = \cup_t S_t$ and the projection $f : S \to \mathbb{C} - \{0,1\}$ defined by t. The family of cohomology spaces $H^1(S_t, \mathbb{Z})$ form a local system on $\mathbb{C} - \{0,1\}$. Since the fundamental group of $\mathbb{C} - \{0,1\}$ is a free group generated by two loops around the two punctures $\{0,1\}$, the local system is completely determined by the two associated monodromy linear operators around 0 and 1.

The monodromy is related to the presence of a critical point $(0,0)$ (resp. (1,0)) for $P(z,w,t)$ for $t = 0$ (resp. $t = 1$). It is possible to check that the Hessian matrix at these critical points is invertible. Such a property of points, known as non degenerate critical points, has been the center of continuous attention by mathematicians and an hypothesis of fundamental results known as Morse theory. We give the basic results in this theory in the next subsection.

We apply these results in our case to the variety $S = \cup_{t \in \mathbb{C}} S_t$. Locally S is defined in $\mathbb{C}^3$ by $P(z,w,t) = w^2 - z(z-1)(z-t) = 0$ and the projection to $\mathbb{C}$ is defined by the projection on the parameter t space.

We conclude from the Picard-Lefschetz transformation below that the monodromy linear operators around 0 and 1 are defined resp. by the matrices

$$A = \begin{pmatrix} 1 & 2 \\ 0 & 1 \end{pmatrix}, \quad B = \begin{pmatrix} 1 & 0 \\ -2 & 1 \end{pmatrix}$$

as computed for example in ([1], thm 1.1.20).

A.2. Non-degenerate critical points

Let $z = (z_1, \dots, z_n) \in U \subset \mathbb{C}^n$ and let $f\colon U \to \mathbb{C}$ be a Morse function defined for $n > 1$ by $f(z) = \Sigma_{i=1}^n z_i^2$. The Hessian of f at zero is the matrix $(\partial^2 f / \partial z_i \partial z_j(0))_{i,j}$. The point 0 is a non-degenerate critical point if the differential $df(0) = 0$ and the determinant of the Hessian matrix is non-zero. These properties together are independent of the coordinates.

The following result shows that the local study of the fibration defined by f near a non-degenerate critical point is isomorphic to the case defined by a Morse function.

Lemma A.1 (Morse). *Let $f : U \to \mathbb{C}$ be a holomorphic map on an open set U in an analytic manifold M with a non-degenerate critical point $a \in U$. Locally near a, f can be written in a suitable set of coordinates $z_1, \dots, z_n$ as $f(z) = f(a) + \sum_{i=1}^n z_j^2$.*

See for example ([3], II, 1.1).

The Morse lemma has the following real form.

Lemma A.2. *Let $f : X \to \mathbf{R}$ be a differentiable map with a non-degenerate critical point $a \in X$. Locally near a, f can be written in a suitable set of coordinates $x_1, \dots, x_n$ as $f(x) = f(a) - \sum_{j=1}^r x_j^2 + \sum_{j=r+1}^n x_j^2$.*

The number r is called the index of f at a and it is independent of the choice of coordinates.

A.2.1. Invariants attached to a non-degenerate critical point. Let $z = (z_1, \dots, z_n) \in U \subset \mathbb{C}^n$ and let $f\colon U \to \mathbb{C}$ be defined for $n > 1$ by $f(z) = \Sigma_{i=1}^n z_i^2$. There exists a unique critical point at the origin since $\partial f/\partial z_i = 2z_i$ vanish for all i only at such a point. The inverse image $Y = f^{-1}(0)$ of zero has an isolated singular point at the origin.

Moreover, this singular point is non-degenerate, since the determinant of the Hessian matrix $(\partial^2 f/\partial z_i \partial z_j)(0)_{i,j}$ of f at zero doesn't vanish. The map f induces a morphism $\overline{B_\varepsilon} \to D_{\varepsilon^2}$ from the closed $2n$-ball of radius ε to the disc of radius ε^2.

Let $z_j = x_j + iy_j$, then $f(z.) = \Sigma_{j=1}^n (x_j^2 - y_j^2) + 2i(\Sigma_{j=1}^n x_j y_j)$.

Lemma A.3. *For $|t|$ small enough, the fiber of f at t meets transversally the boundary S^{2n-1}.*

Corollary A.4. *Let $f : B_\varepsilon - f^{-1}(0) \to D^*_{\varepsilon^2}$ be the restriction of the fibration to the punctured disc. There exists δ small enough such that $f : f^{-1}D^*_\delta \cap (B_\varepsilon - f^{-1}(0)) \to D^*_\delta$ is a differentiable bundle.*

This corollary follows from the theorem on differentiable fibration. It is valid for any morphism f as proved by Milnor.

A.2.2. A *Milnor fiber* is defined for t small enough by

$$F_t = \{z. \in B_\varepsilon\colon f(z.) = t\}.$$

Since it is a differentiable invariant, it is denoted by F instead of F_t. For each t, the fiber F_t has a holomorphic structure depending on t in general.

A.2.3. Vanishing cycle. With the usual norm and scalar product on $\mathbf{R}^n$, let

$$Q = \{(u., v.) \in \mathbf{R}^n \times \mathbf{R}^n\colon \ \| u. \| = 1, \| v. \| \le 1, (u., v.) = 0\}.$$

Q is homeomorphic to the space of tangent vectors to the sphere S^{n-1} of length less than or equal to 1. Since the sphere S^{n-1} is a deformation retract of Q, $H_{n-1}(Q, \mathbb{Z}) \simeq H_{n-1}(S^{n-1}, \mathbb{Z}) \simeq \mathbb{Z}$.

We define an isomorphism from Q onto the Milnor fiber F_ρ for real numbers ρ as follows. Let $z_j = x_j + iy_j$, then F_ρ can be considered as a subset of $\mathbf{R}^n \times \mathbf{R}^n$ defined by three real conditions:

$$F_\rho = \{(x., y.) \in \mathbf{R}^n \times \mathbf{R}^n\colon \| x. \|^2 + \| y. \|^2 \le \varepsilon^2, \| x. \|^2 - \| y. \|^2 = \rho, \ (x., y.) = 0\}.$$

Let $\sigma\colon = (1/2(\varepsilon^2 - \rho))^{1/2}$, then the change of variables

$$u. = x./ \| x. \|, \ v. = y./\sigma, \quad x. = (\sigma^2 \| v. \|^2 + \rho)^{1/2} (u.), y. = \sigma(v.)$$

maps F_ρ to Q isomorphically. The image of S^{n-1} is the set $z.$ in F_ρ with real coordinates $z_j = \rho^{1/2} u_j$ for all j. When ρ varies in a small interval $[0, \rho_0]$, the family of embedded spheres S^{n-1}, each in a fiber at ρ, form a ball B of radius ρ_0

in the total space $\mathbb{C}^n$ such that the spheres collapse to 0 for $\rho = 0$. In this sense the sphere is a non-vanishing cycle on the fiber which vanishes in the total space.

Definition A.5. A generator of $H_{n-1}(F, \mathbb{Z})$ is called a vanishing cycle. It is defined by the homology class of an orientation on the embedded sphere S^{n-1} in F.

The homology of F vanishes in degrees different from 0 and $n-1$. For example, if $n = 2$, by a change of variables, we are reduced to the case of $f = z_1^2 - z_2^2$; then the Milnor fiber F_ρ is a hyperbola rotating along the circle defined by the vanishing cycle. The singular fiber is given by two complex lines intersecting in one point. This local situation apply in general near an isolated singularity as proved by Milnor (only the number of vanishing cycles vary).

A.2.4. Picard-Lefschetz transformation. Let $f : X \to \Delta$ be a projective morphism defined on an analytic manifold with values in a complex disc Δ. Let $a \in X$ be a unique non-degenerate critical point and let γ be a simple loop in the disc with origin a general point p around the critical value $c = f(a)$ (one positive turn). By the above local theory, a vanishing cycle v_a near a has been constructed on the fiber of a point y_c on the loop γ near c; this cycle v_a can be carried, by a trivialization of f restricted to the induced path γ_{p,y_c} from the loop γ, into a cycle δ (depending on γ), called also a vanishing cycle.

Proposition A.6. *The monodromy action is trivial on $H^i(X_p, \mathbb{Q})$ for $i \neq n-1$ and for $i = n-1$,*

$$\gamma x = x + \varepsilon_n(x, \delta)\delta,$$

where $\varepsilon_n = \pm 1$ is a sign depending on n and (x, δ) is the intersection number $Tr(x \smile \delta)$.

See ([3], II, Thm 3.16).

A.3. Picard-Fuchs Equations

Considering again the equation $A1$ in $(z, w) \in \mathbb{C}^2$, the meromorphic differential form $\omega = \frac{dz}{dw}$ on $\mathbb{C}^2$, induces a form on the curve S_t classically written as $\omega_t = \frac{dz}{\sqrt{z(z-1)(z-t)}}$. It can be checked that this form extends for $t \neq 0, 1$ to a holomorphic form on the compact curve, that is ω_t can be written locally as $\omega_t = f(u)du$ where u is a local coordinate and $f(u)$ is holomorphic. For each t, there exists a small ball B_t such that the family is topologically trivial over B_t so that we can choose constant homology vectors δ, γ generating the homology of S_t. The form ω_t is closed and its class $[\omega_t]$ decomposes on the dual basis of the cohomology δ^*, γ^* as

$$[\omega_t] = (\int_\delta \omega_t)\, \delta^* + (\int_\gamma \omega_t)\, \gamma^*.$$

The coefficients are called the periods $A(t) = \int_\delta \omega_t$ and $B(t) = \int_\gamma \omega_t$ of ω_t. By derivation under the integral sign, we can check that the coefficients are in fact holomorphic in t.

Since the basis (δ^*, γ^*) is locally constant, hence horizontal for the Gauss-Manin connection, the derivation of ω_t by the connection is $\omega_t' = A'(t)\delta^* + B'(t)\gamma^*$. It can be checked that $[\omega_t]$ and $[\omega_t']$ form a basis of the cohomology, so that by derivation we obtain $[\omega_t'']$ which decomposes on such a basis, hence we obtain an equation in cohomology classes

$$a(t)\,\omega'' + b(t)\,\omega' + c(t)\,\omega = 0.$$

If we consider a cycle ξ in S_t and the function $h(t) = \int_\xi \omega$, then we deduce a differential equation in the function $h(t)$; it is possible to determine the coefficients $a(t), b(t)$ and $c(t)$ and obtain the equation with regular singular points ([1], 1.1.17)

$$t(t-1)\,h''(t) + (2t-1)\,h'(t) + \frac{1}{4}h(t) = 0.$$

References

[1] Carlson J., Muller S., Peters C.: Period Mappings and Period Domains, *Cambridge Studies in Advanced Mathematics* 85, 2003.

[2] Springer G.: Introduction to Riemann surfaces, *Addison-Wesley*, 1957.

[3] Voisin C.: Hodge theory and Complex Algebraic Geometry, *Cambridge Studies in Advanced Mathematics* 76, 77, 2002.

Fouad El Zein
Institute of Mathematics of Jussieu, Geometry and Dynamics
Case 7012, 2 place Jussieu
75251 Paris Cedex 05
France

Associate member of CAMS, AUB, Beirut
e-mail: `elzein@math.jussieu.fr`

Jawad Snoussi
Instituto de Mathemáticas
Unidad Cuernavaca, UNAM
México
e-mail: `jsnoussi@matcuer.unam.mx`

Progress in Mathematics, Vol. 283, 155–176

Geometry and Combinatorics of Resonant Weights

Michael Falk

Abstract. Let $\mathcal{A}$ be an arrangement of n hyperplanes in $\mathbb{C}^\ell$. Let $\Bbbk$ be a field and $A = \oplus_{p=0}^{\ell} A^p$ the Orlik-Solomon algebra of $\mathcal{A}$ over $\Bbbk$. The pth *resonance variety* of $\mathcal{A}$ over $\Bbbk$ is the set $\mathcal{R}^p(\mathcal{A}, \Bbbk)$ of one-forms $a \in A^1$ annihilated by some $b \in A^p \setminus (a)$. This notion arises naturally from consideration of cohomology of the rank-one local systems generated by single-valued branches of $\mathcal{A}$-master functions Φ_a.

For the most part we focus on the case $p = 1$. We will describe the features of $R^1(\mathcal{A}, \Bbbk)$ for $\Bbbk = \mathbb{C}$ and also for fields of positive characteristic, and their connections with other phenomena. We derive simple necessary and sufficient conditions for an element a to lie in $\mathcal{R}^1(\mathcal{A}, \Bbbk)$, and consequently obtain a precise description of $\mathcal{R}^1(\mathcal{A}, \Bbbk)$ as a ruled variety. We sketch the description of components of $\mathcal{R}^1(\mathcal{A}, \mathbb{C})$ in terms of multinets, and the related Ceva-type pencils of plane curves. We present examples over fields of positive characteristic showing that the ruling may be quite nontrivial. In particular, $\mathcal{R}^1(\mathcal{A}, \Bbbk)$ need not be a union of linear varieties, in contrast to the characteristic-zero case. We also give an example for which components of $\mathcal{R}^1(\mathcal{A}, \Bbbk)$ do not intersect trivially.

We discuss the current state of the classification problem for Orlik-Solomon algebras, and the utility of resonance varieties in this context. Finally we sketch a relationship between one-forms $a \in \mathcal{R}^p(\mathcal{A}, \Bbbk)$ and the critical loci of the corresponding master functions Φ_a. For $p = 1$ we obtain a precise connection using the associated multinet and Ceva-type pencil.

Mathematics Subject Classification (2000). Primary 32S22; Secondary 52C35, 55N25, 14C21.

Keywords. Arrangement, Orlik-Solomon algebra, local system cohomology, resonance variety, master function, net, multinet, pencil.

1. Arrangements and projective arrangements

Resonance varieties are seen in several of the lectures in this volume ([7, 8]), mostly in connection with the fundamental group and characteristic varieties. But the idea had its genesis in the work of Schechtman and Varchenko and their co-workers on the Knizhnik-Zamolodchikov equations, hypergeometric integrals, and local systems [28, 12, 27], and the subsequent paper of Yuzvinsky [36]. In this lecture we will approach resonance varieties from that direction.

Let $\mathcal{A} = \{H_1, \ldots, H_n\}$ be an arrangement of linear hyperplanes in $\mathbb{C}^\ell$, and let $M = \mathbb{C}^\ell - \bigcup_{i=1}^n H_i$ denote the *complement* of $\mathcal{A}$. For each $1 \leq i \leq n$ let $\alpha_i \colon \mathbb{C}^\ell \to \mathbb{C}$ be a linear form with kernel H_i, uniquely determined up to a nonzero scalar factor. Let $\Omega^{\cdot}_M$ denote the sheaf of germs of holomorphic differential forms on M, with complex of global sections $\Omega^{\cdot}(M) = \Omega^{\cdot}_M(M)$. Let $\omega_i = d\log(\alpha_i) = \frac{d\alpha_i}{\alpha_i} \in \Omega^1(M)$.

Definition 1.1. The *Brieskorn algebra* $A = A(\mathcal{A})$ of $\mathcal{A}$ is the $\mathbb{C}$-subalgebra of $\Omega^{\cdot}(M)$ generated by $\{\omega_1, \ldots, \omega_n\}$.

Note that A consists of closed forms.

Theorem 1.2. *The inclusion $A \hookrightarrow \Omega^{\cdot}(M)$ induces an isomorphism on cohomology. In particular $A \cong H^*(M, \mathbb{C})$.*

This is a famous result of Brieskorn [1] – see also [23, 9, 2]. Orlik and Solomon found a presentation for A that depends only on combinatorial data associated with $\mathcal{A}$, as encoded in the underlying matroid (or intersection lattice – see below). For our current purposes we need only the fact that the ω_i are linearly independent, so A^1 is isomorphic to $\mathbb{C}^n$. For $(a_1, \ldots, a_n) \in \mathbb{C}^n$ we denote $\sum_{i=1}^n a_i\omega_i$ by ω_a.

Since the α_i are homogeneous, they define a set of projective hyperplanes $\bar{\mathcal{A}} = \{\bar{H}_1, \ldots, \bar{H}_n\}$ in $\mathbb{P}^{\ell-1} = \mathbb{P}(\mathbb{C}^\ell)$. The projectivized complement $\bar{M} = \mathbb{P}^{\ell-1} - \bigcup_{i=1}^n \bar{H}_i$ is the quotient of M by the (free) diagonal action of $\mathbb{C}^*$. A trivialization of the $\mathbb{C}^*$-bundle $p : M \to \bar{M}$ can be defined once one chooses a hyperplane H_{i_0} to be the "hyperplane at infinity." The map $(\alpha_{i_0}, p) \colon M \to \mathbb{C}^* \times \bar{M}$ is a fiber-preserving diffeomorphism. Thus $M \cong \mathbb{C}^* \times \bar{M}$.

Moreover, $\bar{M}$ is diffeomorphic to the complement of an arrangement $d\mathcal{A}$ of $(n-1)$ affine hyperplanes in $\mathbb{C}^{\ell-1}$, called the *decone* of $\mathcal{A}$. Again $d\mathcal{A}$ depends on a choice of hyperplane at infinity: the decone relative to H_{i_0} consists of hyperplanes $H_i \cap \alpha_{i_0}^{-1}(1), i \neq i_0$, in the ambient space $\alpha_{i_0}^{-1}(1) \cong \mathbb{C}^{\ell-1}$. Usually one changes variables so that $\alpha_{i_0} = x_\ell$ to simplify the calculations. Because of the arbitrary choice of H_{i_0} in this construction, we will avoid reference to affine arrangements when possible, restricting our discussion to central arrangements and their projective images.

The map $p : M \to \bar{M}$ induces a map of sheaves $\Omega^{\cdot}_{\bar{M}} \to \Omega^{\cdot}_M$, and an injection of global sections $\Omega^{\cdot}(\bar{M}) \to \Omega^{\cdot}(M)$. The image is the subalgebra of forms ω satisfying $\omega(\epsilon) = 0$, where ϵ is an Euler vector field $\sum_{i=1}^\ell x_i \frac{\partial}{\partial x_i}$. Since α_i is linear, $d\alpha_i(\epsilon) = \alpha_i$, and $\omega_i(\epsilon) = 1$. Then $\omega_a(\epsilon) = \sum_{i=1}^n a_i$.

Definition 1.3. The *projective Brieskorn algebra* of $\mathcal{A}$ is the subalgebra $\bar{A} = \bar{A}(\mathcal{A})$ of $A = A(\mathcal{A})$ generated by $\{\omega_a \in A^1 \mid \sum_{i=1}^n a_i = 0\}$.

The preceding discussion yields the following corollary of Brieskorn's Theorem.

Corollary 1.4. *The inclusion $\bar{A} \to \Omega^{\cdot}(\bar{M})$ induces an isomorphism on cohomology $\bar{A} \cong H^*(\bar{M}, \mathbb{C})$.*

2. Master functions and local systems

Let $a = (a_1, \ldots, a_n) \in \mathbb{C}^n \setminus \{0\}$, with $\sum_{i=1}^n a_i = 0$, and set $\Phi_a = \prod_{i=1}^n \alpha_i^{a_i}$. If a is not an integer vector, the function Φ_a is multi-valued, with zeros and poles contained in the union $\bigcup_{i=1}^n H_i$. Functions of this form are called $\mathcal{A}$-*master functions.*

Let $\mathcal{L}_a$ be the sheaf whose local sections are constant multiples of single-valued branches of Φ_a. Then $\mathcal{L}_a$ is locally constant, with stalk $\mathbb{C}$, i.e., $\mathcal{L}_a$ is a rank-one complex local system on M. If $a \in \mathbb{Z}^n$, then $\mathcal{L}_a$ is the constant sheaf.

The assumption on a implies Φ_a is homogeneous of degree zero, hence Φ_a descends to a (still multi-valued) function on the projective image $\bar{M}$. Then $\mathcal{L}_a$ pushes forward to a rank-one local system on $\bar{M}$, which we also denote by $\mathcal{L}_a$.

The rank-one local system $\mathcal{L}_a$ on M is determined by its monodromy, a homomorphism $\pi_1(M) \to \mathrm{GL}(\mathbb{C}) = \mathbb{C}^*$. Since the target is abelian, this map factors through $H_1(M, \mathbb{Z}) \cong \mathbb{Z}^n$, so the monodromy is determined by a homomorphism $\mathbb{Z}^n \to \mathbb{C}^*$, or equivalently, a point $t = (t_1, \ldots, t_n) \in (\mathbb{C}^*)^n$. Using the canonical orientation of $\mathbb{C}^\ell$, the monodromy around H_i is given by multiplication by $t_i = exp(2\pi\sqrt{-1}a_i)$; thus $t = \exp(2\pi\sqrt{-1}a)$. In particular, if $a - a' \in \mathbb{Z}^n$, then $\mathcal{L}_a = \mathcal{L}_{a'}$. Here we have identified $\mathbb{C}^n$ with the tangent space at 1 to the complex torus $(\mathbb{C}^*)^n$, and exp is the exponential map of Lie theory, mapping subspaces of $\mathbb{C}^n$ onto one-parameter subgroups of $(\mathbb{C}^*)^n$.

The assumption $\sum_{i=1}^n a_i = 0$ implies the monodromy vector t lies in the subtorus $\prod_{i=1}^n t_i = 1$, and thus the monodromy factors through the homomorphism $p_* \colon \pi_1(M) \to \pi_1(\bar{M})$. Let $\Lambda = \mathbb{Z}[H_1(\bar{M}, \mathbb{Z})] \cong \mathbb{Z}[t_1^{\pm 1}, \ldots, t_n^{\pm 1}]/(t_1 \cdots t_n - 1)$. The integral chain complex $C.$ of the universal abelian cover of $\bar{M}$ is a Λ-module, and the monodromy of $\mathcal{L}_a$ determines a Λ-module structure on $\mathbb{C}$. The cohomology $H^*(\bar{M}, \mathcal{L}_a)$ is the homology of the complex $\mathrm{Hom}_\Lambda(C, \mathbb{C})$. This can be computed from a presentation of $\pi_1(\bar{M})$ using the Fox calculus. Alternatively, the cohomology of $\bar{M}$ with coefficients in $\mathcal{L}_a$ can be computed via the twisted DeRham theorem, as in Theorem 2.1 below.

Note that $\frac{d\Phi_a}{\Phi_a} = d\log(\Phi_a) = \omega_a$. Define $\nabla_a : \Omega^p(\bar{M}) \to \Omega^{p+1}(\bar{M})$ by $\nabla_a(\tau) = d\tau + \omega_a \wedge \tau$.

Theorem 2.1. $H^*(\bar{M}, \mathcal{L}_a) \cong H^*(\Omega^{\cdot}(\bar{M}), \nabla_a)$.

Sketch of proof: The differential ∇_a defines a sheaf map $\Omega^{\cdot}_{\bar{M}} \to \Omega^{\cdot+1}_{\bar{M}}$, with $\nabla_a \circ \nabla_a = 0$, and the kernel of $\nabla_a : \Omega^0_{\bar{M}} \to \Omega^1_{\bar{M}}$ is precisely $\mathcal{L}_a$. Indeed,

$$\begin{aligned}\nabla_a(\Phi_a) &= d\Phi_a + \omega_a \wedge \Phi_a \\ &= \frac{\Phi_a d\Phi_a - d\Phi_a \Phi_a}{\Phi_a} \\ &= 0.\end{aligned}$$

Conversely, if f is a local section of $\Omega^0_{\bar{M}}$ and $\nabla_a(f) = df + f\omega_a = 0$, then $df = -\frac{fd\Phi_a}{\Phi_a}$. Then

$$\begin{aligned}d(\frac{f}{\Phi_a}) &= \frac{\Phi_a df - fd\Phi_a}{\Phi_a^2} \\ &= \frac{-d\Phi_a f - fd\Phi_a}{\Phi_a^2} \\ &= 0,\end{aligned}$$

so f is a constant multiple of Φ_a.

One shows that

$$\mathcal{L}_a \to \Omega^0_{\bar{M}} \xrightarrow{\nabla_a} \Omega^1_{\bar{M}} \xrightarrow{\nabla_a} \cdots \xrightarrow{\nabla_a} \Omega^\ell_{\bar{M}}$$

is an injective resolution of $\mathcal{L}_a$ by flasque sheaves, and the result follows. □

Logarithmic comparison. The preceding result is not so useful for computation. With some restrictions on a, one can replace $\Omega^{\cdot}(\bar{M})$ with the finite-dimensional subcomplex $\bar{A}$, the projective Brieskorn algebra, using Deligne's logarithmic comparison theorem. This was carried out in [12], later generalized in [27].

The restrictions on a are stated in terms of the intersection lattice of $\mathcal{A}$.

Definition 2.2. The *intersection lattice* of $\mathcal{A}$ is the poset

$$L(\mathcal{A}) = \{X = \bigcap_{i \in I} H_i \mid I \subseteq [n]\},$$

with $X \leq X'$ if and only if $X \supseteq X'$.

The poset $L = L(\mathcal{A})$ is a geometric lattice, with rank function $\mathrm{rk}(X) = \mathrm{codim}(X)$. The minimal element of L is $\mathbb{C}^\ell$, denoted $\hat{0}_L$. The maximal element of L is $\bigcap_{i \in [n]} H_i$, denoted by $\hat{1}_L$. Henceforth we assume that $\mathcal{A}$ is an *essential* arrangement, i.e., that $\hat{1}_L = \{0\}$, so L has rank ℓ.

We often identify $X \in L$ with the subset $\mathcal{A}_X = \{H_i \mid H_i \supseteq X\}$ of $\mathcal{A}$. The collection $\{\mathcal{A}_X \mid X \in L(\mathcal{A})\}$ is the set of flats of the matroid determined by dependent subsets of a set $\mathcal{A}^*$ of defining forms for the hyperplanes of $\mathcal{A}$, a vector configuration in the dual space $(\mathbb{C}^\ell)^*$. Often it is simpler to consider this matroid (e.g., the set of minimal dependent subsets (*circuits*) of $\mathcal{A}^*$), rather than the intersection lattice L, to carry the combinatorial structure of $\mathcal{A}$. In particular, the underlying matroid is uniquely determined by its rank function, or independent sets, or essential flats, or any of the wide variety of combinatorial data which serve to characterize matroids.

Definition 2.3. The arrangement $\mathcal{A}$ is *reducible* if there is a partition $\mathcal{A} = \mathcal{A}_1 \sqcup \mathcal{A}_2$ such that every circuit of $\mathcal{A}$ is contained in $\mathcal{A}_1$ or $\mathcal{A}_2$. An element $X \in L(\mathcal{A})$ is *irreducible* if the arrangement $\mathcal{A}_X$ is not reducible.

An arrangement is irreducible if and only if its underlying matroid is connected. The condition is a little easier to state for affine arrangements: an arrangement $\mathcal{A}$ is reducible if and only if some decone $d\mathcal{A}$ is a disjoint union $d\mathcal{A}_1 \sqcup \mathcal{A}_2$ of subarrangements whose defining forms involve disjoint sets of variables – see [23]. This is exploited in [11] to generate families of arrangements with non-isomorphic matroids but diffeomorphic complements – see also [14].

Irreducible lattice elements are also known as *dense edges*. We denote the set of irreducible elements of $L - \{\hat{1}_L\}$ by $\mathcal{X}$. An element $X \in L$ of rank 2 is irreducible if and only if $|\mathcal{A}_X| \geq 3$.

For $X \in L(\mathcal{A})$, set $\partial a_X = \sum_{H_i \in \mathcal{A}_X} a_i$. The following theorem was proved in a slightly weaker form in [12], and in its current form in [27]. Since d vanishes on $\bar{A}$, the restriction of ∇_a to $\bar{A}$ coincides with left-multiplication by ω_a. The resulting complex will be denoted by $(\bar{A}, \omega_a)$.

Theorem 2.4. *Suppose for each $X \in \mathcal{X}$, ∂a_X is not a positive integer. Then the inclusion of complexes $(\bar{A}, \omega_a) \to (\Omega^{\cdot}(\bar{M}), \nabla_a)$ induces an isomorphism on cohomology.*

Sketch of proof: One blows up $\mathbb{P}^{\ell-1}$ along the projective images of irreducible lattice elements, proceeding according to increasing rank. This realizes $\bar{M}$ as the complement $V \setminus D$ of a normal-crossing divisor D in a smooth, simply-connected variety V. The condition on a guarantees that the monodromy of the lifted local system $\widetilde{\mathcal{L}}_a$ around components of D do not have positive-integer residues: $\exp(2\pi\sqrt{-1}\partial a_X)$ is the monodromy of $\widetilde{\mathcal{L}}_a$ around the exceptional divisor arising from X.

Deligne's logarithmic comparison theorem [6] then implies that $H^*(V \setminus D, \widetilde{\mathcal{L}}_a)$ is carried by the complex of forms on V with logarithmic poles along D. This complex coincides with $\bar{A}$ under the identification of $V \setminus D$ with $\bar{M}$. □

See [25] for a detailed proof of Theorem 2.4. We are left with the problem of computing $H^*(\bar{A}, \omega_a)$. The generic case was treated by S. Yuzvinsky in [36]. Let $\chi(\bar{M}) = \sum_{p=0}^{\ell-1} \dim(\bar{A}^p)$ denote the Euler characteristic of $\bar{M}$. By the Hopf trace formula, we have $\sum_{p=0}^{\ell-1} \dim H^p(A, \omega_a) = \chi(\bar{M})$. Note $|\chi(\bar{M})| = (-1)^{\ell-1}\chi(\bar{M})$ – this quantity is called the beta invariant of (the underlying matroid of) $\mathcal{A}$, denoted $\beta(\mathcal{A})$. (See Section 3.)

Theorem 2.5. *Suppose there exists $H_0 \in \mathcal{A}$ such that, for every $X \in \mathcal{X}$ with $X \not\geq H_0$, $a_X \neq 0$. Then $H^p(\bar{A}, \omega_a) = 0$ for $p < \ell - 1$ and $H^{\ell-1}(\bar{A}, \omega_a) \cong \mathbb{C}^{\beta(\mathcal{A})}$.*

This theorem is proved by considering a complex of sheaves on a finite topological space determined by the intersection lattice of $\mathcal{A}$. See [9] in this volume for a complete proof.

Corollary 2.6. *Suppose, for each $X \in \mathcal{X}$, a_X is not a positive integer, and for some $H_0 \in \mathcal{A}$, $a_X = 0$ only if $X \geq H_0$. Then $H^p(\bar{M}, \mathcal{L}_a) = 0$ for $p < \ell - 1$ and $H^{\ell-1}(\bar{M}, \mathcal{L}_a) \cong \mathbb{C}^{\beta(\mathcal{A})}$.*

3. Euler characteristics and Hilbert series

For $X \in L$ write $M_X = M(\mathcal{A}_X)$. Since $\mathcal{A}_X \subseteq \mathcal{A}$ there is an inclusion $M \hookrightarrow M_X$. This inclusion splits, up to homotopy equivalence. Indeed, if B_X is a small ball centered at a generic point of X, then $B_X \cap M = B_X \cap M_X$ and the composite $B \cap M_X \hookrightarrow M \hookrightarrow M_X$ is a homotopy equivalence. These inclusions give rise to the *Brieskorn decomposition* of $H^*(M, \mathbb{C})$ [1].

Theorem 3.1. *For each p, $0 \leq p \leq \ell$,*

$$H^p(M, \mathbb{C})) \cong \bigoplus_{\substack{X \in L \\ \mathrm{rk}(X)=p}} H^p(M_X, \mathbb{C}).$$

The (one-variable) *Möbius function* of L is the function $\mu \colon L \to \mathbb{Z}$ defined recursively by $\mu(\hat{0}_L) = 1$ and, for $X > \hat{0}_L$, $\mu(X) = -\sum_{Y \in L, Y < X} \mu(Y)$. We earlier observed that $M \cong \mathbb{C}^* \times \bar{M}$. It follows that the Euler characteristic of $\chi(M)$ vanishes. Recall our tacit assumption that $\mathcal{A}$ is essential.

Corollary 3.2. $\dim A^\ell = (-1)^\ell \mu(\hat{1}_L)$.

Proof. We proceed by induction on ℓ. If $\ell = 1$ then $M = \mathbb{C}^*$ and the assertion holds. For $\ell > 1$, since $\chi(M) = 0$, we have $(-1)^\ell \dim A^\ell = -\sum_{p=1}^{\ell-1} (-1)^p \dim A^p$. By Theorem 3.1 $\dim A^p = \sum_{X \in L, \mathrm{rk}(X)=p} \dim A^p_X$, where $A_X = A(\mathcal{A}_X) \cong H^*(M_X)$, and $\dim A^p_X = (-1)^p \mu(X)$ by the inductive hypothesis. Then we have $(-1)^\ell \dim A^\ell = \sum_{X \in L, \mathrm{rk}(X) < \ell} \mu(X)$, which equals $\mu(\hat{1}_L)$ by definition of μ. □

Let $\pi(\mathcal{A}, t) = \sum_{p=1}^{\ell} \dim(A^p) t^p$ denote the Hilbert series of the cohomology algebra $H^*(M, \mathbb{C}) \cong A$. Combining the last result with Theorem 3.1 we obtain a combinatorial formula for $\pi(\mathcal{A}, t)$ [22].

Corollary 3.3. $\pi(\mathcal{A}, t) = \sum_{X \in L} \mu(X)(-t)^{\mathrm{rk}(X)}$.

The polynomial on the right-hand side is related to the characteristic polynomial $\chi(L, t)$ of the intersection lattice: $\chi(L, t) = t^{\mathrm{rk}(\mathcal{A})} \pi(\mathcal{A}, -t^{-1})$. $\chi(\mathcal{A}, t)$ is a specialization of the Tutte polynomial of the underlying matroid. If $\mathcal{A}$ is a complexified arrangement then $\chi(\mathcal{A}, -1) = \pi(\mathcal{A}, 1)$ is equal to the number of connected components in the real complement $M \cap \mathbb{R}^\ell$ [39]. If $\mathcal{A}$ is a graphic arrangement (consisting of hyperplanes $x_i = x_j$ for some collection of pairs (i, j)), then $\chi(\mathcal{A}, t)$ is equal to the chromatic polynomial of the associated graph.

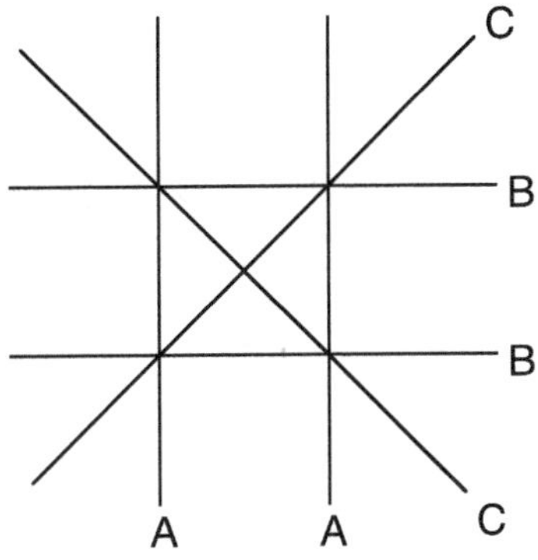

FIGURE 1. The projective D_3 arrangement

Since $M \cong \mathbb{C}^* \times \bar{M}$, $\pi(\mathcal{A},t) = (1+t)\pi(\bar{\mathcal{A}},t)$, where $\pi(\bar{A},t)$ is the Hilbert series of $\bar{A} \cong H^*(\bar{M})$. Then the Euler characteristic $\chi(\bar{M})$ is given by

$$\chi(M(\bar{\mathcal{A}})) = \pi(\bar{\mathcal{A}},-1) = \lim_{t\to -1} \frac{\pi(\mathcal{A},t)}{1+t}. \tag{1}$$

The beta invariant of the underlying matroid of $\mathcal{A}$ is defined by

$$\beta(\mathcal{A}) = (-1)^{\mathrm{rk}(\mathcal{A})-1}\frac{d}{dt}\pi(\mathcal{A},t)|_{t=-1} = \lim_{t\to -1} \frac{\pi(\mathcal{A},t)}{1+t}.$$

Then $(-1)^{\ell-1}\chi(\bar{M}) = \beta(\mathcal{A})$, as above. If $\mathcal{A}$ is a complexified real arrangement then $\beta(\mathcal{A})$ is the number of bounded regions (with any of the hyperplanes of $\mathcal{A}$ at infinity) in the real complement $\bar{M} \cap \mathbb{RP}^{\ell-1}$ [39].

Example 3.4. Let $\mathcal{A}$ be the reflection arrangement of type D_3, whose planes are defined by the linear factors of $Q(x,y,z) = (x^2-y^2)(x^2-z^2)(y^2-z^2)$. The projective arrangement $\bar{\mathcal{A}}$ is pictured in Figure 1. The labels indicate a neighborly partition of $\mathcal{A}$, corresponding to the given factorization of Q, for use in the next section.

One sees that L contains six elements X of rank 1 for which $\mu(X) = -1$, and seven rank-two elements, four with $\mu(X) = 2$ and three with $\mu(X) = 1$. Thus $\pi(\mathcal{A},t) = 1 + 6t + 11t^2 + 6t^3 = (1+t)(1+2t)(1+3t)$. The coefficients $1, 2, 3$ are the exponents of the Weyl group of type D_3. A similar factorization occurs for any reflection arrangement, and, more generally, for any free arrangement.

The Euler characteristic $\chi(\bar{M})$ is then $(1-2)(1-3) = 2$, which is indeed the number of bounded regions relative to any plane $H \in \mathcal{A}$. See Figure 2.

Now consider the arrangement $\mathcal{A}'$ defined by the polynomial $Q(x,y,z) = (x+z)(x-z)(y+z)(y-z)yz$, illustrated in Figure 3.

$\mathcal{A}'$ is a reducible arrangement. The lattice L' of $\mathcal{A}'$ has six rank-one elements ($\mu(X) = -1$) and eight rank-two elements, one with $\mu(X) = 2$, one with $\mu(X) = 3$, and $\mu(X) = 1$ for the six others. Then $\pi(\mathcal{A}',t) = 1 + 6t + 11t^2 + t^3 = \pi(\mathcal{A},t)$.

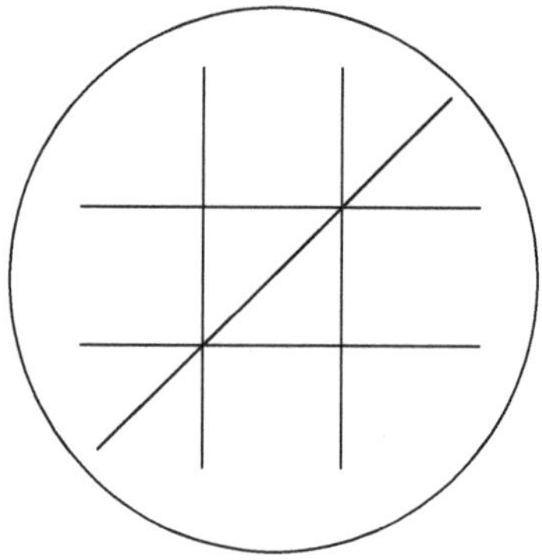

FIGURE 2. The projective D_3 arrangement with one of its lines at infinity

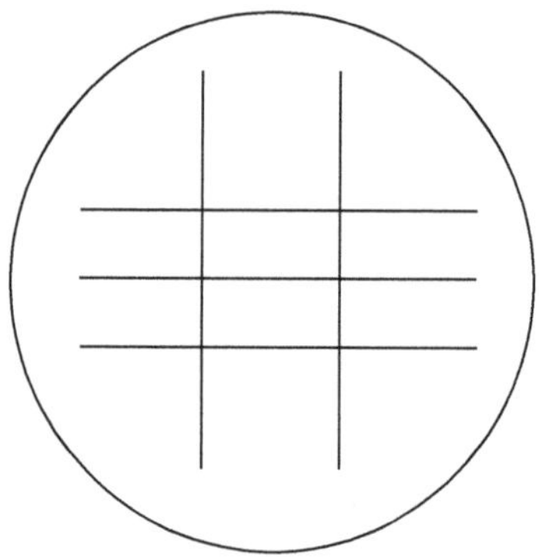

FIGURE 3. A product arrangement

Thus the betti numbers of $M(\mathcal{A}')$ (or $\bar{M}(\mathcal{A})$) agree with those of $M(\mathcal{A})$ (resp., $\bar{M}(\mathcal{A})$) above. The rings $H^*(M(\mathcal{A}))$ and $H^*(M(\mathcal{A}'))$ are easily distinguished using resonance varieties, as we will demonstrate in the next section.

4. Resonance varieties

After Theorem 2.5 the notion of resonance variety appears quite naturally: we ask what happens when the hypothesis of that theorem doesn't hold. The conclusion may or may not hold in that case – the values of a for which it fails make up the resonance varieties of $\mathcal{A}$.

Definition 4.1. Let $1 \leq p < \ell - 1$. The pth *resonance variety* of $\mathcal{A}$ is

$$\mathcal{R}^p(\mathcal{A}) = \{a \in \mathbb{C}^n \setminus \{0\} \mid H^p(\bar{A}, \omega_a) \neq 0\}.$$

These provide powerful invariants of the fundamental group (when $p = 1$) and the cohomology ring of $\bar{M}$. They were first introduced in [13] as a tool in the classification of Orlik-Solomon algebras (see below).

It is not hard to see that $\mathcal{R}^p(\mathcal{A})$ is an algebraic subset of $\mathbb{C}^n$, and is invariant under the diagonal action of $\mathbb{C}^*$. There is a filtration of $\mathcal{R}^p(\mathcal{A})$ by dimension:

$$\mathcal{R}^p_k(\mathcal{A}) = \{a \in \mathcal{R}^p(\mathcal{A}) \mid \dim H^p(\bar{A}, \omega_a) \geq k\}.$$

The Orlik-Solomon algebra. We are interested in extending Definition 4.1 to more general coefficients. To do this most efficiently we introduce the Orlik-Solomon algebra of $\mathcal{A}$. Let $\mathcal{A} = \{H_1, \dots, H_n\}$ with $H_i = \ker(\alpha_i : \mathbb{C}^\ell \to \mathbb{C})$. Let $\Bbbk$ be a field, and let E be the exterior algebra over $\Bbbk$ generated by elements $e_1, \dots, e_n$ of degree 1. Define the derivation $\partial\colon E \to E$ by $\partial(e_i) = 1$ and extending by linearity and the Leibniz rule. Let

$$I = \left(\partial(e_{i_1} \cdots e_{i_p}) \mid \{\alpha_{i_1}, \dots, \alpha_{i_p}\} \text{ is linearly dependent}\right),$$

a homogeneous ideal of E.

Definition 4.2. The *Orlik-Solomon algebra* (or *OS*-algebra) of $\mathcal{A}$ over $\Bbbk$ is the graded algebra $A = A_\Bbbk(\mathcal{A}) = E/I$.

Clearly $E^1 \to A^1$ is an isomorphism – we will abuse notation by using $e_1, \dots, e_n$ to denote the natural basis of A^1 corresponding to the hyperplanes of $\mathcal{A}$. The context should make the meaning clear in any particular situtation.

The connection with the Brieskorn algebra $A(\mathcal{A})$, which consists of differential forms, is well-known [22, 23].

Theorem 4.3. *The Brieskorn algebra $A(\mathcal{A})$ is isomorphic to the OS algebra $A_\mathbb{C}(\mathcal{A})$.*

In lieu of a proof of Theorem 4.3, we show that the *OS* relations hold in $A(\mathcal{A})$ in degree 2. This is all we will require for what is to come. Refer to [23] for the complete argument.

Lemma 4.4. *Suppose $\{\alpha_1, \alpha_2, \alpha_3\}$ is linearly dependent. Then $\omega_2\omega_3 - \omega_1\omega_3 + \omega_1\omega_2 = 0$.*

Proof. We can change variables so that $\alpha_1 = x, \alpha_2 = y$, and $\alpha_3 = x + y$. Then

$$\begin{aligned}\omega_2\omega_3 - \omega_1\omega_3 + \omega_1\omega_2 &= \frac{dy \wedge d(x+y)}{y(x+y)} - \frac{dx \wedge d(x+y)}{x(x+y)} + \frac{dx \wedge dy}{xy} \\ &= \left(\frac{-x - y + (x+y)}{yx(x+y)}\right) dx \wedge dy \\ &= 0.\end{aligned}$$

□

In fact, $A_\mathbb{Z}(\mathcal{A})$ is isomorphic to the $\mathbb{Z}$-subalgebra of $\Omega^\cdot(M)$ generated by $\{\omega_1, \dots, \omega_n\}$, which in turn is isomorphic to $H^*(M, \mathbb{Z})$. For any ring $\Bbbk$, $A_\Bbbk(\mathcal{A})$ is a free $\Bbbk$-module, and its Hilbert series is independent of $\Bbbk$. See [23] for proofs of these results. Henceforth, we let $\Bbbk$ be an arbitrary field, and denote $A_\Bbbk(\mathcal{A})$ by A. (The distinction between A and the (isomorphic) Brieskorn algebra of forms, also denoted by A, should be clear from the context.) We have a canonical isomorphism $A^1 \cong \Bbbk^n$.

Note that $\partial(\sum_{i=1}^n a_i e_i) = \sum_{i=1}^n a_i$. We define the *projective OS algebra* $\bar{A}$ to be the kernel of $\partial\colon A \to A$. It is an exercise to show that $\bar{A}$ is generated by $\bar{A}^1 = \{a \in A^1 \mid \partial a = 0\}$. Let $a \in \bar{A}^1$. Left multiplication by a makes $\bar{A}$ into a cochain complex, with cohomology denoted by $H^*(\bar{A}, a)$. For $1 \le p < \ell - 1$ the pth resonance variety of $\mathcal{A}$ over $\Bbbk$ is defined by

$$\mathcal{R}^p(\mathcal{A}, \Bbbk) = \{a \in \bar{A}^1 \setminus \{0\} \mid H^p(\bar{A}, a) \neq 0\}.$$

It is easy to see that the resonance varieties, up to linear change of coordinates, are invariants of the graded algebra A.

Degree-one resonance varieties. For the remainder of this section we focus on $\mathcal{R}^1(\mathcal{A}, \Bbbk)$. This is much easier than the general case, but nevertheless leads to some beautiful mathematics, with applications to the fundamental group of M. Let $\mathcal{R}$ denote $\mathcal{R}^1(\mathcal{A}, \Bbbk)$. As in the case $\Bbbk = \mathbb{C}$, $\mathcal{R}$ is an affine variety in $A^1 \cong \Bbbk^n$ and is invariant under the diagonal action of $\Bbbk^*$. We will mostly consider its projective image in $\mathbb{P}(\bar{A}^1)$, which we denote by $\bar{\mathcal{R}}$. The image of $a \in \bar{A}^1$ in $\mathbb{P}(\bar{A}^1)$ will be denoted by $\bar{a}$.

Proposition 4.5. *$\bar{a} \in \bar{\mathcal{R}}$ if and only if there exists $\bar{b} \in \bar{\mathcal{R}} \setminus \{\bar{a}\}$ such that the projective line $\bar{a} * \bar{b}$ spanned by $\bar{a}$ and $\bar{b}$ lies in $\bar{\mathcal{R}}$.*

Proof. Since $a \neq 0$, b represents a nontrivial class in $H^1(\bar{A}, a)$ if and only if b is not a scalar multiple of a and $ab = 0$ in $\bar{A}^2$. Then, for any scalars λ and $\eta \neq 0$, $\lambda a + \eta b \neq 0$ and a represents a nontrivial class in $H^1(A, \lambda a + \eta b)$. Then $\lambda a + \eta b \in \mathcal{R}$. Thus $\bar{a} * \bar{b} \subseteq \bar{\mathcal{R}}$. The reverse implication is trivial. □

Corollary 4.6. *The projective resonance variety $\bar{\mathcal{R}}$ is ruled by lines.*

Example 4.7. Suppose $\mathcal{A}$ is an arrangement of rank 2 with $|\mathcal{A}| \ge 3$. Then any three hyperplanes H_1, H_2, H_3 of $\mathcal{A}$ are dependent and we have

$$0 = \partial(e_1 e_2 e_3) = (e_1 - e_2)(e_2 - e_3),$$

whence $e_1 - e_2 \in \mathcal{R}$. It follows that $\mathcal{R} = \bar{A}^1 = \{a \in A^1 \mid \partial a = 0\}$. The natural ruling of $\bar{\mathcal{R}}$ consists of all the lines in $\bar{\mathcal{R}}$.

Using the previous example and some properties of the *OS* algebra, we will be able to describe the ruling of $\bar{\mathcal{R}}$, for an arrangement of arbitrary rank, in a very simple way. The key is the following lemma, a combinatorial version of the Brieskorn decomposition of Theorem 3.1, proven in [22] as part of the argument for Theorem 4.3 – see also [23, 9]. Recall, for $X \in L$, $\mathcal{A}_X = \{H \in \mathcal{A} \mid H \supseteq X\}$. Let A_X denote the *OS* algebra $A(\mathcal{A}_X)$.

Lemma 4.8. *The natural map $A_X \to A$ is a split injection, and*

$$A^p \cong \bigoplus_{\substack{X \in L \\ \text{rk}(X) = p}} A^p_X.$$

Proof. The ordering of the hyperplanes of $\mathcal{A}$ determines a basis of A, the *no broken circuit (nbc)* basis, consisting of the monomials $e_{i_1} \cdots e_{i_p}$ for which $1 \leq i_1 < \cdots < i_p \leq n$ and, for each k, $i_k = \min\{i \mid H_i \supseteq H_{i_{k+1}} \cap \cdots \cap H_{i_n}\}$. Each such *nbc* monomial of degree p lies in A_X for a unique $X \in L$ of rank p, namely $X = H_{i_1} \cap \cdots \cap H_{i_p}$. The construction is local in the sense that *nbc* monomials in A^p_X map to (distinct) *nbc* monomials in A^p. Thus, $A_X \to A$ is injective, and splits. Moreover, the *nbc* monomials in A^p are partitioned by the A_X for $\mathrm{rk}(X) = p$. This yields the direct sum decomposition. □

For $X \in L$ and $a \in A^p$, let a_X denote the image of a under the projection $A^p \to A^p_X$. In particular, if $a = \sum_{i=1}^n a_i e_i \in A^1$, then $a_X = \sum_{H_i \in \mathcal{A}_X} a_i e_i$.

Lemma 4.9. *If* $a, b \in A^1$, *then* $(ab)_X = a_X b_X$.

Proof. An arbitrary product $e_i e_j$ is expressed in terms of degree-two nbc monomials by repeatedly using relations of the form $e_p e_q - e_p e_r + e_q e_r = 0$, with $\{\alpha_p, \alpha_q, \alpha_r\}$ dependent. Then any nbc monomials $e_p e_q$ involved in the resulting expression of $e_i e_j$ satisfies $H_p \cap H_q = H_i \cap H_j$. It follows that the component of $(\sum_{i=1}^n a_i e_i)(\sum_{i=1}^n b_i e_i) = \sum_{i,j=1}^n a_i b_j e_i e_j$ in A^2_X is $\sum_{H_i \cap H_j = X} a_i b_j e_i e_j$, which is $(a_X)(b_X)$. □

The preceding two results immediately lead to a characterization of resonant weights.

Theorem 4.10. *Let* $a, b \in \bar{A}^1$. *Then* $ab = 0$ *if and only if, for every* $X \in L$ *with* $\mathrm{rk}(X) = 2$, *either*

(i) a_X *and* b_X *are proportional, or*
(ii) $|\mathcal{A}_X| \geq 3$ *and* $\partial a_X = \partial b_X = 0$.

If $a = \sum_{i=1}^n a_i e_i$, then $\partial a_X = \sum_{H_i \supseteq X} a_i$. If $a \in \mathcal{R}$ we call $(a_1, \ldots, a_n)$ a *resonant weight*. If a and b satisfy $ab = 0$ and are not proportional, we say (a, b) is a *resonant pair* (or resonant pair of weights under the identification of A^1 with $\mathbb{k}^n$). Condition (ii) of the preceding corollary means that resonant pairs of weights lie in the null space of the incidence matrix of a set $\mathcal{X} \subseteq \{X \in L(\mathcal{A}) \mid \mathrm{rk}(X) = 2 \text{ and } |\mathcal{A}_X| \geq 3\}$ with $\mathcal{A}$.

Neighborly partitions. The dichotomy in Theorem 4.10 leads us to neighborly partitions. For technical reasons it is more convenient to describe these in terms of more general symmetric relations on $\mathcal{A}$, which we think of as undirected simple graphs with vertex set $\mathcal{A}$. A *clique* in a graph is a set of vertices which are pairwise adjacent.

Definition 4.11. A simple graph Γ with vertex set $\mathcal{A}$ is $\mathcal{A}$-neighborly if, for every $X \in L$ with $\mathrm{rk}(X) = 2$, for every $H \in \mathcal{A}_X$, $\mathcal{A}_X \setminus \{H\}$ is a clique of Γ only if $\mathcal{A}_X$ is a clique of Γ.

With the vertex set understood to be $\mathcal{A}$, we write Γ as a set of edges, and abbreviate $\{H_i, H_j\} \in \Gamma$ to $\{i,j\} \in \Gamma$. If $a, b \in A^1$, let

$$\Gamma_{(a,b)} = \left\{ \{i,j\} \mid \det \begin{bmatrix} a_i & a_j \\ b_i & b_j \end{bmatrix} = 0 \right\}. \tag{2}$$

Theorem 4.12. *Suppose (a,b) is a resonant pair. Then the graph $\Gamma_{(a,b)}$ is $\mathcal{A}$-neighborly.*

Proof. Suppose $X \in L$ with $\mathrm{rk}(X) = 2$. Let $H_i \in \mathcal{A}_X$ with $\mathcal{A}_X \setminus \{H_i\}$ a clique of $\Gamma = \Gamma_{(a,b)}$, and suppose $\mathcal{A}_X$ itself is not a clique of Γ. Then a_X and b_X are not proportional, so $\partial a_X = \partial b_X = 0$ by Theorem 4.10. Then, for any $H_k \in \mathcal{A}_X$ with $k \neq i$, $\det \begin{bmatrix} \partial a_X & a_k \\ \partial b_X & b_k \end{bmatrix} = 0$. Since $\{j,k\} \in \Gamma$ for every $H_j \in \mathcal{A}_X$ with $j \neq i$, it follows that $\det \begin{bmatrix} a_i & a_k \\ b_i & b_k \end{bmatrix} = 0$, which implies $\{i,k\} \in \Gamma$. Then $\mathcal{A}_X$ is a clique of Γ, a contradiction. □

Suppose a and b are not proportional, i.e., $\Gamma_{(a,b)}$ is not a clique. Then $a_i = b_i = 0$ if and only if H_i is a cone vertex of $\Gamma_{(a,b)}$, adjacent to every other vertex. If we delete these vertices from $\Gamma_{(a,b)}$, the maximal cliques of the induced subgraph are its connected components. A *neighborly partition* is a neighborly graph whose components are cliques. In this case we refer to the maximal cliques as the blocks of Γ. Every $\mathcal{A}$-neighborly graph induces a neighborly partition of a subarrangement of $\mathcal{A}$, and, conversely, any neighborly partition of a subarrangement of $\mathcal{A}$ extends to an $\mathcal{A}$-neighborly graph by coning on the other vertices of $\mathcal{A}$. The set of non-cone vertices of Γ is called the *support* of Γ.

Let Γ be an $\mathcal{A}$-neighborly graph. Note, if $|\mathcal{A}_X| = 2$ then $\mathcal{A}_X$ is a clique in Γ. Let

$$\mathcal{X}_\Gamma = \{X \in L \mid \mathrm{rk}(X) = 2 \text{ and } \mathcal{A}_X \text{ is not a clique of } \Gamma\}$$

and

$$K_\Gamma = \{a \in \bar{A}^1 \mid \partial a_X = 0 \text{ for all } X \in \mathcal{X}_\Gamma\}.$$

A Γ-*resonant pair* is a pair (a,b) of non-proportional elements of K_Γ with a_S proportional to b_S for all cliques S of Γ.

Let $\mathcal{R}_\Gamma \subseteq \bar{A}^1$ consist of those a in K_Γ for which there exists $b \in K_\Gamma$ such that (a,b) is a Γ-resonant pair. In particular, $\mathcal{R}_\Gamma \neq \emptyset$ only if $\dim(K_\Gamma) \geq 2$. Let $\mathsf{N}(\mathcal{A})$ denote the set of $\mathcal{A}$-neighborly graphs.

Theorem 4.13.

$$\mathcal{R} = \bigcup_{\Gamma \in \mathsf{N}(\mathcal{A})} \mathcal{R}_\Gamma.$$

This result affords a simple description of the ruling of $\bar{\mathcal{R}}$, resulting in a simple geometric construction of $\bar{\mathcal{R}}$. If S is a maximal clique of $\Gamma \in \mathsf{N}(\mathcal{A})$, set

$$D_S = \{a \in K_\Gamma \mid a_i = 0 \text{ for all } i \in S\}.$$

The *arrangement of directrices* associated with Γ is the collection $\mathcal{D}_\Gamma$ of subspaces D_S, where S is a maximal clique of Γ. Note, if i is a cone vertex of Γ, then every $D \in \mathcal{D}_\Gamma$ is contained in the coordinate hyperplane $a_i = 0$.

If $\mathcal{D}$ is an arrangement of subspaces in $\bar{A}^1$, let $\mathsf{L}(\mathcal{D})$ denote the set of lines in $\mathbb{P}(\bar{A}^1)$ which meet the projective image $\bar{D}$ of D for each $D \in \mathcal{D}$. Then $\mathsf{L}(\mathcal{D})$ can be identified with a subset of the Grassmannian $\mathcal{G}_2(\Bbbk^n)$. In fact $\mathsf{L}(D)$ is a linear line complex, and is the intersection of the special Schubert varieties. (See [15, 16] for further discussion.) Using condition (i) of Theorem 4.10 we obtain the following.

Theorem 4.14. *$\bar{\mathcal{R}}_\Gamma$ is the union of the lines in $\mathsf{L}(D_\Gamma)$.*

Proof. If (a, b) is a Γ-resonant pair, and S is a maximal clique of Γ, then a_S is proportional to b_S by definition. Here a_S denotes $(a_i \mid i \in S)$. Then the line $\bar{a} * \bar{b}$ meets $\bar{D}_S$ in $\mathbb{P}(\bar{A}^1)$. Then $\bar{a} * \bar{b} \in \mathsf{L}(D_\Gamma)$. Conversely, if $\bar{a} * \bar{b} \in \mathsf{L}(D_\Gamma)$ then a and b lie in K_Γ and a_S and b_S are proportional for every maximal clique S of Γ. Then (a, b) is a Γ-resonant pair. □

Each element $X \in \mathcal{X}_2$ gives rise to a linear component of $\mathcal{R}$, of dimension $|\mathcal{A}_X| - 1$, as in Example 4.7. These are called *local components* of $\mathcal{R}$.

Example 4.15. Let $\mathcal{A}$ be the D_3 reflection arrangement of Example 3.4. The labels in Figure 1 determine a neighborly partition Γ with support equal to $\mathcal{A}$. The kernel of the 4×6 incidence matrix of $\mathcal{X}_\Gamma = \mathcal{X}_2$ with $\mathcal{A}$ has dimension 2 (over any field $\Bbbk$), with basis $\{(1,1,0,0,-1,-1), (0,0,1,1,-1,-1)$. Here the hyperplanes are labelled in order of their appearance in the defining polynomial Q. Then $\mathcal{R}_\Gamma = K_\Gamma$ is a (non-local, linear) component of $\mathcal{R}(\mathcal{A})$. (The ruling of $\bar{\mathcal{R}}_\Gamma$ consists of a single line.) There are four local components, each of dimension two. Thus $\bar{\mathcal{R}}(\mathcal{A})$ consists of five (pairwise disjoint) lines in $\mathbb{P}(\bar{A}^1) \cong \mathbb{P}^5$.

Multinets and Ceva-type pencils. When $\Bbbk$ has characteristic zero we obtain a more precise description of $\mathcal{R}$. The crucial observation, made in [20], is this: if J is the incidence matrix whose kernel (intersected with $\bar{A}^1$) is K_Γ, and E is the matrix whose every entry is 1, then $Q_\Gamma = J^T J - E$ satisfies $\ker(Q_\Gamma) \cap \ker(E) = K_\Gamma$ and, moreover, Q_Γ is a generalized Cartan matrix. Then one applies the Vinberg-Kac classification of Cartan matrices [18]. One finds that, if $\mathcal{R}_\Gamma$ is nonempty, then K_Γ is spanned by vectors of the form $u_i - u_j$ where the u_i are non-negative integer vectors with disjoint supports. These supports form a neighborly partition of $\mathcal{A}$, and condition (ii) of Theorem 4.10 is satisfied by every pair of vectors in K_Γ. This leads to the following result of [20].

Theorem 4.16. *Suppose $\Bbbk$ has characteristic zero. Then:*

(i) *if $\mathcal{R}_\Gamma$ is a component of $\mathcal{R}$ with support $\mathcal{A}$, each $X \in \mathcal{X}_\Gamma$ meets each block of Γ;*
(ii) *if $\mathcal{R}_\Gamma$ is a component of $\mathcal{R}$ with support $\mathcal{A}$, then $\mathcal{R}_\Gamma = K_\Gamma$;*
(iii) *any two non-proportional elements of K_Γ form a resonant pair;*
(iv) *distinct components of $\mathcal{R}$ intersect trivially.*

As a consequence of (iii) above, the dimension of $H^1(A,a)$ coincides with the dimension of the component of $\bar{\mathcal{R}}(\mathcal{A})$ in which a lies.

Arrangements which support non-local components of $\mathcal{R}$ are rare; when $\Bbbk = \mathbb{C}$ they arise from an additional combinatorial structure on $\mathcal{A}$ called a *multinet*, and comprise the components of completely decomposable curves in Ceva-type pencils on $\mathbb{CP}^2$ - see [17, 9]. We give a brief summary.

Without loss we can assume $\mathcal{A}$ has rank three. Loosely speaking, a *multinet* consists of a map $m\colon \mathcal{A} \to \mathbb{N}$ which assigns a multiplicity to each hyperplane of $\mathcal{A}$, and a partition Γ of $\mathcal{A}$ into $k \geq 3$ blocks, $\mathcal{A} = \mathcal{A}_1 \sqcup \cdots \sqcup \mathcal{A}_k$, satisfying

(i) each block of Γ contains d hyperplanes, counting multiplicity;
(ii) each rank-two lattice element $X \in \mathcal{X}_\Gamma$ is contained in the same number n_X of hyperplanes from each block $\mathcal{A}_i$.

There are further technical conditions: the multiplicities should be mutually relatively prime, and the rank-two lattice elements outside of $\mathcal{X}_\Gamma$, which are intersections of hyperplanes solely from one block, should be "connected" within that block. See [17] for a precise formulation.

A homogeneous polynomial of degree d (up to nonzero constant multiple) is called a *degree d (plane) curve.* A degree d curve is *completely reducible* if its irreducible factors are linear. By multiplying together the defining linear forms of the hyperplanes in each block of a multinet, with repeated factors according to the specified multiplicities of hyperplanes, we obtain $k \geq 3$ completely reducible degree d curves $C_1, \ldots, C_k$. The following theorem is proved in [17] – see also [9] in this volume.

Theorem 4.17. *Suppose $\Bbbk = \mathbb{C}$. Then $\mathcal{A}$ supports a component of dimension $k-2$ of $\bar{\mathcal{R}}(\mathcal{A},\Bbbk)$ if and only if $\mathcal{A}$ supports a multinet with k blocks. In this case the curves $C_1, \ldots, C_k$ are collinear in the projective space of degree d curves.*

The partition Γ associated with a multinet is neighborly, and the multiplicities are determined by the non-negative integer spanning vectors u_i in K_Γ mentioned above.

A line in the $\binom{d+2}{2}$-dimensional projective space of degree d plane curves which includes three or more completely reducible curves is called a *pencil of Ceva type.* (Again there are some technical conditions: the pencil can have no embedded components, and should be "connected" in a certain sense.) Thus Theorem 4.17 shows that arrangements that support resonance components give rise to Ceva-type pencils. Conversely, in [17] it is also shown that the sets of linear components of completely reducible polynomials in a Ceva-type pencil, assigned their multiplicities as factors of C_i, form a multinet.

Example 4.18. For the D_3 reflection arrangement of Example 3.4, the partition of Figure 1 forms a net on $\mathcal{A}$ (all multiplicities =1). The corresponding Ceva pencil has completely reducible elements x^2-y^2, y^2-z^2, and z^2-x^2.

The Ceva arrangements, defined for any $d \geq 2$ by the linear factors of the polynomial $(x^d - y^d)(y^d - z^d)(z^d - x^d)$, consists of the components of three completely reducible curves in a pencil of Ceva type, hence the name. The previous example is the case $d = 2$.

Example 4.19. Let $\mathcal{A}$ be the reflection arrangement of type B_3, obtained by adding the coordinate planes $x = 0, y = 0$, and $z = 0$ to the D_3 arrangement of the preceding example. Assign to each of the coordinate planes multiplicity two, and all other planes in $\mathcal{A}$ multiplicity one, and consider the partition of $\mathcal{A}$ defined by the factors of $x(y^2 - z^2), y(z^2 - x^2)$, and $z(x^2 - y^2)$ respectively. It is easy to check that conditions (i) and (ii) in the definition of multinet given above are satisfied. The completely reducible curves in the resulting Ceva-type pencil are $x^2(y^2 - z^2), y^2(z^2 - x^2)$, and $z^2(x^2 - y^2)$. See [17] for a more detailed discussion.

Theorem 4.17 motivates a search for multinets, or equivalently, Ceva-type pencils on $\mathbb{CP}^2$. The Hurwitz formula imposes some restrictions [20, 17]. Further restrictions arise from the theory of foliations [26]. For instance, the number k of completely reducible fibers in a Ceva-type pencil is at most 4 [38]. Thus projective resonance components have dimension at most 3. Using Theorem 4.16(iii), this yields the following corollary.

Corollary 4.20. *For any arrangement $\mathcal{A}$ and any $a \in \bar{A}^1_{\mathbb{C}}(\mathcal{A})$, $\dim H^1(A_{\mathbb{C}}(\mathcal{A}), a) \leq 3$.*

There are no known examples of nets or multinets with $k = 5$ and, as mentioned above, only one known example, a net, with $k = 4$ – see Example 4.24 below. In recent work S. Yuzvinsky has shown that any multinet with $k = 4$ blocks must be a net [38]. A k-net determines $k - 2$ mutually orthogonal latin squares [20]. This connection is exploited in [19, 10] – related work appears in [37, 30, 32].

Classification of OS algebras. Resonance varieties can be used to distinguish the ring structure of graded algebras, as in the next example.

Example 4.21. Let $\mathcal{A}'$ be the product arrangement of Example 3.4. We saw in that example that the *OS* algebra of $\mathcal{A}'$ is isomorphic as a graded vector space to that of the D_3 reflection arrangement $\mathcal{A}$. But $\mathcal{A}'$ doesn't support any neighborly partition, and $\bar{\mathcal{R}}(\mathcal{A}')$ consists of two linear (local) components, of dimensions 1 and 2, in $\mathbb{P}^5$. We saw in Example 4.15 that the $\bar{\mathcal{R}}(\mathcal{A})$ consists of five lines in $\mathbb{P}(\bar{A}^1) \cong \mathbb{P}^4$. Thus $A(\mathcal{A}')$ is not isomorphic to $A(\mathcal{A})$ as a graded ring.

Indeed we have the following theorem from [13].

Theorem 4.22. *$\mathcal{R}(\mathcal{A})$ is a complete invariant of the quadratic closure of A as a graded algebra.*

In [13] there are examples of pairs of arrangements whose resonance varieties have the same number of components (all local) of each dimension, but yet are not equivalent under a linear change of coordinates. Then the *OS* algebras are not isomorphic. The resonance varieties are distinguished by their associated polymatroids.

It is natural to ask whether $\mathcal{R}(\mathcal{A})$ determines the lattice L of $\mathcal{A}$, at least through rank two. But the construction of [11] mentioned earlier yields arrangements with different numbers of rank-two lattice elements and diffeomorphic complements, hence isomorphic *OS* algebras. Aside from examples arising from this construction (e.g., by truncation) there are no known examples of rank-three arrangements with non-isomorphic lattices but linearly equivalent degree-one resonance varieties.

Linearity and disjointness. When $\Bbbk$ has characteristic zero, the components of $\mathcal{R}$ are linear and intersect trivially, by Theorem 4.16(ii) and (iv). The next examples show that these conclusions do not hold over fields of positive characteristic [16].

Example 4.23. Consider the complexified arrangement $\mathcal{A}$ defined by

$$Q(x,y,z) = (x+y)(x-y)(x+z)(x-z)(y+z)(y-z)z.$$

The underlying matroid of $\mathcal{A}$ is the non-Fano plane, with set of nontrivial irreducible rank-two flats

$$\mathcal{X}_2 = \{136,\ 145,\ 235,\ 246,\ 347,\ 567\}.$$

Here we have identified hyperplanes with their labels, and labelled according to the order of factors in Q.

If $\Bbbk$ is a field of characteristic zero, no neighborly partition of $\mathcal{A}$ satisfies $\dim K_\Gamma > 1$, which is necessary for $\mathcal{A}$ to support a resonant pair. Suppose $\Bbbk$ is a field of characteristic 2 and $\Gamma = 127|3|4|5|6$. Then Γ is a neighborly partition, and $\mathcal{X}_\Gamma = \mathcal{X}_2$. The 6×7 incidence matrix of $\mathcal{X}_2$ with $\mathcal{A}$ has rank 4 over $\Bbbk$. Hence K_Γ has dimension 3, and $\bar{K}_\Gamma$ is a plane in $\mathbb{P}(\bar{A}^1) \cong \mathbb{P}^5$. The directrices corresponding to the singleton blocks of Γ are lines in this plane, while $\bar{D}_{127}$ is the point $\bar{a}$, $a = 0011110$. Thus L_Γ consists of the lines in $\bar{K}_\Gamma$ passing through the point $\bar{a}$. Then $\bar{\mathcal{R}}_\Gamma = \bar{K}_\Gamma$, a plane in $\mathbb{P}(\bar{A}^1) \cong \mathbb{P}^5$.

The previous example gives rise to an example for which components of $\mathcal{R}$ intersect nontrivially [15, 16]. The example is the "deleted B_3 arrangement," obtained by deleting the plane $x - y = 0$ from the B_3 arrangement of Example 4.19. There are two non-Fano subarrangements (Example 4.23) which meet in a copy of the D_3 arrangement of Example 3.4. If $\Bbbk$ has characteristic 2 these give rise to two planes in $\bar{\mathcal{R}}$ which intersect in a line. This example also shows that components of $\mathcal{R}$ need not be isotropic (as in Theorem 4.16(iii)) over fields of positive characteristic. There is an apparent relationship, perhaps coincidental, between this structure in $\mathcal{R}(\mathcal{A}, \bar{\mathbb{Z}}_2)$ and the positive-dimensional translated component of the first characteristic variety [31, 9].

The following example shows that $\mathcal{R}$ may have nonlinear components if $\Bbbk$ has positive characteristic.

Example 4.24. Let $\mathcal{A}$ be the Hessian arrangement in $\mathbb{C}^3$, corresponding to the set of twelve lines passing through the nine inflection points of a nonsingular cubic in $\mathbb{P}^2(\mathbb{C})$ [23, Example 6.30]. The underlying matroid is the deletion of one point from $PG(2,3)$, the projective plane over $\mathbb{Z}_3$. We label the hyperplanes of $\mathcal{A}$ with the numerals $1,\ldots,9$ and the letters α,β,γ. We choose a labelling so that the set of nontrivial rank-two flats is

$$\mathcal{X}_2 = \{148\gamma, 159\alpha, 167\beta, 247\alpha, 258\beta, 269\gamma, 349\beta, 357\gamma, 368\alpha\}.$$

Let $\Bbbk$ be an algebraically-closed field of characteristic 3, and let $\Gamma = 123|456|789|\alpha\beta\gamma$. Then we have $\mathcal{X}_\Gamma = \mathcal{X}_2$. The 9×12 incidence matrix of $\mathcal{X}_2$ with $\mathcal{A}$ has rank 6 over $\Bbbk$, so $\dim K_\Gamma = 6$. For each block S of Γ, the corresponding directrix has dimension 3. Then the projectivized arrangement of directrices consists of four planes in $\bar{K}_\Gamma \cong \mathbb{P}^5$.

The placement of these four planes is special: each meets the other three in three collinear points. It follows that the six points of intersection are coplanar, and are the six points of intersection of four lines in general position in that plane. Also the four planes of $\bar{\mathcal{D}}_\Gamma$ lie in a hyperplane in $\mathbb{P}^5$. $\bar{\mathcal{R}}_\Gamma$ is the union of the lines meeting each of the four planes of $\bar{\mathcal{D}}_\Gamma$.

A *Macaulay2* computation tells us that $\bar{\mathcal{R}}_\Gamma$ is an irreducible cubic hypersurface in $\mathbb{P}^4$. This can be confirmed by hand by analyzing the line complex $\mathsf{L}(\mathcal{D}_\Gamma)$ using Schubert calculus in $\mathcal{G}_2(\Bbbk^5)$ – see [15].

A more detailed computation (thanks to H. Schenck) shows that the plane containing the six intersection points of the directrices is the singular locus of $\mathcal{R}$. This plane consists of the points that have "depth" 2 in L_Γ. For those points $\dim H^1(A,a) = 2$ – see [15, 16].

For $\Bbbk = \mathbb{C}$ the dimension of K_Γ is 3, the directrices are four lines in $\mathbb{P}^2$, and $\bar{\mathcal{R}}_\Gamma$ is a 2-dimensional linear component in $\bar{\mathcal{R}}$. The partition Γ defines a net on $\mathcal{A}$. In fact the Hessian is the only known example of a multinet with more than three blocks.

5. Critical loci of master functions

We now return to the case $\Bbbk = \mathbb{C}$. There is a relationship between the resonance properties of $a = \sum_{i=1}^n a_i e_i \in A^1$ and the topology of the critical locus $\mathrm{crit}(\Phi_a)$ of the $\mathcal{A}$-master function Φ_a on the projective complement $\bar{M}$. (Recall $\Phi_a = \prod_{i=1}^n \alpha_i^{a_i}$ where $H_i = \ker(\alpha_i\colon \mathbb{C}^\ell \to \mathbb{C}$.) To compute critical points on $\bar{M}$ one must choose a coordinate chart, which we take to be the complement of $\bar{H}_n$. We assume that H_n is given by $x_\ell = 0$, and hence consider the critical points of $\Phi_a(x_1,\ldots,x_{\ell-1},1)$ on $\underset{\sim}{M} \subset \mathbb{C}^{\ell-1}$. Since different branches of Φ_a differ by a constant, the critical locus is well-defined, even when Φ_a is multi-valued. Moreover, $\mathrm{crit}(\Phi_a)$ coincides with the set of zeros of the one-form $\frac{d\Phi_a}{\Phi_a} = d\log(\Phi_a) = \sum_{i=1}^n a_i d\log(\alpha_i) = \omega_a$.

The generic case. This line of research began with a theorem of A. Varchenko, and his conjectured generalization later proven by P. Orlik and H. Terao.

Theorem 5.1. *If $a \in \bar{A}^1$ is generic, then Φ_a has $\beta(\mathcal{A})$ nondegenerate isolated critical points.*

This was established for complexified arrangements and conjectured for arbitrary arrangements in [33]. If $\mathcal{A}$ is a complexified arrangement and a_i is a positive real for $1 \leq i < n$, then Φ_a has one nondegenerate critical point in each bounded (relative to H_n) component of the real complement $\bar{M} \cap \mathbb{RP}^{n-1}$. (For instance the function $\Phi_{(1,1,-2)}(x_1, x_2) = \frac{(x_1+1)(x_1-1)}{x_2^2}$ has one critical point $x_1 = 0$ in the single bounded chamber $-1 < x_1 < 1$ in the affine chart $x_2 = 1$.) The general statement for complexified arrangements follows from this. Varchenko's conjecture was proven in [24]. The conclusion holds for a outside of some proper algebraic subset of $\bar{A}^1$, but the proofs do not yield precise conditions on a that guarantee $\mathrm{crit}(\Phi_a)$ is as stated. Note that, for generic a, the number of components of $\mathrm{crit}(\Phi_a)$ coincides with the dimension of $H^{\ell-1}(\bar{A}, a)$. Also the codimension of $\mathrm{crit}(\Phi_a)$ in $\bar{M}$ is $\ell - 1$, while $H^p(\bar{A}, a) = 0$ for $0 \leq p < \ell - 1$.

The motivation for the preceding result comes from the Bethe Ansatz to generate solutions of the $\mathfrak{sl}_k$ Knizhnik-Zamolodchikov equation (for conformal blocks on the sphere) in the quasi-classical limit. One assumes $\mathcal{A} = \mathcal{A}_{k,\ell-1}$ is an affine *discriminantal arrangement*, given by the defining polynomial

$$Q_{k,\ell-1}(x_1, \dots x_\ell) = x_\ell \prod_{i=1}^{\ell-1} \prod_{j=1}^{k} (x_i - z_j x_\ell) \prod_{1 \leq i < j \leq \ell-1} (x_i - x_j),$$

where $z_1, \dots, z_k$ are distinct complex numbers. The weight vector a corresponds to a set of representations of $\mathfrak{sl}_k(\mathbb{C})$. Each isolated, nondegenerate critical point of Φ_a determines an eigenfunction of the associated KZ operator. (The Bethe Ansatz conjecture is that a complete set of solutions is obtained this way; it is true for $k = 2$ but false in general [21].) See [34] for an extensive treatment. (There $\mathcal{A}_{k,\ell-1}$ is defined to be an affine arrangement, obtained from our version by setting $x_\ell = 1$.)

Exotic critical sets. In the same context it is shown in [29] that, for a discriminantal arrangement $\mathcal{A}$ and certain integer weights a, $\mathrm{crit}(\Phi_a)$ is one dimensional. In [5] these weights are shown to lie in $\mathcal{R}^{\ell-2}(\mathcal{A})$.

Theorem 5.2. *Let $\mathcal{A} = \mathcal{A}_{k,\ell-1}$ and $A = A(\mathcal{A}, \mathbb{C})$. Let $m = (m_1, \dots, m_k) \in \mathbb{N}^k$ be a k-tuple of non-negative integers, and set $|m| = \sum_{i=1}^k m_i$. Suppose m satisfies $0 \leq |m| - (\ell - 1) + 1 < \ell - 1$. Define a weight vector $a \in \bar{A}^1$ by assigning weight $-m_j$ to each of the hyperplanes $x_i - z_j x_\ell = 0, 1 \leq i \leq \ell - 1$, weight 2 to each of the hyperplanes $x_i - x_j, 1 \leq i < j \leq \ell - 1$, and weight $(\ell - 1)(|m| - \ell + 2 - \ell)$ to the hyperplane $x_\ell = 0$. Then $a \in \mathcal{R}^{\ell-2}(\mathcal{A}, \mathbb{C})$ with $H^p(\bar{A}, a) = 0$ for $p < \ell - 2$, and, for generic $z_1, \dots, z_k$, $\mathrm{crit}(\Phi_a)$ has codimension 1 in $\bar{M}$.*

Thus, again, we have that the codimension of $\mathrm{crit}(\Phi_a)$ in $\bar{M}$ matches the degree of the first nonvanishing cohomology group of $(\bar{A}, a)$. However, in this case the number of components of $\mathrm{crit}(\Phi_a)$ is equal to the dimension of the skew-symmetric part of $H^{\ell-2}(\bar{A}, a)$, which in some cases is a proper submodule of $H^{\ell-2}(\bar{A}, a)$. (Both the arrangement and the system of weights are preserved under permutation of the first $\ell - 1$ coordinates, giving rise to a $\Sigma_{\ell-1}$-module structure on $H^*(A, a)$.)

Example 5.3. Let $\mathcal{A} = \mathcal{A}_{2,2}$ with $z_1 = 0$ and $z_2 = 1$. Let $m = (1,1)$. Then m satisfies the hypothesis of Theorem 5.2. Figure 4 below exhibits the projective arrangement $\bar{\mathcal{A}}$, with the weight vector $a = (-1,-1,-1,-1,2,2)$ indicated. The critical locus of Φ_a is illustrated with dotted lines.

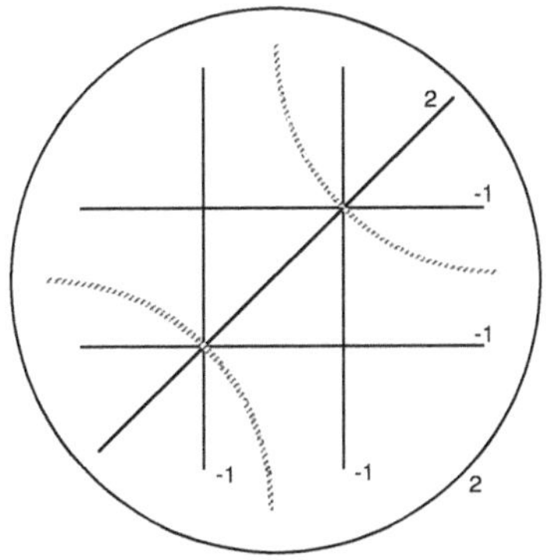

FIGURE 4. The discriminantal arrangement $\mathcal{A}_{2,2}$ and crit(Φ_a)

In recent joint work [4] we have proved partial generalizations of these results.

Theorem 5.4. *Suppose $\mathcal{A}$ is a free arrangement, and suppose $H^q(\bar{A},a) \neq 0$, while $H^p(\bar{A},a) = 0$ for $0 \le p < q$. Then* crit(Φ_a) *has codimension at least q in $\bar{M}$.*

See [23] for a definition of free arrangement. The theorem holds in the more general setting of *tame* arrangements [35].

The rank-three case. We close with a brief discussion of the rank-three case, using Theorem 4.17. A generalization of this argument will appear in [3]. Suppose $\mathcal{A}$ is an arrangement of rank 3, and let $a \in \mathcal{R}_\Gamma$, with the support of Γ equal to $\mathcal{A}$. For simplicity assume $\dim H^1(A,a) = 1$. Then $\mathcal{A}$ supports a multinet with three blocks. According to the discussion of multinets in the previous section, $\mathcal{R}_\Gamma = K_\Gamma$ is spanned by $s_1 = u_1 - u_3$ and $s_2 = u_2 - u_3$, where the u_i are non-negative integer vectors (supported on the three blocks of Γ) forming a basis for $\ker(Q_\Gamma)$. The master functions Φ_{u_i} are precisely the curves C_i of Theorem 4.17. Then, without loss of generality, $\Phi_{u_1} + \Phi_{u_2} = \Phi_{u_3}$. This implies $\Phi_{s_1} + \Phi_{s_2} = 1$.

We have $a = r_1 s_1 + r_2 s_2$ for some $r_1, r_2 \in \Bbbk$, so $\Phi_a = \Phi_{s_1}^{r_1}\Phi_{s_2}^{r_2}$. Then the critical locus of Φ_a is the inverse image under the (single-valued) mapping $(\Phi_{s_1}, \Phi_{s_2})\colon \mathbb{CP}^2 \to \mathbb{C}^2$ of the critical locus of the restriction of $f(y_1,y_2) = y_1^{r_1} y_2^{r_2}$ to the line $y_1 + y_2 = 1$.

The critical points of f restricted to $y_1 + y_2 = 1$ can be found by the method of Lagrange (or rather its holomorphic analogue). One finds a single isolated non-degenerate critical point if and only if r_1, r_2, and $r_1 + r_2$ are nonzero. It follows from the connectivity condition for Ceva-type pencils that the preimage of such a point is connected. We conclude that, aside from the exceptional cases $a = s_1, a = s_2$, and $a = s_1 - s_2 = u_1 - u_2$, $\mathrm{crit}(\Phi_a)$ has codimension 1 in $\mathbb{CP}^2$, and has one irreducible component, matching $\dim H^1(A,a)$. Note that the exceptional weights are precisely those weights in $\mathcal{R}_\Gamma$ supported on a proper subset of $\mathcal{A}$. A similar argument works when $k = 4$, in which case $\mathrm{crit}(\Phi_a)$ has codimension 1 and $\dim H^1(A,a) = 2$ components, with a few exceptional cases supported on proper subarrangements.

In Example 5.3, we have $u_1 = (1,0,0,1,0,0), u_2 = (0,1,1,0,0,0)$, and $u_3 = (0,0,0,0,1,1)$. Then $\Phi_{s_1}(x,y,z) = -\frac{x(y-z)}{(x-y)z}$ and $\Phi_{s_2}(x,y,z) = \frac{(x-z)y}{(x-y)z}$. The sign is chosen so that $\Phi_{s_1} + \Phi_{s_2} = 1$. We have $a = -s_1 - s_2$, so $r_1 = r_2 = -1$. The function $f(y_1, y_2) = (y_1 y_2)^{-1}$, restricted to $y_1 + y_2 = 1$, has one critical point $(1/2, 1/2)$. The critical locus of Φ_a is given by the equation $\Phi_{s_1} = 1/2$.

References

[1] E. Brieskorn. *Sur les groupes de tresses*, volume 317 of *Lecture Notes in Mathematics*, pages 21–44. Springer-Verlag, Berlin, Heidelberg, New York, 1973.

[2] D. Cohen, G. Denham, M. Falk, G.H. Schenck, A. Suciu, H. Terao, and S. Yuzvinsky. Complex Arrangements: Algebra, Geometry, Topology. In preparation.

[3] D. Cohen, G. Denham, M. Falk, and A. Varchenko. Critical loci of $\mathcal{A}$-master functions and special resonant weights. In preparation.

[4] D. Cohen, G. Denham, M. Falk, and A. Varchenko. Critical points of master functions and resonance of hyperplane arrangements. *Canadian J. of Mathematics.* To appear.

[5] D. Cohen and A. Varchenko. Resonant local systems and representations of $\mathfrak{sl}_2$. *Geom. Dedicata*, 101:217–233, 2003.

[6] P. Deligne. *Equations differentielles á points singuliers reguliers*, volume 163 of *Lecture Notes in Mathematics.* Springer Verlag, Berlin Heidelberg New York, 1970.

[7] G. Denham. Homological Aspects of Hyperplane Arrangements. This volume, 39–58.

[8] A. Dimca. Pencils of Plane Curves and Characteristic Varieties. This volume, 59–82.

[9] A. Dimca and S. Yuzvinsky. Lectures on Orlik-Solomon algebras. In F. El Zein, A. Suciu, M. Tosun, A.M. Uludag, and S. Yuzvinsky, editors, *Lectures on Arrangements, Local Systems, and Singularities.* Birkhäuser Boston, 2009.

[10] C. Dunn, M. Miller, M. Wakefield, and S. Zwicknagl. Equivalence classes of latin squares and nets in $\mathbb{C}P^2$. arxiv.org/abs/math/0703142, 2007.

[11] C. Eschenbrenner and M. Falk. Tutte polynomials and Orlik-Solomon algebras. *Journal of Algebraic Combinatorics*, 10:189–199, 1999.

[12] H. Esnault, V. Schechtman, and E. Viehweg. Cohomology of local systems on the complement of hyperplanes. *Inventiones Mathematicae*, 109:557–561, 1992. erratum: ibid. 112 (1993), 447.

[13] M. Falk. Arrangements and cohomology. *Annals of Combinatorics*, 1:135–157, 1997.

[14] M. Falk. Combinatorial and algebraic structure in Orlik-Solomon algebras. *European Journal of Combinatorics*, 22:687–698, 2001.

[15] M. Falk. The line geometry of resonance varieties. arxiv.org/abs/math/0405210, 2004.

[16] M. Falk. Resonance varieties over fields of positive characteristic. *International Mathematics Research Notices*, 2007:25 pages, 2007. doi:10.1093/imrn/rnm009.

[17] M. Falk and S. Yuzvinsky. Multinets, resonance varieties, and pencils of plane curves. *Compositio mathematica*, 143(4):1069–1088, 2007.

[18] V. Kac. *Infinite Dimensional Lie Algebras.* Cambridge University Press, 1990.

[19] Y. Kawahara. The non-vanishing cohomology of Orlik-Solomon algebras. preprint, 2005.

[20] A. Libgober and S. Yuzvinsky. Cohomology of the Orlik-Solomon algebras and local systems. *Compositio mathematica*, 121:337–361, 2000.

[21] E. Mukhin and A. Varchenko. Multiple orthogonal polynomials and a counterexample to the Gaudin Bethe Ansatz conjecture. *Transactions of the American Mathematical Society*, 359(11):5383–5418, 2007.

[22] P. Orlik and L. Solomon. Topology and combinatorics of complements of hyperplanes. *Inventiones mathematicae*, 56:167–189, 1980.

[23] P. Orlik and H. Terao. *Arrangements of Hyperplanes.* Springer-Verlag, Berlin Heidelberg New York, 1992.

[24] P. Orlik and H. Terao. The number of critical points of a product of powers of linear functions. *Inventiones mathematicae*, 120(1):1–14, 1995.

[25] P. Orlik and H. Terao. *Arrangements and Hypergeometric Integrals*, volume 9 of *MSJ Memoirs.* Japan Mathematical Society, Tokyo, 2001.

[26] J. V. Pereira and S. Yuzvinsky. Completely reducible hypersurfaces in a pencil. *Advances in Mathematics*, 219:672–688 (2008).

[27] V.V. Schechtman, H. Terao, and A.N. Varchenko. Cohomology of local systems and Kac-Kazhdan condition for singular vectors. *Journal of Pure and Applied Algebra*, 100:93–102, 1995.

[28] V.V. Schechtman and A.N. Varchenko. Arrangements of hyperplanes and Lie algebra homology. *Inventiones mathematicae*, 106:139–194, 1991.

[29] I. Scherbak and A. Varchenko. Critical points of functions, $\mathfrak{sl}_2$ representations, and Fuchsian differential equations with only univalued solutions. *Moscow Mathematics Journal*, 3:621–645, 2003.

[30] J. Stipins. *On Finite k-nets in the Complex Projective Plane.* PhD thesis, University of Michigan, 2007.

[31] A. Suciu. Translated tori in the characteristic varieties of complex hyperplane arrangements. *Topology and Its Applications*, 118:209–224, 2002. Special Issue - Arrangements in Boston, A Conference on Hyperplane Arrangements, D. Cohen and A. Suciu, editors.

[32] G. Urzua. On line arrangements with applications to 3-nets. arxiv.org/abs/math/0704.0469, 2007.

[33] A. Varchenko. Critical points of the product of powers of linear functions and families of bases of singular vectors. *Compositio mathematica*, 97(3):385–401, 1995.

[34] A. Varchenko. *Special functions, KZ type equations, and representation theory*, volume 98 of *CBMS Regional Conference Series in Mathematics*. Conference Board of the Mathematical Sciences, 2003.

[35] J. Wiens and S. Yuzvinsky. De Rham cohomology of logarithmic forms on arrangements of hyperplanes. *Transactions of the American Mathematical Society*, 349(4):1653–1662, 1997.

[36] S. Yuzvinsky. Cohomology of Brieskorn-Orlik-Solomon algebras. *Communications in Algebra*, 23:5339–5354, 1995.

[37] S. Yuzvinsky. Realization of finite abelian groups by nets in $\mathbb{P}^2$. *Compositio mathematica*, 140(6):1614–1624, 2004.

[38] S. Yuzvinsky. A new bound on the number of special fibers in a pencil of curves. arxiv.org/abs/math/0801.1521, 2008.

[39] T. Zaslavsky. *Facing Up to Arrangements – Face Count Formulas for Partition of Space by Hyperplanes*, volume 154 of *Memoirs of the American Mathematical Society*. American Mathematical Society, Providence, 1975.

Michael Falk
Department of Mathematics and Statistics
Northern Arizona University
Flagstaff, AZ 86011-5717
USA
e-mail: `michael.falk@nau.edu`

Progress in Mathematics, Vol. 283, 177–190

The Characteristic Quasi-Polynomials of the Arrangements of Root Systems and Mid-Hyperplane Arrangements

Hidehiko Kamiya[1], Akimichi Takemura and Hiroaki Terao

Abstract. Let q be a positive integer. In [8], we proved that the cardinality of the complement of an integral arrangement, after the modulo q reduction, is a quasi-polynomial of q, which we call the characteristic quasi-polynomial. In this paper, we study general properties of the characteristic quasi-polynomial as well as discuss two important examples: the arrangements of reflecting hyperplanes arising from irreducible root systems and the mid-hyperplane arrangements. In the root system case, we present a beautiful formula for the generating function of the characteristic quasi-polynomial which has been essentially obtained by Ch. Athanasiadis [2] and by A. Blass and B. Sagan [3]. On the other hand, it is hard to find the generating function of the characteristic quasi-polynomial in the mid-hyperplane arrangement case. We determine them when the dimension is less than six.

Mathematics Subject Classification (2000). 32S22, 05B35, 17B20.

Keywords. Characteristic quasi-polynomial, elementary divisor, hyperplane arrangement, root system, mid-hyperplane arrangement.

1. Introduction

Let S be an arbitrary $m \times n$ integral matrix without zero columns. For each positive integer $q \in \mathbb{Z}_{>0}$, denote $\mathbb{Z}_q = \mathbb{Z}/q\mathbb{Z}$ and $\mathbb{Z}_q^\times = \mathbb{Z}_q \setminus \{0\}$. Consider the set

$$M_q(S) := \{\mathbf{z} = (z_1, \ldots, z_m) \in \mathbb{Z}_q^m : \mathbf{z}S \in (\mathbb{Z}_q^\times)^n\},$$

and its cardinality $|M_q(S)|$. In our recent paper [8], we showed that there exists a monic **quasi-polynomial** (periodic polynomial) $\chi_S(q)$ with integral coefficients of degree m such that

$$\chi_S(q) = |M_q(S)|, \quad q \in \mathbb{Z}_{>0}.$$

[1] This work was partially supported by JSPS KAKENHI (19540131).

Note that the set $M_q(S)$ is the complement of an arrangement of hyperplanes in the following sense: Let $S_1, S_2, \dots, S_n$ be the columns of S. Each set

$$H_{i,q} := \{\mathbf{z} = (z_1, \dots, z_m) \in \mathbb{Z}_q^m : \mathbf{z}S_i = 0\}, \quad 1 \le i \le n,$$

can be called a "hyperplane" in $\mathbb{Z}_q^m$ by a slight abuse of terminology. Then

$$M_q(S) = \mathbb{Z}_q^m \setminus \bigcup_{i=1}^{n} H_{i,q}.$$

For a sufficiently large prime number q, $\chi_S(q)$ is known [2] to be equal to the **characteristic polynomial** [9, Def. 2.52] of the real arrangement consisting of the following hyperplanes (ignoring possible repetitions):

$$H_{i,\mathbb{R}} := \{\mathbf{z} = (z_1, \dots, z_m) \in \mathbb{R}^m : \mathbf{z}S_i = 0\}, \quad 1 \le i \le n.$$

It is thus natural to call the quasi-polynomial $\chi_S(q)$ the **characteristic quasi-polynomial** of S as in [8]. Let us define its **generating function**

$$\Phi_S(t) := \sum_{q=1}^{\infty} \chi_S(q) t^q.$$

We understand that $M_1(S) = \emptyset$ for $q = 1$ and hence the summation is in effect for $q \ge 2$.

In this paper, we study the characteristic quasi-polynomial $\chi_S(q)$ or equivalently its generating function $\Phi_S(t)$. In Section 2, we discuss general properties of the characteristic quasi-polynomials and their generating functions. In the subsequent chapters, we deal with two kinds of specific arrangements defined over $\mathbb{Z}$: the arrangements of reflecting hyperplanes arising from irreducible root systems (Section 3) and the mid-hyperplane arrangements (Section 4). Let R be an irreducible root system of rank m and $n = |R|/2$. We assume that an $m \times n$ integral matrix $S = S(R) = [S_{ij}]$ satisfies

$$R_+ = \{\sum_{i=1}^{m} S_{ij}\alpha_i : j = 1, \dots, n\},$$

where R_+ is a set of positive roots and $B(R) = \{\alpha_1, \alpha_2, \dots, \alpha_m\}$ is the set of **simple roots** associated with R_+. In other words, S is a coefficient matrix of R_+ with respect to the basis $B(R)$. Define the characteristic quasi-polynomial $\chi_R(q) := \chi_S(q)$ and the generating function $\Phi_R(t) := \Phi_S(t)$ for each irreducible root system R. Then $\chi_R(q)$ and $\Phi_R(t) := \Phi_S(t)$ depend only upon R. In Section 3, we present a beautiful formula for the generating function $\Phi_R(t)$ for every irreducible root system R. This formula has been essentially proved by Ch. Athanasiadis in [2] and A. Blass and B. Sagan in [3]. (See [6] also.) In Theorem 3.1, we will state the formula in our language and include a proof following [2, 3, 6] for completeness. In Section 4, we will give a formula for the generating function $\Phi_S(t)$ when S is equal to the coefficient matrix for the mid-hyperplane arrangement of dimension less than 6.

We are aided by the computer package PARI/GP [10].

2. Results on the characteristic quasi-polynomial of an integral matrix

Let $\chi_S(t)$ be the characteristic quasi-polynomial of an $m \times n$ integral matrix S without zero columns. Fix a nonempty $J \subseteq [n] := \{1, 2, \ldots, n\}$ and define an $m \times |J|$ matrix S_J consisting of the columns of S corresponding to the set J. Let $e_{J,1}, \ldots, e_{J,\ell(J)} \in \mathbb{Z}_{>0}$ be the elementary divisors of S_J numbered so that $e_{J,1}|e_{J,2}|\cdots|e_{J,\ell(J)}$, where $\ell(J) := \operatorname{rank} S_J$. Write $e(J) := e_{J,\ell(J)}$, and define the **lcm period** $\rho_0(S)$ of S by

$$\begin{aligned}\rho_0 = \rho_0(S) &:= \operatorname{lcm}\{e(J) : J \subseteq [n],\ J \neq \emptyset\} \\ &= \operatorname{lcm}\{e(J) : J \subseteq [n],\ 1 \leq |J| \leq \min\{m, n\}\}.\end{aligned}$$

Then it is known ([8, Theorem 2.4]) that the lcm period ρ_0 is a period of $\chi_S(t)$.

It is further shown in [8] that the constituents of the quasi-polynomial $\chi_S(t)$ are the same for all q's with the same value of $\gcd\{\rho_0, q\}$. Let d be a positive integer which divides ρ_0, and define a monic polynomial $P_d(t) = P_{S,d}(t)$ with integral coefficients of degree m by

$$\chi_S(q) = P_d(q) \quad \text{for all} \quad q \in d + \rho_0\mathbb{Z}_{\geq 0}. \tag{1}$$

Put

$$e(J, d) := \prod_{j=1}^{\ell(J)} \gcd\{e_{J,j}, d\}.$$

Then the following formula was essentially proved in our previous paper [8].

Theorem 2.1. *For each $d \in \mathbb{Z}_{>0}$ with $d|\rho_0$, the polynomial $P_d(t)$ is given by*

$$P_d(t) = \sum_{J \subseteq [n]} (-1)^{|J|} e(J, d) t^{m-\ell(J)},$$

where for $J = \emptyset$, we understand that $\ell(\emptyset) = 0$ and that $e(\emptyset, d) = 1$.

Proof. Obtained from [8, (10)] and the inclusion-exclusion principle. □

Theorem 2.2 ([8] **Theorem 2.5**). *The polynomial*

$$P_1(t) = \sum_{J \subseteq [n]} (-1)^{|J|} t^{m-\ell(J)}$$

is equal to the ordinary characteristic polynomial [9, Def. 2.52] *of the real arrangement consisting of the hyperplanes (ignoring possible repetitions) $H_{1,\mathbb{R}}, H_{2,\mathbb{R}}, \ldots, H_{n,\mathbb{R}}$.*

Corollary 2.3. *Suppose $d, d' \in \mathbb{Z}_{>0}$ both divide ρ_0, and assume the following condition holds true for some positive integer s:* $\gcd\{e(J), d\} = \gcd\{e(J), d'\}$ *for all $J \subseteq [n]$ with $|J| \leq s$. Then*

$$\deg\{P_d(t) - P_{d'}(t)\} < m - s.$$

In particular, we have $\deg\{P_d(t) - P_1(t)\} < m - s$ *if* $\gcd\{e(J), d\} = 1$ *for all $J \subseteq [n]$ with $|J| \leq s$.*

Proof. We apply Theorems 2.1 and 2.2. It is enough to show $e(J, d) = e(J, d')$ for $J \subseteq [n]$ with $\ell(J) \leq s$. We can choose a subset $J' \subseteq J$ such that $\ell(J') = |J'| = \ell(J) \leq s$. Then $\gcd\{e(J'), d\} = \gcd\{e(J'), d'\}$. Since $e(J)|e(J')$ [8, Lemma 2.3], $\gcd\{e(J), d\} = \gcd\{e(J), d'\}$. This shows $e(J, d) = \prod_{j=1}^{\ell(J)} \gcd\{e_{J,j}, d\} = \prod_{j=1}^{\ell(J)} \gcd\{e_{J,j}, e(J), d\} = \prod_{j=1}^{\ell(J)} \gcd\{e_{J,j}, e(J), d'\} = \prod_{j=1}^{\ell(J)} \gcd\{e_{J,j}, d'\} = e(J, d')$. □

Corollary 2.4. *Suppose that $d \in \mathbb{Z}_{>0}$ and $d' \in \mathbb{Z}_{>0}$ both divide ρ_0 and that* $\gcd\{d, d'\} = 1$. *In addition, we assume the following condition holds true for some positive integer s:*

$$\gcd\{e(J), d\} = 1 \quad \textit{or} \quad \gcd\{e(J), d'\} = 1 \tag{2}$$

for all $J \subseteq [n]$ with $|J| \leq s$. Then

$$\deg\{P_1(t) + P_{dd'}(t) - P_d(t) - P_{d'}(t)\} < m - s.$$

Proof. Suppose $J \subseteq [n]$ with $\ell(J) \leq s$. It is enough to show

$$1 + e(J, dd') - e(J, d) - e(J, d') = 0.$$

We can choose a subset $J' \subseteq J$ such that $\ell(J') = |J'| = \ell(J) \leq s$. Then either $\gcd\{e(J'), d\} = 1$ or $\gcd\{e(J'), d'\} = 1$ by (2). Since $e(J)|e(J')$,

$$\gcd\{e(J), d\} = 1 \quad \text{or} \quad \gcd\{e(J), d'\} = 1.$$

This shows that either $e(J, d) = 1$ or $e(J, d') = 1$. We finally have

$$\begin{aligned} 0 &= \{1 - e(J, d)\}\{1 - e(J, d')\} = 1 - e(J, d) - e(J, d') + e(J, d)e(J, d') \\ &= 1 - e(J, d) - e(J, d') + e(J, dd'). \end{aligned}$$

□

Corollary 2.5. *Suppose that $d \in \mathbb{Z}_{>0}$ and $d' \in \mathbb{Z}_{>0}$ both divide ρ_0 and that* $\gcd\{d, d'\} = 1$. *If $e(J)$ are prime powers or 1 for all J, we have $P_{dd'}(t) = P_d(t) + P_{d'}(t) - P_1(t)$.*

Proof. Easily follows from Corollary 2.4. □

For the rest of this section we discuss general properties of the generating functions $\Phi_S(t)$ of characteristic quasi-polynomials. Let $\omega = \exp(2\pi i/\rho_0)$ which is a primitive ρ_0's root of unity. By (1)

$$\Phi_S(t) = \sum_{d=1}^{\rho_0} \Phi_{S,d}(t), \qquad \Phi_{S,d}(t) = \sum_{s=0}^{\infty} P_d(d + \rho_0 s) t^{d+\rho_0 s}.$$

Note that $\Phi_S(\omega^k t) = \sum_{d=1}^{\rho_0} \omega^{kd}\Phi_{S,d}(t)$. Therefore from the orthogonality relations among powers of ω, i.e., by the Fourier inversion, $\Phi_{S,d}(t)$ for each d can be recovered from $\Phi_S(t)$ by

$$\Phi_{S,d}(t) = \frac{1}{\rho_0}\sum_{k=1}^{\rho_0} \omega^{-kd}\Phi_S(\omega^k t). \tag{3}$$

This relation will be used in Example 3.5 below.

Taking a common denominator, we can express $\Phi_S(t)$ as a rational function

$$\Phi_S(t) = \frac{Q(t)}{(1-t^{\rho_0})^{m+1}}, \qquad \deg Q < (m+1)\rho_0.$$

In the numerator $Q(t)$ the powers $t^{d+\rho_0 s}$, $s = 0, 1, \dots$, correspond to P_d. Therefore as in (3) for each d we can extract these powers as

$$\Phi_{S,d}(t) = \frac{Q_d(t)}{(1-t^{\rho_0})^{m+1}}, \qquad Q_d(t) = \frac{1}{\rho_0}\sum_{k=1}^{\rho_0}\omega^{-kd}Q(\omega^k t). \tag{4}$$

Let

$$P_d(q) = \sum_{k=0}^{m} c_{d,k}q^k \ \ (c_{d,k} \in \mathbb{Z}).$$

Then

$$\begin{aligned} Q_d(t) &= (1-t^{\rho_0})^{m+1}\sum_{s=0}^{\infty} P_d(d+\rho_0 s)t^{d+\rho_0 s} \\ &= (1-t^{\rho_0})^{m+1}\sum_{k=0}^{m} c_{d,k}\sum_{s=0}^{\infty}(d+\rho_0 s)^k t^{d+\rho_0 s}. \end{aligned}$$

Define polynomials $q_{d,k}(t)$ by

$$\sum_{s=0}^{\infty}(d+\rho_0 s)^k t^{d+\rho_0 s} = \frac{q_{d,k}(t)}{(1-t^{\rho_0})^{k+1}} \qquad (d = 1, \dots, \rho_0). \tag{5}$$

Then we obtain

$$Q_d(t) = \sum_{k=0}^{m}(1-t^{\rho_0})^{m-k}c_{d,k}q_{d,k}(t). \tag{6}$$

Now we present the following proposition, to which we give a proof because we were not able to find an appropriate reference in the literature.

Proposition 2.6. *Define $q_{d,k}(t)$ by (5). Let $q_{d,k}^{(j)}(1)$ be their j-th derivatives at $t=1$. Then*

$$0 \neq q_{1,k}^{(j)}(1) = q_{2,k}^{(j)}(1) = \cdots = q_{\rho_0,k}^{(j)}(1) \qquad (j = 0, \dots, k). \tag{7}$$

Proof. For notational simplicity write $q_d(t) = q_{d,k}(t)$ and let

$$\tilde{q}_l(t) = \sum_{d=1}^{\rho_0}\omega^{ld}q_d(t) \qquad (l = 1, \dots, \rho_0).$$

Then the inverse Fourier transform is

$$q_d(t) = \frac{1}{\rho_0}\sum_{l=1}^{\rho_0} \omega^{-ld}\tilde{q}_l(t) \qquad (d = 1, \dots, \rho_0).$$

The j-th derivative of this at $t = 1$ is

$$q_d^{(j)}(1) = \frac{1}{\rho_0}\sum_{l=1}^{\rho_0} \omega^{-ld}\tilde{q}_l^{(j)}(1).$$

It follows that $q_d^{(j)}(1)$ does not depend on d if and only if

$$\tilde{q}_l^{(j)}(1) = 0 \qquad (l = 1, \dots, \rho_0 - 1). \tag{8}$$

By the use of Eulerian numbers $W(k, h)$ (see Chapter III of [1]) we can write

$$\tilde{q}_l(t) = \left(\sum_{h=0}^{k-1} W(k,h)(\omega^l t)^{k-h}\right)\left(1 + \omega^l t + \omega^{2l}t^2 + \dots + \omega^{(\rho_0-1)l}t^{\rho_0-1}\right)^{k+1}. \tag{9}$$

Note that $0 = 1 + \omega^l + \omega^{2l} + \dots + \omega^{(\rho_0-1)l}$ for $1 \le l < \rho_0$. Therefore differentiating (9) with respect to t, we have $\tilde{q}_l^{(j)}(1) = 0$ for $1 \le l < \rho_0$ and for $0 \le j \le k$. Thus $q_{1,k}^{(j)}(1) = q_{2,k}^{(j)}(1) = \dots = q_{\rho_0,k}^{(j)}(1)$. By summing up (5) we have

$$\sum_{d=1}^{\rho_0} q_{d,k}(t) = (1 - t^{\rho_0})^{k+1}\sum_{q=1}^{\infty} q^k t^q.$$

Since the Eulerian numbers are positive integers, it is not hard to see that the right-hand side is a polynomial of degree $\ge k$ with positive integer coefficients. Thus $q_{d,k}^{(j)}(1)$ is not zero for $0 \le j \le k$, $1 \le d \le \rho_0$. □

Proposition 2.6 and (6) imply that

$$P_d(t) = P_{d'}(t) \Leftrightarrow c_{d,k} = c_{d',k} \text{ for } 0 \le k \le m \Leftrightarrow Q_d^{(j)}(1) = Q_{d'}^{(j)}(1) \text{ for } 0 \le j \le m.$$

Furthermore note that lower order derivatives of Q_d at $t = 1$ determine coefficients of higher degree terms in $P_d(t)$. Therefore in terms of the generating function the relations in Corollaries 2.3 and 2.4 can be written as follows:

$$\deg\{P_d(t) - P_{d'}(t)\} < m - s \Leftrightarrow Q_d^{(j)}(1) = Q_{d'}^{(j)}(1) \quad (j = 0, 1, \dots, s),$$

$$\begin{aligned}&\deg\{P_1(t) + P_{dd'}(t) - P_d(t) - P_{d'}(t)\} < m - s \\ &\qquad \Leftrightarrow Q_1^{(j)}(1) + Q_{dd'}^{(j)}(1) - Q_d^{(j)}(1) - Q_{d'}^{(j)}(1) = 0 \qquad (j = 0, 1, \dots, s).\end{aligned}$$

3. Arrangements of root systems

Let V be an m-dimensional Euclidean space and E be the affine space underlying V. Let R be an irreducible root system in V of rank m and $n = |R|/2$. Suppose that R_+ is a set of positive roots and $B = \{\alpha_1, \dots, \alpha_m\}$ is the set of simple roots associated with R_+. Denote the coefficient matrix (with an arbitrary order of columns) of the positive roots R_+ with respect to B by $S = [S_{ij}]$, which is an $m \times n$ matrix:

$$R_+ = \{\sum_{i=1}^{m} S_{ij}\alpha_i : j = 1, \dots, n\}.$$

In this section we give an explicit formula for the generating function $\Phi_R(t) := \Phi_S(t)$. The formula was essentially proved by Ch. Athanasiadis [2] and by A. Blass and B. Sagan [3]. (See [6] also.) Let

$$\tilde{\alpha} = \sum_{i=1}^{m} n_i \alpha_i$$

be the highest root. Then it is well-known that $h := 1 + \sum_{i=1}^{m} n_i$ is equal to the **Coxeter number**; see [4, Ch. VI, §1, 11. Prop. 31].

Theorem 3.1.

$$\Phi_R(t) = \frac{(n_1 \cdots n_m)(m!)t^h}{(1-t)\prod_{i=1}^{m}(1-t^{n_i})}.$$

Before proving this formula after [2, 3, 6], we introduce basic concepts. Let $\beta_1, \dots, \beta_m$ be the basis for V which is dual to the basis B: $(\alpha_i, \beta_j) = \delta_{ij}$. Define a free abelian group

$$P(R^\vee) = \mathbb{Z}\beta_1 + \mathbb{Z}\beta_2 + \cdots + \mathbb{Z}\beta_m$$

of rank m. Let

$$H_{i,k} = \{x \in E : (x, \alpha_i) = k\}$$

for $1 \le i \le n$ and $k \in \mathbb{Z}$. Then $\mathcal{A}_a = \{H_{i,k} : 1 \le i \le n,\ k \in \mathbb{Z}\}$ is an arrangement of (infinitely many) affine hyperplanes in E. The reflection with respect to $H_{i,k}$ is denoted by $s_{i,k}$:

$$s_{i,k}(x) = x - 2\frac{(x, \alpha_i) - k}{(\alpha_i, \alpha_i)}\alpha_i \qquad (x \in E).$$

The **affine Weyl group** W_a is the group generated by $\{s_{i,k} : 1 \le i \le n,\ k \in \mathbb{Z}\}$. Each connected component of $E \backslash \bigcup \mathcal{A}_a$ is called an **alcove**. The closure of an alcove is a fundamental domain of the group W_a acting on E [4, Ch. VI, §2, 1]. Consider a special alcove

$$\begin{aligned} C &= \{x \in E : 0 < (x, \alpha) < 1 \ (\alpha \in R_+)\} \\ &= \{x \in E : 0 < (x, \alpha_i) < 1 \ (i = 1, \dots, m), \ \ (x, \tilde{\alpha}) < 1\} \\ &= \{\sum_{i=1}^{m} c_i \beta_i : 0 < c_i \ (i = 1, \dots, m), \ \sum_{i=1}^{m} c_i n_i < 1\} \end{aligned}$$

which is an open simplex with vertices 0 and the β_i/n_i $(1 \le i \le m)$.

Proof of Theorem 3.1. ([2, 3, 6]) Let A be the parallelepiped defined by

$$A = \{\sum_{i=1}^{m} c_i \beta_i : 0 < c_i < 1 \ (i = 1, \dots, m)\}.$$

Then A is, by definition, a union of alcoves. The number of alcoves in A is equal to

$$\frac{\mathrm{vol}(A)}{\mathrm{vol}(C)} = (n_1 \cdots n_m)(m!);$$

see [4, Ch. VI, §2, 4. (5)]. Compute

$$\begin{aligned} \chi_R(t) &= |M_q(S)| = |\{\mathbf{z} = (z_1, \dots, z_m) \in \mathbb{Z}_q^m : \mathbf{z}S \in (\mathbb{Z}_q^{\times})^n\}| \\ &= |\{\gamma \in A \cap \frac{1}{q} P(R^{\vee}) : \gamma \notin \bigcup \mathcal{A}_a\}| \\ &= |(A \cap \frac{1}{q} P(R^{\vee})) \setminus \bigcup \mathcal{A}_a| \\ &= (n_1 \cdots n_m)(m!) \, |C \cap \frac{1}{q} P(R^{\vee})|. \end{aligned}$$

On the other hand,

$$\begin{aligned} |C \cap \frac{1}{q} P(R^{\vee})| &= |\{\sum_{i=1}^{m} c_i \beta_i : c_i \in \{\frac{1}{q}, \frac{2}{q}, \dots, \frac{q-1}{q}\} \ (i = 1, \dots, m), \ \sum_{i=1}^{m} c_i n_i < 1\}| \\ &= |\{\sum_{i=1}^{m} c_i \beta_i : c_i \in \mathbb{Z}_{>0} \ (i = 1, \dots, m), \ \sum_{i=1}^{m} c_i n_i < q\}|. \end{aligned}$$

This function is known as the Ehrhart quasi-polynomial of the open simplex bounded by the coordinate hyperplanes and the hyperplane $\sum_{i=1}^{m} c_i x_i = q$; see [11, page 235ff]. Thus

$$\begin{aligned} \Phi_R(t) &= (n_1 \cdots n_m)(m!) \sum_{q=1}^{\infty} |C \cap \frac{1}{q} P(R^{\vee})| t^q \\ &= \frac{(n_1 \cdots n_m)(m!) t^{n_1 + \cdots + n_m + 1}}{(1-t) \prod_{i=1}^{m} (1 - t^{n_i})} = \frac{(n_1 \cdots n_m)(m!) t^h}{(1-t) \prod_{i=1}^{m} (1 - t^{n_i})}. \end{aligned}$$

This completes the proof. □

Corollary 3.2. *The minimum period of the characteristic quasi-polynomial for an irreducible root system is equal to* $\mathrm{lcm}(n_1, \dots, n_m)$.

Proof. The assertion holds true because by Proposition 4.4.1 of [11] the minimum period of the Ehrhart quasi-polynomial of the open simplex bounded by the coordinate hyperplanes and the hyperplane $\sum_{i=1}^m n_i x_i = q$ is equal to $\mathrm{lcm}(n_1, \dots, n_m)$. □

The minimum periods for all irreducible root systems are shown in the following table:

root system	$n_1, n_2, \dots, n_m$	$h = 1 + n_1 + n_2 + \cdots + n_m$	minimum period
A_m	$1, 1, \dots, 1, 1$	$m + 1$	1
B_m	$1, 2, 2, \dots, 2$	$2m$	2
C_m	$2, 2, \dots, 2, 1$	$2m$	2
D_m	$1, 2, 2, \dots, 2, 1, 1$	$2m - 2$	2
E_6	$1, 2, 2, 3, 2, 1$	12	6
E_7	$2, 2, 3, 4, 3, 2, 1$	18	12
E_8	$2, 3, 4, 6, 5, 4, 3, 2$	30	60
F_4	$2, 3, 4, 2$	12	12
G_2	$2, 3$	6	6

Remark 3.3. With PARI/GP we checked that for every irreducible root system the minimum period coincides with the lcm period $\rho_0(S)$.

Corollary 3.4. *Let q be a positive integer. For an irreducible root system R with its Coxeter number h, $\chi_R(q) > 0$ if and only if $q \geq h$.*

Proof. The lowest non-zero term of $\Phi_R(t)$ is equal to $(n_1 \cdots n_m)(m!)t^h$. □

Example 3.5. (B_m, C_m, D_m) By Theorem 3.1 and the table above, we have the generating functions of type B_m, C_m, D_m:

$$\Phi_{B_m}(t) = \Phi_{C_m}(t) = \frac{2^{m-1}(m!)t^{2m}}{(1-t)^2(1-t^2)^{m-1}}, \quad \Phi_{D_m}(t) = \frac{2^{m-3}(m!)t^{2m-2}}{(1-t)^4(1-t^2)^{m-3}}.$$

Thus the characteristic quasi-polynomials of B_m is the same as the characteristic quasi-polynomials of C_m. Since the minimum periods of these three root systems are all equal to 2, there exist four polynomials $P_1(q), P_2(q), Q_1(q), Q_2(q)$ satisfying

$$\Phi_1(t) := \Phi_{B_m}(t) = \Phi_{C_m}(t) = \sum_{i=0}^{\infty} P_1(2i+1)t^{2i+1} + \sum_{i=1}^{\infty} P_2(2i)t^{2i},$$

$$\Phi_2(t) := \Phi_{D_m}(t) = \sum_{i=0}^{\infty} Q_1(2i+1)t^{2i+1} + \sum_{i=1}^{\infty} Q_2(2i)t^{2i}.$$

Compute

$$\begin{aligned}2\sum_{i=1}^{\infty} P_2(2i)t^{2i} &= \Phi_1(t)+\Phi_1(-t)\\ &= \frac{2^{m-1}(m!)t^{2m}}{(1-t)^2(1-t^2)^{m-1}}+\frac{2^{m-1}(m!)t^{2m}}{(1+t)^2(1-t^2)^{m-1}}\\ &= \frac{2^m(m!)t^{2m}(1+t^2)}{(1-t^2)^{m+1}}\end{aligned}$$

and

$$\begin{aligned}2\sum_{i=1}^{\infty} Q_1(2i+1)t^{2i+1} &= \Phi_2(t)-\Phi_2(-t)\\ &= \frac{2^{m-3}(m!)t^{2m-2}}{(1-t)^4(1-t^2)^{m-3}}-\frac{2^{m-3}(m!)t^{2m-2}}{(1+t)^4(1-t^2)^{m-3}}\\ &= \frac{2^m(m!)t^{2m+1}(1+t^2)}{(1-t^2)^{m+1}} = 2\sum_{i=1}^{\infty} P_2(2i)t^{2i+1}.\end{aligned}$$

This implies

$$P_2(q) = Q_1(q-1).$$

Since $Q_1(q)$ is equal to the ordinary characteristic polynomial of D_m by Theorem 2.2, we obtain

$$P_2(q) = Q_1(q-1) = (q-2)(q-4)\dots(q-2m+2)(q-m).$$

Actually we may derive the following characteristic quasi-polynomials from the generating functions $\Phi_1(t)$ and $\Phi_2(t)$:

$$\chi_{B_m}(q) = \chi_{C_m}(q) = \begin{cases}(q-1)(q-3)\cdots(q-2m+1) & \text{if } q \text{ is odd,}\\ (q-2)(q-4)\cdots(q-2m+2)(q-m) & \text{if } q \text{ is even,}\end{cases}$$

$$\chi_{D_m}(q) = \begin{cases}(q-1)(q-3)\cdots(q-2m+3)(q-m+1) & \text{if } q \text{ is odd,}\\ (q-2)(q-4)\cdots(q-2m+4)\left\{q^2-2(m-1)q+\frac{m(m-1)}{2}\right\} & \text{if } q \text{ is even.}\end{cases}$$

Remark 3.6. We may also prove $\chi_{B_m}(2q) = \chi_{D_m}(2q-1)$ by constructing a one-to-one correspondence between $M_{2q}(S(B_m))$ and $M_{2q-1}(S(D_m))$.

Example 3.7. Let R be a root system of type E_6. We use PLATE V in [4] to get the 6×36 matrix $S = S(E_6)$:

$$S(E_6)=\left[\begin{array}{ccccccccccccccccccccc}
1&0&0&0&0&0&1&0&0&0&0&1&0&0&0&0&1&1&0&0&0\\
0&1&0&0&0&0&0&1&0&0&0&0&1&1&0&0&1&0&1&1&0\\
0&0&1&0&0&0&1&0&1&0&0&1&1&0&1&0&1&1&1&0&1\\
0&0&0&1&0&0&0&1&1&1&0&1&1&1&1&1&1&1&1&1&1\\
0&0&0&0&1&0&0&0&0&1&1&0&0&1&1&1&0&1&1&1&1\\
0&0&0&0&0&1&0&0&0&0&1&0&0&0&0&1&0&0&0&1&1
\end{array}\right.$$

$$\left.\begin{array}{ccccccccccccccc}
1&1&0&0&1&1&0&1&1&0&1&1&1&1&1\\
1&0&1&1&1&1&1&1&1&1&1&1&1&1&2\\
1&1&1&1&1&1&1&2&1&1&2&1&2&2&2\\
1&1&1&2&2&1&2&2&2&2&2&2&2&3&3\\
1&1&1&1&1&1&1&1&2&1&2&2&2&2&2\\
0&1&1&0&0&1&1&0&1&1&1&1&1&1&1
\end{array}\right].$$

Thus $(n_1, n_2, n_3, n_4, n_5, n_6) = (1, 2, 2, 3, 2, 1)$ and $h = 12$. By Theorem 3.1, we have the generating function of type E_6:

$$\Phi_{E_6}(t) = \frac{24 \cdot (6!)t^{12}}{(1-t)^3(1-t^2)^3(1-t^3)}.$$

By expanding this formal power series we have the characteristic quasi-polynomial of E_6:

$$\chi_{E_6}(q) = \begin{cases}
q^6 - 36q^5 + 510q^4 - 3600q^3 + 13089q^2 - 22284q + 12320 \\
\quad = (q-1)(q-4)(q-5)(q-7)(q-8)(q-11), \\
\qquad \gcd\{6, q\} = 1, \\
q^6 - 36q^5 + 510q^4 - 3600q^3 + 13224q^2 - 23904q + 16640 \\
\quad = (q-2)(q-4)(q-8)(q-10)(q^2 - 12q + 26), \\
\qquad \gcd\{6, q\} = 2, \\
q^6 - 36q^5 + 510q^4 - 3600q^3 + 13089q^2 - 22284q + 12960 \\
\quad = (q-3)(q-9)(q^4 - 24q^3 + 195q^2 - 612q + 480), \\
\qquad \gcd\{6, q\} = 3, \\
q^6 - 36q^5 + 510q^4 - 3600q^3 + 13224q^2 - 23904q + 17280 \\
\quad = (q-6)^2(q^4 - 24q^3 + 186q^2 - 504q + 480), \\
\qquad \gcd\{6, q\} = 6.
\end{cases}$$

We have computed $\{e(J) : |J| \leq 1\} = \{e(J) : |J| \leq 2\} = \{e(J) : |J| \leq 3\} = \{1\}$, $\{e(J) : |J| \leq 4\} = \{e(J) : |J| \leq 5\} = \{1, 2\}$, $\{e(J) : |J| \leq 4\} = \{1, 2, 3\}$ and the constituents of the quasi-polynomial are consistent with Corollaries 2.3 and 2.4.

Remark. R. Suter [12] gave essentially the same calculation for every irreducible root system.

4. Mid-hyperplane arrangement

The mid-hyperplane arrangement was defined in [7] to find the number of "ranking patterns" generated by the unidimensional unfolding model in mathematical psychology ([5]). This arrangement is defined as follows.

Let $m \geq 4$ be an integer. We define two kinds of hyperplanes as follows:

$$\begin{aligned} H_{ij} &:= \{(\alpha_1,\dots,\alpha_m)\in\mathbb{R}^m : \alpha_i=\alpha_j\}, \quad 1\leq i<j\leq m, \\ H_{ijkl} &:= \{(\alpha_1,\dots,\alpha_m)\in\mathbb{R}^m : \alpha_i+\alpha_j=\alpha_k+\alpha_l\}, \quad (i,j,k,l)\in I_4, \end{aligned}$$

where $I_4 := \{(i,j,k,l)\colon 1\leq i<j\leq m,\ i<k<l\leq m,\ j \text{ is different from } k \text{ and } l\}$. Then the mid-hyperplane arrangement $\mathcal{M}_m$ is defined as

$$\mathcal{M}_m := \{H_{ij}\ (1\leq i<j\leq m),\ H_{ijkl}\ ((i,j,k,l)\in I_4)\}.$$

Now, let $T(\mathcal{M}_m) : m\times n$ be the coefficient matrix of $\mathcal{M}_m$, where $n=|\mathcal{M}_m| = \binom{m}{2}+3\binom{m}{4}$.

4.1. Characteristic quasi-polynomial and generating function of $\mathcal{M}_4$

When $m=4$, we have $\{e(J) : |J|\leq 1\}=\{1\}$, $\{e(J) : |J|\leq 2\}=\{e(J) : |J|\leq 3\}=\{e(J) : |J|\leq 4\}=\{1,2\}$, and thus $\rho_0=2$. The characteristic quasi-polynomial is

$$\chi_{T(\mathcal{M}_4)}(q) = \begin{cases} q^4-9q^3+23q^2-15q = q(q-1)(q-3)(q-5) & \text{if } q \text{ is odd}, \\ q^4-9q^3+26q^2-24q = q(q-2)(q-3)(q-4) & \text{if } q \text{ is even}. \end{cases}$$

From this characteristic quasi-polynomial, we obtain

$$\Phi_{T(\mathcal{M}_4)}(t) = \frac{48t^6(t^3+5t^2+7t+3)}{(1-t^2)^5} = \frac{48t^6(t+3)}{(1-t)^5(1+t)^3}.$$

4.2. Characteristic quasi-polynomial and generating function of $\mathcal{M}_5$

When $m=5$, we have $\{e(J) : |J|\leq 1\}=\{1\}$, $\{e(J) : |J|\leq 2\}=\{1,2\}$, $\{e(J) : |J|\leq 3\}=\{1,2,3\}$, $\{e(J) : |J|\leq 4\}=\{e(J) : |J|\leq 5\}=\{1,2,3,4,5,6\}$, and thus $\rho_0=60$. With the help of the relations

$$P_d(t)+P_{d'}(t)-P_{dd'}(t)-P_1(t) = \begin{cases} 0 & \text{for } (d,d')=(2,5),(3,5),(4,5),(5,6), \\ & \qquad (5,12), \\ -120t & \text{for } (d,d')=(2,3),(2,15),(3,4),(3,10), \\ & \qquad (3,20),(4,15) \end{cases}$$

(see Corollary 2.4), we obtained the constituents of $\chi_{T(\mathcal{M}_5)}(q)$ as

$$\begin{aligned} P_1(q) &= q^5-25q^4+215q^3-695q^2+504q \\ &= q(q-1)(q-7)(q-8)(q-9), \\ P_2(q) &= q^5-25q^4+230q^3-920q^2+1104q \\ &= q(q-2)(q^3-23q^2+184q-552), \end{aligned}$$

$$
\begin{aligned}
P_3(q) &= q^5 - 25q^4 + 215q^3 - 735q^2 + 864q \\
&= q(q-3)(q-9)(q^2 - 13q + 32), \\
P_4(q) &= q^5 - 25q^4 + 230q^3 - 920q^2 + 1344q \\
&= q(q-4)(q-6)(q-7)(q-8), \\
P_5(q) &= q^5 - 25q^4 + 215q^3 - 695q^2 + 600q \\
&= q(q-5)(q^3 - 20q^2 + 115q - 120), \\
P_6(q) &= q^5 - 25q^4 + 230q^3 - 960q^2 + 1584q \\
&= q(q-6)(q^3 - 19q^2 + 116q - 264), \\
P_{10}(q) &= q^5 - 25q^4 + 230q^3 - 920q^2 + 1200q \\
&= q(q-10)(q^3 - 15q^2 + 80q - 120), \\
P_{12}(q) &= q^5 - 25q^4 + 230q^3 - 960q^2 + 1824q \\
&= q(q^4 - 25q^3 + 230q^2 - 960q + 1824), \\
P_{15}(q) &= q^5 - 25q^4 + 215q^3 - 735q^2 + 960q \\
&= q(q^4 - 25q^3 + 215q^2 - 735q + 960), \\
P_{20}(q) &= q^5 - 25q^4 + 230q^3 - 920q^2 + 1440q \\
&= q(q^4 - 25q^3 + 230q^2 - 920q + 1440), \\
P_{30}(q) &= q^5 - 25q^4 + 230q^3 - 960q^2 + 1680q \\
&= q(q^4 - 25q^3 + 230q^2 - 960q + 1680), \\
P_{60}(q) &= q^5 - 25q^4 + 230q^3 - 960q^2 + 1920q \\
&= q(q^4 - 25q^3 + 230q^2 - 960q + 1920).
\end{aligned}
$$

The generating function is

$$
\begin{aligned}
\Phi_{T(\mathcal{M}_5)}(t) = \; & 240t^{11}(6t^{20} + 40t^{19} + 112t^{18} + 282t^{17} + 511t^{16} \\
& +917t^{15} + 1301t^{14} + 1818t^{13} + 2163t^{12} + 2493t^{11} \\
& +2479t^{10} + 2462t^9 + 2078t^8 + 1734t^7 + 1263t^6 \\
& +903t^5 + 523t^4 + 308t^3 + 137t^2 + 59t + 11) \\
& /\{(1-t)^6(1+t)^4(1-t+t^2)^2(1+t+t^2)^3 \\
& (1+t^2)^2(1+t+t^2+t^3+t^4)^2\}.
\end{aligned}
$$

Remark 4.1. For $\mathcal{M}_6$, we have $\{e(J) : |J| \le 1\} = \{1\}$, $\{e(J) : |J| \le 2\} = \{1,2\}$, $\{e(J) : |J| \le 3\} = \{1,2,3\}$, $\{e(J) : |J| \le 4\} = \{1,2,3,4,5,6\}$, $\{e(J) : |J| \le 5\} = \{e(J) : |J| \le 6\} = \{1,2,3,4,5,6,7,8,9,10,11\}$, and thus $\rho_0 = 27720$. It was computationally infeasible for us to obtain $\chi_{T(\mathcal{M}_6)}(q)$, so we cannot tell whether this ρ_0 is the minimum period or not.

References

[1] M. Aigner, *Combinatorial Theory*, Springer, Berlin, 1979.

[2] C. A. Athanasiadis, Characteristic polynomials of subspace arrangements and finite fields, *Adv. Math.* **122** (1996), 193–233.

[3] A. Blass and B. Sagan, Characteristic and Ehrhart polynomials, *J. Algebraic Combin.* **7** (1998), 115–126.

[4] N. Bourbaki, *Lie Groups and Lie Algebras: Chapters 4-6,* Springer-Verlag, Berlin-Heidelberg-New York, 2002

[5] C. H. Coombs, *A Theory of Data,* John Wiley & Sons, New York, 1964.

[6] M. Haiman, Conjectures on the quotient ring of diagonal invariants, *J. Algebraic Combin.* **3** (1994), 17–76.

[7] H. Kamiya, P. Orlik, A. Takemura and H. Terao, Arrangements and ranking patterns, *Ann. Comb.* **10** (2006), 219-235.

[8] H. Kamiya, A. Takemura and H. Terao, Periodicity of hyperplane arrangements with integral coefficients modulo positive integers, *J. Alg. Combin.* **27** (2008), 317–330.

[9] P. Orlik and H. Terao, *Arrangements of Hyperplanes,* Springer-Verlag, Berlin, 1992.

[10] PARI/GP (http://pari.math.u-bordeaux.fr).

[11] R. Stanley, *Enumerative Combinatorics,* vol. I, Cambridge University Press, Cambridge, 1997.

[12] R. Suter, The number of lattice points in alcoves and the exponents of the finite Weyl group, *Math. Comp.*, **67** (1998), 751-758.

Hidehiko Kamiya
Faculty of Economics
Okayama University
Okayama 700-8530
Japan
e-mail: `kamiya@e.okayama-u.ac.jp`

Akimichi Takemura
Graduate School of Information Science and Technology
University of Tokyo
Tokyo 113-0033
Japan
e-mail: `takemura@stat.t.u-tokyo.ac.jp`

Hiroaki Terao
Department of Mathematics
Hokkaido University
Sapporo 060-0810
Japan
e-mail: `terao@math.sci.hokudai.ac.jp`

Progress in Mathematics, Vol. 283, 191–207

Toric Varieties and the Diagonal Property

Ali Ulaş Özgür Kişisel and Özer Öztürk

Abstract. A smooth variety X of dimension n is said to satisfy the diagonal property if there exists a vector bundle $\mathcal{E}$ of rank n on $X \times X$ and a section s of $\mathcal{E}$ such that the image $\Delta(X)$ of the diagonal embedding of X into $X \times X$ is the zero scheme of s. A study of varieties satisfying the diagonal property was begun by Pragacz, Srinivas and Pati, in [8]. Even though there are many cases where the answer is affirmative, only in a few examples an explicit description of the vector bundle is known. After an exposition of toric varieties, we discuss this question in the particular case when X is a toric surface, in search for such examples.

Keywords. Toric variety, toric surface, torus action, convex cone, Hirzebruch surface, diagonal property, vector bundle.

1. Introduction

These notes are based partly on a lecture given by the first author in the CIMPA Summer School on Hyperplane Arrangements and Singularities that took place between June 11-22, 2007 at Galatasaray University in İstanbul, and partly on the research of the authors on the diagonal property on toric surfaces. Section 2 is of expository character, and can be regarded as a standard introduction to toric varieties. In section 3, we introduce the diagonal property and present some of the results in [8] without proofs. Section 4 contains a discussion of the diagonal property on toric surfaces through examples, based on the work of the authors. A complete discussion for all smooth toric surfaces will be presented in [7].

An algebraic variety X is said to satisfy the **diagonal property** if there exists a vector bundle $\mathcal{E}$ on $X \times X$ of rank equal to the dimension of X, and a section $s : X \times X \to \mathcal{E}$ of $\mathcal{E}$ such that the zero scheme of s coincides with $\Delta(X)$. After several appearances in the literature in relation to other problems in algebraic geometry, a systematic study of the diagonal property was started by Pragacz, Srinivas, and Pati, in [8] (see also [9]). Among other results, in [8] one can find a detailed discussion of algebraic surfaces satisfying the diagonal property. However, even in the cases when the surface is known to satisfy the property, the proof is often not constructive. In fact, varieties known to satisfy the diagonal property via an explicit construction of the required bundle are rather rare. The known examples are curves, flag varieties of the form SL_n/P and their products.

A natural task then is to expand the list of examples for which the vector bundle in question can be determined explicitly. Toric varieties are natural candidates due to their combinatorial nature, and also due to the fact that there seem to be no rational varieties known not to satisfy the diagonal property as yet. In this paper we construct the bundles in question for some smooth toric surfaces; namely $\mathbb{P}^2$, $\mathbb{P}^1 \times \mathbb{P}^1$ and $\mathbb{F}_1$ under the same footing. It should be noted that by the results in [8] it follows that any smooth toric surface satisfies the diagonal property. The explicit construction for a general smooth toric surface will appear in [7].

Acknowledgments

We thank Prof. Piotr Pragacz for introducing us to the subject, and for helpful discussions. We also thank the organizers of the CIMPA school and the Galatasaray University for their hospitality and the stimulating atmosphere. We thank the referee for providing many helpful corrections and comments.

2. Toric Varieties

2.1. Definition and examples

Definition 1. An irreducible variety X over $\mathbb{C}$ of dimension n is said to be a **toric variety** if

- X is normal,
- there exists an embedding of $(\mathbb{C}^*)^n$ into X as a Zariski open subset,
- there is an action of $(\mathbb{C}^*)^n$ on X which agrees with the canonical action of the algebraic torus $(\mathbb{C}^*)^n$ on $(\mathbb{C}^*)^n \subset X$.

Remark. The data for a toric variety includes X, a choice of embedding $i : (\mathbb{C}^*)^n \hookrightarrow X$, and the torus action all together. But with abuse of terminology, one usually just says that "X is a toric variety". In some contexts, especially those concerning local algebraic geometry, the assumption about normality is dropped.

Since X is irreducible, the map $i : (\mathbb{C}^*)^n \hookrightarrow X$ has an inverse on the Zariski open set $i((\mathbb{C}^*)^n)$, and this shows that any toric variety is rational.

Examples. $(\mathbb{C}^*)^n$ is clearly a toric variety. $\mathbb{C}^n$ is a toric variety if we extend the action of $(\mathbb{C}^*)^n$ to $\mathbb{C}^n$ by the formula $(\lambda_1, \dots, \lambda_n) \cdot (x_1, \dots, x_n) = (\lambda_1 x_1, \dots, \lambda_n x_n)$. $\mathbb{P}^n$ is a toric variety as well: Fix an embedding of $(\mathbb{C}^*)^n$ into $\mathbb{P}^n$, say $(\lambda_1, \dots, \lambda_n) \mapsto [\lambda_1 : \dots : \lambda_n : 1]$. Then the action of $(\mathbb{C}^*)^n$ extends to $\mathbb{P}^n$ by the formula $(\lambda_1, \dots, \lambda_n) \cdot [X_0 : \dots : X_{n-1} : X_n] = [\lambda_1 X_0 : \dots : \lambda_n X_{n-1} : X_n]$. The product of any two toric varieties is a toric variety.

Proposition 1. *Let X be a toric variety. Then $(\mathbb{C}^*)^n \subset X$ is the unique Zariski open orbit of the torus action on X.*

Proof. The intersection of any two distinct orbits for a group action is the empty set. Thus, if there exists another Zariski open orbit for the torus action, then it is disjoint from $(\mathbb{C}^*)^n \subset X$. But this would contradict the irreducibility of X. □

2.2. Affine toric varieties

Denote the $\mathbb{C}$-algebra of Laurent polynomials $\mathbb{C}[z_1, z_2, \dots, z_n, z_1^{-1}, z_2^{-1}, \dots, z_n^{-1}]$ by $\mathbb{C}[\mathbb{Z}^n]$. Notice that

$$(\mathbb{C}^*)^n = Spec(\mathbb{C}[\mathbb{Z}^n]).$$

Suppose that X is an affine toric variety of dimension n. Then $X = Spec(A)$ for some finitely generated, reduced $\mathbb{C}$-algebra A. Then $A = \mathcal{O}(X)$, the $\mathbb{C}$ algebra of regular functions on X. Since $i : (\mathbb{C}^*)^n \hookrightarrow X$ is a morphism of affine algebraic varieties, there is an induced homomorphism of $\mathbb{C}$-algebras

$$i^* : A \to \mathbb{C}[\mathbb{Z}^n]$$

where i^* is characterized by the relation $i^*(f)(c) = f(i(c))$ for any $f \in A, c \in (\mathbb{C}^*)^n$ (see [5], Ch. I, Prop 3.5).

Proposition 2. *i^* is a monomorphism of $\mathbb{C}$-algebras.*

Proof. This immediately follows from the hypothesis that i is a dominant map. □

Therefore, we may identify A with a $\mathbb{C}$-subalgebra of $\mathbb{C}[\mathbb{Z}^n]$. From now on, we will often make this identification and ignore the map i^*.

Proposition 3. *A can be generated by monomials in $\mathbb{C}[\mathbb{Z}^n]$.*

Proof. By definition of a toric variety, there is a compatible $(\mathbb{C}^*)^n$ action on the flag of varieties $(\mathbb{C}^*)^n \subset X$, which induces a compatible dual action on the pair of algebras of regular functions $A \subset \mathbb{C}[\mathbb{Z}^n]$ on these varieties. More precisely, if $\lambda \in (\mathbb{C}^*)^n$ and $f \in A$, define $\lambda \cdot f$ by the relation $(\lambda \cdot f)(x) = f(\lambda^{-1} \cdot x)$ for $x \in X$. Then $(\lambda \cdot f)(\lambda \cdot x) = f(x)$, so $\lambda \cdot f$ is regular on X and therefore must belong to A. But it is easy to see that any subalgebra of $\mathbb{C}[\mathbb{Z}^n]$ which is closed under this action must contain the monomial summands of each of its elements. □

Since $A = \mathcal{O}(X)$ is a priori finitely generated, the monomial summands of any finite set of generators would generate A. Therefore A can be generated by finitely many monomials.

Given any subalgebra B of $\mathbb{C}[\mathbb{Z}^n]$ generated by monomials, let

$$S_B = \{(a_1, \dots, a_n) \in \mathbb{Z}^n | z_1^{a_1} \dots z_n^{a_n} \in B\}.$$

It is easy to see that S_B is a sub-semigroup of $\mathbb{Z}^n$. We will also use the notation $B = \mathbb{C}[S_B]$ of which $\mathbb{C}[\mathbb{Z}^n]$ is an example. The discussion above shows that, given any affine toric variety X of dimension n, we can associate to it a sub-semigroup $S_{\mathcal{O}(X)}$ of $\mathbb{Z}^n$. Denote $S_{\mathcal{O}(X)}$ by S_X for brevity.

Definition 2. A sub-semigroup S of $\mathbb{Z}^n$ is called **saturated** if $ms \in S$ with m a positive integer and $s \in \mathbb{Z}^n$ implies $s \in S$.

Example. The sub-semigroup S of $\mathbb{Z}$ generated by 2 and 3 is not saturated since 1 is not in S.

Definition 3. The group **generated** by a sub-semigroup S of $\mathbb{Z}^n$ is the intersection of all subgroups of $\mathbb{Z}^n$ containing S, and will be denoted by $G(S)$.

Theorem 1. *Suppose that X is an n-dimensional affine toric variety. Then S_X is a saturated, finitely generated sub-semigroup of $\mathbb{Z}^n$ such that $G(S) = \mathbb{Z}^n$. Conversely, given any saturated, finitely generated sub-semigroup S of $\mathbb{Z}^n$ such that $G(S) = \mathbb{Z}^n$, there exists an n-dimensional affine toric variety X such that $S = S_X$.*

Proof. Say X is an n-dimensional affine toric variety. Since X is rational, the function field $k(X)$ of X is isomorphic to $k((\mathbb{C}^*)^n) = \mathbb{C}(x_1, \dots, x_n)$. In particular $G(S_X) = \mathbb{Z}^n$. Also, $\mathcal{O}(X)$ can be generated by finitely many monomials, therefore S_X is a finitely generated semigroup. Suppose, contrary to the assertion, that S_X is not saturated. Then there exists $s = x_1^{a_1} \dots x_n^{a_n} \in \mathbb{Z}^n$ such that $ms \in S_X$ but s is not in S_X. However, $s \in k(X)$ and the relation $s^m - x^{ma_1} \dots x^{ma_n} = 0$ shows that s is integral over $\mathcal{O}(X)$. This contradicts the normality of X.

Conversely, suppose that S is a saturated, finitely generated sub-semigroup of $\mathbb{Z}^n$ such that $G(S) = \mathbb{Z}^n$. Let $\mathbb{C}[S]$ denote the algebra generated by the monomials corresponding to the elements of S. It is a finitely generated $\mathbb{C}$-algebra since S is finitely generated. $\mathbb{C}[S]$ is an integral domain, so X is irreducible. Therefore, $X = Spec(\mathbb{C}[S])$ is an affine variety. Since $G(S) = \mathbb{Z}^n$, $k(X) = k(x_1, \dots, x_n)$, which shows that X is n-dimensional. The inclusion $\mathbb{C}[S] \subset \mathbb{C}[x_1, x_1^{-1}, \dots, x_n, x_n^{-1}]$ induces a regular map $i : (\mathbb{C}^*)^n \to X$. The equality of function fields $k(X) = k(x_1, \dots, x_n)$ implies that i is injective. So i embeds $(\mathbb{C}^*)^n$ in X as a Zariski open subset. Finally, define a $\mathbb{C}$-algebra homomorphism

$$\Psi : \mathbb{C}[S] \to \mathbb{C}[\lambda_1, \lambda_1^{-1}, \dots, \lambda_n, \lambda_n^{-1}] \otimes \mathbb{C}[X]$$

by

$$\Psi(x_1^{a_1} \dots x_n^{a_n}) = \lambda_1^{a_1} \dots \lambda_n^{a_n} \otimes x_1^{a_1} \dots x_n^{a_n}.$$

Taking *Spec*'s, we see that this induces an action of $(\mathbb{C}^*)^n$ on X compatible with the natural action of $(\mathbb{C}^*)^n$ on itself. Therefore, X is an n-dimensional affine toric variety. □

Definition 4. Let $M = \mathbb{Z}^n \otimes \mathbb{R} \cong \mathbb{R}^n$. Given $v_1, \dots, v_m \in \mathbb{R}^n$, the **cone** generated by $v_1, \dots, v_m$ in M is the set

$$\{c_1v_1 + c_2v_2 + \dots + c_mv_m | c_1, \dots, c_m \in \mathbb{R}_{\geq 0}\}$$

(which will also be called the **positive hull** [1] of $v_1, \dots, v_m$). If furthermore $v_1, \dots, v_m \in \mathbb{Z}^n$, then σ will be called a **lattice cone.**

---

[1] The term "nonnegative hull" would be more appropriate, but it is customary to make this abuse of language

Lemma 1 (Gordan's lemma). *If $\sigma \subset M$ is a lattice cone, then $\sigma \cap \mathbb{Z}^n$ is a finitely generated sub-semigroup of $\mathbb{Z}^n$.*

Proof. Suppose that σ is the positive hull of $v_1, \ldots, v_m \in \mathbb{Z}^n$, and that $\{v_1, \ldots, v_m\}$ is linearly independent. Let

$$F_\sigma = \{c_1 v_1 + \ldots + c_m v_m | 0 \leq c_i \leq 1 \forall i\}.$$

Then F_σ is compact, therefore $F_\sigma \cap \mathbb{Z}^n$ is a finite set. Note that $v_1, \ldots, v_m \in F_\sigma \cap \mathbb{Z}^n$. We claim that $F_\sigma \cap \mathbb{Z}^n$ generates $\sigma \cap \mathbb{Z}^n$ as a semigroup. Indeed, let $x = c_1 v_1 + \ldots + c_m v_m \in \sigma \cap \mathbb{Z}^n$. Then, $(c_1 - \lfloor c_1 \rfloor) v_1 + \ldots + (c_m - \lfloor c_m \rfloor) v_m \in \sigma \cap \mathbb{Z}^n$ where $\lfloor c \rfloor$ denotes the greatest integer less than or equal to c. But $(c_1 - \lfloor c_1 \rfloor) v_1 + \ldots + (c_m - \lfloor c_m \rfloor) v_m \in F_\sigma \cap \mathbb{Z}^n$. This shows that $F_\sigma \cap \mathbb{Z}^n$ generates $\sigma \cap \mathbb{Z}^n$ as a semigroup. □

Theorem 2. *Let $S \subset \mathbb{Z}^n$ be a saturated, finitely generated semigroup. Let σ_S denote the lattice cone generated by the elements of S. Then, $S = \sigma_S \cap \mathbb{Z}^n$. Conversely, given any lattice cone $\sigma \subset \mathbb{R}^n$, $\sigma \cap \mathbb{Z}^n$ is a saturated, finitely generated semigroup.*

Proof. $S \subset \sigma_S \cap \mathbb{Z}^n$ is clear. Say $v = c_1 v_1 + \ldots + c_m v_m \in \sigma_S \cap \mathbb{Z}^n$ where $c_1, \ldots, c_m \in \mathbb{R}$. We may assume that $\{v_1, \ldots, v_m\}$ is a basis for the $\mathbb{Q}$-linear space generated by S without loss of generality. v belongs to this $\mathbb{Q}$-linear space, thus $c_1, \ldots, c_m$, being the coordinates of v with respect to this basis, are uniquely determined, and are in $\mathbb{Q}$. Multiplying with a common denominator, we see that there exists a positive integer l such that $lv \in S$. But S is saturated, therefore $v \in S$. Hence $\sigma_S \cap \mathbb{Z}^n \subset S$, so $\sigma_S \cap \mathbb{Z}^n = S$

Conversely, suppose that $\sigma \subset \mathbb{R}^n$ is a lattice cone. Then $S = \sigma \cap \mathbb{Z}^n$ is clearly a saturated semigroup, and it is finitely generated by Gordan's lemma. Hence the claim is established. □

Theorems 1 and 2 imply that, to any n-dimensional affine toric variety X, there corresponds an n-dimensional lattice cone σ_{S_X}, which will be denoted by σ_X for brevity. Conversely, given any n-dimensional lattice cone $\sigma \subset \mathbb{R}^n$, there exists an n-dimensional affine toric variety X such that $\sigma = \sigma_X$.

Suppose that X_1, X_2 are n-dimensional affine toric varieties, and $\sigma_{X_1}, \sigma_{X_2}$ corresponding n-dimensional lattice cones. Suppose that there exists $A \in SL(n, \mathbb{Z})$ such that $A\sigma_{X_1} = \sigma_{X_2}$. Then A induces a morphism of algebras $\varphi_A : \mathbb{C}[S_{X_1}] \to \mathbb{C}[S_{X_2}]$, and consequently a morphism of varieties $\Psi_A : X_2 \to X_1$. Clearly, $\Psi_{A^{-1}} = (\Psi_A)^{-1}$, and this implies that X_1 and X_2 are isomorphic.

Lemma 2. *Let X be an n-dimensional affine toric variety, and $\sigma_X \subset \mathbb{R}^n$ be the corresponding n-dimensional lattice cone. If there exists $v \neq 0$ such that $v \in \sigma_X$ and $-v \in \sigma_X$, then X is isomorphic to $\mathbb{C}^* \times Y$ where Y is an $n-1$-dimensional affine toric variety.*

Proof. Since σ_X is saturated, without loss of generality we may assume that v is a primitive vector, i.e., $rv \in \mathbb{Z}^n$ with $r \in \mathbb{Z}$ if and only if $v \in \mathbb{Z}^n$. Then, there exists $A \in SL(n, \mathbb{Z})$ such that $Av = (1, 0, \ldots, 0)$. Let us apply the isomorphism

Ψ_A to X and work with the new variety. Let $H \subset \mathbb{R}^n$ be the hyperplane $a_1 = 0$, and $\sigma' = \sigma_X \cap H$. Then σ' is a lattice cone and since σ_X is saturated, so is σ'. By Theorems 1 and 2, there exists an affine $n-1$-dimensional toric variety Y such that $\sigma' = \sigma_Y$. Since $\sigma_X \cap \mathbb{Z}^n = (\sigma' \cap \mathbb{Z}^{n-1}) \oplus \mathbb{Z}v$, we have $X \cong \mathbb{C}^* \times Y$. □

How can we test if a given affine toric variety X is singular or not by looking at σ_X? In view of the previous lemma, it is enough to consider the case where σ_X does not contain any linear subspace.

Theorem 3. *Let X be an n-dimensional affine toric variety, $\sigma_X \subset \mathbb{R}^n$ be the corresponding n-dimensional lattice cone, and assume that $v \in \sigma_X$ and $-v \in \sigma_X$ implies that $v = 0$. Then, X is nonsingular if and only if $\sigma_X \cap \mathbb{Z}^n$ is generated by n lattice vectors $v_1, \ldots, v_n$.*

Proof. First, suppose that $\sigma_X \cap \mathbb{Z}^n$ is generated by n lattice vectors $v_1, \ldots, v_n$. Then $|\det(v_1, \ldots, v_n)| = 1$, otherwise the parallelpiped with sides $v_1, \ldots, v_n$ would contain an interior integral point which would belong to $\sigma_X \cap \mathbb{Z}^n$ but not to $\mathbb{Z}[v_1, \ldots, v_n]$, which is a contradiction. Assume that $\det(v_1, \ldots, v_n) = 1$, by reordering the v_i's if necessary. Then there exists $A \in SL(n, \mathbb{Z})$ such that Av_i is the i-th standard basis vector for $\mathbb{R}^n$. This proves that $X \cong \mathbb{C}^n$, in particular X is nonsingular.

Now assume that X is nonsingular. Consider the ideal $\mathbf{m}$ of $\mathbb{C}[\sigma_X \cap \mathbb{Z}^n]$ generated by all monomials $x_1^{a_1} \ldots x_n^{a_n}$ in this ring where at least one of the a_i's is not 0. This ideal is proper since the cone σ_X does not contain any nontrivial linear subspaces. Then it is easy to see that $\mathbf{m}$ is a maximal ideal. By Nullstellensatz, there exists a closed point $z \in X$ corresponding to this maximal ideal. Since X is nonsingular, in particular z is a nonsingular point, and therefore we must have $dim_{\mathbb{C}}(\mathbf{m}/\mathbf{m}^2) = n$. But σ_X is an n-dimensional cone which doesn't contain nontrivial linear subspaces, thus it must have at least n one-dimensional faces. Monomials corresponding to primitive vectors along one-dimensional faces are independent elements in the vector space $\mathbf{m}/\mathbf{m}^2$, therefore, σ_X must have exactly n one-dimensional faces. The primitive vectors along them must generate $\sigma_X \cap \mathbb{Z}^n$, otherwise one would need extra generators for $\mathbf{m}/\mathbf{m}^2$. This proves the claim. □

2.3. Fans

Let $(\mathbb{R}^n)^*$ denote the dual vector space to $\mathbb{R}^n$, and let $< \ | \ >$ denote the natural pairing between $\mathbb{R}^n$ and $(\mathbb{R}^n)^*$.

Definition 5. Suppose that $\sigma \subset \mathbb{R}^n$ is a cone. Let

$$\check{\sigma} = \{y \in (\mathbb{R}^n)^* | < x|y > \geq 0 \ \forall x \in \sigma\} \subset (\mathbb{R}^n)^*.$$

Then, $\check{\sigma}$ is called the **dual cone** to σ.

It is easy to see that if σ is a lattice cone, then $\check{\sigma}$ is a lattice cone in $(\mathbb{R}^n)^*$.

Lemma 3. *Identifying $(\mathbb{R}^n)^{**}$ with $\mathbb{R}^n$ via the canonical homomorphism, $\check{\check{\sigma}} = \sigma$.*

Proof. $x \in \check{\check{\sigma}}$ if and only if $< x|y > \geq 0$ for all $y \in \check{\sigma}$. But this happens if and only if $x \in \sigma$. □

Lemma 4. $\sigma_1 \subset \sigma_2 \Leftrightarrow \check{\sigma}_2 \subset \check{\sigma}_1$.

Proof. Say $y \in \check{\sigma}_2$. Then $< x|y > \geq 0$ for all $x \in \sigma_2$, in particular for all $x \in \sigma_1$. So $y \in \check{\sigma}_1$, and consequently $\check{\sigma}_2 \subset \check{\sigma}_1$. The converse can be proven by the same argument, using Lemma 3. □

If σ_1 and σ_2 are two cones in $\mathbb{R}^n$, let $\sigma_1 + \sigma_2 = \{x + y | x \in \sigma_1, y \in \sigma_2\}$. Then $\sigma_1 + \sigma_2$ is a cone.

Lemma 5. $(\sigma_1 + \sigma_2)\check{} = \check{\sigma}_1 \cap \check{\sigma}_2$.

Proof.

$$\begin{aligned} y \in (\sigma_1 + \sigma_2)\check{} &\Leftrightarrow\ < x|y > \geq 0 \quad \forall x \in \sigma_1 + \sigma_2 \\ &\Leftrightarrow\ < x|y > \geq 0 \quad \forall x \in \sigma_1, \quad \forall x \in \sigma_2 \\ &\Leftrightarrow\ y \in \check{\sigma}_1 \cap \check{\sigma}_2. \end{aligned}$$ □

Definition 6. A cone $\sigma \subset \mathbb{R}^n$ is said to be **strongly convex** if $\sigma \cap (-\sigma) = \{0\}$.

Lemma 6. *σ is strongly convex if and only if $\check{\sigma}$ has dimension n.*

Proof. $\check{\sigma}$ has dimension less than n if and only if it is contained in a hyperplane in $(\mathbb{R}^n)^*$. But this happens if and only if there exists $0 \neq x \in \sigma$ such that $< x|y >= 0$ for all $y \in \check{\sigma}$. And the latter is equivalent to the assertion that there exists $0 \neq x \in \sigma$ such that $-x \in \sigma$. This finishes the proof. □

Definition 7. Suppose that σ is a cone generated by $\mathcal{V} = \{v_1, \dots, v_m\}$, and assume that none of the v_i's is contained in the positive hull of $\mathcal{V} - \{v_i\}$ (i.e., $\mathcal{V}$ is a *minimal* set of vectors generating the cone σ). Suppose that $\mathcal{V}' \subset \mathcal{V}$. Then the cone generated by the elements of $\mathcal{V}'$ is said to be a **face** of σ. (The cone generated by the empty subset is assumed to consist of just the origin.)

Definition 8. A nonempty finite set Σ of strongly convex lattice cones in $\mathbb{R}^n$ is called a **fan** if the following conditions are satisfied:

- $\sigma \in \Sigma$ implies that all faces of σ are in Σ,
- for all $\sigma_1, \sigma_2 \in \Sigma$, $\sigma_1 \cap \sigma_2$ is a common face of σ_1 and σ_2.

Notice that since 0 is a face of every lattice cone, it belongs to any fan Σ.

Given any fan $\Sigma \subset \mathbb{R}^n$, we would like to construct a corresponding toric variety X_Σ. If the fan consists in a unique cone σ of maximal dimension n and its proper faces, then X will be the affine toric variety corresponding to $\check{\sigma}$. For a general fan, the corresponding toric variety is obtained by gluing the affine varieties obtained from the maximal cones. We will describe this procedure now.

Given $\sigma \in \Sigma$, let $U_\sigma = Spec(\mathbb{C}[\check{\sigma} \cap \mathbb{Z}^n])$. We proved in the previous section that U_σ is an affine toric variety. Since σ is strongly convex, $\check{\sigma}$ is an n-dimensional cone. This implies that U_σ has dimension n. Notice that $U_{\{0\}} = (\mathbb{C}^*)^n$.

Proposition 4. *Suppose that $\sigma \in \Sigma$ and τ is a face of σ. Then $U_\tau \subset U_\sigma$ is a Zariski open subset.*

Proof. $\check{\sigma} \subset \check{\tau}$ since $\tau \subset \sigma$. Therefore $\mathbb{C}[\check{\sigma} \cap \mathbb{Z}^n]$ is a subalgebra of $\mathbb{C}[\check{\tau} \cap \mathbb{Z}^n]$. This induces a regular map i from $U_\tau = Spec(\mathbb{C}[\check{\tau} \cap \mathbb{Z}^n])$ to $U_\sigma = Spec(\mathbb{C}[\check{\sigma} \cap \mathbb{Z}^n])$. Since $\check{\sigma}$ and $\check{\tau}$ are both n-dimensional, U_τ and U_σ both have $k(x_1, \ldots, x_n)$ as their function fields. This implies that i is injective. □

Now, consider the collection of affine toric varieties U_σ as σ ranges over Σ. For any $\sigma_1, \sigma_2 \in \Sigma$, $\sigma_1 \cap \sigma_2 \in \Sigma$ is a common face of σ_1 and σ_2. Therefore there exist injective regular maps $i_1 : U_{\sigma_1 \cap \sigma_2} \hookrightarrow U_{\sigma_1}$ and $i_2 : U_{\sigma_1 \cap \sigma_2} \hookrightarrow U_{\sigma_2}$. Furthermore, if $\delta \subset \tau \subset \sigma$ are three cones in Σ, then it is clear that the composition of the inclusions $U_\delta \hookrightarrow U_\tau$ and $U_\tau \hookrightarrow U_\sigma$ agrees with the inclusion $U_\delta \hookrightarrow U_\sigma$. Therefore, it is possible to glue all U_σ for $\sigma \in \Sigma$ together along their common intersections in order to obtain a scheme X_Σ (see [5] pg. 87, thm. 3.3 for a more general discussion of the existence of a fibred product, which in particular implies the existence of the scheme X_Σ).

Theorem 4. *X_Σ is a toric variety.*

Proof. (see also [2]) First of all, let us show that X_Σ is a variety. This amounts to showing that X_Σ is an integral, separated scheme of finite type. By [5], proposition 3.1, page 82, a scheme is integral if and only if it is both reduced and irreducible. A variety is reduced if there are no nilpotent elements in its structure sheaf. Since this is a local property, and is true for each U_σ, it is true for X_Σ. To see that X_Σ is irreducible, note that every U_σ contains $(\mathbb{C}^*)^n$, and that every U_σ is irreducible. X_Σ is clearly of finite type since Σ is a finite set and the ring of regular functions of each U_σ is a finitely generated $\mathbb{C}-$algebra.

It remains to prove that X_Σ is a separated scheme. By definition, this amounts to showing that the diagonal morphism $\Delta : X_\Sigma \to X_\Sigma \times X_\Sigma$ is a closed immersion, i.e., that its image is closed and the induced map $i^\# : \mathcal{O}_{X_\Sigma \times X_\Sigma} \to \mathcal{O}_{X_\Sigma}$ of structure sheaves is surjective. To see this, note that the Zariski topology on X_Σ is generated by those on U_σ's. Therefore, it is enough to show that the image of $\Delta : U_{\sigma_1 \cap \sigma_2} \to U_{\sigma_1} \times U_{\sigma_2}$ is closed for all $\sigma_1, \sigma_2 \in \Sigma$. But this is equivalent to the claim that the multiplication map

$$\varphi : \mathbb{C}[\check{\sigma}_1 \cap \mathbb{Z}^n] \otimes \mathbb{C}[\check{\sigma}_2 \cap \mathbb{Z}^n] \to \mathbb{C}[(\sigma_1 \cap \sigma_2)\check{} \cap \mathbb{Z}^n]$$

is surjective. But by Lemma 5, $(\sigma_1 \cap \sigma_2)\check{} = \check{\sigma}_1 + \check{\sigma}_2$, so $\mathbb{C}[(\sigma_1 \cap \sigma_2)\check{} \cap \mathbb{Z}^n] = \mathbb{C}[(\check{\sigma}_1 \cap \mathbb{Z}^n) + (\check{\sigma}_2 \cap \mathbb{Z}^n)]$. Then it is clear that φ is surjective. Therefore X_Σ is a variety.

Let us now show that X is a toric variety. By composing $i : (\mathbb{C}^*)^n \hookrightarrow U_\sigma$ and $U_\sigma \hookrightarrow X_\Sigma$ for any $\sigma \in \Sigma$, we get an injective regular map from $(\mathbb{C}^*)^n$ to X_Σ. Since the torus $(\mathbb{C}^*)^n$ acts on each U_σ, this gives an action on X_Σ. Finally, X_Σ is normal since every local ring of each U_σ and therefore every local ring of X_Σ, is normal. □

Theorem 4 has a converse, whose proof we will omit.

Theorem 5. *Let X be an arbitrary n-dimensional toric variety. Then there exists a fan Σ of cones in $\mathbb{R}^n$ such that $X \cong X_\Sigma$.*

Proof. See [6], I.2, theorem 6. □

Example. In $\mathbb{R}^2$, let $\sigma_0 = PHull((1,0),(0,1))$, $\sigma_1 = PHull((0,1),(-1,-1))$, and $\sigma_2 = PHull((1,0),(-1,-1))$ where $PHull(v_1,\dots,v_n)$ denotes the positive hull of $v_1,\dots,v_n$. Then, let Σ be the fan consisting in $\sigma_0,\sigma_1,\sigma_2$ and their proper faces $\sigma_{01} = \sigma_0 \cap \sigma_1$, $\sigma_{02} = \sigma_0 \cap \sigma_2$, $\sigma_{12} = \sigma_1 \cap \sigma_2$ and $\sigma_{012} = \sigma_0 \cap \sigma_1 \cap \sigma_2 = \{0\}$. Then, the following holds:

$$\begin{aligned}
U_{\sigma_0} &= Spec(\mathbb{C}[x,y]) \cong \mathbb{C}^2\\
U_{\sigma_1} &= Spec(\mathbb{C}[x^{-1},x^{-1}y]) \cong \mathbb{C}^2\\
U_{\sigma_2} &= Spec(\mathbb{C}[xy^{-1},y^{-1}]) \cong \mathbb{C}^2\\
U_{\sigma_{01}} &= Spec(\mathbb{C}[x,x^{-1},y]) \cong \mathbb{C}\times\mathbb{C}^*\\
U_{\sigma_{02}} &= Spec(\mathbb{C}[x,y,y^{-1}]) \cong \mathbb{C}\times\mathbb{C}^*\\
U_{\sigma_{12}} &= Spec(\mathbb{C}[xy^{-1},x^{-1}y,y^{-1}]) \cong \mathbb{C}\times\mathbb{C}^*\\
U_{\sigma_{123}} &= Spec(\mathbb{C}[x,x^{-1},y,y^{-1}]) \cong (\mathbb{C}^*)^2.
\end{aligned}$$

Introduce the homogenous coordinates $[X:Y:Z]$ subject to the relations $x = \frac{X}{Z}$, $y = \frac{Y}{Z}$. Then, $\mathbb{C}[x,y] = \mathbb{C}[\frac{X}{Z},\frac{Y}{Z}]$, $\mathbb{C}[xy^{-1},y^{-1}] = \mathbb{C}[\frac{X}{Y},\frac{Z}{Y}]$, and $\mathbb{C}[x^{-1},x^{-1}y] = \mathbb{C}[\frac{Z}{X},\frac{Y}{X}]$. The gluing in this case coincides with the standard gluing of three copies of $\mathbb{A}^2$ in the process of obtaining $\mathbb{P}^2$. Therefore, $X_\Sigma \cong \mathbb{P}^2$.

Example. Let $\sigma_0 = \mathbb{R}_{\geq 0}$, $\sigma_1 = \mathbb{R}_{\leq 0}$ and $\sigma_{01} = \{0\}$ and $\Sigma = \{\sigma_0,\sigma_1,\sigma_{01}\}$. Then, $U_{\sigma_0} = Spec(\mathbb{C}[x])$, $U_{\sigma_1} = Spec(\mathbb{C}[x^{-1}])$, and $U_{\sigma_{01}} = Spec(\mathbb{C}[x,x^{-1}])$. It is easy to see that $X_\Sigma \cong \mathbb{P}^1$. Let us consider the fan $\Sigma\times\Sigma$ consisting of the cones $\sigma\times\sigma' \subset \mathbb{R}^2$ where σ,σ' range over Σ. The affine subvarieties corresponding to the maximal cones are as follows.

$$\begin{aligned}
U_{\sigma_0\times\sigma_0} &= Spec(\mathbb{C}[x,y]) \cong \mathbb{C}^2\\
U_{\sigma_0\times\sigma_1} &= Spec(\mathbb{C}[x,y^{-1}]) \cong \mathbb{C}^2\\
U_{\sigma_1\times\sigma_0} &= Spec(\mathbb{C}[x^{-1},y]) \cong \mathbb{C}^2\\
U_{\sigma_1\times\sigma_1} &= Spec(\mathbb{C}[x^{-1},y^{-1}]) \cong \mathbb{C}^2.
\end{aligned}$$

The affine subvarieties corresponding to the proper faces of these cones can be computed analogously. If we introduce two pairs of homogenous coordinates $[X_0 : X_1]$ and $[Y_0 : Y_1]$ so that $x = \frac{X_1}{X_0}$ and $y = \frac{Y_1}{Y_0}$, we see that the gluing can be performed separately in each coordinate, hence $X_{\Sigma\times\Sigma} \cong \mathbb{P}^1\times\mathbb{P}^1$.

Definition 9. Let Σ_1 and Σ_2 be two fans consisting of cones in $\mathbb{R}^n$ and $\mathbb{R}^m$ respectively. Then the **product fan** $\Sigma_1\times\Sigma_2$ of cones in $\mathbb{R}^{n+m}$ is defined as

$$\Sigma_1\times\Sigma_2 = \{\sigma\times\sigma' \mid \sigma\in\Sigma_1, \sigma'\in\Sigma_2\}.$$

Proposition 5. $X_{\Sigma_1\times\Sigma_2} \cong X_{\Sigma_1}\times X_{\Sigma_2}$.

Proof. $(\sigma \times \sigma')\check{} = \check{\sigma} \times \check{\sigma}' \subset (\mathbb{R}^{n+m})^*$. Therefore, $S_{\sigma\times\sigma'} \cong S_\sigma \oplus S_{\sigma'}$. This implies that $\mathbb{C}[S_{\sigma\times\sigma'}] \cong \mathbb{C}[S_\sigma \oplus S'_\sigma] \cong \mathbb{C}[S_\sigma]\otimes\mathbb{C}[S_{\sigma'}]$, and consequently $U_{\sigma\times\sigma'} \cong U_\sigma \times U_{\sigma'}$. Since the gluing can be performed first for intersections within one of the fans and then the other, the gluing process for the separate factors do not intervene each other, and one obtains $X_{\Sigma\times\Sigma'} \cong X_\Sigma \times X_{\Sigma'}$. □

Example. Suppose that $k \in \mathbb{Z}$. Let $\sigma_0 = PHull((1,0),(0,1))$, $\sigma_1 = PHull((-1,0),(0,1))$, $\sigma_2 = PHull((-1,0),(k,-1))$ and $\sigma_3 = PHull((1,0),(k,-1))$. Suppose that Σ consists of these four maximal cones and their proper faces. We then have the following.

$$\begin{aligned} U_{\sigma_0} &= Spec(\mathbb{C}[x,y]) \cong \mathbb{C}^2 \\ U_{\sigma_1} &= Spec(\mathbb{C}[x^{-1},y]) \cong \mathbb{C}^2 \\ U_{\sigma_2} &= Spec(\mathbb{C}[x^{-1}y^{-k},y^{-1}]) \cong \mathbb{C}^2 \\ U_{\sigma_3} &= Spec(\mathbb{C}[xy^{k},y^{-1}]) \cong \mathbb{C}^2. \end{aligned}$$

Gluing U_{σ_0} and U_{σ_1} we get a variety isomorphic to $\mathbb{P}^1 \times \mathbb{C}$, and similarly for U_{σ_2} and U_{σ_3}. Finally, the two copies of $\mathbb{P}^1 \times \mathbb{C}$ are glued along their second factors. The second factors, which are both copies of $\mathbb{C}$, glue to form $\mathbb{P}^1$. The resulting surface X_Σ is a Hirzebruch surface, which is a ruled surface. Alternatively, X_Σ is the total space of the projectivization of the bundle $\mathcal{O}_{\mathbf{P}^1} \oplus \mathcal{O}_{\mathbf{P}^1}(k)$. This surface will be denoted by $\mathbb{F}_k$.

Definition 10. Let Σ be a fan in $\mathbb{R}^n$. The **support** of Σ is the union of all $\sigma \in \Sigma$, and is denoted by $|\Sigma|$.

Theorem 6. *An n-dimensional toric variety X is complete if and only if $|\Sigma_X| = \mathbb{R}^n$.*

Proof. See [4], pages 36-41. □

3. The Diagonal Property

Let X be a smooth, irreducible variety over an algebraically closed field, and let $\Delta : X \mapsto X \times X$ be the diagonal embedding. X is separated since it is a variety, and therefore the image $\Delta(X) \subset X \times X$ is a closed subvariety.

Definition 11. X is said to satisfy the **diagonal property** if there exists a vector bundle $\mathcal{E}$ on $X \times X$ of rank equal to $dim(X)$, and a section $s \in H^0(X \times X, \mathcal{E})$ such that the zero scheme of s coincides with $\Delta(X)$.

In [8], Pragacz, Srinivas and Pati started a program concerning the classification of varieties which satisfy the diagonal property. If X is a curve, it evidently satisfies the diagonal property, since then $\Delta(X) \subset X \times X$ is a hypersurface, therefore there exists a line bundle $\mathcal{L}$ on $X \times X$ and a section s of $\mathcal{L}$ such that $\Delta(X)$ is the zero scheme of s (see [5], Ch. II, Cor. 6.16). In the case that X is a surface, Pragacz, Srinivas and Pati were able to obtain detailed results. We present the statements from [8] about surfaces in the theorem below:

Theorem 7 (Pragacz, Srinivas, Pati [8]). *Let X be a smooth, projective surface over an algebraically closed field.*

(a) *There exists a smooth surface Y and $f : Y \to X$, where f is a birational morphism, such that Y satisfies the diagonal property.*
(b) *If $Y \to X$ is a birational morphism, X satisfies the diagonal property, and if $Pic(X)$ is finitely generated, then Y satisfies the diagonal property.*
(c) *Kodaira dimension$= -\infty$ case: all of these surfaces (namely rational and ruled surfaces) satisfy the diagonal property.*
(d) *Kodaira dimension$= 0$ case:*
   - *If X is an Abelian surface, then it satisfies the diagonal property,*
   - *if X is an Enriques surface, then it satisfies the diagonal property,*
   - *if X is a hyperelliptic surface, then it satisfies the diagonal property,*
   - *if X is a $K3$ surface with two rational curves, then it satisfies the diagonal property, however a general $K3$ surface does not satisfy the diagonal property* [2].
(e) *Kodaira dimension$= 1$ case: if X is an elliptic fibration which admits a section, then it satisfies the diagonal property.*
(f) *A general hypersurface of degree ≥ 5 does not satisfy the diagonal property.*

Even in the cases above where X satisfies the diagonal property, the proofs in [8] are not constructive. Rather, the proof of the fact is reduced to the existence of a *cohomologically trivial line bundle* on X, namely a line bundle $\mathcal{L}$ on X such that $H^i(X, \mathcal{L}) = 0$ for all i. Then for each individual case, the existence of $\mathcal{L}$ is proved separately. [8] also contains some partial results for higher dimensional varieties, and a discussion of the diagonal property for vector bundles in other categories than the algebraic category.

One case in which it is known that X satisfies the diagonal property, and at the same time the vector bundle $\mathcal{E}$ on $X \times X$ and its section s can be explicitly identified, is when X is a Grassmannian (of A-type). Suppose that $X = Gr(k, n)$, parameterizing complex k-dimensional subspaces of $\mathbb{C}^n$. Then, X is a smooth, rational algebraic variety of dimension $k(n-k)$. There is a tautological rank k sub-bundle S of the trivial rank n bundle on X where

$$S = \{(x, v) | v \in x\} \subset X \times \mathbb{C}^n.$$

Here, x is thought of as a point in X in the expression (x, v), and a k-plane in $\mathbb{C}^n$ in the expression $v \in x$. Let Q denote the rank $n-k$ quotient bundle of the trivial bundle $V = X \times \mathbb{C}^n$ by S. Then we have a short exact sequence of bundles

$$0 \longrightarrow S \longrightarrow V \longrightarrow Q \longrightarrow 0$$

on X. Now, let us return to $X \times X$. Denote the projections on the first and second factors by π_1 and π_2. We can pull-back any vector bundle on the target of a morphism to obtain a vector bundle on the domain, and this process does not

[2] It is a theorem that every K3 surface contains at least one rational curve ([1],page 364).

disturb the exactness. Therefore we have two exact sequences of vector bundles

$$0 \longrightarrow \pi_1^* S \xrightarrow{\varphi} \pi_1^* V \longrightarrow \pi_1^* Q \longrightarrow 0$$

$$0 \longrightarrow \pi_2^* S \longrightarrow \pi_2^* V \xrightarrow{\psi} \pi_2^* Q \longrightarrow 0$$

on $X \times X$. Since V is the trivial bundle on X, $\pi_1^* V$ and $\pi_2^* V$ are trivial bundles on $X \times X$, and they are naturally isomorphic. Consider the vector bundle $\mathcal{E} = Hom(\pi_1^* S, \pi_2^* Q)$, which is a rank $k(n-k)$ vector bundle on $\mathcal{E}$. Consider the section $s = \psi \circ \varphi$ of $\mathcal{E}$, where the isomorphism mentioned in this paragraph is suppressed in the notation. We claim that the zero scheme $Z(s) \subset X \times X$ of s is equal to $\Delta(X)$. Indeed, $(x, y) \in Z(s)$ if and only if $s((x,y),v) = \psi(\varphi((x,y),v)) = \psi((x,y),v) = 0$ for all $((x,y),v) \in \pi_1^* S$. By the exactness of the second sequence, $\psi((x,y),v) = 0$ if and only if $((x,y),v) \in \pi_2^* S$. But $((x,y),v) \in \pi_1^* S$ if and only if $v \in x$ and $((x,y),v) \in \pi_2^* S$ if and only if $v \in y$. This brings us to the equivalent condition: $((x,y),v) \in Z(s)$ if and only if $v \in x \Leftrightarrow v \in y$. But the latter is equivalent to $x = y$, so $Z(s) = \Delta(X)$. Therefore, in this case we see that $X = Gr(k,n)$ satisfies the diagonal property, and furthermore we have an explicit description of a vector bundle $\mathcal{E}$ and section s.

4. The Diagonal Property on Toric Surfaces

Let X be a complete, nonsingular toric surface. Then X is rational, therefore by Theorem 7 part (c), X satisfies the diagonal property. However, it is natural to ask if one can write the vector bundle in question explicitly in terms of the fan Σ_X. This is indeed possible. We will discuss three examples $\mathbb{P}^2$, $\mathbb{P}^1 \times \mathbb{P}^1$ and $\mathbb{F}_1$ below. The general case will be described in [7]. Before passing to these examples, let us establish some basic facts. Suppose that the maximal cones of Σ_X are cyclically numbered as $\sigma_0, \sigma_2, \ldots, \sigma_{m-1}$, and $U_i = Spec(\mathbb{C}[\check{\sigma}_i \cap \mathbb{Z}^2])$. Thus, σ_i and σ_j intersect on a 1-dimensional face if and only if $i - j \equiv \pm 1 \mod m$.

Lemma 7. $U_{i\times j} \cap U_{k\times l} = (U_{i\times j} \cap U_{k\times j}) \cap (U_{k\times j} \cap U_{k\times l})$.

Proof. This is equivalent to proving $\sigma_{i\times j} \cap \sigma_{k\times l} = \sigma_{i\times j} \cap \sigma_{k\times j} \cap \sigma_{k\times l}$ which in turn is equivalent to $\sigma_{i\times j} \cap \sigma_{k\times l} \subset \sigma_{k\times j}$. This is clear since $\sigma_{s\times t} = \sigma_s \times \sigma_t$. □

The lemma implies that it is enough to specify $g_{i\times j,k\times j}$ and $g_{i\times j,i\times k}$ for all i, j, k, since the other transition functions can be determined by the cocycle condition, without having to pass to any proper subset of the domains of definition of these functions. We can simplify the situation further, using the natural cyclic ordering of the maximal cones of the fan for the case of a surface:

Lemma 8. *If $i \neq j$, then $U_{i\times k} \cap U_{j\times k} \subset U_{i\times k} \cap U_{i+1\times k}$ or $U_{i\times k} \cap U_{j\times k} \subset U_{i\times k} \cap U_{i-1\times k}$ (the increment and decrement being considered $\mod m$).*

Proof. Indeed, unless $j \in \{i-1, i, i+1\}$, we have $\sigma_i \cap \sigma_j = \{0\}$. This implies the assertion. □

Similarly, we can formulate the lemma for the second index. Therefore, it suffices to specify the transition functions $g_{i\times j,i+1\times j}$ and $g_{i\times j,i\times j+1}$ for all i,j mod m, and check only the "small square" cocycle conditions of the form

$$g_{i+1\times j,i+1\times j+1}g_{i\times j,i+1\times j} = g_{i\times j+1,i+1\times j+1}g_{i\times j,i\times j+1} \tag{1}$$

and the "loop" cocycle conditions of the forms

$$\begin{aligned} g_{m-1\times j,0\times j}\cdots g_{1\times j,2\times j}g_{0\times j,1\times j} &= 1, \\ g_{i\times m-1,i\times 0}\cdots g_{i\times 1,i\times 2}g_{i\times 0,i\times 1} &= 1 \end{aligned} \tag{2}$$

in order to describe the vector bundle.

(i) $\mathbb{P}^2$

In $\mathbb{R}^2$, let $\sigma_0 = PHull((1,0),(0,1))$, $\sigma_1 = PHull((0,1),(-1,-1))$ and $\sigma_2 = PHull((1,0),(-1,-1))$. The fan Σ with these maximal cones gives us $\mathbb{P}^2$ as a toric variety. Let $U_i = Spec(\mathbb{C}[\check{\sigma}_i \cap \mathbb{Z}^2])$ as before. The fan corresponding to $\mathbb{P}^2 \times \mathbb{P}^2$ is $\Sigma \times \Sigma$. $\mathbb{P}^2 \times \mathbb{P}^2$ is covered by $U_i \times U_j$ where $i,j \in \{0,1,2\}$. Denote $U_i \times U_j$ by $U_{i\times j}$ as above. We wish to specify a vector bundle of rank 2 on $\mathbb{P}^2 \times \mathbb{P}^2$, and a section of this vector bundle whose zero scheme coincides with $\Delta(\mathbb{P}^2)$, the diagonal image of $\mathbb{P}^2$ in $\mathbb{P}^2 \times \mathbb{P}^2$. We will do this by explicitly specifying the transition functions $g_{i\times j,i+1\times j} : U_{i\times j} \to U_{i+1\times j}$ and $g_{i\times j,i\times j+1} : U_{i\times j} \to U_{i\times j+1}$ for all i,j. Let us fix notation for the variables. Let the first copy of $\mathbb{C}[\mathbb{Z}^2]$ be denoted by $\mathbb{C}[x,y]$ and the second copy of $\mathbb{C}[\mathbb{Z}^2]$ by $\mathbb{C}[z,w]$. Then, for instance, $U_{0\times 1} = Spec(\mathbb{C}[x,y,z^{-1}w,z^{-1}])$, etc. Let

$$g_{0\times j,1\times j} = \begin{bmatrix} -x^{-1}y & 1 \\ -x^{-1} & 0 \end{bmatrix}, \quad g_{1\times j,2\times j} = \begin{bmatrix} -y^{-1} & 1 \\ -xy^{-1} & 0 \end{bmatrix}, \quad g_{2\times j,0\times j} = \begin{bmatrix} -x & 1 \\ -y & 0 \end{bmatrix}$$

regardless of the value of j. And,

$$g_{i\times 0,i\times 1} = \begin{bmatrix} z^{-1} & 0 \\ 0 & z^{-1} \end{bmatrix}, \quad g_{i\times 1,i\times 2} = \begin{bmatrix} zw^{-1} & 0 \\ 0 & zw^{-1} \end{bmatrix}, \quad g_{i\times 2,i\times 0} = \begin{bmatrix} w & 0 \\ 0 & w \end{bmatrix}$$

regardless of the value of i. One can check directly the regularity of these transition functions and their inverses on their respective domains of definition and that the cocycle conditions (1) and (2) are satisfied (for invertibility, it suffices to check that the determinant is invertible). Thus we have a rank 2 vector bundle on $\mathbb{P}^2 \times \mathbb{P}^2$. The section s is defined via the following local expressions:

$$s_{0\times 0} = \begin{bmatrix} x-z \\ y-w \end{bmatrix}, \quad s_{1\times 0} = \begin{bmatrix} x^{-1}yz-w \\ x^{-1}z-1 \end{bmatrix}, \quad s_{2\times 0} = \begin{bmatrix} y^{-1}w-1 \\ xy^{-1}w-z \end{bmatrix},$$

$$s_{0\times 1} = \begin{bmatrix} xz^{-1}-1 \\ yz^{-1}-wz^{-1} \end{bmatrix}, \quad s_{1\times 1} = \begin{bmatrix} x^{-1}y-z^{-1}w \\ x^{-1}-z^{-1} \end{bmatrix}, \quad s_{2\times 1} = \begin{bmatrix} z^{-1}y^{-1}w-z^{-1} \\ z^{-1}xy^{-1}w-1 \end{bmatrix},$$

$$s_{0\times 2} = \begin{bmatrix} xw^{-1}-zw^{-1} \\ yw^{-1}-1 \end{bmatrix}, \quad s_{1\times 2} = \begin{bmatrix} zw^{-1}x^{-1}y-1 \\ zw^{-1}x^{-1}-w^{-1} \end{bmatrix}, \quad s_{2\times 2} = \begin{bmatrix} y^{-1}-w^{-1} \\ xy^{-1}-zw^{-1} \end{bmatrix}.$$

Again, the regularity of s, and the fact that it is a section of this vector bundle can be checked directly. Finally we prove:

Theorem 8. *The zero scheme of s in $\mathbb{P}^2 \times \mathbb{P}^2$ coincides with the diagonal, $\Delta(\mathbb{P}^2)$.*

Proof. First, consider $U_{i\times i}$. From the local expressions of s, it is clear that the zero scheme of $s_{i\times i}$ coincides with $\Delta(\mathbb{P}^2) \cap U_{i\times i}$ for $i \in \{0,1,2\}$. Now, $\Delta(\mathbb{P}^2) \subset U_{0\times 0} \cup U_{1\times 1} \cup U_{2\times 2}$. Since the zero scheme of s is connected, it must coincide with $\Delta(\mathbb{P}^2)$ on all of $\mathbb{P}^2 \times \mathbb{P}^2$. □

For the other two examples, the line of reasoning is identical: one checks regularity and cocycle conditions directly, and proves that the zero scheme of s coincides with $\Delta(X)$ by looking at $U_{i\times i}$. Therefore, we will only list $g_{i\times j,i+1\times j}, g_{i\times j,i\times j+1}$ and $s_{i\times j}$ for all i,j. Before passing to these examples, let us relate the vector bundle constructed here to the bundle described at the end of section 3. We remark that $\mathbb{P}^2 \cong Gr(1,3)$. In order to match the two results, suppose that the first copy of $\mathbb{P}^2$ has function field $\mathbb{C}(z,w)$, and the second copy, $\mathbb{C}(x,y)$. Keeping the notation of section 3, trivialize $\pi_1^*(S)$ on U_0, U_1, U_2 respectively as

$$((z,w),\lambda) \mapsto (\lambda z, \lambda w, \lambda) \in \mathbb{C}^3,$$
$$((z^{-1}w, z^{-1}),\mu) \mapsto (\mu, \mu z w^{-1}, \mu z^{-1}) \in \mathbb{C}^3,$$
$$((w^{-1}, zw^{-1}),\nu) \mapsto (\nu z w^{-1}, \nu, \nu w^{-1}) \in \mathbb{C}^3,$$

therefore the transition functions of this line bundle under this trivialization are $h_{0,1} = z, h_{1,2} = z^{-1}w, h_{2,3} = w^{-1}$ (incidentally, this line bundle is isomorphic to $\mathcal{O}(-1)$). Trivialize π_2^*Q respectively on U_0, U_1, U_2 by taking orthogonal complements to fibers of π_2^*S with respect to the standard dot product on $\mathbb{C}^3$:

$$((x,y),(\lambda_1,\lambda_2)) \mapsto (\lambda_1, \lambda_2, -\lambda_1 x - \lambda_2 y) \in \mathbb{C}^3,$$
$$((x^{-1}y, x^{-1}),(\mu_1,\mu_2)) \mapsto (-\mu_1 x^{-1}y - \mu_2 x^{-1}, \mu_1, \mu_2) \in \mathbb{C}^3$$
$$((y^{-1}, xy^{-1}),(\nu_1,\nu_2)) \mapsto (\nu_2, -\nu_1 y^{-1} - \nu_2 x y^{-1}, \nu_1) \in \mathbb{C}^3,$$

from which we can read off the transition functions as

$$k_{0,1} = \begin{bmatrix} 0 & 1 \\ -x & -y \end{bmatrix}, k_{1,2} = \begin{bmatrix} 0 & 1 \\ -x^{-1}y & -x^{-1} \end{bmatrix}, k_{2,0} = \begin{bmatrix} 0 & 1 \\ -y^{-1} & -xy^{-1} \end{bmatrix}.$$

The bundle constructed in section 3 is $\mathcal{E} = Hom(\pi_1^*S, \pi_2^*Q) \cong (\pi_1^*S)^* \otimes \pi_2^*Q$. We can directly check the equality

$$g_{i\times j,i\times j+1} = (h_{j,j+1} \otimes I)^{-1},$$

therefore the first factors agree on the nose. On the other hand,

$$g_{i\times j,i+1\times j} = (k_{i,i+1}^t)^{-1},$$

so the bundle constructed in this section is isomorphic to $(\pi_1^*S)^* \otimes (\pi_2^*Q)^*$ instead.

(ii) $\mathbb{P}^1 \times \mathbb{P}^1$

Let $\sigma_0 = PHull((1,0),(0,1))$,$\sigma_1 = PHull((0,1),(-1,0))$,$\sigma_2 = PHull((-1,0),(0,-1))$ and $\sigma_3 = PHull((0,-1),(1,0))$. We set

$$g_{0\times j,1\times j} = \begin{bmatrix} 0 & 1 \\ -x^{-1} & 0 \end{bmatrix}, \qquad g_{1\times j,2\times j} = \begin{bmatrix} 0 & 1 \\ -y^{-1} & 0 \end{bmatrix},$$

$$g_{2\times j,3\times j} = \begin{bmatrix} 0 & 1 \\ -x & 0 \end{bmatrix}, \qquad g_{3\times j,0\times j} = \begin{bmatrix} 0 & 1 \\ -y & 0 \end{bmatrix}$$

regardless of j. On the other hand, $g_{i\times j,i\times j+1}$ will depend nontrivially on both i and j, so we list them all:

$$g_{0\times 0,0\times 1} = g_{2\times 0,2\times 1} = \begin{bmatrix} z^{-1} & 0 \\ 0 & 1 \end{bmatrix}, g_{1\times 0,1\times 1} = g_{3\times 0,3\times 1} = \begin{bmatrix} 1 & 0 \\ 0 & z^{-1} \end{bmatrix},$$

$$g_{0\times 1,0\times 2} = g_{2\times 1,2\times 2} = \begin{bmatrix} 1 & 0 \\ 0 & w^{-1} \end{bmatrix}, g_{1\times 1,1\times 2} = g_{3\times 1,3\times 2} = \begin{bmatrix} w^{-1} & 0 \\ 0 & 1 \end{bmatrix},$$

$$g_{0\times 2,0\times 3} = g_{2\times 2,2\times 3} = \begin{bmatrix} z & 0 \\ 0 & 1 \end{bmatrix}, g_{1\times 2,1\times 3} = g_{3\times 2,3\times 3} = \begin{bmatrix} 1 & 0 \\ 0 & z \end{bmatrix},$$

$$g_{0\times 3,0\times 0} = g_{2\times 3,2\times 0} = \begin{bmatrix} 1 & 0 \\ 0 & w \end{bmatrix}, g_{1\times 3,1\times 0} = g_{3\times 3,3\times 0} = \begin{bmatrix} w & 0 \\ 0 & 1 \end{bmatrix}.$$

The local expressions for the section s are:

$$s_{0\times 0} = \begin{bmatrix} x-z \\ y-w \end{bmatrix}, \quad s_{1\times 0} = \begin{bmatrix} y-w \\ x^{-1}z-1 \end{bmatrix}, s_{2\times 0} = \begin{bmatrix} x^{-1}z-1 \\ y^{-1}w-1 \end{bmatrix}, s_{3\times 0} = \begin{bmatrix} y^{-1}w-1 \\ x-z \end{bmatrix}$$

$$s_{0\times 1} = \begin{bmatrix} xz^{-1}-1 \\ y-w \end{bmatrix}, s_{1\times 1} = \begin{bmatrix} y-w \\ x^{-1}-z^{-1} \end{bmatrix}, s_{2\times 1} = \begin{bmatrix} x^{-1}-z^{-1} \\ y^{-1}w-1 \end{bmatrix}, s_{3\times 1} = \begin{bmatrix} y^{-1}w-1 \\ xz^{-1}-1 \end{bmatrix}$$

$$s_{0\times 2} = \begin{bmatrix} xz^{-1}-1 \\ yw^{-1}-1 \end{bmatrix}, s_{1\times 2} = \begin{bmatrix} yw^{-1}-1 \\ x^{-1}-z^{-1} \end{bmatrix}, s_{2\times 2} = \begin{bmatrix} x^{-1}-z^{-1} \\ y^{-1}-w^{-1} \end{bmatrix}, s_{3\times 2} = \begin{bmatrix} y^{-1}-w^{-1} \\ xz^{-1}-1 \end{bmatrix}$$

$$s_{0\times 3} = \begin{bmatrix} x-z \\ yw^{-1}-1 \end{bmatrix}, s_{1\times 3} = \begin{bmatrix} yw^{-1}-1 \\ x^{-1}z-1 \end{bmatrix}, s_{2\times 3} = \begin{bmatrix} x^{-1}z-1 \\ y^{-1}-w^{-1} \end{bmatrix}, s_{3\times 3} = \begin{bmatrix} y^{-1}-w^{-1} \\ x-z \end{bmatrix}$$

(iii) $\mathbb{F}_1$

Considering the maximal cones of the fan of $\mathbb{F}_1$, let $\sigma_0 = PHull((1,0),(0,1))$, $\sigma_1 = PHull((0,1),(-1,0))$, $\sigma_2 = PHull((-1,0),(1,-1))$, and $\sigma_3 = PHull((1,-1),(1,0))$. Let

$$g_{0\times j,1\times j} = \begin{bmatrix} 0 & 1 \\ -x^{-1} & 0 \end{bmatrix}, \qquad g_{1\times j,2\times j} = \begin{bmatrix} 0 & y^{-1} \\ -y^{-1} & 0 \end{bmatrix},$$

$$g_{2\times j,3\times j} = \begin{bmatrix} 0 & 1 \\ -xy & 0 \end{bmatrix}, \qquad g_{3\times j,0\times j} = \begin{bmatrix} 0 & 1 \\ -y & 0 \end{bmatrix}.$$

regardless of j. And,

$$g_{0\times 0,0\times 1}=g_{2\times 0,2\times 1}=\begin{bmatrix} z^{-1} & 0\\ 0 & 1\end{bmatrix}, g_{1\times 0,1\times 1}=g_{3\times 0,3\times 1}=\begin{bmatrix} 1 & 0\\ 0 & z^{-1}\end{bmatrix},$$

$$g_{0\times 1,0\times 2}=\begin{bmatrix} 1 & xz^{-1}w^{-1}\\ 0 & w^{-1}\end{bmatrix}, g_{1\times 1,1\times 2}=\begin{bmatrix} w^{-1} & 0\\ -z^{-1}w^{-1} & 1\end{bmatrix},$$

$$g_{2\times 1,2\times 2}=\begin{bmatrix} 1 & z^{-1}w^{-1}\\ 0 & w^{-1}\end{bmatrix}, g_{3\times 1,3\times 2}=\begin{bmatrix} w^{-1} & 0\\ -xyz^{-1}w^{-1} & 1\end{bmatrix},$$

$$g_{0\times 2,0\times 3}=g_{2\times 2,2\times 3}=\begin{bmatrix} zw & 0\\ 0 & 1\end{bmatrix}, g_{1\times 2,1\times 3}=g_{3\times 2,3\times 3}=\begin{bmatrix} 1 & 0\\ 0 & zw\end{bmatrix},$$

$$g_{0\times 3,0\times 0}=\begin{bmatrix} w^{-1} & -x\\ 0 & w\end{bmatrix}, g_{1\times 3,1\times 0}=\begin{bmatrix} w & 0\\ 1 & w^{-1}\end{bmatrix}$$

$$g_{2\times 3,2\times 0}=\begin{bmatrix} w^{-1} & -1\\ 0 & w\end{bmatrix}, g_{3\times 3,3\times 0}=\begin{bmatrix} w & 0\\ xy & w^{-1}\end{bmatrix}.$$

The section s is determined by the local expressions:

$$s_{0\times 0}=\begin{bmatrix} x-z\\ y-w\end{bmatrix},\quad s_{1\times 0}=\begin{bmatrix} y-w\\ x^{-1}z-1\end{bmatrix}, s_{2\times 0}=\begin{bmatrix} x^{-1}y^{-1}z-y^{-1}\\ y^{-1}w-1\end{bmatrix}, s_{3\times 0}=\begin{bmatrix} y^{-1}w-1\\ x-z\end{bmatrix}$$

$$s_{0\times 1}=\begin{bmatrix} xz^{-1}-1\\ y-w\end{bmatrix}, s_{1\times 1}=\begin{bmatrix} y-w\\ x^{-1}-z^{-1}\end{bmatrix}, s_{2\times 1}=\begin{bmatrix} x^{-1}y^{-1}-y^{-1}z^{-1}\\ y^{-1}w-1\end{bmatrix}, s_{3\times 1}=\begin{bmatrix} y^{-1}w-1\\ xz^{-1}-1\end{bmatrix}$$

$$s_{0\times 2}=\begin{bmatrix} xz^{-1}yw^{-1}-1\\ yw^{-1}-1\end{bmatrix}, s_{1\times 2}=\begin{bmatrix} yw^{-1}-1\\ x^{-1}-z^{-1}yw^{-1}\end{bmatrix}$$

$$s_{2\times 2}=\begin{bmatrix} x^{-1}y^{-1}-w^{-1}z^{-1}\\ y^{-1}-w^{-1}\end{bmatrix}, s_{3\times 2}=\begin{bmatrix} y^{-1}-w^{-1}\\ xz^{-1}yw^{-1}-1\end{bmatrix}$$

$$s_{0\times 3}=\begin{bmatrix} xy-zw\\ yw^{-1}-1\end{bmatrix}, s_{1\times 3}=\begin{bmatrix} yw^{-1}-1\\ x^{-1}zw-y\end{bmatrix},$$

$$s_{2\times 3}=\begin{bmatrix} x^{-1}zy^{-1}w-1\\ y^{-1}-w^{-1}\end{bmatrix}, s_{3\times 3}=\begin{bmatrix} y^{-1}-w^{-1}\\ xy-zw\end{bmatrix}$$

References

[1] Barth W.P., Hulek, K., Peters, C.A.M., Van de Ven, A., *Compact complex surfaces*, Ergebnisse der Math. **4**. Springer Verlag, Berlin, 2004.

[2] Cox, D., *Lectures on toric varieties*, CIMPA Lecture Notes.

[3] Ewald, G., *Combinatorial convexity and algebraic geometry*, Graduate Texts in Mathematics **168**. Springer Verlag, New York, 1996.

[4] Fulton, W., *Introduction to toric varieties*, Annals of Math. Studies **131**. Princeton University Press, New Jersey, 1993.

[5] Hartshorne, R., *Algebraic Geometry*, Graduate Texts in Mathematics **168**. Springer Verlag, New York, 1977.

[6] Kempf, G. , Knudsen, F. , Mumford, D. , Saint-Donat, B., *Toroidal Embeddings I*, Lecture Notes in Mathematics **339**. Springer Verlag, Berlin, 1973.

[7] Kişisel, A.U.Ö, Öztürk, Ö., *The diagonal property on toric surfaces*, in preparation.

[8] Pragacz, P., Srinivas, V., Pati, V., *Diagonal subschemes and vector bundles*, preprint, arXiv: math.AG/0609381.

[9] Pragacz, P. *Miscellany on the zero schemes of sections of vector bundles*, (notes by Ö. Öztürk) in: "Algebraic Cycles, Sheaves, Shtukas, and Moduli" – ILN, "Trends in Mathematics", Birkhäuser, 2007, 105–116.

Ali Ulaş Özgür Kişisel

Özer Öztürk
Middle East Technical University
Department of Mathematics
06531 Ankara
Turkey
e-mail: akisisel@metu.edu.tr
ozero@metu.edu.tr

Progress in Mathematics, Vol. 283, 209–245

Introduction to Plane Curve Singularities. Toric Resolution Tower and Puiseux Pairs

Mutsuo Oka

Abstract. We give an elementary introduction of the toric modification, using an irreducible plane curve germ. We explain also the relation between the tower of the toric modifications which gives a resolution of the curve and the Puiseux pairs.

2000 Mathematics Subject Classification. 14H20.

Keywords. Resolution, Puiseux characteristics.

1. Introduction

Let (C,O), $O=(0,0)$ be a plane curve which is defined by $\{(x,y)\in U;\, f(x,y)=0\}$ where U is an open neighborhood of O and $f(x,y)$ is a holomorphic function defined on U. The purpose of this survey is to give an elementary proof of an embedded resolution of a curve germ (C,O), first by ordinary blowing-ups and then by toric modifications. This note is prepared for the lecture at CIMPA school at Istanbul, June 2007. We hope that the readers will learn basic ideas of toric modifications through a very simple setting of plane curves. We explain that Puiseux characteristics are better understood through minimal toric resolution towers. In fact, we prove that there exists a bijective correspondence between minimal toric resolution towers and the set of Puiseux characteristics. See Theorem 32, Theorem 34 and Theorem 35 in §5.

2. Preliminaries

2.1. Presentations of a germ of a curve

Usually a presentation of a plane curve is done in one of the following ways.

1. (A smooth curve) by a graph of a holomorphic function

$$\Gamma=\{(x,y)\in\mathbb{C}^2\,|\,y=\phi(x)\},\quad \phi\in\mathbb{C}\{x\}.$$

2. A good parametrization

$$\phi : \mathbb{C} \supset U \to \mathbb{C}^2,\ t \mapsto (x(t), y(t)),\ x(t), y(t) \in \mathbb{C}\{t\}$$

   where ϕ is generically injective. For example, the presentation $t \mapsto (t^2, t^4)$ is a 2:1 mapping onto the curve $y - x^2 = 0$ and not a good parametrization.
3. Zero locus of a function:

$$\Gamma : \{(x, y) \in U \subset \mathbb{C}^2 \mid f(x, y) = 0\}, \quad f \in \mathbb{C}\{x, y\}.$$

4. An algebraic curve (global)

$$C : \ \{(x, y) \in \mathbb{C}^2; f(x, y) = 0\}, \quad f \in \mathbb{C}[x, y].$$

In this note, we focus our study on the cases (2) and (3).

2.2. Intersection number

Suppose we have two curves defined in an open neighborhood U of the origin $O = (0, 0)$,

$$C = \{(x, y) \in U \mid f(x, y) = 0\}, \quad C' = \{(x, y) \in U \mid g(x, y) = 0\}.$$

The *local intersection number* $I(C, C'; O)$, or simply $C \cdot C'$ is defined as

$$C \cdot C' := \dim_{\mathbb{C}} \mathcal{O}/(f, g).$$

If C is given by a good parametrization $\{(x(t), y(t)),\ t \in \mathbb{C}\}$ with $(x(0), y(0)) = (0, 0)$, we can also define

$$C \cdot C' := \mathrm{order}_t\, g(x(t), y(t)).$$

We say that two smooth curves C, C' are *transverse at* O if $I(C, C'; O)) = 1$. We say O is a *smooth* (or *simple*) point of C if $(\frac{\partial f}{\partial x}(O), \frac{\partial f}{\partial y}(O)) \neq (0, 0)$. Otherwise, O is called a *singular* point. The tangent line $T_P C$ at a smooth point $P = (\alpha, \beta) \in U \cap C$ is given as

$$T_P C : \quad \frac{\partial f}{\partial x}(P)\,(x - \alpha) + \frac{\partial f}{\partial y}(P)\,(y - \beta) = 0.$$

We can see by the definition that $I(C, T_O C; O) \geq 2$. Note that $I(C, C'; O) = 1$ iff C, C' are smooth at O and their tangent lines are transverse.

Let $f(x, y) = \sum_{i,j} a_{ij} x^i y^j$ be the Taylor expansion. By the assumption $f(0, 0) = 0$, we have $a_{0,0} = 0$. Now C is smooth at O if and only if $(a_{10}, a_{01}) \neq (0, 0)$. *The multiplicity* of (C, O), which we denote as $m(C, O)$, is defined by the minimum of the integer $i + j$ such that $a_{ij} \neq 0$. This is equal to the minimum of $I(L, C; O)$ where L moves in the set of lines passing through O.

Lemma 1. *The following properties are satisfied.*

1. $C \cdot C' \in \mathbb{N}$,

$$C \cdot (D + D') = C \cdot D + C \cdot D'.$$

2. *Symmetry:* $C \cdot D = D \cdot C$.

2.3. Newton boundary

Fix an analytic coordinate system (x,y) and let $f(x,y) = \sum_\nu c_\nu x^{\nu_1}y^{\nu_2}$ be the Taylor expansion. The *Newton polygon*
$\Gamma_+(f;x,y)$ is the convex hull of

$$\bigcup_{\nu,c_\nu\neq 0} \{\nu+\mathbb{R}^2_+\}$$

and *the Newton boundary* $\Gamma(f;x,y)$ is the union of compact faces or vertices of $\Gamma_+(f;x,y)$. We define $\Gamma_-(f;x,y)$ by the cone over $\Gamma(f;x,y)$ with the origin as the vertex of the cone. Note that $\Gamma_+(f;x,y)\cap\Gamma_-(f;x,y) = \Gamma(f;x,y)$. Take a 1-face Δ of $\Gamma(f)$. The *face function* f_Δ is defined as

$$f_\Delta(x,y) := \sum_{\nu\in\Delta} c_\nu x^{\nu_1}y^{\nu_2}.$$

There exists a unique positive primitive vector $P = {}^t(a,b)$, $a,b>0$, which is called *the weight vector*, such that $f_\Delta(x,y)$ is a weighted homogeneous polynomial with weights P. Namely we have the equality

$$f_\Delta(t^a\,x, t^b\,y) = t^d f_\Delta(x,y)$$

where the integer d is characterized by $\Delta\subset\{a\nu_1+b\nu_2=d\}$.

2.3.1. Example. The following Figure 1 is the Newton boundary of $f(x,y) = (x^2+y^3)(x^4+y^3)$. The vertices of the Newton boundary are $(6,0)$, $(2,3)$, $(0,6)$. $\Gamma(f)$ has two faces Δ_1,Δ_2 with

$$f_{\Delta_1}(x,y) = y^3(x^2+y^3),\quad f_{\Delta_2}(x,y) = x^2(x^4+y^3).$$

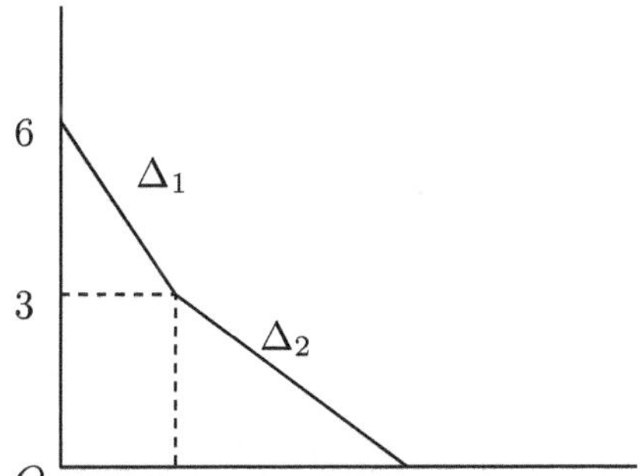

FIGURE 1. Newton boundary of $(x^2+y^3)(x^4+y^3)$

Remark 2. Let $f_1(x,y) = y^2x + x^n$ with $n \geq 2$ and $f_2(x,y) = xy$. Then the Newton boundary $\Gamma(f_1)$ has a unique face $\overline{AB}$ with $A = (n,0)$ and $B = (1,2)$. The Newton boundary of $\Gamma(f_2)$ consists of a single vertex $(1,1)$ and it has no 1-face. Both of f_1, f_2 are not "convenient" and $\Gamma(f_1)$, $\Gamma(f_2)$ do not intersect with the y-axis. See §3.6 for a definition of convenience.

2.4. Weierstrass preparation theorem

Let $\mathcal{O}_O$ be the ring of germs of holomorphic functions at the origin and let $\mathcal{O}(U)$ be the ring of holomorphic functions defined on an open set U. It is well-known that $\mathcal{O}_O$ is isomorphic to the ring of convergent power series $\mathbb{C}\{x,y\}$. It is a factorial and Noetherian local ring. We refer, for example, to Gunning-Rossi [4] or the book of C.T.C. Wall [14] for this property. Consider a curve $C = \{(x,y) \in U; f(x,y) = 0\}$ which is reduced. Assuming O is a smooth point and for example $\frac{\partial f}{\partial y}(0,0) \neq 0$, by the implicit function theorem ([4]) we can solve $f(x,y)$ in y as $y = \phi(x)$, where $\phi(x)$ is a holomorphic function of x.

If O is a singular point, we have no solution $y = \phi(x)$ with a holomorphic function ϕ. However if (C,O) is irreducible at O, we can find a holomorphic function $\phi(t)$ defined on $|t| < \delta$ so that $f(x,y) = 0$ can be solved as $y = \phi(x^{1/n})$ for some positive integer $n \geq 2$. (This implies $f(t^n, \phi(t)) \equiv 0$ for any t, $|t| < \delta$.) This is called a *Puiseux expansion* which we will explain later. First we recall the following basic fact, which is called the *Weierstrass preparation theorem*. Assume that $f(x,y) \in \mathbb{C}\{x,y\}$ has an isolated critical point at O, namely the origin is an isolated point of

$$\{(x,y) \in U; \frac{\partial f}{\partial x}(x,y) = \frac{\partial f}{\partial y}(x,y) = 0\}.$$

This is equivalent to $f(x,y)$ is reduced in $\mathcal{O}_O$. $f(x,y)$ is called *regular in y of order s* if $f(0,y) \not\equiv 0$ and $\mathrm{ord_y\, f(0,y) = s}$. Then the Weierstrass preparation theorem says that

Theorem 3. ([4]) *Under the above assumption, there exists a unit $u \in \mathbb{C}\{x,y\}$ and analytic functions $a_j \in \mathbb{C}\{x\}$, $j = 1,\dots,s$ such that*

$$f(x,y) = u(x,y) \times P(x,y),\ P(x,y) = y^s + a_1(x)y^{s-1} + \cdots + a_s(x)$$

with $a_i(0) = 0, i = 1,\dots,s$.

The polynomial $P(x,y) \in \mathbb{C}\{x\}[y]$ is called *the Weierstrass polynomial* of $f(x,y)$ in y at the origin. Suppose that $f(x,y)$ has a factorization $f = f_1^{\nu_1} \cdots f_k^{\nu_k}$ in $\mathcal{O}_O$. If $f(x,y)$ is regular in y, each $f_i(x,y)$ is also regular in y and the Weierstrass polynomial of $f(x,y)$ is simply a product of the Weierstrass polynomial $P_i(x,y)$ of $f_i(x,y)$. It is known that $f(x,y)$ is irreducible if and only if $P(x,y)$ is irreducible in $\mathbb{C}\{x\}[y]$ ([14]).

2.5. Resultant and discriminant

We consider two polynomials in $K[y]$ with degrees m, n, where K is an algebraically closed field of characteristic 0,

$$f(y) = a_m y^m + \cdots + a_0 x^m, \ g(y) = b_n y^n + \cdots + b_0 x^m.$$

Then we consider the following expression.

$$\begin{pmatrix} y^{n-1}f(y) \\ \vdots \\ f(y) \\ y^{m-1}g(y) \\ \vdots \\ g(y) \end{pmatrix} = \begin{pmatrix} a_m & \cdots & a_0 & 0 & \cdots & 0 \\ 0 & \ddots & 0 & \ddots & & \\ 0 & \cdots & a_m & \cdots & a_0 & \\ b_n & \cdots & & b_0 & 0 & \ldots 0 \\ 0 & \ddots & & \cdots & \ddots & 0 \\ 0 & \cdots & b_n & & \cdots & b_0 \end{pmatrix} \begin{pmatrix} y^{n+m-1} \\ y^{n+m-2} \\ \vdots \\ \vdots \\ 1 \end{pmatrix}.$$

$R(f,g)$ is defined by the determinant of the right-hand $(n+m)\times(n+m)$-matrix and we call it the *resultant* of f, g.

Lemma 4. *Assume*

$$f(y) = c\prod_1^m (y-\alpha_i), \quad g(y) = c'\prod_1^n (y-\beta_j),\ c, c' \neq 0.$$

Then

1. $R(f,g) = c^n c'^m \prod_{i,j}(\alpha_i - \beta_j)$.
2. $R(f,g) = (-1)^{mn} R(g,f)$.
3. $R(f,gh) = R(f,g)R(f,h)$.
4. $R(f,g) = c^n \prod_j g(\alpha_j)$.
5. *If* $\deg g < \deg f$, *for any* ϕ *with* $\deg \phi = \deg f - \deg g$, *we have the equality* $R(f + g\phi, g) = R(f,g)$.
6. $R(f,g) = 0$ *if and only if* $\deg \gcd(f,g) > 0$ *i.e., a common root* $\alpha_i = \beta_j$.

See Lang [6] or [14] for the proof.
Now the *discriminant* of $f(y) \in K[y]$ in y is defined as

$$D(f,y) := R(f, \frac{df}{dy}).$$

Corollary 5. $D(f) = 0$ *iff* $f = 0$ *has a multiple root or* $c = 0$.

The above definition can be applied to a Weierstrass polynomial in $P(x,y) \in \mathbb{C}\{x\}[y]$ considering it a polynomial y. The discriminant $D(P,y)$ then belongs to $\mathbb{C}\{x\}$.

2.6. Weierstrass polynomial and discriminant

Assume that $f \in \mathcal{O}_O$ with $f(0,0) = 0$, regular in y of order n and reduced with factorization $f = f_1 \cdots f_k$. Let $P_j(x,y) = y^{m_j} + a_{j1}(x)y^{m_j-1} + \cdots + a_{jm_j}(x)$ be the Weierstrass polynomial of f_j so that $P(x,y) = P_1(x,y)\cdots P_k(x,y)$ is the Weierstrass polynomial of $f(x,y)$. Note that $n = \sum_{j=1}^k m_j$.

Lemma 6. *Assume that $P(x,y)$ is reduced. Then the discriminant $d(x)$ of $P(x,y)$ as a polynomial of y is not constantly zero in $\mathbb{C}\{x\}$.*

Proof. Consider the variety $D := \{(x,y);\ P(x,y) = \frac{\partial P}{\partial y}(x,y) = 0\}$ at the origin. If $d(x) \equiv 0$, it implies $\dim D = 1$. Let $C_1,\dots,C_k$ be the components of (C,O). Then (as a germ), there exists a component $C_i = \{P_i(x,y) = 0\}$ of C such that $C_i \subset D$. We may assume that $C_i \cap C_j = \{O\}$ for $j \neq i$. Thus $P_i(x,y) = 0$ implies $P(x,y) = \frac{\partial P}{\partial y}(x,y) = 0$. As $\frac{\partial P}{\partial y}(x,y) = \frac{\partial P_i}{\partial y}(x,y) \times \prod_{j\neq i} P_j(x,y)$ on C_i, this implies P_i divides $\partial P_i/\partial y$. However this is not possible as $\deg_y \partial P_i/\partial y = \deg_y P_i - 1$. □

2.7. Puiseux expansion

Using Lemma 6, we obtain:

Theorem 7 (Puiseux expansion theorem). *Assume that $f(x,y) \in \mathbb{C}\{x,y\}$ is irreducible, $f(0,0) = 0$ and $f(x,y)$ is regular in y of order n and let $P(x,y)$ be the Weierstrass polynomial. Then there exists a holomorphic function $\phi(t)$ defined on a small open disk $\Delta_\varepsilon = \{t; |t| < \sqrt[n]{\varepsilon}\}$ and an open neighborhood $U' = \Delta_\varepsilon \times \Delta_\delta \subset U$ such that $f(x,y) = 0$ for some $(x,y) \in U'$ if and only if $y = \phi(t)$ and $x = t^n$ for some t with $t^n \in \Delta_\varepsilon$. Let $\omega = \exp(2\pi i/n)$. Then*

$$P(x,y) = \prod_{j=0}^{n-1} (y - \phi(t\omega^j))$$

where the expansion of the right-hand side involves only exponents of type $t^{n\nu}$ so that we replace $y^j t^{n\nu}$ by $y^j x^\nu$.

Proof. Put

$$P(x,y) = y^n + a_1(x)y^{n-1} + \cdots + a_n(x).$$

We assume that the coefficients $a_j(x)$, $1 \le j \le n$ are defined on Δ_ε. As $a_j(0) = 0$, we may assume that any root y of $P(x,y) = 0$ is $|y| < \delta$ for any $x \in \Delta_\varepsilon$. We may also assume that the discriminant $d(x)$ of $P(x,y)$ in y vanishes only at $x = 0$.

Step 1. For any $x \in \Delta_\varepsilon^* := \Delta_\varepsilon - \{O\}$, let $g_1(x),\dots,g_n(x)$ be the roots of $P(x,y) = 0$ in y. By the implicit function theorem, $g_i(x)$ is locally holomorphic in $x \in \Delta_\varepsilon^*$, possibly multi-valued but it is single-valued for $\Delta_\varepsilon - \{x; x \le 0\}$.

Step 2. By analytic continuations, along $|x| = \varepsilon'$, $x = \varepsilon' \exp(i\theta)$, $0 \le \theta \le 2\pi$, for any $\varepsilon' < \varepsilon$, we can divide $\{g_i\}$ into several orbits: $S_1,\dots,S_k$. We assert:

Claim 8. *If (C,O) is irreducible, we have only one orbit, i.e., $k = 1$ and after a suitable ordering, the analytic continuation gives a cyclic permutation $(g_1,\dots,g_n)$.*

Proof. Put

$$P_1(x,y) := \prod_{i\in S_1} (y - g_i(x)).$$

By the assumption, $P_i(x,y)$ is holomorphic, single-valued on $\Delta_\delta^* \times \mathbb{C}$. By Riemann's extension theorem ([4]), P_i is holomorphic on $\Delta_\delta \times \mathbb{C}$. Then the assumption implies that $P_1(x,y)$ divides $P(x,y)$. This is possible only if $k = 1$ and $P_1 = P$. □

Step 3. Consider the n-fold branched cyclic mapping $p : \mathbb{C} \to \mathbb{C}$, defined by $p(x) = x^n$. Then as the order of the cyclic permutation $(1, 2, \ldots, n)$ is n, each pull-back function $p^* g_i(x) = g_i(x^n)$ is a single-valued holomorphic function. Fix an analytic branch of $x^{1/n}$ on $\mathbb{C} \setminus \{z \le 0\}$. This implies that $g_i(x)$ is an analytic function of $t = x^{1/n}$. Namely g_i is a convergent power series in t. Put $\phi(t) = g_1(t^n)$. Then $\phi(t) \in \mathbb{C}\{t\}$. Note that by the analytic continuation of the function $x^{1/n}$ over $|x| = \varepsilon$, $x^{1/n}$ is transformed into $x^{1/n}\omega$. Consider the Taylor expansion $\phi(t) = \sum_{i=1}^{\infty} c_i t^i$. Then the monodromy of $p : \mathbb{C} \to \mathbb{C}$ by the analytic continuation is simply given by $t \mapsto t \times \omega$. Thus this implies that we may assume that

$$g_{j+1}(x) = \phi(t\omega^j) = \sum_{i=1}^{\infty} c_i \omega^{ji} t^i, \quad j = 1, \ldots, n-1.$$

As $P(x, y) = \prod_{j=0}^{n-1}(y - \phi(t\omega^j))$, the assertion follows. □

2.8. Puiseux characteristics

Consider a curve C with a good parametrization given by Theorem 7:

$$C : x = t^m, \quad y = \sum_{j=m}^{\infty} c_j t^j \tag{1}$$

such that $m = m(C, O)$, multiplicity of C at O. (Note that (C, O) is non-singular if $m = 1$). Assuming $m \ge 2$, we define inductively integers $\beta_1, e_1, \beta_2, e_2, \ldots$:

$$\beta_1 = \min\{k \mid c_k \ne 0,\ m \not| k\}, \quad e_1 = \gcd(m, \beta_1)$$

and if $e_i > 1$,

$$\beta_{i+1} = \min\{k \mid c_k \ne 0,\ e_i \not| k\}, \quad e_{i+1} = \gcd(e_i, \beta_{i+1}).$$

This operation ends when we arrive at $e_g = 1$ or β_g is not defined. The latter case does not happen. In fact, if this is the case, $c_k = 0$ for any k such that $k \not\equiv 0$ modulo e_{g-1}. This implies that y is a convergent series in $t^{e_{g-1}}$, and then the parametrization (1) is $e_{g-1} : 1$ and it is not a good parametrization which is a contradiction for (1) to be good.

We denote by $(m; \beta_1, \ldots, \beta_g)$ *the Puiseux characteristics* of C. Put $\beta_0 = e_0 = m$. *The Puiseux pairs* of C is defined by $\{(m_i, n_i) \mid i = 1, \ldots, g\}$ where

$$\gcd(m_i, n_i) = 1, \ \frac{\beta_i}{m} = \frac{n_i}{m_1 \cdots m_i}, \ i = 1, \ldots, g.$$

The uniqueness of the Puiseux characteristics will be discussed later.

3. Resolution of Singularities

3.1. Blowing-up

Let U be an open neighborhood of the origin $O \in \mathbb{C}^2$. An ordinary blowing-up centered at O is defined as follows. Let (x, y) be fixed analytic coordinates and consider the canonical embedding

$$\iota : U - \{O\} \to U \times \mathbb{P}^1, \quad (x, y) \mapsto ((x, y), (x : y)).$$

Let $\overline{U}$ be the closure of the image of $\iota : U - \{O\} \hookrightarrow U \times \mathbb{P}^1$ and let $p_1 : \overline{U} \to U$ be the restriction of the following canonical projection mapping.

$$\begin{array}{ccc} U - \{O\} & \xrightarrow{\iota} & \overline{U} \subset U \times \mathbb{P}^1 \\ \downarrow i & & \downarrow p_1 \\ U & = & U \end{array}$$

Proposition 9. 1. *$\overline{U}$ is a smooth complex manifold of dimension 2 and $p_1 : \overline{U} \to U$ is a proper map.*
2. *$p_1 : \overline{U} - p_1^{-1}(O) \to U - \{O\}$ is biholomorphic.*
3. *The exceptional divisor $E := p_1^{-1}(O)$ is isomorphic to $\mathbb{P}^1$ as a complex manifold.*

Proof. Consider the canonical charts $\mathbb{P}^1 = U_0 \cup U_1$, $U_0 = \{(x : y) | x \neq 0\}$, $U_1 = \{(x : y) | y \neq 0\}$ which have coordinate functions $t = y/x$, $s = x/y$ respectively.

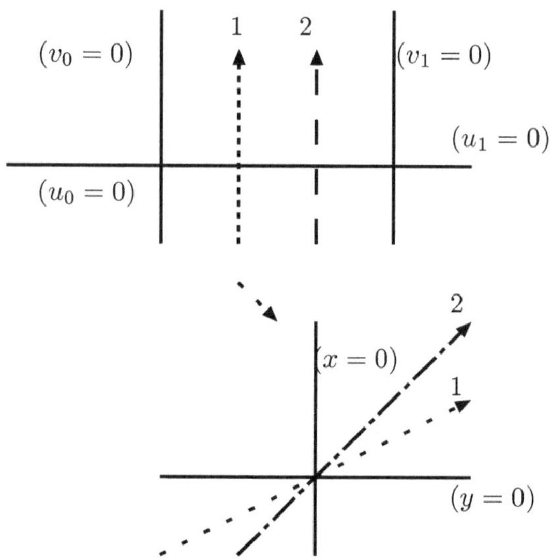

FIGURE 2. Blowing-up

Put $U^* := U - \{O\}$. For simplicity, we assume that $U = \mathbb{C}^2$. We observe that

$$\overline{U} \cap (U^* \times U_0) = \{(x, y, t); x \neq 0,\ y = xt\} \cong \mathbb{C}^* \times \mathbb{C}$$

by the canonical projection $(x, y, t) \mapsto (x, t)$. Therefore

$$\widetilde{U}_0 := \overline{U} \cap (U \times U_0) = \{y = xt\} \subset U \times U_0$$

which is the graph of the function $(x,t) \mapsto xt$ and thus $\overline{U}$ is a smooth manifold and we can take (x,t) as a system of coordinates on $\overline{U} \cap (U \times U_0)$. Similarly $\widetilde{U}_1 := \overline{U} \cap U_1 = \{((x,y),s); x = ys\} \cong \mathbb{C}^2$ and we take the canonical coordinates (y,s). The properness of the mapping p_1 is obvious as p_1 is the restriction of the projection $U \times \mathbb{P}^1 \to U$ to the closed subset $\overline{U}$. The assertion (2) is immediate as the canonical inverse is given by $\iota : U^* \to \overline{U}$.

Proof of assertion 3. $U_0' := \overset{-1}{p_1}(O) \cap (\{O\} \times U_0) = \{(0,t) | t \in \mathbb{C}\}$ and $U_1' := \overset{-1}{p_1}(O) \cap (\{O\} \times U_1) = \{(0,s) | s \in \mathbb{C}\}$. The gluing is given by $U_0' \ni (0,t) \iff (0,1/s) \in U_1'$. This is the usual gluing of $\mathbb{P}^1$ which proves the assertion $\overset{-1}{p_1}(O) \cong \mathbb{P}^1$. □

The *canonical coordinates* (u_0, v_0) on $\widetilde{U}_0$ and (u_1, v_1) on $\widetilde{U}_1$ are defined as $u_0 = x, v_0 = t$, $u_1 = s$, $v_1 = y$. The projections are described by the relations

$$\begin{pmatrix} x \\ y \end{pmatrix} = \begin{pmatrix} u_0 \\ u_0 v_0 \end{pmatrix} = \begin{pmatrix} u_1 v_1 \\ u_1 \end{pmatrix}.$$

Figure 2 shows the geometry of the blowing-up.

3.2. Tangent cone

Let $C = \{f(x,y) = 0\}$ and let $m = m(C,O)$ and

$$f(x,y) = f_m(x,y) + f_{m+1}(x,y) + \cdots$$

be the graduation by degree. We may assume (after a liner change of coordinates)

$$f_m(x,y) = (y - \alpha_1 x)^{\nu_1} \cdots (y - \alpha_k x)^{\nu_k}. \tag{2}$$

The equation $f_m(x,y) = 0$ gives k lines $L_j : y - \alpha_j x = 0$, $j = 1, \ldots, k$ passing through the origin. We call these lines *the tangent cone* of C at the origin. Note that a line L is one of the tangent lines $L_1, \ldots, L_k$ if and only if $I(C, L; O) > m$.

Consider a blowing-up $\pi : (X, E) \to (\mathbb{C}^2, O)$. Let C' be the closure of $\pi^{-1}(C - \{O\}) \subset X$ and we call C' *the strict transform* of C. Under the assumption 2, we see that $C' \cap E$ is in the coordinates chart $(U_0; (u_0, v_0))$. In the coordinate $(\widetilde{U}_0; (u_0, v_0))$, the pull-back of f is defined as

$$\begin{aligned} \pi^* f(u_0, v_0) &= f_m(u_0, u_0 v_0) + f_{m+1}(u_0, u_0 v_0) + \cdots \\ &= u_0^m \times \hat{f}(u_0, v_0), \\ \hat{f}(u_0, v_0) &:= f_m(1, v_0) + u_0 f_{m+1}(1, v_0) + \cdots. \end{aligned}$$

Thus C' is contained in $\widetilde{U}_0$ where it defined by $\hat{f}(u_0, v_0) = 0$ and note that

$$C' \cap E = \{(0, v_0); f_m(1, v_0) = \prod_{j=1}^{k} (v_0 - \alpha_j)^{\nu_j} = 0\}.$$

This expression implies that

Proposition 10. *The intersection $E \cap C'$ is topologically k points $\{(u_0, v_0) = (0, \alpha_1), \dots, (0, \alpha_k)\} \subset U_0$. Let $\xi_j = (0, \alpha_j)$ and take the coordinate (u_0, v_j) with $v_j = v_0 - \alpha_j$. Then $\hat{f}$ is regular of order ν_j in v_j and $m(C', \xi_j) \le \nu_j$. In particular, $I(E, C'; \xi_j) = \nu_j$ and $m(C', \xi_j) \le m$. If the multiplicity does not drop (so $m(C', \xi_j) = m$ holds), then $k = 1$.*

3.3. Resolution of Singularities

A modification $\pi : (X_k, E) \to (\mathbb{C}^2, O)$ with $E = \pi^{-1}(O)$ is called *a weak resolution of* (C, O) if the strict transform C' of C in X_k is non-singular. Usually π is given as the composition of a sequence of blowing-ups:

$$X_k \xrightarrow{\pi_k} X_{k-1} \longrightarrow \cdots \longrightarrow X_1 \xrightarrow{\pi_1} X_0 = \mathbb{C}^2.$$

Theorem 11. *For a given (C, O), there exists a weak resolution.*

Proof 1. We first follow the proof by Wall [14], assuming (C, O) is irreducible. Assume

$$x = t^m, \quad y = b_1 t^m + b_2 t^{2m} + \cdots + c t^{\beta_1} + \cdots, \quad m = \text{multiplicity}\,(\mathrm{C}, \mathrm{O})$$

where β_1 is the first Puiseux characteristic. We use a double induction on (m, β_1) with the dictionary order. After one blowing-up, we get the parametrization

$$u_0 = t^m, \quad v_0 - b_1 = b_2 t^m + \cdots + c t^{\beta_1 - m} + \dots.$$

Case 1: If $\beta_1 - m < m$, $v_0 - b_1 = c t^{\beta_1 - m} + \dots$, and let $C^{(1)}$ be the strict transform of C on X_1 and $\xi = C'^{(1)} \cap E$. Then $m\,(C^{(1)}, \xi) = \min\,(m, \beta_1 - m) = \beta_1 - m < m$.
Case2: $\beta_1 - m \ge m$. Taking local coordinates (u_0, v_0'), $v_0' = v_0 - b_1$,

$$(m(C^{(1)}), \beta_1(C^{(1)})) = (m, \beta_1 - m) < (m, \beta_1). \qquad \square$$

Proof 2, from the Newton boundary point of view. We give another proof which works for any (not necessarily irreducible) curve. We use double induction on m and $\mathrm{Vol}\,\Gamma_-(\mathrm{f}; (\mathrm{x}, \mathrm{y}))$ (with respect to suitable coordinates (x, y)).
Case 1. If a tangent cone contains k lines with $k \ge 2$, we have seen that the strict transform C' splits into k germs of curves in X_1, say $\xi_1, \dots, \xi_k \in E$ with $m(C', \xi_j) \le \nu_j < m$ by Proposition 10.
Case 2. Assume that $k = 1$. (Note that (C, O) may be still reducible.) Then we will see that $\mathrm{Vol}\,\Gamma_-(\mathrm{f}; (\mathrm{x}, \mathrm{y}))$ strictly decreases. First we can assume that $f_m(x, y) = y^m$ (after a change of coordinates). Then we look at the Newton boundary. Let $Q = (j, \ell)$, $\ell < m$, $j + \ell > m$ be a vertex of $\Gamma(f)$ which corresponds to a monomial $x^j y^\ell$. Then after a blowing-up and using the coordinate (u_0, v_0) with $y = u_0 v_0$, $x = u_0$, $x^j y^\ell \mapsto u_0^{j+\ell-m} v_0^\ell$ which makes the area $\Gamma_-(f; u_0, v_0)$ strictly smaller than before. Note that $\deg u_0^{j+\ell-m} v_0^\ell = (j + \ell) - (m - \ell) < \deg x^j y^\ell$. Every vertex of $\Gamma(f; (x, y))$ except $(0, m)$ moves to the left side, keeping the second coordinate. Let $A = (0, m)$ and let $P = (m_1, \ell)$ be the vertex such that $\{A, P\}$ makes the first face Δ of the Newton boundary from the left. After a blowing-up, P changes into $P' = (m_1 - m + \ell, \ell)$ and thus the degree P' is $m_1 + \ell - (m - \ell) < m_1 + \ell$. Then

it is easy to see that after a finite number of blowing-ups, the multiplicity will be strictly smaller than m. □

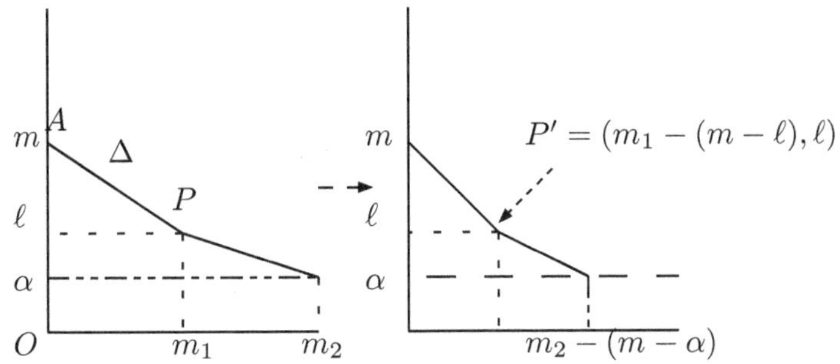

FIGURE 3. Newton boundary after a blowing-up

In the above example, f is not convenient if $\alpha \geq 1$.

3.4. Geometry of the resolution

Lemma 12. *Assume $C \subset S$ is a curve in a complex surface S. Take $P \in C \subset S$ and consider a blowing-up centered at P: $\pi : T \to S$. Let C' be the strict transform of C and let E be the exceptional divisor. If $m(C,P) = m$, $I(C', E) = m$. In particular, if (C, P) is non-singular, $m = 1$ and this implies that C' is non-singular and C' intersects E transversely.*

Proof. The assertion is immediate from Proposition 10. □

A holomorphic mapping $\pi : X \to (U, O)$ ($U \subset \mathbb{C}^2$: an open neighborhood of $O \in \mathbb{C}^2$) is called a *modification* at $O \in U$ if:

- X is a smooth complex manifold of dimension 2, π is a proper holomorphic mapping and
- $\pi : X - \pi^{-1}(O) \to U - O$ is a biholomorphic mapping.

A modification $\pi : X \to U$ is called a *good resolution* of (C, O) if each component of $\pi^{-1}(C)$ is smooth and $\pi^{-1}(C)$ has at most normal crossing singularities. Here "normal crossing" means that for any singular point $P \in \pi^{-1}(C)$ there exists an analytic coordinate system (u, v) centered at P so that

$$\pi^* f(u, v) = V\, u^a\, v^b,\ a, b \geq 0$$

where V is a unit. A good resolution is *minimal* if there is no possible exceptional divisor with self-intersection number -1 and can be contracted keeping the goodness of the resolution.

Remark 13. The self-intersection number of a smooth divisor C in a smooth surface M is defined as follows. Take an isotopy h_t, $-1 \leq t \leq 1$ of M so that $h_t(C)$ $t \neq 0$

intersects C transversely. Then the self-intersection C^2 is defined by the intersection number of C and $h_t(C)$, $t \neq 0$. It can be also defined as the first Chen number of the restriction to C of the line bundle $[C]$ defined by C. See [7, 3] for example.

Theorem 14. *There exists a good resolution.*

Proof. First we may assume by Theorem 11 that there exists a composition of blowing-ups $\pi : (X, E) \to (\mathbb{C}^2, O)$ so that each component of $\pi^* f = 0$ is smooth. Next suppose that two smooth curves C_1, C_2 are tangent at O and put $s := I(C_1, C_2; O) = s \geq 2$. We may assume that

$$C_1 : x = 0, \quad C_2 : x = \phi(y), \ \phi(y) = y^s + (\text{higher terms}).$$

Then after one blowing-up, $x = uv$, $y = u$,

$$C_1' : v = 0, \quad C_2 : v = u^{s-1} + \text{higher terms}.$$

Thus we observe that $I(C_1', C_2'; O') = s - 1$ where $O' = (0, 0)$ in the coordinates (u, v). Thus we observe that after a finite number of blowing-ups $\pi' : X' \to X$, we can make components of $(\pi \circ \pi')^* f = 0$ that are transversal. Thirdly if there are more than three components of $(\pi \circ \pi')^* f = 0$ which intersect at a point, we proceed one more blowing-up to separate them. □

3.4.1. Example. Let us consider a cusp $C : y^3 = x^2$. First blowing-up $\pi_1 : (X_1, E_1) \to (\mathbb{C}^2, O)$,

$$x = uv, \ y = v, \ \pi^* f = v^2(u^2 - v).$$

This is a weak resolution but not a good resolution.

Second blowing-up $\pi_2 : (X_2, E_2) \to (X_1, O_1)$,

$$v = pq, \ u = p, \ f \mapsto p^3 q^2 (p - q).$$

Third blowing-up $\pi_3 : (X_3, E_3) \to (X_2, O_2)$,

$$p = t, \ q = st, \ f \mapsto t^6 s^2 (1 - s).$$

(There are four coordinate charts on X_3.) Now the composition of three projections is given as

$$(s, t) \mapsto (p, q) = (t, st) \mapsto (u, v) = (t, st^2) \mapsto (x, y) = (st^3, st^2).$$

This is a typical example of a *monomial mapping*, which is true for any toric modifications which we explain in the next section.

3.5. Toric modification

Consider the space of weight vectors $N \cong \mathbb{Z}^2$ of the fixed variables, say x, y. It has a canonical basis $E_1 = {}^t(1, 0)$, $E_2 = {}^t(0, 1)$. $N \otimes \mathbb{R}$ can be identified with $\mathbb{R}^2$. Consider a cone subdivision Σ^* of $\mathbb{R}_+^2$ by half lines generated by primitive integer vectors

$$P_0 = E_1, \ P_1 = \begin{pmatrix} a_1 \\ b_1 \end{pmatrix}, \ldots, P_m = \begin{pmatrix} a_m \\ b_m \end{pmatrix}, \ P_{m+1} = E_2.$$

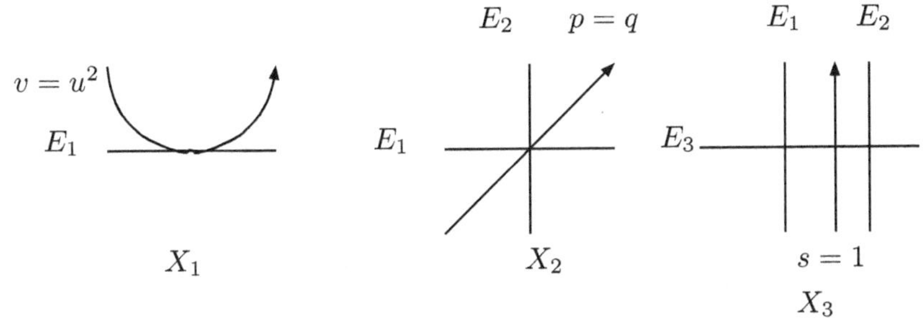

FIGURE 4. Resolution Tower

For each cone Cone$(\mathrm{P_i}, \mathrm{P_{i+1}})$, we associate a matrix

$$\sigma_i := \begin{pmatrix} a_i & a_{i+1} \\ b_i & b_{i+1} \end{pmatrix}.$$

The cone subdivision (*fan*) is *regular* if $\det \sigma_i = 1$ for $i = 0, \dots, m$.

For a regular subdivision Σ^*, we define a modification $\pi : X \to \mathbb{C}^2$ as follows. First, we associate a coordinate chart $(\mathbb{C}_i^2, (u_i, v_i))$ for each cone σ_i, $i = 0, \dots, m$ and we consider monomial mapping

$$\pi_{\sigma_i} : \mathbb{C}_i^2 \to \mathbb{C}^2, \quad (u_i, v_i) \mapsto (x, y) = (u_i^{a_i} v_i^{a_{i+1}}, u_i^{b_i} v_i^{b_{i+1}}).$$

Here by an abuse of notation, we identify the cone Cone$(\mathrm{P_i}, \mathrm{P_{i+1}})$ and the corresponding unimodular matrix $\sigma_i = (P_i, P_{i+1})$. This morphism has an obvious property:

$$\pi_\sigma \circ \pi_\tau = \pi_{\sigma\tau}, \; \pi_\sigma^{-1} = \pi_{\sigma^{-1}}.$$

Thus $(\pi_\sigma)^{-1} = \pi_{\sigma^{-1}}$ and π_σ is a *birational map*. Note that σ^{-1} may have negative coefficients and therefore it is only defined on $\mathbb{C}^{*2}$ in general.

Example 15. Consider the cone subdivision $\Gamma^* = \{E_1, P, E_2\}$ where $P = {}^t(3, 2)$ and $\Sigma^* = \{E_1, Q, P, R, E_2\}$ where $Q = {}^t(2, 1)$ and $R = {}^t(1, 1)$. We observe that Γ^* is not regular as $\det(E_1, P) = 2$ for example, but Σ^* is a regular cone subdivision of Γ^*. Note that Γ^* is the dual Newton diagram (see §3.6 below for definition) of the cusp singularity $f(x, y) = \{y^3 - x^2 = 0\}$.

We define a complex surface X as follows. First take the disjoint sum $\tilde{X} := \amalg_{i=0}^m \mathbb{C}_i^2$ and then we glue each coordinate chart as follows. Two points $\mathbf{u}_i = (u_{i1}, u_{i2}) \in \mathbb{C}_i^2$ and $\mathbf{u}_j = (u_{j1}, u_{j2}) \in \mathbb{C}_j^2$ are identified if $\pi_{\sigma_i^{-1}\sigma_j} : \mathbb{C}_j^2 \to \mathbb{C}_i^2$ is well-defined at $\mathbf{u}_j$ and $\pi_{\sigma_i}^{-1} \circ \pi_{\sigma_j}(\mathbf{u}_j) = \mathbf{u}_i$. A complex surface X is defined as the quotient space of $\tilde{X}$ by this identification.

Proposition 16. *By the definition, we have the following properties.*

1. *X is a complex surface with affine coordinate charts $(\mathbb{C}_i^2, (u_i, v_i))$, $i = 0, \dots, m$.*

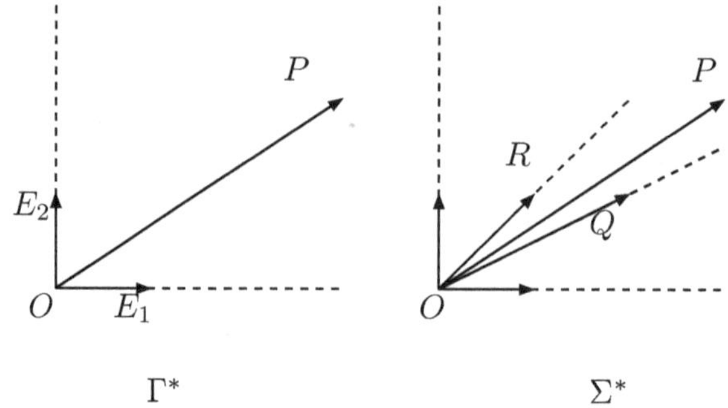

FIGURE 5. Γ^* (left) and Σ^* (right)

2. *There exists a canonical mapping* $\pi : X \to \mathbb{C}^2$ *which is compatible with* $\pi_{\sigma_i} : \mathbb{C}_i^2 \to \mathbb{C}^2$, *namely* $\pi \circ \iota_i = \pi_{\sigma_i}$ *with* $\iota_i : \mathbb{C}_i^2 \to X$ *being the canonical inclusion.*
3. *The two divisors* $\{v_i = 0\} \subset \mathbb{C}_i^2$ *and* $\{u_{i+1} = 0\} \subset \mathbb{C}_{i+1}^2$ *glue together to make a compact divisor isomorphic to* $\mathbb{P}^1$ *for* $i = 1, \dots, m$. *We denote this divisor by* $\widehat{E}(P_i)$. *Then* $\pi^{-1}(O) = \bigcup_{i=1}^m \widehat{E}(P_i)$ *and* $\pi : X - \pi^{-1}(O) \to \mathbb{C}^2 - \{O\}$ *is biholomorphic.*

Proof. First we observe that the tori $\mathbb{C}_i^{*2} \subset \mathbb{C}_i^2$ and $\mathbb{C}_j^{*2} \subset \mathbb{C}_j^2$ are identified by the isomorphism $\pi_{\sigma_i^{-1}\sigma_j} : \mathbb{C}_j^{*2} \to \mathbb{C}_i^{*2}$. If $|i-j| = 1$, this identification is extended on one common coordinate axis as we see below. Take a basis $\{P_i, P_{i+1}\}$ of the lattice $\mathbb{Z}^2 \subset \mathbb{R}^2$. For $Q \in \mathbb{R}_+^2$, we can write $Q = aP_i + bP_{i+1}$. Then a is positive if Q is on the same side as P_i with respect to the line OP_{i+1}. If P, Q and R, S are a basis of $\mathbb{Z}^2$, note that

$$(P,Q)^{-1}(R,S) = \begin{bmatrix} a & b \\ c & d \end{bmatrix} \iff R = aP + cQ,\ S = bP + dQ.$$

Thus we have

$$\sigma_i^{-1}\sigma_{i+1} = (P_i, P_{i+1})^{-1}(P_{i+1}, P_{i+2}) = \begin{bmatrix} 0 & -1 \\ 1 & d \end{bmatrix}, \quad d > 0.$$

This implies

$$u_i = v_{i+1}^{-1}, \quad v_i = u_{i+1} v_{i+1}^d$$

and

$$\{u_{i+1} = 0\} \leftrightarrow \{v_i = 0\}, \quad u_i = v_{i+1}^{-1} \iff \{u_{i+1} = 0\} \cup \{v_i = 0\} \cong \mathbb{P}^1.$$

This shows the compactness of the divisor

$$\widehat{E}(P_{i+1}) = \{v_i = 0\} \cup \{u_{i+1} = 0\}.$$

As $a_i, b_i > 0$ for $i = 1, \dots, m$, it is easy to see that $\pi(\widehat{E}(P_i)) = \{O\}$. Note also that for $i = 0$ (respectively for $i = m$), we have $b_1 = 1$ (resp. $a_m = 1$) and

$$\pi_{\sigma_0}(u_0, v_0) = (x, y) = (u_0 v_0^{a_1}, v_0),\ \pi_{\sigma_m}(u_m, v_m) = (x, y) = (u_m, u_m^{b_m} v_m)$$

and the non-compact divisor $\{u_0 = 0\}$ (resp. $v_m = 0$) is isomorphically mapped onto the axis $x = 0$ (resp. $y = 0$). Thus $\pi^{-1}(O) = \widehat{E}(P_1) \cup \dots \cup \widehat{E}(P_m)$. □

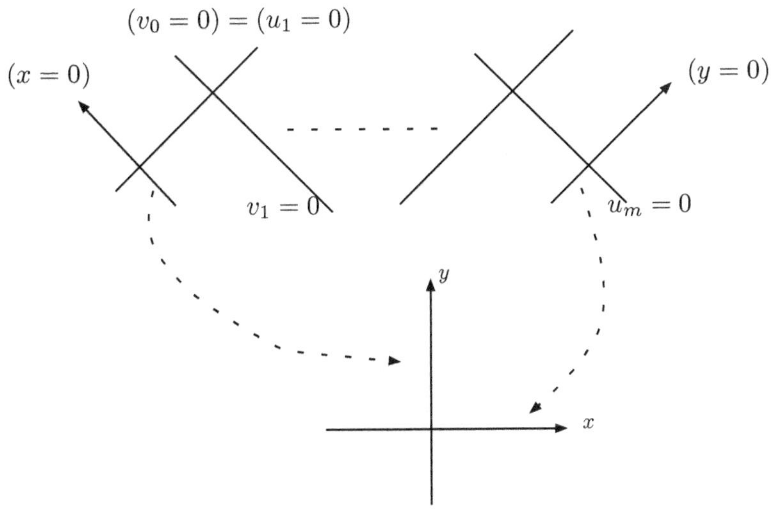

FIGURE 6. Toric modification ($E_i = \widehat{E}(P_i)$)

The dual graph of a modification $p : Y \to \mathbb{C}^2$ is defined as follows. Let $E_1, \dots, E_r$ be the exceptional divisors so that $p^{-1}(O) = \bigcup_{i=1}^r E_i$. For each E_j, we associate a vertex v_j. Two vertices v_i, v_j are joined by an edge iff E_i and E_j intersect. Let $C_1, \dots, C_s$ be irreducible components and let $p : Y \to \mathbb{C}^2$ be a good resolution of C. *An extended dual graph* is defined by adding s white vertices $w_1, \dots, w_s$ beside $v_1, \dots, v_r$ (they are usually denoted by bullets) and w_i and v_j is joined by an edge if the strict transform C_i' of C_i and E_j intersect.

Note that the dual graph of a toric modification $\pi : X \to \mathbb{C}^2$ is a bamboo.

Example 17. Consider a cusp singularity $C := \{y^3 - x^2 = 0\}$. The toric modification $\pi : X \to \mathbb{C}^2$ with respect to the regular subdivision $\Sigma^* = \{E_1, Q, P, R, E_2\}$ where $Q = {}^t(2,1)$, $P = {}^t(3,2)$ and $R = {}^t(1,1)$ gives a good resolution of C and the dual resolution diagram is given as in Figure 7.

3.6. Non-degeneracy and a canonical resolution

Let $f(x, y)$ be a germ of functions such that $f(O) = 0$. Let $f(x, y) = \sum_\nu c_\nu x^{\nu_1} y^{\nu_2}$ be the Taylor expansion. The Newton polygon $\Gamma_+(f; x, y)$ is the convex hull of $\bigcup_{\nu, c_\nu \neq 0} \{\nu + \mathbb{R}_+^2\}$ and the Newton boundary $\Gamma(f; x, y)$ is the union of compact

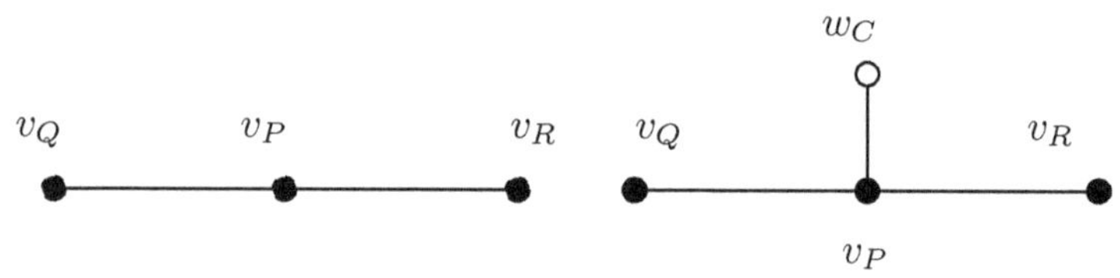

$$v_Q \iff \hat{E}(Q),\ v_P \iff \hat{E}(P),\ v_R \iff \hat{E}(R),\ w_C \iff C$$

FIGURE 7. Dual resolution graph Γ (left) and the extended dual graph (right)

faces of $\Gamma_+(f; x, y)$. Take a 1-dimensional face Δ of $\Gamma(f)$. The *face function* f_Δ is defined as

$$f_\Delta(x, y) := \sum_{\nu \in \Delta} c_\nu x^{\nu_1} y^{\nu_2}.$$

There exists a unique positive primitive vector (weight vector) $P = {}^t(a, b)$ such that $f_\Delta(x, y)$ is a weighted homogeneous polynomial with respect to P so that

$$f_\Delta(t^a x, t^b y) = t^d f_\Delta(x, y), \quad \text{or } \Delta \subset \{a\nu_1 + b\nu_2 = d\}.$$

We can factorize as

$$f_\Delta(x, y) = c\, x^\alpha y^\beta \prod_{j=1}^{\delta} (y^a - \gamma_j x^b)^{\mu_j},\ \gamma_i \neq \gamma_j,\ (i \neq j). \tag{3}$$

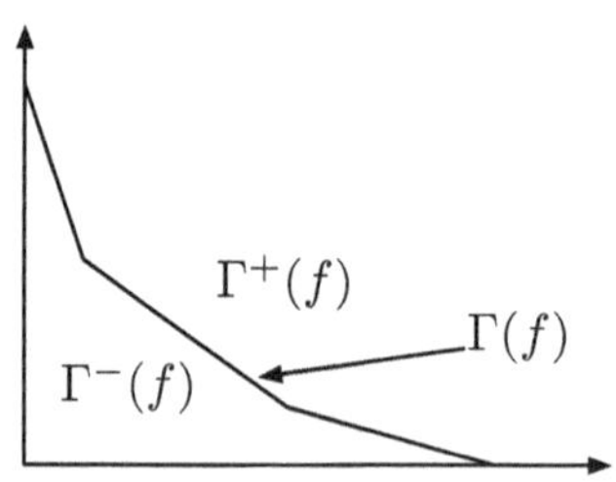

FIGURE 8. Newton Diagram

We say that f is *non-degenerate* on a face $\Delta \in \Gamma(f)$ if $\mu_1 = \cdots = \mu_\delta = 1$ and f is called *non-degenerate* (in the sense of a Newton boundary) if f is non-degenerate on every 1-face Δ. Note that the non-degeneracy is a notion which depends on the choice of the coordinates (x, y).

Proposition 18. *If f is non-degenerate on a 1-face $\Delta \in \Gamma(f)$ for which the face function is factored in (3), $\delta + 1$ is equal to the number of integral points on Δ.*

The proof follows from the observation that the number of integral points is invariant under the linear transformation defined by a unimodular matrix. See also the proof of Theorem 22.

For any positive primitive weight vector $P = {}^t(a,b)$, we consider a linear function on $\Gamma(f)$ defined by P. Let $d(P;f)$ (or simply $d(P)$) be the minimal value of $P|\Gamma(f)$ and let $\Delta(P;f)$ (or simply $\Delta(P)$) be the face where it takes the minimal value. Here $\Delta(P)$ can be a vertex. Let $f_P = f_{\Delta(P)}$. Then f_P is a weighted homogeneous polynomial of type $(a,b;d)$ with $d = d(P;f)$. Let N be the space of positive weight vectors. We introduce an equivalence relation as follows.

$$P \sim Q \iff \Delta(P;f) = \Delta(Q;f).$$

Then this defines a fan (a conical subdivision) of N and we denote it as $\Gamma^*(f)$ and we call it *the dual Newton diagram.* Let $\Delta_1, \dots, \Delta_s$ be the 1-faces of $\Gamma(f;x,y)$ and let $P_1, \dots, P_s$ be the corresponding weight vectors. Let $E_1 = {}^t(1,0)$, $E_2 = {}^t(0,1)$ be as before. Then $\Gamma^*(f)$ consists of $s+1$ cones $\mathrm{Cone}(\mathrm{P_i}, \mathrm{P_{i+1}})$ where $P_0 = E_1$, $P_{s+1} = E_2$.

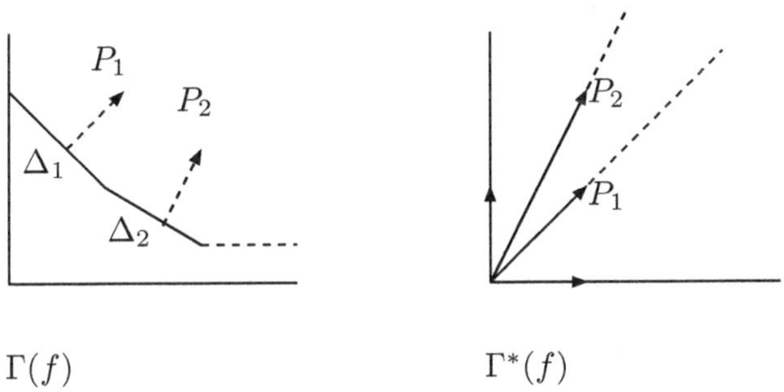

FIGURE 9. (Dual) Newton boundary of $f = y(y^2 + x^2)(y + x^2)$

A regular fan Σ^* is called *admissible* for $\Gamma^*(f)$ if it is a subdivision of $\Gamma^*(f)$. The function germ f is called *convenient* if $f(0,y) \not\equiv 0$ and $f(x,0) \not\equiv 0$.

Example 19. For example, $f(x,y) = y(y^2 + x^2)(y + x^2)$ is not convenient. See Figure 9 for its Newton boundary.

Lemma 20. ([11]) *Suppose that* $\det(P,Q) = d > 1$. *Then there exists a unique integer* $0 < d_1 < d$ *such that* $T := (Q + d_1 P)/d$ *is an integer vector and* $\det(T,Q) = d_1$.

Applying this lemma inductively on the cone $\mathrm{Cone}(\mathrm{T},\mathrm{Q})$ if $d_1 > 1$, we obtain:

Corollary 21. ([11]) *There is a canonical subdivision $\Sigma_{can} := \{P, T_1, \dots, T_s\}$ of* $\mathrm{Cone}(\mathrm{P},\mathrm{Q})$ *so that any regular subdivision of* $\mathrm{Cone}(\mathrm{P},\mathrm{Q})$ *is a subdivision of Σ_{can}. The number s is determined by expanding d/d_1 in a continued fraction*

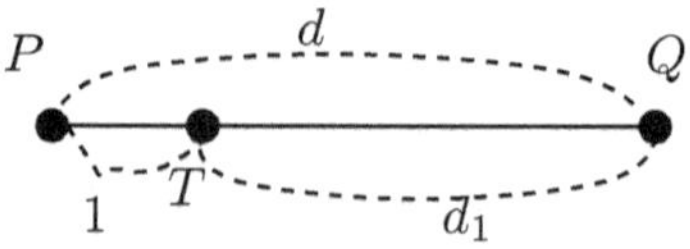

FIGURE 10. First subdvision of Cone(P, Q)

$[m_1, \dots, m_s]$, $m_i \geq 2$ where the continued fraction is defined inductively by $[m_1, \dots, m_s] = m_1 - \frac{1}{[m_2, \dots, m_s]}$ and $[m_s] = m_s$. □

3.6.1. Problem. Let $f(x, y) = x^2 + y^{2k+1}$. Compute the dual Newton diagram $\Gamma^*(f)$ and the canonical regular subdivision Σ^*.
Answer: $\Gamma^*(f)$ has three vertice $\{E_1, P, E_2\}$ where $P = {}^t(2k+1, 2)$. Σ^* is the cone with vertices

$$\Sigma^* : \ \{E_1, T, P, S_k, S_{k-1} \dots, S_1, E_2\} \quad \text{where}$$

$$T = \begin{bmatrix} k+1 \\ 1 \end{bmatrix}, \ P = \begin{bmatrix} 2k+1 \\ 2 \end{bmatrix}, \ S_k = \begin{bmatrix} k \\ 1 \end{bmatrix}, \ S_{k-1} = \begin{bmatrix} k-1 \\ 1 \end{bmatrix}, \dots, S_1 = \begin{bmatrix} 1 \\ 1 \end{bmatrix}.$$

Theorem 22. *Assume that $f(x, y)$ is a convenient polynomial with 1-faces $\Delta_1, \dots, \Delta_s$ and let $P_1, \dots, P_s$ be the corresponding weight vectors. Put $P_i = {}^t(a_i, b_i)$ for $i = 1, \dots, s$. Let Σ^* be an admissible regular fan and let $\pi : (X, E) \to (\mathbb{C}^2, O)$ be the corresponding toric modification with vertices $Q_0, Q_1, \dots, Q_m$ with $Q_0 = E_1$, $Q_m = E_2$. Then $E = \pi^{-1}(O) = \bigcup_{j=1}^{m-1} \widehat{E}(Q_j)$ and*

$$(\pi^* f) = C' + \sum_{i=1}^{m-1} d(Q_i) \widehat{E}(Q_i)$$

where $d(Q_i) = d(Q_i; f)$ and C' is the strict transform of C. Assume that $P_i = Q_{\nu_i}$, $i = 1, \dots, s$ and consider the factorization

$$f_{P_i}(x, y) = c_i x^{\alpha_i} y^{\beta_i} \prod_{j=1}^{\ell_i} (y^{a_i} - \gamma_{ij} x^{b_i})^{\nu_{ij}}, \ i = 1, \dots, m.$$

Then

1. *(C, O) has at least $\sum_i \ell_i$ irreducible components. More precisely, $(C', C' \cap E) \subset (X, E)$ consists of $\sum_{i=1}^{s} \ell_i$ disjoint components, where each component may not be irreducible.*
2. *The strict transform C' intersects with only exceptional divisors $\widehat{E}(Q_{\nu_i}) = \widehat{E}(P_i)$, $i = 1, \dots, s$ and $C' \cap \widehat{E}(P_i) = \{(0, \gamma_{ij}); j = 1, \dots, \ell_i\}$. Put $\xi_{ij} = (0, \gamma_{ij}) \in \widehat{E}(P_i) \cap C'$. Then the intersection number is given by $I(C', \widehat{E}(P_i); \xi_{ij}) = \nu_{ij}$. If Σ^* is the canonical toric subdivision, the self intersection number satisfies the inequality: $\widehat{E}(Q_i)^2 \leq -2$ for any $Q_i \neq P_1, \dots, P_s$.*
3. *Assume that $\nu_{ij} = 1$ for $j = 1, \dots, \ell_i$. Then C' intersects with $\widehat{E}(Q_{\nu_i})$ at ℓ_i points transversely.*

4. *In particular, if $f(x,y)$ is non-degenerate, $\pi : (X,E) \to (\mathbb{C}^2, O)$ gives a good resolution of (C,O) and the number of irreducible components is equal to the number of integral points on $\Gamma(f;x,y)$ minus 1.*

Proof. Let us consider the toric chart $\sigma_\iota = \mathrm{Cone}(Q_\iota, Q_{\iota+1})$ and write it $Q_\iota = {}^t(p_\iota, q_\iota)$ and we identify $\mathrm{Cone}(Q_\iota, Q_{\iota+1})$ and the unimodular matrix

$$\sigma_\iota = \begin{pmatrix} p_\iota & p_{\iota+1} \\ q_\iota & q_{\iota+1} \end{pmatrix}.$$

We assume that $f(x,y) = \sum_\nu c_\nu x^{\nu_1} y^{\nu_2}$. First we note that

$$\pi_{\sigma_\iota}^*(x^\alpha y^\beta) = u_\iota^{d(Q_\iota; x^\alpha y^\beta)} v_\iota^{d(Q_{\iota+1}; x^\alpha y^\beta)} = u_\iota^{p_\iota \alpha + q_\iota \beta} v_\iota^{p_{\iota+1}\alpha + q_{\iota+1}\beta}.$$

In this chart, the strict transform C' is defined by $f_{\sigma\iota}(u_\iota, v_\iota) = 0$ where

$$\pi_\iota^* f(u_\iota, v_\iota) = u_\iota^{d(Q_\iota; f)} v_\iota^{d(Q_{\iota+1}; f)} f_{\sigma_\iota}(u_\iota, v_\iota).$$

We put also

$$f_{Q_\iota, \sigma_\iota}(u_\iota, v_\iota) = \frac{1}{u_\iota^{d(Q_\iota; f)} v_\iota^{d(Q_{\iota+1})}} \pi_\iota^* f_{Q_\iota}(u_\iota, v_\iota).$$

Now we observe that

$$\hat{E}(Q_\iota) \cap C' := \{(0, v_\iota) \mid f_{\sigma_\iota}(0, v_\iota) = \textstyle\sum_{\nu \in \Delta(Q_\iota)} c_\nu v_\iota^{Q_{\iota+1}(\nu) - d(Q_{\iota+1}; f)} = 0\}.$$

Thus if $\Delta(Q_\iota) = \{\nu\}$, $\widehat{E}(Q_\iota) \cap C' = \emptyset$ as $f_{\sigma_\iota}(0,v) = c_\nu$. Similarly we have $\widehat{E}(Q_\iota) \cap \widehat{E}(Q_{\iota+1}) \cap C' = \emptyset$. Thus we see that $\widehat{E}(Q_\iota) \cap C' \neq \emptyset$ iff $Q_\iota = P_i$ for some i. Let us assume that $\iota = n_i$. Thus using the equality $p_\iota q_{\iota+1} - p_{\iota+1} q_\iota = 1$, we see that

$$y^{p_\iota} - \gamma x^{q_\iota} = u_\iota^{p_\iota q_\iota} v_\iota^{p_{\iota+1} q_\iota}(v_\iota - \gamma).$$

Therefore

$$f_{\sigma_\iota}(0, v_\iota) \equiv f_{Q_\iota, \sigma_\iota}(0, v_\iota) \equiv c_i \prod_{j=1}^{\ell_i} (v_\iota - \gamma_j)^{\nu_{ij}}.$$

Putting $\xi_{ij} := (0, \gamma_{ij}) \in \widehat{E}(P_i)$, this implies that

$$\widehat{E}(Q_\iota) \cap C' = \{\xi_{ij} | j = 1, \dots, \ell_i\}$$

and $I(\widehat{E}(P_i), C'; \xi_{ij}) = \nu_{ij}$. Taking a translated coordinate $(u_\iota, v_{\iota j})$, $v_{\iota j} = v_\iota - \gamma_{ij}$, we see that f_{σ_ι} is regular of order ν_{ij} in $v_{\iota j}$. The union of the images of the germ (C', ξ_{ij}) by π for $i = 1, \dots, s$, $j = 1, \dots, \ell_i$ is nothing but (C,O). Thus the number of irreducible components of (C,O) is at least $\sum_{i=1}^s \ell_i$. In particular, if $\nu_{ij} = 1$, (C', ξ_{ij}) is irreducible and smooth and transverse to $\widehat{E}(P_i)$ as

$$\frac{\partial f_{\sigma_\iota}}{\partial v_\iota}(0, \gamma_j) = \prod_{k \neq j} (\gamma_j - \gamma_k) \neq 0.$$

For the assertion about the intersection numbers, we refer to [9, 11] □

Corollary 23. *Assume that (C,O) is irreducible. Then there is only one face Δ on $\Gamma(f;x,y)$ for any given coordinate system (x,y).*

To study the local geometry of C' at ξ_{ij}, we first take the corresponding toric coordinate chart $\mathbb{C}^2_{\sigma_i}$ with coordinates (u_i, v_i) where $\widehat{E}(P_i)$ is defined by $u_i = 0$. Then put $v_{ij} = v_i - \gamma_{ij}$. We call (u_i, v_{ij}) a *translated toric coordinate*. To proceed to further toric modification, we may need to make a little change of coordinates of the type (u_i, v'_{ij}) where $v'_{ij} = v_{ij} + h(u_i)$, with $h(u_i)$ being a suitable polynomial. We call such a system of coordinates (u_i, v'_{ij}) an *admissible toric coordinate system* at ξ_{ij}.

3.7. Relation between a tower of ordinary blowing-ups and a toric blowing-up

Theorem 24. 1. *A toric modification $\pi : (X, E) \to \mathbb{C}^2$ is a finite composition of ordinary blowing-ups .*

2. *Conversely let*

$$X_k \xrightarrow{p_k} X_{k-1} \xrightarrow{p_{k-1}} \cdots \xrightarrow{p_2} X_1 \xrightarrow{p_1} X_0 = \mathbb{C}^2$$

be a tower of ordinary blowing-ups such that the center of $p_i : X_i \to X_{i-1}$ is the origin of two canonical coordinate charts in the preceding chart. Then the composition $\pi_k : X_k \to X_0$ is a toric modification with $k+1$ regular cones.

Proof. The first assertion is proved in Lê-Oka [12]. We will show the second assertion by an induction on k. On X_1, we have two canonical charts $(\widetilde{U}_0, (u_0, v_0))$ and $(\widetilde{U}_1, (u_1, v_1))$ so that

$$p_1 : (u_0, v_0) \mapsto (x, y) = (u_0, u_0 v_0), \quad (u_1, v_1) \mapsto (x, y) = (u_1 v_1, v_1).$$

Assume that the center of p_2 is $(u_0, v_0) = (0, 0)$ in $\widetilde{U}_0$. (The case that the center is $(u_1, v_1) = (0, 0)$ is also proved exactly in a similar discussion.) By the induction's assumption, there exists a regular cone subdivision with vertices

$$\left\{ P_i = \begin{pmatrix} a_i \\ b_i \end{pmatrix}, \, i = 0, 1, \dots, k \right\} \text{ with } P_0 = \begin{pmatrix} 1 \\ 0 \end{pmatrix}, \, P_k = \begin{pmatrix} 0 \\ 1 \end{pmatrix}.$$

Then X_k has $(k+1)$ canonical coordinate charts $(W_i, (s_i, t_i))$ which corresponds to the toric coordinate chart associated with $\mathrm{Cone(P_i, P_{i+1}),\ i = 0, 1, \dots, k-1}$ and $(\widetilde{U}_1, (u_1, v_1))$. Now we compute the composition $\xi_j : W_j \xrightarrow{\sigma_j} \widetilde{U}_0 \xrightarrow{p_1} X_0 = \mathbb{C}^2$ as

$$\begin{aligned}(s_j, t_j) \xrightarrow{\sigma_j} (u_0, v_0) &= (s_j^{a_j} t_j^{a_{j+1}}, s_j^{b_j} t_j^{b_{j+1}}) \\ &\xrightarrow{p_1} (x, y) = (u_0, u_0 v_0) = (s_j^{a_j} t_j^{a_{j+1}}, s_j^{a_j + b_j} t_j^{a_{j+1} + b_{j+1}}).\end{aligned}$$

Put

$$\tilde{P}_j = \begin{pmatrix} a_j \\ a_j + b_j \end{pmatrix}, \; j = 0, \dots, k.$$

Then we see that $\tilde{P}_0 = T = \begin{pmatrix} 1 \\ 1 \end{pmatrix}$ and $\tilde{P}_k = E_2$ and $\det(\tilde{P}_j, \tilde{P}_{j+1})) = 1$. Thus by taking a regular fan with vertices

$$\{E_1, \tilde{P}_0, \tilde{P}_1, \dots, \tilde{P}_k\},$$

we see immediately that the composition ξ_j is equal to π_{σ_j} where $\sigma_j = (\tilde{P}_j, \tilde{P}_{j+1})$). The projection $\tilde{U}_1 \xrightarrow{p_1} X_0$ is associated with $\mathrm{Cone}(\mathrm{E}_1, \tilde{\mathrm{P}}_0)$. □

3.7.1. Example. Consider the sequence

$$p : X_3 \xrightarrow{p_3} X_2 \xrightarrow{p_2} X_1 \xrightarrow{p_1} \mathbb{C}^2$$

which we have constructed in Example 3.4.1 to resolve the cusp singularity C : $y^3 - x^2 = 0$. This is equivalent to the toric modification with respect to the regular fan Σ^* which is generated by

$$\Sigma^* : \left\{ E_1, Q = \begin{pmatrix} 2 \\ 1 \end{pmatrix}, P = \begin{pmatrix} 3 \\ 2 \end{pmatrix}, R = \begin{pmatrix} 1 \\ 1 \end{pmatrix}, E_2 \right\}$$

The strict transform C' intersects with $\hat{E}(P)$ and, using the toric coordinates chart $\sigma = \mathrm{Cone}(P, R)$ with coordinates (u, v), the projection π_σ is defined as

$$(u_1, u_2) \mapsto (x, y), \quad \begin{cases} x = u_1^3 u_2 \\ y = u_1^2 u_2 \end{cases}.$$

This corresponds to $p_1 \circ p_2 \circ p_3$ with identification $(u_1, u_2) = (t, s)$ in Example 3.4.1.

4. Milnor fibration

4.1. Milnor fibration

Let $C : f(x, y) = 0$ where $f(x, y)$ is reduced and $O \in C$. More generally let $V = \{f(z_1, \ldots, z_n) = 0\}$ be a reduced hypersurface with isolated singularity at $O \in V$.

Lemma 25. *There exists $\varepsilon_0 > 0$ such that $S_\varepsilon^3 \pitchfork C$ (or $S_\varepsilon^{2n-1} \pitchfork V$) for any* $0 < \varepsilon \leq \varepsilon_0$.

Proof. In general dimension, we use the curve selection lemma ([8]). We give an easy proof for curves. Assume that C has irreducible components $C_1, \ldots, C_k$ and assume that C_i is parametrized as

$$C_i : x = t^{m_i}, y = \sum_{i=m_i}^{\infty} a_i t^i.$$

Then the tangent line at $(x(t), y(t))$ is given by

$$y'(t)(x - x(t)) - x'(t)(y - y(t)) = 0.$$

Thus $T_{P(t)} C_i \to L : a_{m_i} x - y = 0 \, (t \to 0)$. On the other hand,

$$\overline{OP(t)} \to \mathbb{C}(1, a_{m_i}) \iff y = a_{m_i} x$$

where $\overline{OP(t)}$ is the line passing through two points $O, P(t)$ and $\mathbb{C}(1, a_{m_i})$ is the line $\{(\tau, \tau a_{m_i}); \tau \in \mathbb{C}\}$. The convergence is in the space of lines in $\mathbb{C}^2$. So the assertion follows immediately. □

Hereafter we restrict our consideration to the case of curves.

Theorem 26 (Milnor Fibration, Second description). *For a fixed $\varepsilon \le \varepsilon_0$, take $\delta > 0$ so that the level hypersurface $f^{-1}(\eta)$ intersects transversely with S_ε for any η such that $|\eta| \le \delta$. Then*

$$f : B_\varepsilon \cap f^{-1}(D_\delta^*) \to D_\delta^*$$

is a locally trivial fibration.

Proof. The assertion is immediate from the Ehresmann fibration theorem [15] □

We call $F = f^{-1}(\eta) \cap B_\varepsilon$, a *Milnor fiber of f*. A Milnor fiber is a 1-dimensional complex open manifold and it is known that F has a homotopy type of 1-dimensional CW-complex ([8]). The first Betti number $b_1(F)$ is called a *Milnor number of f* at the origin O and usually denoted as $\mu(f, O)$.

Consider the restriction of the fibration $f : B_\varepsilon \cap f^{-1}(S_\delta^1) \to S_\delta^1$. As the fibration is locally trivial, the total space $E(\varepsilon, \delta) := B_\varepsilon \cap f^{-1}(S_\delta^1)$ can be identified with the quotient space $F \times [0, 2\pi]/h$ where $h : F \to F$ is the monodromy diffeomorphism and the identification is $(\mathbf{z}, 2\pi) \sim (h(\mathbf{z}), 0)$ for $\mathbf{z} \in F$. It is well-known that h is well-defined up to an isotopy. To construct an $h : F \to F$, we can use a horizontal vector field $\mathcal{V}$ on $B_\varepsilon \cap f^{-1}(S_\delta^1)$ such that its image by df is the unit tangent vector $\frac{\partial}{\partial \theta}$ of the circle S_δ^1. We assume also that $\mathcal{V}(\mathbf{z}) \in T_{\mathbf{z}} \partial E(\varepsilon, \delta)$ if $\mathbf{z} \in \partial E(\varepsilon, \delta)$. Then take integral curves of $\mathcal{V}$, $\hat{h} : E(\varepsilon, \delta) \times \mathbb{R} \to E(\varepsilon, \delta)$ so that $\hat{h}(\mathbf{z}, 0) = \mathbf{z}$ for any $\mathbf{z} \in E(\varepsilon, \delta)$ and $f(\hat{h}(\mathbf{z}, t)) = f(\mathbf{z}) \exp(it)$. Then the monodromy map h is defined as $h(\mathbf{z}) = \hat{h}(\mathbf{z}, 2\pi)$.

Let $P_i(t)$ be the characteristic polynomial of the monodromy map: $h_{*,i} : H_i(F; \mathbb{Q}) \to H_i(F; \mathbb{Q})$ $i = 0, 1$. The zeta function for the Milnor fibration is defined as

$$\zeta(t) = P_0(t)^{-1} P_1(t) = P_1(t)/(t - 1).$$

In particular, we have

$$\mu(f, O) = \deg \zeta(t) + 1.$$

Suppose that $\pi : X \to \mathbb{C}^2$ is a good resolution of $C = \{f(x, y) = 0\}$ and let C' be the strict transform of C and let $E_1, \dots, E_s$ be the exceptional divisors. Let m_j be the multiplicity of $\pi^* f$ along E_j and let $E_j^* = E_j - C' \bigcup_{i \ne j} E_i$. Then the following formula is fundamental.

Theorem 27 (A'Campo [1]).

$$\zeta(t) = \prod_j (1 - t^{m_j})^{-\chi(E_j^*)}.$$

This follows essentially from the additivity of the Euler characteristics:

Lemma 28. ([11]) *Let $h : F \to F$ be the monodromy map and let $F = F_1 \cup F_2$ be the decomposition of subspaces, which are stable by h, $h(F_i) \subset F_i$, $i = 1, 2$. Put*

$h_i := h|F_i$ and $h_0 := h|F_0$, where $F_0 = F_1 \cap F_2$. Then

$$\chi(F) = \chi(F_1) + \chi(F_2) - \chi(F_0), \quad \text{and}$$
$$\zeta(t;h) = \zeta(t;h_1)\zeta(t;h_2)/\zeta(t;h_0).$$

4.2. Non-degenerate plane curves

Theorem 29. ([5, 13]) *Let $C : f(x,y) = 0$ be a non-degenerate plane curve, with a convenient Newton boundary. Assume that $\Gamma(f) \cap \{v = 0\} = \{(a,0)\}$ and $\Gamma(f) \cap \{u = 0\} = \{(0,b)\}$. Let Δ_j, $j = 1,\dots,k$ be the 1-faces of $\Gamma(f;x,y)$, let ℓ_j be the number of factors in $f_{\Delta_j}(x,y)$ and $P_j = {}^t(a_j,b_j)$ be the corresponding weight vector. Thus we can factorize $f_{P_j}(x,y) = c_j x^{\alpha_i} y^{\beta_i} \prod_{i=1}^{\ell_i}(y^{a_j} - \gamma_{ji}x^{b_j})$. Then the zeta function is given by the formula:*

$$\zeta(t) = \frac{\prod_{j=1}^{k}(1-t^{d(P_j)})^{\ell_j}}{(1-t^a)(1-t^b)}, \ \mu(f;O) = \sum_{j=1}^{k} d(P_j)\ell_j - (a+b) + 1. \tag{4}$$

The number $d(P_j)\ell_j$ is equal to $2\operatorname{Vol}\Delta_j(O)$ where $\Delta_j(O)$ is the cone of Δ_j with the origin O and $d(P_j) = d(P_j, f)$.

The right-hand side of the last equality (4) is called *the Newton Number of* $\Gamma_-(f)$ [5].

Proof. Assume that $\Sigma^* = \{Q_0, Q_1, \dots, Q_m, Q_{m+1}\}$ is a regular cone subdivision of the dual Newton diagram $\Gamma^*(f)$ where $Q_0 = E_1$, $Q_{m+1} = E_2$. Let $P_j = Q_{\iota_j}$, $j = 1,\dots,k$. By Theorem 22, the graph of the exceptional divisors is a bamboo (a linear graph without any branches). By Theorem 27, the zeta function is determined by the information of exceptional divisors $\widehat{E}(Q_i)$ with $\chi(\widehat{E}(Q_t)^*) \neq 0$, namely those divisors which intersect with either only one divisor or more than or equal to three other divisors among $\widehat{E}(Q_t)$, $t \neq i$ and C', namely exceptional divisors $E(Q_i)^*$ which have more than or equal to three holes. They are exactly $\widehat{E}(Q_1)$, $\widehat{E}(Q_m)$ (two end divisors) and $\widehat{E}(P_j) = \widehat{E}(Q_{\iota_j})$, $j = 1,\dots,k$. As $\det(Q_0, Q_1) = 1$, we can write Q_1 as ${}^t(\alpha_1, 1)$. By the admissibility of Σ^*, Q_1 takes its minimum at $(0,b)$ (corresponds to y^b) and thus $d(Q_1) = b$. Similarly $d(Q_m) = a$. As $\chi(\widehat{E}(P_j)^*) = -\ell_j$ and $m_j = d(P_j)$, the assertion follows. The last equality $\ell_j\, d(P_j) = 2\mathrm{Vol}\,\Delta_j(O)$ follows from the Figure 11 ([11]): We put $\sigma = (Q_{\iota_j}, Q_{\iota_j+1})$. We use the fact that a unimodular matrix preserves the volume and $\sigma(\Delta_j)$(the image of Δ_j by the linear map σ) is a vertical edge of length ℓ_j. □

4.2.1. Problem. Let $f(x,y) = x^{n+4} + x^2y^2(x^2 + 2axy + y^2) + y^{n+4}$ with $n \geq 3$. Show that $f(x,y)$ is non-degenerate if and only if $a^2 \neq 1$. If this is the case, compute the canonical regular subdivision Σ^* of the dual Newton diagram $\Gamma^*(f)$ and then show that the extended dual diagram is given as in Figure 12.

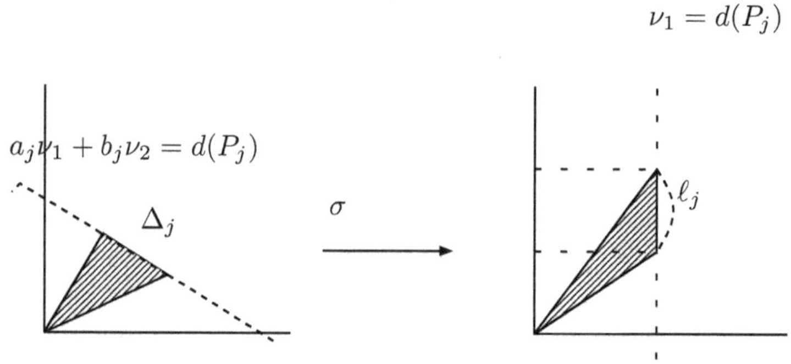

FIGURE 11. σ preserves volume

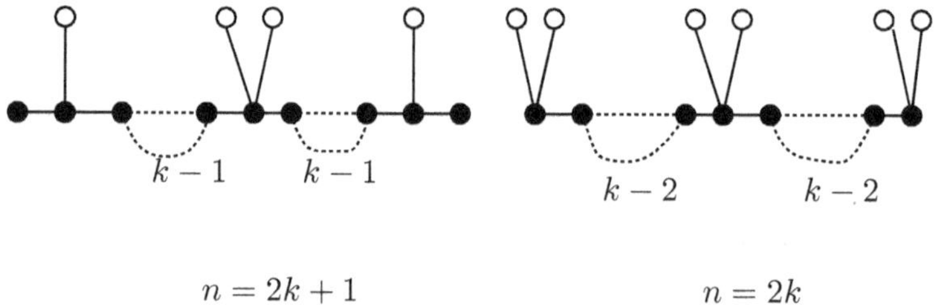

FIGURE 12. Extended dual graph

5. Puiseux characteristics and their geometry

5.1. Change of parameters

Let $h(t) = \sum_i a_i t^i$ and $n \in \mathbb{N}$ be a given positive integer. Let us define

$$P_1(h;n) := \min\{j; a_j \neq 0,\ j \not\equiv 0 \mod n\},\ D_1(h;n) := \gcd(n, P_1(h;n)).$$

Using these integers, we define inductively the following integers:

$$\begin{Bmatrix} \beta_1 = P_1(h;n) \\ e_1 = D_1(h;n) \end{Bmatrix} \to \begin{Bmatrix} \beta_2 = P_1(h;e_1) \\ e_2 = D_1(h;e_1) \end{Bmatrix} \to \cdots \to \begin{Bmatrix} \beta_k = P_1(h;e_{k-1}) \\ e_k = D_1(h;e_{k-1}) \end{Bmatrix}.$$

Thus this process stops when β_{k+1} can not be defined. Namely

$$j \equiv 0 \text{ modulo } e_k, \text{ for any } j \text{ with } a_j \neq 0.$$

This is the case if for example $e_k = 1$. If $h(t)$ involves only powers of t^n from the beginning, we write $\mathcal{P}(h;n) = \emptyset$. Note that $e_k = 1$ for some k, if $x = t^n$, $y = h(t)$ is a good parametrization of an irreducible plane curve. Put

$$\mathcal{P}(h;n) := \{\beta_1, \ldots, \beta_k\}, \quad \mathcal{D}(h;n) := \{e_1, \ldots, e_k\}.$$

Note that by the definition,

$$e_j \mid e_{j-1} \quad \text{and} \quad e_j \mid \beta_j.$$

For a good parametrization of an irreducible curve

$$C : x = t^n,\ y = \phi(t),\quad n = m(C, O),$$

we call $(n, \mathcal{P}(\phi, n))$ *the Puiseux characteristics of* (C, O). The following is easily shown.

Lemma 30. ([10], Lemma 5-1, 5-2)

1. *For any* $r \in \mathbb{Q}$ *and* $h \in \mathcal{O}^*$, $\mathcal{P}(h; n) = \mathcal{P}(h^r; n)$.
2. *If* $h_1, h_2 \in \mathcal{O}^*$, $P_1(h_1h_2; n) \geq \min(P_1(h_1; n), P_1(h_2; n))$ *and the equality holds if* $P_1(h_1; n) \neq P_1(h_2; n)$.
3. *Assume that* $n|m$. *Then* $\mathcal{P}(t^m h; n) = \mathcal{P}(h; n) + m$.
4. (Change of parameter) *Suppose* $\tau = t \cdot \tau_0(t)$, $t = \tau \cdot t_0(\tau)$ *with* $t_0, \tau_0 \in \mathcal{O}^*$. *Then* $\mathcal{P}(\tau_0(t); n) = \mathcal{P}(t_0(\tau); n)$.

   If we assume further $P_1(\tau_0(t); \ell) \geq P_1(h(t); \ell)$, *then*

   $$P_1(h(t(\tau)); \ell) = P_1(h(t); \ell)$$

   where the left side is defined as a power series in τ. □

For an integer α with $\alpha \equiv 0$ modulo n, we define

$$\alpha + \{\beta_1, \ldots, \beta_k\} := \{\alpha + \beta_1, \ldots, \alpha + \beta_k\}.$$

The following is immediate from Lemma 30.

Lemma 31. *Let* $x(t) = t^n$, $y = y(t)$ *be a good parametrization of an irreducible curve* (C, O) *with* $n = m(C, O)$ *and let* $\mathcal{P}(y(t); n) = \{\beta_1, \ldots, \beta_k\}$ *and* $\mathcal{D}(y(t); n) = \{e_1, \ldots, e_k\}$ *with* $e_0 = n$ *and* $e_k = 1$.

1. $P_1(x^p(t)\, y^q(t); e_j) \geq P_1(y(t); e_j) + n$ *for any* j, *if* $q > 1$.
2. $\mathcal{P}(a\, x(t) + b\, y(t); n) = \mathcal{P}(y(t); n)$ *if* $b \neq 0$.
3. *Let* $g(x, y) = \sum_\nu c_{\nu_1\nu_2} x^{\nu_1} y^{\nu_2}$ *be a convergent power series such that* $c_{0,0} = 0$ *and* $c_{0,1} \neq 0$. *Then* $\mathcal{P}(g(x(t), y(t)); n) = \mathcal{P}(y(t); n)$.

Proof. Put $y(t) = t^{\beta_1'} y_0(t)$ with $y_0(t) \in \mathcal{O}^*$. Here $\beta_1' \leq \beta_1$ and $\beta_1' \equiv 0$ modulo n if $\beta_1' < \beta_1$. Then

$$P_1(y_0(t); e_j) = \beta_{j+1} - \beta_1' \text{ for } j > 0$$

and thus

$$P_1(y(t)^q; e_j) \ = q\beta_1' + (\beta_{j+1} - \beta_1'),\ j \geq 0;$$

therefore

$$P_1(x(t)^p y(t)^q; e_j) = pn + (q-1)\beta_1' + \beta_{j+1} \geq \beta_{j+1} + n.$$

The other assertion follows from this observation. □

Using Lemma 31, we will show that the characteristic series will be independent of the choice of the coordinate function y when we fix a generic line $x = 0$ for the parametrization $x(t) = t^m$ with $m = m(C, O)$. It is also independent of the choice of the coordinate system (x, y), as is shown in the following.

Theorem 32. *Let $x(t) = t^n$, $y = y(t)$ be a good parametrization of an irreducible curve (C, O) with $m = m(C, O)$ and let $\mathcal{P}(y(t); n) = (\beta_1, \dots, \beta_k)$ and $\mathcal{D}(y(t); n) = (e_1, \dots, e_k)$ with $e_0 = n$ and $e_k = 1$. Let m be the multiplicity of (C, O).*

1. (Independence of Puiseux characteristics) *If $n = m$, $(n; \beta_1, \dots, \beta_k)$ and $(e_1, \dots, e_k)$ does not depend on the choice of the coordinate system (x, y).*
2. (Non-admissible Puiseux characteristics) *Assume that $n > m$. Take a new parameter τ so that $t = t(\tau)$ and $y(t(\tau)) = \tau^m$. Then the Puiseux characteristics are given by*

$$\mathcal{P}(x(t(\tau)); m) = \begin{cases} (n, \beta_2 + n - m, \dots, \beta_k + n - m), & n \not\equiv 0 \mod m, \\ (\beta_2 + n - m, \dots, \beta_k + n - m), & n \equiv 0 \mod m. \end{cases}$$

Proof. We first consider the case 1., $n = m$. We consider another set of coordinates (x_1, y_1). We assume that

$$x_1 = \sum_{i+j \geq 1} c_{i,j} x^i y^j, \quad y_1 = \sum_{i+j \geq 1} d_{i,j} x^i y^j.$$

For any non-zero $h(t) \in \mathbb{C}\{t\}$, we can write in a unique way $h(t) = t^d h_0(t)$ with $h_0(t) \in \mathcal{O}^*$.

(a) Assume that $c_{0,1} \neq 0$. We see that (x_1, x) is also a coordinate. Then we see that $x_1(t) = t^n x_{10}(t)$ with $x_{10}(t) \in \mathcal{O}^*$ and $\mathcal{P}(x_1(t); n) = \{\beta_1, \dots, \beta_k\}$ by (3) of Lemma 31. Now we take a new parameter τ so that $x_1(t(\tau)) = \tau^n$ and $t = \tau \cdot t_0(\tau)$, $\tau = t \cdot t_0(\tau)$. By Lemma 30 and Lemma 31, we see that

$$\begin{aligned} \mathcal{P}(x(t(\tau)); n) &= \mathcal{P}(\tau^n\, t_0(\tau); n) = n + \mathcal{P}(t_0(\tau); n) \\ &= n + \mathcal{P}(\tau_0(t); n) = \mathcal{P}(\tau(t)^n; n) \\ &= \mathcal{P}(x_1(t); n) = \{\beta_1, \dots, \beta_k\}. \end{aligned}$$

Now we consider y_1. Expand y_1 in a convergent power series of variables x, x_1: $y_1 = \sum_{i+j>0} d'_{i,j} x^i (x_1)^j$. Then $d'_{1,0} \neq 0$. Then we apply Lemma 31 to see that

$$\mathcal{P}(y_1(\tau); n) = \mathcal{P}(x(t(\tau)); n) = \{\beta_1, \dots, \beta_k\}.$$

(b) Assume now $c_{0,1} = 0$ and $c_{1,0} \neq 0$. Then $d_{01} \neq 0$. This implies that (x_1, y) is also a local set of coordinates. Then by Lemma 31, we have

$$x_1(t) = t^m x_{10}(t), \; x_{10}(t) \in \mathcal{O}^*.$$

By (1) of Lemma 31, $P_1(x_1(t) - c_{1,0}x(t); e_j) \geq \beta_{j+1} + m$. Thus $P_1(x_{10}(t); e_j) \geq \beta_{j+1}$. Take a new parameter τ so that $x_1 = \tau^m$. $t = \tau t_0(\tau)$ and $\tau = t\tau_0(t)$. Then $\tau_0(t) = x_0(t)^{1/m}$. By (1) and (4) of Lemma 30, we see that

$$P_1(y(t(\tau)); e_j) = P_1(y(t); e_j), \quad j = 0, \dots, k-1.$$

This implies that $\mathcal{P}(y(t(\tau)); m) = \mathcal{P}(y(t); m)$. Now write y_1 as

$$y_1 = \sum_{a+b\geq 1} d'_{a,b} {x_1}^a y^b$$

and we consider $y_1(\tau)$. As $d'_{0,1} \neq 0$, we conclude by Lemma 31 that $\mathcal{P}(y_1(t(\tau)); m) = \mathcal{P}(y(t(\tau)); m) = \{\beta_1, \dots, \beta_k\}$. This proves the assertion (1) of Theorem 32.

Now we prove the assertion 2. of Theorem 32. Thus assume that $n > m$. Geometrically this case corresponds to $x = 0$ is the tangent cone of (C, O). Note that $\mathrm{ord_t\, y(t) = m}$ in this case. Thus $\beta_1 = m$ and $e_1 = \gcd(n, m)$. Then writing $\tau = t \cdot \tau_0(t)$, $t = \tau \cdot t_0(\tau)$ so that $y(t(\tau)) = \tau^m$, $x(t(\tau)) = t^n(\tau)$.

Assume that $e_1 < m$ or equivalently $n \not\equiv 0 \mod m$. Then $P_1(t^n(\tau); m) = n$ and $e_1 = \gcd(n, m)$. For e_j, $j \geq 1$, we have $n \equiv 0$ modulo e_j and

$$\begin{aligned} P_1(t^n(\tau); e_j) &= n + P_1(t_0(\tau)^n; e_j) = n + P_1(t_0(\tau); e_j) \\ &= n + P_1(\tau_0(t); e_j) = n + P_1(\tau_0^m(t); e_j) \\ &= (n - m) + P_1(y(t); e_j) = \beta_{j+1} + (n - m). \end{aligned}$$

Thus we conclude that $\mathcal{P}(x(t(\tau)); m) = (n, \beta_2 + n - m, \dots, \beta_k + n - m)$.

Assume now $e_1 = m$. Then $\beta_2 = P_1(y(t); m)$. Thus

$$P_1(t^n(\tau); m) = P_1(t^n(\tau); e_1) = \beta_2 + n - m$$

by the above calculation and the other argument is the same. □

In the case of 2., we call the $(n; \beta_1, \dots, \beta_k)$ *non-admissible Puiseux pairs.*

5.1.1. Example. Consider the parametrization

$$C: \quad \begin{cases} x = t^4, \\ y = t^8 + t^{10} + t^{13}. \end{cases}$$

It is a good parametrization and the Puiseux characteristics are given as $(4; 10, 13)$. Consider the coordinates $(x_1, y_1) = (y, x)$ and take the parameter τ so that $y(t) = \tau^8$. Then $t = \tau t_0(\tau)$ is given as

$$\tau - \frac{1}{8}\tau^3 + \frac{13}{128}\tau^5 - \frac{1}{8}\tau^6 - \frac{115}{1024}\tau^7 + O(8).$$

The corresponding parametrization is given as

$$\begin{aligned} x_1 &= \tau^8, \\ y_1 &= \tau^4\left(1 - \frac{1}{8}\tau^2 + \frac{13}{128}\tau^4 - \frac{1}{8}\tau^5 - \frac{115}{1024}\tau^6 + O(7)\right)^4 \end{aligned}$$

which gives a non-admissible Puiseux characteristics $(8; 4, 6, 9)$. Thus $(m; \beta_2 + (n - m), \beta_3 + (n - m)) = (4; 10, 13)$.

5.2. Tower of toric modifications and Puiseux characteristics

We refer to [10, 2] for this subsection. In this section, we study the geometry of the resolution of irreducible curves.

Let $C : f(x,y) = 0$ be an irreducible germ of a curve and let Γ be the minimal resolution graph. Let $r(v)$ be the number of edges touching the vertex v. The *complexity of* (C,O) is defined using the minimal resolution graph Γ as $\rho(C,O) = 1 + \sum_{v\in\{\mathrm{Vertex}(\Gamma)\}} \max\{r(v)-2,0\}$.

Theorem 33. ([12]) *The complexity is equal to the minimal number of toric modifications to resolve C, if C is irreducible.*

Suppose that we have a tower of toric modification:

$$X_j \xrightarrow{\pi_j} X_{j-1} \to \cdots \xrightarrow{\pi_2} X_1 \xrightarrow{\pi_1} X_0 = \mathbb{C}^2$$

where each π_{j+1} is a toric modification on the center $(0,\lambda_j) \in \widehat{E}(P_j)$, $P_j = {}^t(a_j,b_j)$ which is the weight vector of the unique face of $\Gamma(\Pi_{j-1}^* f; U_{j-1}, V_{j-1})$ and $\Pi_j : X_j \to X_0$ is the composition $\pi_1 \circ \cdots \circ \pi_j$. In X_j, we take the toric coordinate (u_j,v_j) so that $\widehat{E}(P_j) = \{u_j = 0\}$ and $C_j \cap \widehat{E}(P_j) = \{(0,\lambda_j)\}$. Here C_j is the strict transform of C to X_j. For the next toric modification, assuming $\Pi_j : X_j \to \mathbb{C}^2$ is not yet a good resolution, we take some admissible change of coordinates:

$$U_j = u_j, \quad V_j = v_j - \lambda_j + h_j(u_j), \ h_j \in \mathbb{C}[u_j]$$

so that the unique face $\Gamma(\Pi_j^* f; (U_j,V_j))$ of the pull-back $\Pi_j^* f$ has a weight vector $P_{j+1} = {}^t(a_{j+1},b_{j+1})$ with $a_{j+1} > 1$. We take a regular cone subdivision of $\Gamma^*(\Pi_j^* f; U_j, V_j)$, say $\Sigma_j^* = \{Q_{j,0} = E_1, Q_{j,1}, \dots, Q_{j,\nu_j} = E_2\}$ and we take the corresponding toric modification $\pi_{j+1} : X_{j+1} \to X_j$. In this way, we do the same construction until we arrive at X_k so that $\Pi_k : X_k \to X_0$, which is the composition

$$\Pi_k : X_k \xrightarrow{\pi_k} X_{k-1} \to \cdots \xrightarrow{\pi_2} X_1 \xrightarrow{\pi_1} X_0 = \mathbb{C}^2,$$

is a good resolution. We may assume that $a_j \geq 2$ for any j by taking a good coordinate change ([12]). We assume also $2 \leq a_1 < b_1$, exchanging x,y if necessary. Consider the integers

$$A_i := a_i \cdots a_k, \quad i = 1,\dots,k,\ A_{k+1} := 1,$$
$$\beta_i := b_i A_{i+1} + b_{i-1} A_i + \cdots + b_1 A_2,\ i = 1,\dots,k.$$

Theorem 34. ([10]) *Suppose that C has the toric modification tower as above. The parametrization of C is given as*

$$\begin{cases} x = \psi(t) = t^{A_1}, \\ y = \phi(t) = t^{b_1 A_2}\phi_0(t),\ \phi_0(t) \in \mathcal{O}^*, \end{cases}$$

so that their Puiseux characteristics are given as

$$\mathcal{P}(\phi; A_1) = \{\beta_1,\dots,\beta_k\}, \quad \mathcal{D}(\phi; A_1) = \{A_2,\dots,A_{k+1}\}.$$

The multiplicity of C is A_1 under the assumption $2 \leq a_1 < b_1$.

Proof. We prove the assertion by induction on k. Assume that $C_1 \subset X_1$ is parametrized as

$$U_1 = t^{A_2}, \quad V_1 = \phi_1(t),\ \phi_1(t) = t^{b_2 A_3}\phi_{10}(t),\ \phi_{10} \in \mathcal{O}^*,$$
$$\mathcal{P}(V_1(t); A_2) = \{\bar\beta_2, \dots, \bar\beta_k\}, \quad \mathcal{D}(V_1(t); A_2) = \{A_3, \dots, A_{k+1}\},$$
$$\bar\beta_i := \beta_i - b_1 A_2 = b_i A_{i+1} + \cdots + b_2 A_3.$$

We recall that the toric coordinates are related to the admissible toric coordinates (U_1, V_1) by

$$u_1 = U_1, \quad v_1 = \gamma_1 + V_1 - h_1(u_1),\ \exists h_1 \in \mathcal{O},\ h_1(0) = 0.$$

First we note that $\gamma_1 - h_1(u_1) = \gamma_1 - h_1(t^{A_2})$ is expanded in t^{A_2}. Thus $P_1(v_1(t); A_j) = P_1(V_1(t); A_j)$ for $j \geq 2$. Now we compute the parametrization of C as the composition with $\pi_1 : X_1 \to \mathbb{C}^2$, $\pi_1(u_1, v_1) = (x, y)$ where

$$x = u_1^{a_1} v_1^{a_1'} = t^{A_1} v_1(t)^{a_1'},$$
$$y = u_1^{b_1} v_1^{b_1'} = t^{b_1 A_2} v_1^{b_1'}, \quad v_1(t) \in \mathcal{O}^*.$$

Now as $v_1(0) = \gamma_1 \neq 0$, we take the new parameter s so that

$$x = s^{A_1} = t^{A_1} v_1(t)^{a_1'}, \quad s = t s_0(t),\ t = s t_0(s),\ t_0, s_0 \in \mathcal{O}^*.$$

Then $s_0(t) = v_1(t)^{a_1'/A_1}$ and

$$y = x^{b_1/a_1} v_1^{b_1' - b_1 a_1'/a_1} = s^{b_1 A_2} v_1^{1/a_1}.$$

Thus first, we have

$$P_1(y(s); A_1) = b_1 A_2, \ D_1(y(s); A_1) = A_2.$$

As $v_1(t) = \gamma_1 + \phi_1(t) - h_1(t^{A_2})$, we observe that

$$P_1(t_0(s); A_2) = P_1(s_0(t); A_2) = P_1(s_0(t)^{A_1}; A_2) = P_1(s(t)^{A_1}; A_2) - A_1$$
$$= P_1(x(t); A_2) - A_1 = P_1(v_1(t); A_2) = P_1(V_1(t); A_2) = \bar\beta_2$$

Therefore

$$P_1(y(t(s)); A_2) = P_1(s^{b_1 A_2} v_1(t(s))^{1/a_1}; A_2)$$
$$= \bar\beta_2 + b_1 A_2 = P_1(v_1(s); A_2) + b_1 A_2 = \beta_2.$$

The same argument works for $P_1(y(t(s)); e_i)$. □

5.3. Toric resolution tower from a Puiseux expansion

We consider the inverse problem of the previous section. Let (C, O) be an irreducible curve defined on a neighborhood of the origin U and assume that it has a Puiseux parametrization

$$x(t) = t^m, \quad y(t) = \sum_{j=m}^{\infty} c_j t^j$$

and let $\mathcal{P}(y(t); m) = (\beta_1, \ldots, \beta_k)$ be the Puiseux characteristics and let $\mathcal{D}(y(t); m) = (e_1, \ldots, e_k)$. Let us recall that

$$e_1 = \gcd(\beta_1, m),\ e_2 = \gcd(\beta_2, e_1), \ldots, e_k = \gcd(\beta_k, e_{k-1}) = 1.$$

We define integers a_j, b_j, $j = 1, \ldots, k$ inductively:

$$\begin{cases} a_1 := m/e_1,\ b_1 := \beta_1/e_1, \\ a_j := e_{j-1}/e_j,\ b_j := (\beta_j - \beta_{j-1})/e_j, & j = 2, \ldots, k, \\ P_j := \begin{pmatrix} a_j \\ b_j \end{pmatrix}, \quad j = 2, \ldots, k. \end{cases}$$

Then we define integer $A_j = a_j \cdots a_k$ as before. Note that $A_1 = a_1 \cdots a_k = (m/e_1)(e_1/e_2) \cdots (e_{k-1}/e_k) = m$. As $e_0 = m > e_1 > \cdots > e_k = 1$ and $\beta_1 < \beta_2 < \cdots < \beta_k$, we see that a_j, b_j are integers with $a_j \geq 2$ and $\gcd(a_j, b_j) = 1$. Now we assert the inverse of Theorem 34:

Theorem 35. *Under the above assumption, taking a suitable choice of coordinates at each stage, (C, O) is resolved by a tower of toric modifications of length k with corresponding weight vector P_j:*

$$\Pi_k : X_k \xrightarrow{\pi_k} X_{k-1} \xrightarrow{\pi_{k-1}} \cdots \xrightarrow{\pi_2} X_1 \xrightarrow{\pi_1} X_0 = \mathbb{C}^2.$$

Proof. We prove the assertion by the induction on k. First we take coordinates (x_0, y_0) so that $x_0 = x$, $y_0 = y - \sum_{j=1}^{\infty} c_{mj} x^j$. Then $\mathrm{ord_t}\, \mathrm{y_0(t)} = \beta_1$. Thus we assume from the beginning that $y(t) = \sum_{j=\beta_1} c_j t^j$ and $c_j = 0$ if $j \equiv 0$ modulo m. We use (x, y) instead of (x_0, y_0) for simplicity. Put $\phi(t) = \sum_{j \geq \beta_1} c_j t^j$. The defining function $f(x, y)$ is given as

$$f(x, y) = \prod_{i=0}^{m-1} (y - \phi(t\omega^i)), \quad \omega = \exp(2\pi\sqrt{-1}/m)$$

where t^m is replaced by x in the right-hand side.

Assertion 36. *$f(x, y)$ is expressed as*

$$f(x, y) = (y^{a_1} - c_{\beta_1}^{a_1} x^{b_1})^{A_2} + \textit{(higher terms)}.$$

Proof. Consider the weights $\deg y = \beta_1$ and $\deg t = 1$ and take the lowest term in y, t. Note that $b_1 = \beta_1/A_2$ and $m = a_1 A_2$. Then the lowest term of the term $(y - \phi(t\omega^i))$ is given by $y - c_{\beta_j} t^{\beta_1} \omega^{\beta_1 i} = y - c_{\beta_j} \eta^{b_1 i} t^{b_1 A_2}$ where $\eta = \exp(2\pi i/a_1)$. Thus the lowest term of $f(x, y)$ is

$$\begin{aligned} \prod_{i=1}^{m-1} (y - c_{\beta_1} \omega^{i\beta_1} t^{\beta_1}) &= \left(\prod_{i=1}^{a_1 - 1} (y - c_{\beta_1} \eta^{ib_1} (t^{A_2})^{b_1}) \right)^{A_2} \\ &= (y^{a_1} - (c_{\beta_1} t^{\beta_1})^{a_1})^{A_2} \\ &= (y^{a_1} - c_{\beta_1}^{a_1} x^{b_1})^{A_2}. \end{aligned}$$

□

Step 1. Assume that $k = 1$. Then $e_1 = 1$ and $m = a_1$. The defining function $f(x, y)$ is given as

$$f(x,y) \quad = (y^{a_1} - c_{\beta_1}^{a_1} x^{b_1}) + (\text{higher terms})$$

and the assertion is obvious.

Step 2. Assume that $k \geq 2$ and we assume the assertion for $k-1$ characteristics. We take a regular cone $\Sigma^* = \{Q_0, \dots, Q_{\ell_0+1}\}$ so that $P_1 = Q_\iota$ and we take the corresponding toric modification $\pi_1 : X \to \mathbb{C}^2$. Take the cone $\sigma = \mathrm{Cone}(\mathrm{Q}_\iota, \mathrm{Q}_{\iota+1})$ and we assume that

$$(Q_\iota, Q_{\iota+1}) = \begin{pmatrix} a_1 & p_1 \\ b_1 & q_1 \end{pmatrix}, \quad a_1 q_1 - p_1 b_1 = 1$$

Let (u_1, v_1) be the toric coordinate associated with σ. They are related to (x, y) by

$$\begin{pmatrix} x \\ y \end{pmatrix} = \begin{pmatrix} u_1^{a_1} v_1^{p_1} \\ u_1^{b_1} v_1^{q_1} \end{pmatrix}, \quad \begin{pmatrix} u_1 \\ v_1 \end{pmatrix} = \begin{pmatrix} x^{q_1} y^{-p_1} \\ x^{-b_1} y^{a_1} \end{pmatrix}.$$

Then we see that

$$\mathrm{ord}_t u_1(t) = q_1 A_1 - p_1 b_1 A_2 = A_2, \ \mathrm{ord}_t v_1(t) = 0.$$

Write $y(t) = t^{b_1 A_2} y_0(t)$. Then $\mathcal{P}(y_0(t); A_2) = (\beta_2 - \beta_1, \dots, \beta_k - \beta_1)$ where $\beta_1 = b_1 A_2$. Thus we have

$$u_1(t) = t^{A_2} y_0(t)^{-p_1}, \ v_1 = y_0(t)^{a_1}.$$

Take a new parameter τ so that

$$t = \tau\, t_0(\tau), \ u_1(t(\tau))) = \tau^{A_2}, \quad t_0(\tau) \in \mathcal{O}^*.$$

Then $\tau_0(t) = y_0(t)^{-p_1/A_2}$ and $\tau = t\, y_0(t)^{-p_1/A_2}$ and thus

$$\mathcal{P}(\tau_0(t); A_2) = (\beta_2 - \beta_1, \dots, \beta_k - \beta_1), \quad y_0(0) = c_{\beta_1}.$$

As $v_1(t(\tau)) = y_0(t(\tau)))^{a_1}$, this implies that

$$\mathcal{P}(v_1(t(\tau)); A_2) = (\beta_2 - \beta_1, \dots, \beta_k - \beta_1), \ v_1(t(0)) = c_{\beta_1}^{a_1}.$$

Now expand $v_1(t(\tau))$ as $v_1(t(\tau)) = c_{\beta_1}^{a_1} + \sum_{i>0} d_i \tau^i$. As $P_1(v_1(t(\tau)); A_2) = \beta_2 - \beta_1$, we see that $d_i \equiv 0$ modulo A_2 for any $i < \beta_2 - \beta_1$. Consider the coordinates $(u_1, \hat{v}_1)$ where

$$\hat{v}_1 = v - c_{\beta_1}^{a_1} - \sum_{\nu < \beta_2 - \beta_1} d_\nu u_1^{\nu/A_2}.$$

Now we see that

$$\mathcal{P}(\hat{v}_1(t(\tau)); A_2) = \mathcal{P}(v_1(t(\tau)); A_2), \quad \mathrm{ord}_\tau \hat{\mathrm{v}}_1(\mathrm{t}(\tau)) = \beta_2 - \beta_1.$$

Thus the assertion follows by the induction's assumption of the strict transform $C' \subset X_1$ of C. □

5.4. Resolution graph and the zeta function of the Milnor fibration

We assume that the situation is the same as in §5.2. Let $C : f(x,y) = 0$ be an irreducible germ of a curve. Suppose that we have a tower of toric modifications:

$$X_k \xrightarrow{\pi_k} X_{k-1} \to \cdots \xrightarrow{\pi_2} X_1 \xrightarrow{\pi_1} X_0 = \mathbb{C}^2.$$

We recall the process. Each π_{j+1} is a toric modification on the center $\Lambda_j = (0, \lambda_j) \in \widehat{E}(P_j)$ in the canonical toric coordinates (u_j, v_j) where $\widehat{E}(P_j)$ is defined by $\{u_j = 0\}$. We call $\widehat{E}(P_j)$ *the support divisor* of the strict transform $C^{(j)}$. Then we choose a suitable admissible toric coordinate (U_j, V_j) centered at Λ_j and $U_j = u_j$, $V_j = v_j + h_j(u_j)$ with $h_j(0) = \lambda_j$. Let $P_{j+1} = {}^t(a_{j+1}, b_{j+1})$ which is the weight vector of the unique face of $\Gamma(\Pi_j^* f; U_j, V_j)$ (see Corollary 23) where $\Pi_j : X_j \to X_0$ is the composition $\pi_1 \circ \cdots \circ \pi_j$. Let $\Sigma_j^* = \{Q_0^{(i)}, \ldots, Q_{\ell_j+1}^{(i)}\}$ be the canonical regular subdivision of the dual Newton diagram $\Gamma^*(\Pi_j^* f) = \{E_1, P_j, E_2\}$. Here $E_1 = Q_0^{(i)}$, $E_2 = Q_{\ell_j+1}^{(i)}$. Assume that $P_j = Q_{\xi_j}^{(i)}$. Let Γ_j be the bamboo of the exceptional divisor with ℓ_j vertices. The left-end and the right-end vertices correspond to $\widehat{E}(Q_1^{(i)})$ and $\widehat{E}(Q_{\ell_j}^{(i)})$ respectively. Then the resolution $\Pi_k : X_k \to X_0$ is a minimal good resolution and the resolution graph is given as

1. First starting with Γ_1, then by connecting Γ_1, Γ_2 by an edge which connects the vertex $\widehat{E}(Q_{\xi_1}^{(1)})$ and the left-end vertex $\widehat{E}(Q_1^{(2)})$.
2. By induction we add Γ_j by connecting its left-end vertex corresponding to $\widehat{E}(Q_1^{(j)})$ with the vertex corresponding to $\widehat{E}(Q_{\xi_{j-1}}^{(j-1)})$. In Figure 13, we denoted the vertex corresponding to $\widehat{E}(Q)$ simply by Q.

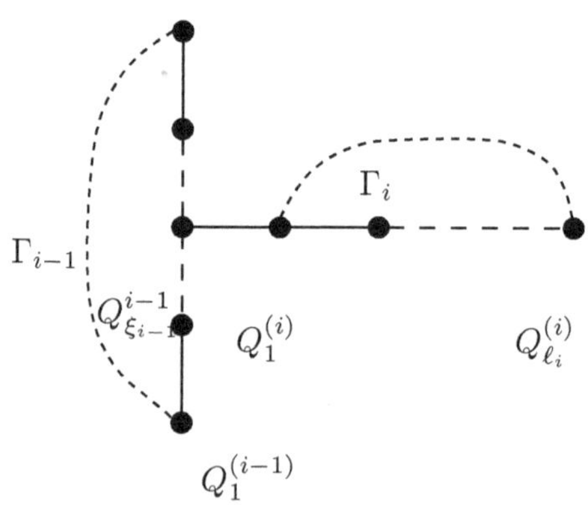

FIGURE 13. Joining Γ_i to Γ_{i-1}

Now for the computation of the zeta-function, we need to know the multiplicity n_1' of $\widehat{E}(Q_1^{(1)})$ and the multiplicity n_j of $\Pi_j^* f$ on the right-hand edge divisor

$\widehat{E}(Q_{\ell_j}^{(j)})$ and the multiplicity m_j on the support divisor $\widehat{E}(Q_{\xi_j}^{(j)})$ respectively for $j = 1, \dots, k$. Note that $\widehat{E}(Q_1^{(1)})$, $\widehat{E}(Q_{\ell_j}^{(j)})$, $1 \le j \le k$ correspond to the end vertices and the vertex corresponding to $\widehat{E}(Q_{\xi_j}^{(j)})$ has three branches. By A'Campo's theorem, the zeta-function of the Milnor filbration is given as

$$\zeta(t) = \frac{\prod_{i=1}^k (1 - t^{m_i})}{(1 - t^{n'_1}) \prod_{i=1}^k (1 - t^{n_i})}, \quad \mu(f, O) = \sum_{i=1}^k (m_i - n_i) - n'_1 + 1.$$

Thus the next task is to determine the multiplicities. First we observe that

$$f(x, y) = (y^{a_1} - \lambda_1 x^{b_1})^{A_2} + (\text{higher terms})$$

which implies that

$$n'_1 = A_1,\ n_1 = b_1 A_2,\ m_1 = a_1 b_1 A_2.$$

Here we have used the fact that $Q_1^{(1)} = {}^t(\nu_1, 1)$, $Q_{\ell_1}^{(1)}$ and $Q_1^{(1)}$, $Q_{\ell_1}^{(1)}$ take their minimum values A_1, $b_1 A_2$ respectively. Then we use induction to see that

$$\Pi_i^* f(U_i, V_i) = U_i^{m_i} \{ (V_i^{a_{i+1}} - \lambda_{i+1} U_i^{b_{i+1}})^{A_{i+2}} + (\text{higher terms}) \}$$

which implies

$$m_{i+1} = a_{i+1} m_i + a_{i+1} b_{i+1} A_{i+2}, \quad n_{i+1} = m_i + b_{i+1} A_{i+2}.$$

Using this , we obtain

Theorem 37. *Under the above condition, the multiplicities, Milnor number and the zeta-function are given by the following* ([2]).

$$m_i = \left(\sum_{j=1}^i a_j b_j A_{j+1}^2 \right) / A_{i+1}$$

$$n_i = \left(\sum_{j=1}^i a_j b_j A_{j+1}^2 \right) / A_i$$

$$\mu(f, O) = 1 - A_1 + \sum_{j=1}^k (1 - \frac{1}{a_i}) \left(\sum_{j=1}^i a_j b_j A_{j+1}^2 \right) / A_{i+1}$$

$$\zeta(t) = \frac{\prod_{i=1}^k (1 - t^{m_i})}{(1 - t^{A_1}) \prod_{i=1}^k (1 - t^{n_i})}.$$

Example. Consider an irreducible curve C parametrized as

$$x = t^6, \quad y = \phi(t),\ \phi(t) = t^6 + t^9 + t^{12} + t^{14} + t^{18}.$$

First Puiseux characteristics are given as

$$n = 6,\ \beta_1 = 9,\ e_1 = 3,\ a_1 = 2,\ b_1 = 3,\ \beta_2 = 14,\ e_2 = 1,\ a_2 = 3,\ b_2 = 5.$$

So $n_1' = 6$, $n_1 = 9$, $m_1 = 18$, $m_2 = 69$. Thus Theorem 37 says that

$$\zeta(t) = \frac{(1-t^{18})(1-t^{69})}{(1-t^6)(1-t^9)(1-t^{23})}, \ \mu(f) = 50.$$

Graphs Γ_1 and Γ_2 are associated with the canonical subdivisions:

$$\Sigma_1^* = \left\{E_1 = \begin{pmatrix}1\\0\end{pmatrix}, Q_1^{(1)} = \begin{pmatrix}1\\1\end{pmatrix}, P_1 = \begin{pmatrix}2\\3\end{pmatrix}, Q_3^{(1)} = \begin{pmatrix}1\\2\end{pmatrix}, E_2 = \begin{pmatrix}0\\1\end{pmatrix}\right\},$$

$$\Sigma_2^* = \left\{E_1, Q_1^{(2)} = \begin{pmatrix}1\\1\end{pmatrix}, Q_2^{(2)} = \begin{pmatrix}2\\3\end{pmatrix}, P_2 = \begin{pmatrix}3\\5\end{pmatrix}, Q_3^{(2)} = \begin{pmatrix}1\\2\end{pmatrix}, E_2\right\}.$$

Thus the extended resolution graph is given as Figure 14. The white circle denotes the strict transform C_2.

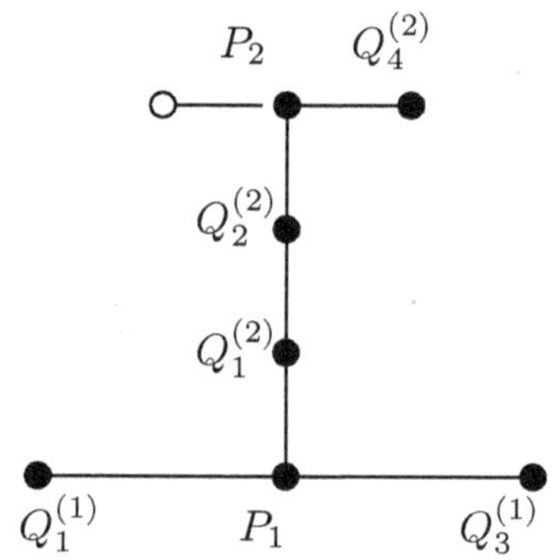

FIGURE 14. Extended resolution graph

We will see this directly. First, we take new coordinates (x_1, y_1) where $x_1 = x$, $y_1 = y - x - x^2 - x^3$. Then the parametrization changes to

$$x_1 = t^6, \ y_1 = t^9 + t^{14}.$$

The defining function $f(x_1, y_1)$ is given as

$$\begin{aligned} f(x_1, y_1) &= \prod_{j=1}^{6} (y_1 - \phi(\omega_6^j t)) \\ &= -{x_1}^9 + 3\,{x_1}^6{y_1}^2 - 6\,{x_1}^{10}y_1 - 3\,{x_1}^3{y_1}^4 - 2\,{y_1}^3{x_1}^7 + {y_1}^6 + {x_1}^{14} \end{aligned}$$

where $\omega_6 = \exp(\pi\sqrt{-1}/3)$. Note that

$$f(x_1, y_1) = (x^3 - y^2)^3 + (\text{higher terms}).$$

We take first a toric modification associated with Σ_1^* and consider the pull-back in $\sigma = \mathrm{Cone}(\mathrm{P}_1, \mathrm{Q}_3^{(1)})$. Let (u, v) be the toric coordinates of the chart $\mathbb{C}_\sigma^2$. Then as $(x_1, y_1) = (u^2v, u^3v^2)$, using admissible coordinates $(u_1, v_1) = (u, v-1)$, we get

$$\pi_1^*(u_1, v_1) = u_1^{18}(v_1+1)^9\{v_1^3 - 8\,u_1^5 + (\text{higher terms})\}.$$

Thus taking another toric modification with respect to Σ_2^*, we obtain a good minimal resolution which is described in Theorem 37.

5.5. The compound torus structure

Let (C,O) be a germ of an irreducible plane curve at the origin O. We will explain the structure of the knot $C \cap S_\varepsilon^3$ where $S_\varepsilon^3 = \{(x,y) \in \mathbb{C}^2 \mid |x|^2 + |y|^2 = \varepsilon\|\}$. First we consider the case

$$C^{(1)} : \; x = t^{a_1}, \; y = t^{b_1}.$$

Consider the knot $K_\varepsilon^{(1)} = C^{(1)} \cap S_\varepsilon^3$. It is easy to see that there exists a positive number $\rho(\varepsilon)$ such that

$$\rho(\varepsilon)^{2a_1} + \rho(\varepsilon)^{2b_1} = \varepsilon^2$$

so that K_ε is a knot embedded in the torus $S^1(\rho(\varepsilon)^{a_1}) \times S^1(\rho(\varepsilon)^{b_1})$ parametrized as

$$C^{(1)} : x = \rho(\varepsilon)^{a_1} \exp(a_1\theta\sqrt{-1}), \; y = \rho(\varepsilon)^{b_1} \exp(b_1\theta\sqrt{-1}), \; 0 \le \theta \le 2\pi.$$

Here $S^1(a) = \{\eta \in \mathbb{C} \mid |\eta| = a\}$. Put $D^2(a) = \{\eta \in \mathbb{C} \mid |\eta| \le a\}$. Consider the solid torus

$$B(\varepsilon) := S^1(\rho(\varepsilon)^{a_1}) \times D^2(3\rho(\varepsilon^{b_1})/2).$$

Note that $K_\varepsilon^{(1)}$ is a torus knot embedded in $S^1(\rho(\varepsilon)^{a_1}) \times S^1(\rho(\varepsilon^{b_1}))$. Consider the canonical homotopy equivalence

$$\psi_\varepsilon : \mathbb{C}^2 \setminus \{O\} \to S_\varepsilon^3, \quad (x,y) \mapsto \varepsilon(x,y)/\sqrt{|x|^2 + |y|^2}.$$

The restriction of ψ_ε to $B(\varepsilon)$ gives a canonical embedding: $\psi_\varepsilon : B(\varepsilon) \hookrightarrow S_\varepsilon^3$. Now we consider the case with two characteristics $(n; \beta_1, \beta_2)$:

$$C^{(2)} : \; x = t^n, \; y = t^{\beta_1} + t^{\beta_2}, \; \beta_1 > n.$$

Recall that $e_1 = \gcd(n, \beta_1) = a_2$, $n = a_1 a_2$, $\beta_2 = b_1 a_2 + b_2$. Now we see that the absolute value $|t|$ on $K_\varepsilon^{(2)} = C^{(2)} \cap S_\varepsilon^3$ is no longer constant but the knot $K_\varepsilon^{(2)}$ is parametrized as

$$x(\theta) = (r(\varepsilon,\theta)\exp(\theta i))^{a_1 a_2}, \quad y(\theta) = (r(\varepsilon,\theta)\exp(\theta i))^{\beta_1} + (r(\varepsilon,\theta)\exp(\theta i))^{\beta_2}.$$

Though $r(\varepsilon,\theta)$ is not constant, we can see easily that $\rho(\varepsilon)/r(\varepsilon,\theta)^{a_2} \to 1$ when $\varepsilon \to 0$. It is more convenient to consider the knot

$$L^{(2)} : \; x(t) = t^{a_1 a_2}, \; y(t) = t^{b_1 a_2} + t^{b_1 a_2 + b_2}, \; t = \rho(\varepsilon)^{1/a_2} \exp(\theta\sqrt{-1})$$

with $0 \le \theta \le 2\pi$. Put

$$B(\varepsilon,\delta) = \{(x, y + y') \mid (x,y) \in S^1(\rho(\varepsilon)^{a_1}) \times S^1(\rho(\varepsilon)^{b_1}), \; |y'| \le \rho(\varepsilon)^{b_1}\delta\}.$$

We can see that $B(\varepsilon,\delta)$ is diffeomorphic to a tubular neighborhood of $K^{(1)}$ if δ is sufficiently small. Now it is easy to see that there exists a positive number ε_0 so that if $\varepsilon < \varepsilon_0$, $|t^{\beta_2}/t^{\beta_1}| < \delta$ on $L^{(2)}$ and

$$L^{(2)} \subset B(\varepsilon,\delta) \subset B(\varepsilon)$$

thus $L^{(2)}$ is a compound torus along $K^{(1)}$. Now we consider the image $\hat{L}^{(2)}$ of $L^{(2)}$ in S_ε by ψ_ε.

Assertion 38. *$\hat{L}^{(2)}$ is isotopic to $K^{(2)}(\varepsilon)$ in S^3_ε.*

Proof. First we consider a deformation from $L^{(2)}$ to $K^{(2)}$ in C. Let us consider a homotopy $h : [0, 2\pi] \times [0, 1] \to B(\varepsilon) \cap C$ defined by

$$h_s(\theta) = h(\theta, s) = (x(\theta, s), y(\theta, s)),$$
$$\begin{cases} x(\theta, s) &= (\rho(\theta, s)\exp(\theta i))^{a_1 a_2}, \\ y(\theta, s) &= (\rho(\theta, s)\exp(\theta i))^{\beta_1} + (\rho(\theta, s)\exp(\theta i))^{\beta_2} \end{cases}, \; 0 \le \theta \le 2\pi,$$
$$\rho(\theta, s) = (1 - s)\rho(\varepsilon)^{1/a_1} + s\, r(\varepsilon, \theta).$$

It is easy to see that this is a deformation inside C and $h_0([0, 2\pi]) = L^{(2)}$ and $h_1([0, 2\pi]) = K^{(2)}$. Denote by $L^{(2)}(s)$ the image knot $h_s([0, 2\pi])$. Now take the image of this homotopy by ψ_ε. Then the image gives an isotopy of our knot $\hat{L}^{(2)}$ and $K^{(2)}$. In fact, it is easy to see that $\psi_\varepsilon : L^{(2)}(s) \to S^3_\varepsilon$ is submersion. Then Assertion 38 follows easily from the following. □

Assertion 39. *The restriction $\psi_\varepsilon : L^{(2)}(s) \to S^3_\varepsilon$ is an embedding.*

Proof. Assume that there exists θ_1, θ_2 such that $\psi_\varepsilon(h_s(\theta_1)) = \psi_\varepsilon(h_s(\theta_2))$. Put $\psi_\varepsilon(h_s(\theta_j)) = (x_j, y_j)$. This implies that $\arg x_1 = \arg x_2$ and $\arg y_1 = \arg y_2$. As ψ_ε is argument preserving for each set of coordinates, $a_1 a_2(\theta_1 - \theta_2) \equiv 0\,(2\pi)$ and we may assume that $\theta_2 = \theta_1 + \frac{2k\pi}{a_1 a_2}$ for some integer k, $0 \le k \le a_1 a_2$. As

$$y_j = \rho(\theta_j, s)^{\beta_1}\exp(b_1 a_2 \theta_j)(1 + \rho(\theta, s)^{\beta_2 - \beta_1}\exp((\beta_2 - \beta_1)\theta_j),$$

we have the equality

$$\arg y_j = b_1 a_2 \theta_j + \arg(1 + \rho(\theta, s)^{\beta_2 - \beta_1}\exp((\beta_2 - \beta_1)\theta_j)$$

and the second quantity is as small as we wish if ε is chosen sufficiently small. Therefore we need to have $b_1 a_2 \theta_1 \equiv b_1 a_2 \theta_2\,(2\pi)$. This implies that $k \equiv 0\,(a_1)$. Thus put $k = a_1 k_1$. Then

$$\arg y_2 = \arg y_1 \iff \theta_1 = \theta_2.$$ □

This shows that $K^{(2)}$ is a compount torus knot along $K^{(1)}$.

This analysis works for an arbitrary number of characteristics. Let (C, O) be an irreducible curve germ with characteristic $(n; \beta_1, \dots, \beta_k)$ and n be the multiplicity of (C, O). Let us assume for simplicity that

$$L^{(k-1)} : x(t) = t^n, \; y(t) = t^{\beta_1} + t^{\beta_2} + \cdots + t^{\beta_{k-1}},$$
$$L^{(k)} : x(t) = t^n, \; y(t) = t^{\beta_1} + t^{\beta_2} + \cdots + t^{\beta_k}$$

where $t = \rho^{1/A_2}\exp(\theta\sqrt{-1})$, $0 \le \theta \le 2\pi$. By the same argument, we see that $L^{(k)}$ is on the boundary torus of $L^{(k-1)}$ which is the boundary of a sufficiently small tubular neighborhood of $L^{(k-1)}$. It turns a_k times along the longitude and β_k times along the meridian.

Acknowledgment

I would like to thank the referee for his comments which made this lecture more understandable.

References

[1] N. A'Campo. La fonction zeta d'une monodromie. *Comm. Math. Helv.*, 50:539–580, 1975.

[2] N. A'Campo and M. Oka. Geometry of plane curves via Tchirnhausen resolution tower. *Osaka J. Math.*, 33:1003–1033, 1996.

[3] P. Griffiths and J. Harris. *Principles of algebraic geometry.* Wiley Classics Library. John Wiley & Sons Inc., New York, 1994. Reprint of the 1978 original.

[4] R. C. Gunning and H. Rossi. *Analytic functions of several complex variables.* Prentice-Hall Inc., Englewood Cliffs, N.J., 1965.

[5] A. G. Kouchnirenko. Polyèdres de Newton et nombres de Milnor. *Invent. Math.*, 32:1–31, 1976.

[6] S. Lang. *Algebra.* Addison-Wesley, Reading, Mass., second edition, 1984.

[7] J. Milnor. *Lectures on the h-cobordism theorem.* Notes by L. Siebenmann and J. Sondow. Princeton University Press, Princeton, N.J., 1965.

[8] J. Milnor. *Singular Points of Complex Hypersurface*, volume 61 of *Annals Math. Studies.* Princeton Univ. Press, 1968.

[9] M. Oka. On the resolution of the hypersurface singularities. In *Complex analytic singularities*, volume 8 of *Adv. Stud. Pure Math.*, pages 405–436. North-Holland, Amsterdam, 1987.

[10] M. Oka. Geometry of plane curves via toroidal resolution. In *Algebraic geometry and singularities (La Rábida, 1991)*, pages 95–121. Birkhäuser, Basel, 1996.

[11] M. Oka. *Non-degenerate complete intersection singularity.* Hermann, Paris, 1997.

[12] L. D. Tráng and M. Oka. On resolution complexity of plane curves. *Kodai Math. J.*, 18(1):1–36, 1995.

[13] A. N. Varchenko. Zeta-function of monodromy and Newton's diagram. *Invent. Math.*, 37(3):253–262, 1976.

[14] C. T. C. Wall. *Singular points of plane curves*, volume 63 of *London Mathematical Society Student Texts.* Cambridge University Press, Cambridge, 2004.

[15] J. A. Wolf. Differentiable fibre spaces and mappings compatible with Riemannian metrics. *Michigan Math. J.*, 11:65–70, 1964.

Mutsuo Oka
Tokyo University of Science
Wakamiyacho 26, Shinjuku-ku
Tokyo
Japan

Progress in Mathematics, Vol. 283, 247–272

Surface Singularities Appeared in the Hyperbolic Schwarz Map for the Hypergeometric Equation

Takeshi Sasaki and Masaaki Yoshida

Dedicated to Professor Fritz Hirzebruch on his eightieth birthday

Abstract. Surface singularities, swallowtail and cuspidal edge, appear in the hyperbolic Schwarz map for the hypergeometric differential equation. Such singularities are studied in detail. After an overview of classical staffs, the hypergeometric equation and the Schwarz map, the hyperbolic Schwarz map is introduced. We study the singularities of this map, whose target is the hyperbolic 3-space, and visualize its image when the monodromy group is a finite group or a typical Fuchsian group. Several confluences of swallowtails are also observed.

Mathematics Subject Classification (2000). 33C05, 53C42.

Keywords. Cuspidal edge, swallowtail, hypergeometric functions, Schwarz map, hyperbolic Schwarz map, derived Schwarz map, hyperbolic space.

1. Introduction

Surface singularities such as cuspidal edge and swallowtail are interesting objects. One can visualize them and can easily make paper-models. However, there are still many things to study about them (see for example [1], [6]). In this note, we show that they appear also in the hypergeometric world.

We first present the swallowtail as the discriminant of the quartic polynomial. The description is very elementary, but highly non-trivial.

We next recall the classical hypergeometric staffs: The *hypergeometric differential equation*

$$x(1-x)u'' + \{c-(a+b+1)x\}u' - abu = 0, \qquad E(a,b,c)$$

and its *Schwarz map* by

$$s : X = \boldsymbol{C} - \{0,1\} \ni x \longmapsto u_0(x) : u_1(x) \in Z \cong \boldsymbol{P}^1, \tag{1.1}$$

where u_0 and u_1 are linearly independent solutions of $E(a,b,c)$ and $\boldsymbol{P}^1$ is the complex projective line. The Schwarz map of the hypergeometric differential equation was studied by Schwarz, around 1870, when the parameters (a,b,c) are real.

We then proceed to the *hyperbolic Schwarz map*, which is introduced in [8]. It is defined as follows: Change the equation $E(a,b,c)$ into the so-called SL-form:

$$E^{\rm SL}: \quad u'' - q(x)u = 0, \tag{1.2}$$

and transform it to the matrix equation

$$\frac{d}{dx}(u,u') = (u,u')\Omega, \quad \Omega = \begin{pmatrix} 0 & q(x) \\ 1 & 0 \end{pmatrix}. \tag{1.3}$$

We now define the hyperbolic Schwarz map, denoted by $\mathscr{S}$, as the composition of the (multi-valued) map

$$X \ni x \longmapsto H = U(x)\ {}^t\overline{U}(x) \in \mathrm{Her}^+(2) \tag{1.4}$$

and the natural projection $\mathrm{Her}^+(2) \to \boldsymbol{H}^3 := \mathrm{Her}^+(2)/\boldsymbol{R}^+$, where $U(x)$ is a fundamental solution of the system, $\mathrm{Her}^+(2)$ the space of positive-definite Hermitian matrices of size 2, and $\boldsymbol{R}^+$ the multiplicative group of positive real numbers; the space $\boldsymbol{H}^3$ is called the *hyperbolic 3-space*.

Note that the target of the hyperbolic Schwarz map is $\boldsymbol{H}^3$, whose boundary is $\boldsymbol{P}^1$, which is the target of the Schwarz map. In this sense, our hyperbolic Schwarz map is a lift-to-the-air of the Schwarz map. The image surface (of X under $\mathscr{S}$) has the following geometrically nice property: The classical Schwarz map s is recovered as one of the two *hyperbolic Gauss maps* of the image surface. The image surface admits singularities; generically they are cuspidal edges and swallowtails.

In this note, we study the hyperbolic Schwarz map $\mathscr{S}$ of the equation $E(a,b,c)$ mainly when the parameters (a,b,c) are real, especially when its monodromy group is a finite (polyhedral) group or a Fuchsian group. We did our best to visualize the image surfaces.

In a computational aspect of this visualization, we use the composition of the hyperbolic Schwarz map $\mathscr{S}$ and the inverse of the Schwarz map s, $\Phi = \mathscr{S} \circ s^{-1}$, especially when the inverse of the Schwarz map is single-valued globally; refer to Section 6.2. This choice is very useful, because the inverse map is often given explicitly as an automorphic function for the monodromy group acting properly discontinuously on the image of the Schwarz map. Moreover, in one of the cases where we treat the lambda function for drawing pictures, it is indispensable, because we have a series that converges very fast.

In the end, we introduce the derived Schwarz map $ds : X \to \boldsymbol{P}^1$, present an associated (parallel) 1-parameter family of surfaces in $\boldsymbol{H}^3$ having $\mathscr{S}$ as a generic member, and having s and ds as the two extremes. We also study the confluence of swallowtail singularities; Refer to §7.3 and §7.4.

Basic ingredients of the hypergeometric function and its Schwarz map can be found in [2] and [10]. For this article only, the first half of [11] is enough.

2. Swallowtail as a discriminant

In this section, we describe surface singularities: cuspidal edges and a swallowtail. We start from the well-known discriminant of a quadratic polynomial.

2.1. Discriminant of a quadratic polynomial

Consider a quadratic polynomial

$$F := t^2 + at + b$$

in t with real coefficients (a, b). An irreducible polynomial in these coefficients which vanishes only when F has a multiple root is called a discriminant; it is unique up to a multiplicative constant and is given as

$$D := a^2 - 4b.$$

When $D > 0$, F has two distinct real roots, and when $D < 0$, F has a pair of complex conjugate roots. There are at least three ways to derive D.

2.1.1. Difference of two roots. Let α and β be two roots of F. Since

$$(t-\alpha)(t-\beta) = t^2 - (\alpha+\beta)t + \alpha\beta,$$

we have

$$\alpha + \beta = -a, \quad \alpha\beta = b.$$

Thus we have

$$(\alpha-\beta)^2 = (\alpha+\beta)^2 - 4\alpha\beta = a^2 - 4b = D.$$

2.1.2. Parallel displacement. A change of variable $t \to t + c$ does not affect the status of roots. Choose c so that the resulting polynomial has no t-term:

$$t^2 + at + b = \left(t + \frac{a}{2}\right)^2 - \frac{a^2 - 4b}{4} = \left(t + \frac{a}{2}\right)^2 - \frac{D}{4}.$$

2.1.3. Usage of differentiation. The polynomial F has a multiple root at a point c if and only if

$$R: \quad F(c) = 0 \quad \text{and} \quad F'(c) = 0.$$

So the point is whether the system $F(c) = F'(c) = 0$ is solvable. Eliminating c from

$$R: \quad c^2 + ac + b = 0, \quad 2c + a = 0,$$

we have

$$0 = \left(-\frac{a}{2}\right)^2 + a\left(-\frac{a}{2}\right) + b = -\frac{D}{4}.$$

This method does not tell us the difference of the status when $D > 0$ and that when $D < 0$; this is the weak point.

2.2. Discriminant of a cubic plynomial

We consider

$$F := t^3 + pt^2 + qt + r;$$

by performing a suitable parallel displacement, we (can) assume $p = 0$.

2.2.1. Difference of three roots. Let α, β, γ be the three roots of $F = 0$. Since

$$(t-\alpha)(t-\beta)(t-\gamma) = t^3 - (\alpha+\beta+\gamma)t^2 + (\alpha\beta+\beta\gamma+\gamma\alpha)t - \alpha\beta\gamma,$$

we have

$$\alpha+\beta+\gamma = 0, \quad \alpha\beta+\beta\gamma+\gamma\alpha = q, \quad \alpha\beta\gamma = -r.$$

The discriminant should be given as a constant multiple of

$$D = \{(\alpha-\beta)(\beta-\gamma)(\gamma-\alpha)\}^2,$$

which is not so easy to write in terms of q and r.

2.2.2. Usage of differentiation. We eliminate t from

$$F = t^3 + qt + r = 0, \quad F' = 3t^2 + q = 0.$$

(The second equation gives an expression of t^2. Substitute this expression into the first equation and you get an expression of t. Substitute this expression into the second equation.) Anyway, we get the expression $4q^3 + 27r^2$, which is now straightforward to see,

$$D = -(4q^3 + 27r^2).$$

2.2.3. Visualization. Draw in the (q, r)-plane the curve $D = 4q^3 + 27r^2 = 0$. Since

$$F = (t-\alpha)^2(t-\beta), \quad 2\alpha + \beta = 0,$$

the curve can be parameterized as

$$q = -3\alpha^2, \quad r = -2\alpha^3, \quad r \in \boldsymbol{R};$$

by this expression, you can draw its image (see Figure 1).

In this occasion, let us recall that if a polynomial with real coefficients has a complex root α, then its complex conjugate $\bar{\alpha}$ is also a root, where the complex conjugate of $z = x + iy$ is defined as $\bar{z} = x - iy$; this can be proved by repeated use of

$$\overline{z+w} = \bar{z} + \bar{w}, \quad \overline{zw} = \bar{z}\bar{w}.$$

Convention:

- $1+1+1$: polynomial with three distinct real roots
- $1+(1+\bar{1})$: polynomial with a real root and a pair of conjugate imaginary roots
- $2+1$: polynomial with double root on the left of a simple root
- $1+2$: polynomial with double root on the right of a simple root
- 3: polynomial with triple root

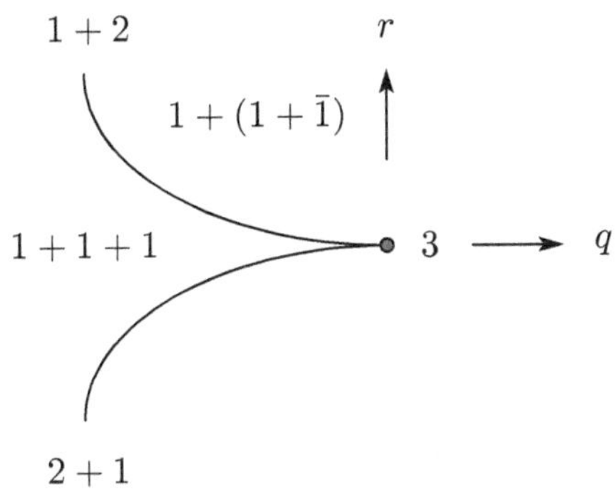

FIGURE 1. The curve $D = 0$ in the qr-plane

The (q, r)-space, the totality of the cubic polynomials F, is divided by the curve $D = 0$ into two parts: the smaller one consists of the polynomials of type $1+1+1$, and the bigger one $1 + (1 + \bar{1})$; the curve $D = 0$ consists of two parts $1 + 2$ and $2 + 1$ with the common point 3. See the diagram in Figure 2.

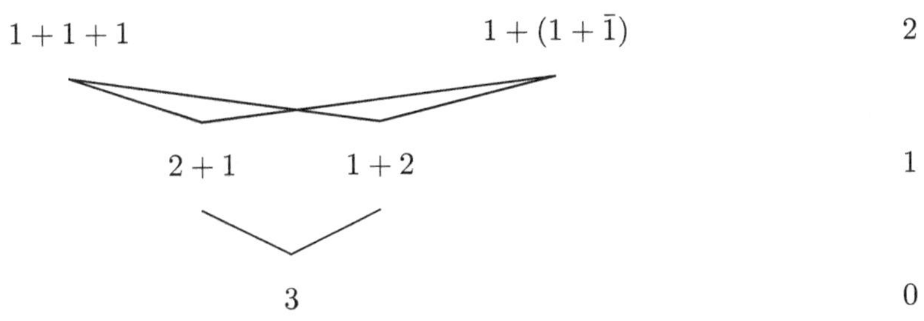

FIGURE 2. Degeneration diagram of the status of three roots

When two are connected by a segment, it means that the lower one is a specialization of the above one. The numerals on the right denote the freedom/dimension. From this diagram, you can draw Figure 3, which is topologically equivalent to Figure 1.

2.2.4. Relation between roots and coefficients. Recall that the coefficients q, r of the polynomial $F = t^3 + qt + r$ and the roots $\alpha, \beta, \gamma = -\alpha - \beta$ are related as

$$q = \alpha\beta + \beta\gamma + \gamma\alpha = \alpha\beta - (\alpha + \beta)^2,$$
$$r = -\alpha\beta\gamma = \alpha\beta(\alpha + \beta).$$

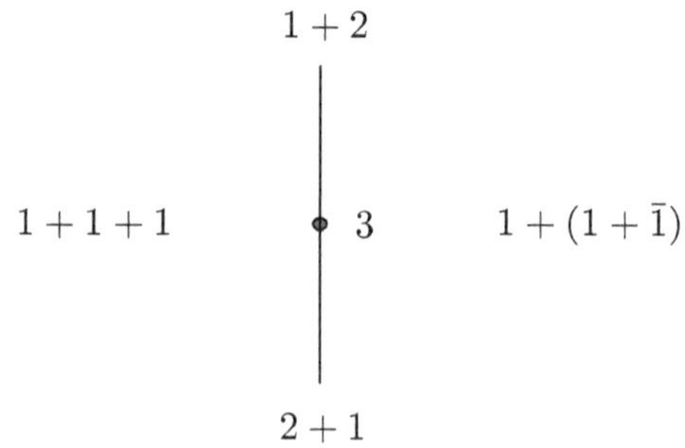

FIGURE 3. The curve $D = 0$ in qr-space, simplified

Since these expressions are symmetric in α and β, we put

$$t = \alpha + \beta, \quad s = \alpha\beta;$$

then the expressions above become

$$q = s - t^2, \quad r = st.$$

It is interesting to study the map

$$\varphi : (s, t) \mapsto (q, r) = (s - t^2, st).$$

For generic (q, r), i.e., $D(q, r) = 4q^3 + 27r^2 \neq 0$, its preimage consists of three points, since the polynomial F has three distinct roots. The preimage of the discriminant curve $D = 0$ consists of two parabolas; indeed we have

$$4q^3 + 27r^2 = (s + 2t^2)^2(4s - t^2).$$

Since we have

$$\frac{\partial(q, r)}{\partial(s, t)} = \begin{vmatrix} 1 & -2t \\ t & s \end{vmatrix} = s + 2t^2,$$

the map is degenerate along the parabola $s + 2t^2 = 0$. These two parabolas are shown in Figure 4. To help understand this map, the image of a hemicircle centered at the origin in the (s, t)-space is also shown.

2.3. Discriminant of a quartic polynomial

We consider a quartic plynomial

$$F = t^4 + xt^2 + yt + z.$$

2.3.1. Discriminant. Let $\alpha, \beta, \gamma, \delta$ be the four roots of F. By the relations

$$\alpha + \beta + \gamma + \delta = 0, \quad \alpha\beta + \alpha\gamma + \alpha\delta + \beta\gamma + \beta\delta + \gamma\delta = x,$$
$$\beta\gamma\delta + \alpha\gamma\delta + \alpha\beta\delta + \alpha\beta\gamma = -y, \quad \alpha\beta\gamma\delta = z,$$

we can express

$$D = \{(\alpha - \beta)(\alpha - \gamma)(\alpha - \delta) \cdot (\beta - \gamma)(\beta - \delta) \cdot (\gamma - \delta)\}^2$$

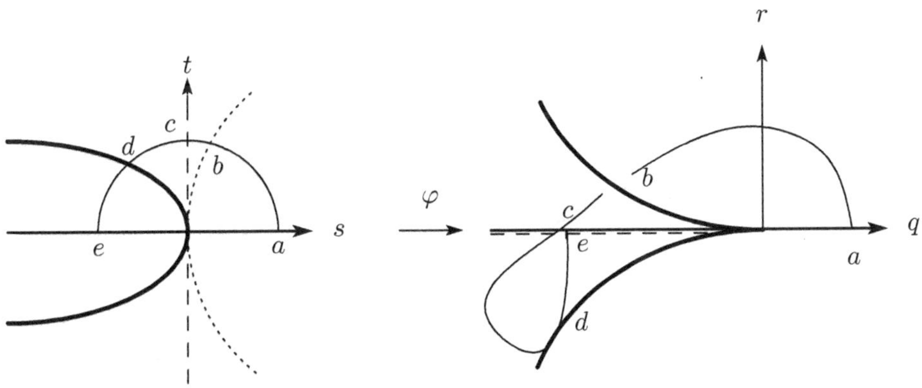

FIGURE 4. Image under the map φ

in terms of x, y, z. But no one would like to carry out the computation.

We eliminate t from $F(t) = F'(t) = 0$:

$$t^4 + xt^2 + yt + z = 0, \quad 4t^3 + 2xt + y = 0.$$

(The second equation gives an expression of t^3. Substitute this into the first and you get an expression of t^2. Substitute this into the second, you get an expression of t,....) Eventually, we get the expression

$$256z^3 - 128x^2z^2 + (144y^2x + 16x^4)z - 4y^2x^3 - 27y^4;$$

you can easily check that it coincides with D.

For a cosmetic reason, we change z into $z/4$:

$$F = t^4 + xt^2 + yt + \frac{z}{4},$$

$$D := 4z^3 - 8x^2z^2 + (36y^2x + 4x^4)z - (27y^2 + 4x^3)y^2.$$

Can you guess what the set $D = 0$ looks like?

2.3.2. Various status of the four roots. Three-dimensional happenings are difficult to understand. The set $D = 0$ will divide the xyz-space into several parts. In some part, the corresponding quartic polynomial $t^4 + xt^2 + yt + z/4 = 0$ has four distinct real roots, in some other parts two real roots and a pair of complex conjugate roots, ...:

We make a degeneration diagram in Figure 5, generalizing that in Figure 2. When two are connected by a segment, it means that the lower one is a specialization of the above one. The numerals on the right denote the freedom/dimension. Convention:

- $1 + 1 + 1 + 1$: 4 distinct real roots

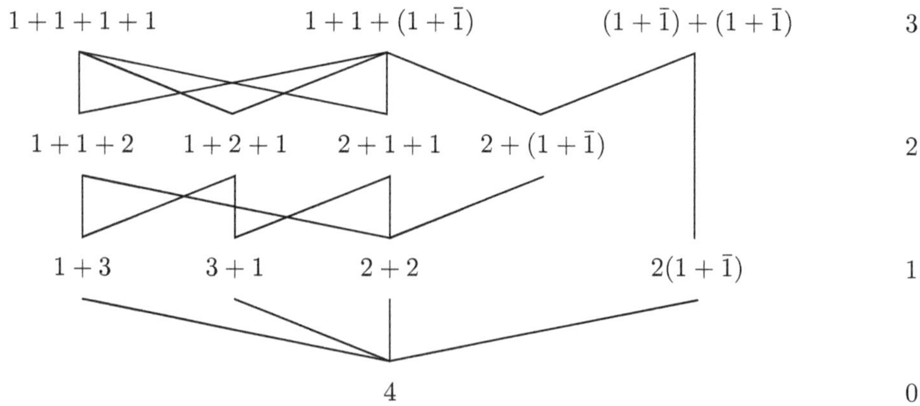

FIGURE 5. Degeneration diagram of the status of four roots

- $1+1+(1+\bar{1})$: 2 distinct real, and a pair of conjugate roots
- $(1+\bar{1})+(1+\bar{1})$: 2 distinct pairs of conjugate roots
- $1+1+2$: double real root on the right of 2 distinct real roots
- $1+2+1$: double real root at the middle
- $2+1+1$: double real root on the left of 2 distinct real roots
- $2+(1+\bar{1})$: double real root and a pair of conjugate roots
- $1+3$: triple root on the right of a real root
- $3+1$: triple root on the left of a real root
- $2+2$: distinct 2 double roots
- $2(1+\bar{1})$: a pair of conjugate double roots
- 4: 4-ple root

From only these data we can draw a (topological) picture of the surface S consisting of polynomials of types

$$1+1+2, \qquad 1+2+1, \qquad 2+1+1, \qquad 2+(1+\bar{1}).$$

This surface divides the space into three parts:

$$1+1+1+1, \qquad 1+1+(1+\bar{1}), \qquad (1+\bar{1})+(1+\bar{1}),$$

which are the inside of the triangular cone, the inside of the circular cone, and the complement of the two cones, respectively. Note that the set $D=0$ consists of S and a curve corresponding to $2(1+\bar{1})$, which lies inside the circular cone $(1+\bar{1})+(1+\bar{1})$.

2.4. Parametrization of the surface S

Let γ be a double root. Then we have

$$t^4+xt^2+yt+z/4=(t-\alpha)(t-\beta)(t-\gamma)^2, \quad \alpha+\beta+2\gamma=0,$$

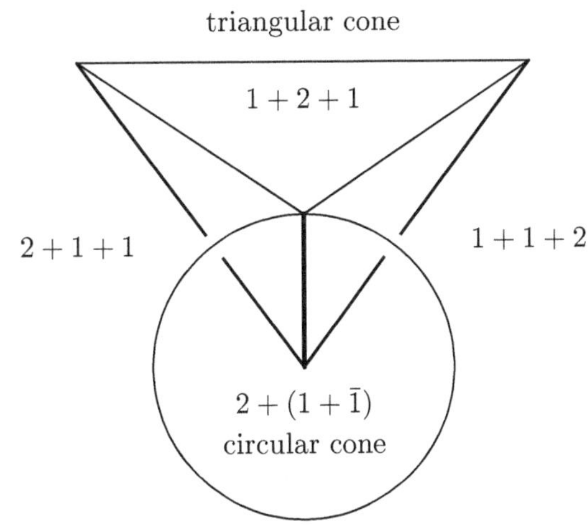

FIGURE 6. A surface homeomorphic to the surface S in xyz-space

and so

$$x=-\frac{3}{4}\alpha^2-\frac{1}{2}\alpha\beta-\frac{3}{4}\beta^2,\quad y=-\frac{1}{4}(\alpha+\beta)(-\beta+\alpha)^2,\quad z=(\alpha+\beta)^2\alpha\beta,$$

and

$$x=-\frac{3}{4}a^2+b,\quad y=\frac{1}{4}a(a^2-4b),\quad z=a^2b,$$

where $\alpha+\beta=-a,\ \alpha\beta=b$. The surface S can be parametrized by $a,b\in\boldsymbol{R}$ (see Figure 7); note that

$$\alpha=\beta\Leftrightarrow a^2-4b=0,\qquad \alpha=\gamma\Leftrightarrow 3a^2+4b=0.$$

Now it is not difficult for a machine to draw S (see Figure 8). It is not difficult also for human beings to see that the surface S has cuspidal edges along the curves $3+1$ and $1+3$, and has normal-crossings along the curve $2+2$; we make a calculation with the expression of x,y,z in (a,b). We get a better picture: see Figure 9.

3. Hypergeometric differential equation

The *hypergeometric differential equation*

$$E(a,b,c): x(1-x)u''+\{c-(a+b+1)x\}u'-abu=0$$

and its solutions, the hypergeometric functions, were found and studied by Euler and Gauss more than two hundred years ago. It is linear, of second order

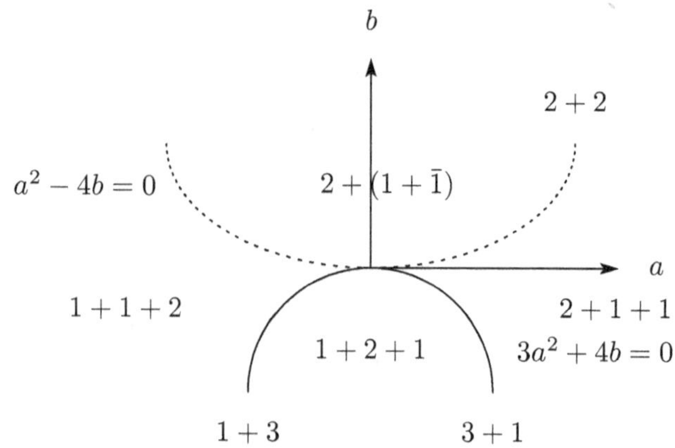

FIGURE 7. (a, b)-space parameterizing S

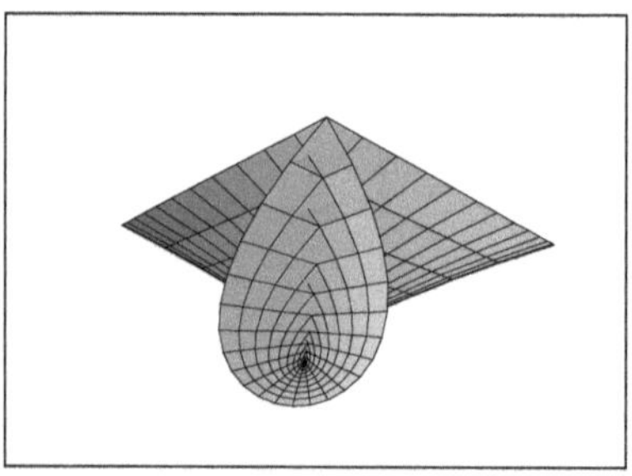

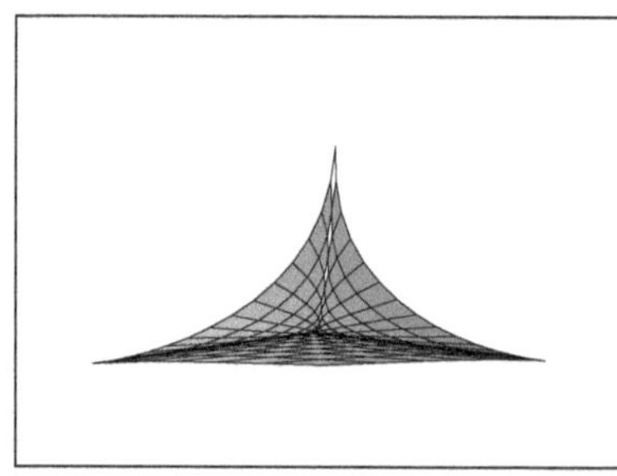

FIGURE 8. Maple figures

with singularities at $x = 0, 1$ and ∞. It is the simplest differential equation not solvable by highschool mathematics. The fundamental theorem of Cauchy asserts that

- at any point $x_0 \neq 0, 1, \infty$, the local solutions at x_0 (they are holomorphic) form a 2-dimensional linear space over $\boldsymbol{C}$, and
- any solution at x_0 can be continued holomorphically along any curve starting at x_0 and not passing through $0, 1$ or ∞.

3.1. Local properties around the singular points

By substituting a local expression $u = x^\alpha(1 + c_1x + c_2x^2 + \cdots)$ into the equation, we find two solutions around the origin of the form

$$1 + O(x) \quad \text{and} \quad x^{1-c}(1 + O(x)).$$

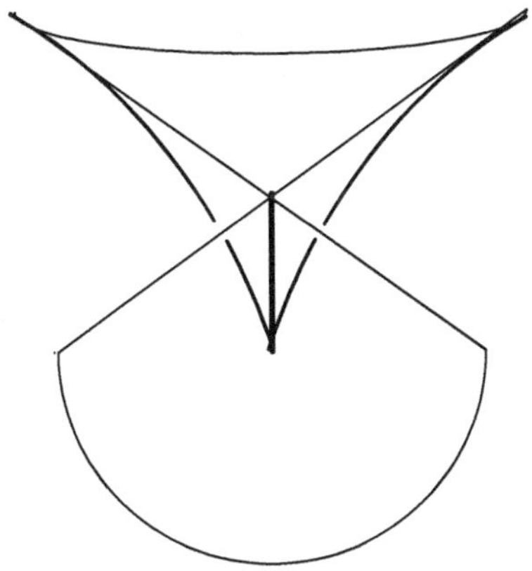

FIGURE 9. A better picture of the surface S with a swallowtail

In the same way we get two solutions around $x = 1$ of the form

$$1 + O(1-x) \quad \text{and} \quad (1-x)^{c-a-b}(1 + O(1-x)),$$

and around $x = \infty$,

$$\xi^a(1 + O(\xi)) \quad \text{and} \quad \xi^b(1 + O(\xi)), \qquad \xi = 1/x.$$

The pairs

$$\{0, 1-c\}, \quad \{0, c-a-b\}, \quad \{a, b\}$$

are called the (local) exponents around $0, 1$ and ∞, respectively. Set the exponent differences as

$$\mu_0 = 1 - c, \quad \mu_1 = c - a - b, \quad \mu_\infty = a - b.$$

3.2. Monodromy group

The fundamental group of

$$X := \boldsymbol{C} - \{0, 1\}$$

with a base point, say $x_0 \in X$, can be generated by a loop ρ_0 around 0, and a loop ρ_1 around 1. If you continue analytically the pair (u_0, u_1) of linear independent solutions at x_0 along the loop ρ_0, then it changes linearly as

$$\begin{pmatrix} u_0 \\ u_1 \end{pmatrix} \longrightarrow M_0 \begin{pmatrix} u_0 \\ u_1 \end{pmatrix},$$

for a matrix $M_0 \in GL_2(\boldsymbol{C})$. Along ρ_1, we get M_1. The group generated by these two matrices is called the *monodromy group* of the differential equation $E(a, b, c)$ (with respect to a pair of linearly independent solutions (u_0, u_1) with base x_0). If you make use of another pair of solutions at another base point, the resulting group is conjugate in $GL_2(\boldsymbol{C})$ to this group. So the differential equation determines a conjugacy class represented by this group. Any representative, or its conjugacy class, is called the monodromy group of the equation.

There are many important subgroups of $GL_2(\boldsymbol{C})$ generated by two elements. Almost all such groups (including all the irreducible ones) can be realized as the monodromy group of the hypergeometric equation $E(a,b,c)$ for some (a,b,c).

4. The Schwarz map of the hypergeometric differential equation

The Schwarz map is defined as

$$s : X \ni x \mapsto z = u_0(x) : u_1(x) \in \boldsymbol{P}^1_z \ (\cong \boldsymbol{C} \cup \{\infty\}),$$

where u_0 and u_1 are linearly independent solutions of the hypergeometric equation $E(a,b,c)$.

4.1. Schwarz triangle when exponents are real

In this section, we assume that the parameters a, b and c are real; so the coefficients of the equation $E(a,b,c)$ are real, if x is real. Along each of the three intervals

$$(-\infty, 0), \quad (0,1), \quad (1,+\infty),$$

there are two linearly independent solutions which are real-valued on the interval. Note that a real-valued solution along an interval may not be real-valued along another interval. Since the Schwarz map is in general multi-valued, we restrict it on the upper half-plane X_+, and study the shape of its image in the target space $\boldsymbol{P}^1_z$, the projective line with coordinate z. The Schwarz map depends on the choice of the two linearly independent solutions; if we choose other two such solutions, then the new and the old Schwarz maps relate with a linear fractional transformation, which is an automorphism of $\boldsymbol{P}^1_z$.

Recall the following fundamental fact: A linear fractional transformation takes a circle to a circle; here a line is considered to be a circle which passes through ∞.

Around the singular points $x = 0, 1$ and ∞, the Schwarz map s is, after performing suitable linear fractional transformations, near to the power functions $x \to x^{|\mu_0|}, (1-x)^{|\mu_1|}$ and $x^{|\mu_\infty|}$. Thus small hemi-disks with center $x = 0, 1$ and ∞ are mapped to horns with angle $\pi|\mu_0|, \pi|\mu_1|$ and $\pi|\mu_\infty|$. See Figure 10.

Summing up, the image of X_+ under s is bounded by three arcs (parts of circles), and the three arcs meet with angle $\pi|\mu_0|, \pi|\mu_1|$ and $\pi|\mu_\infty|$. If

$$|\mu_0| < 1, \quad |\mu_1| < 1, \quad |\mu_\infty| < 1,$$

then the image is an arc-triangle, which is called the *Schwarz triangle*. In this case, the Schwarz map gives a conformal equivalence between X_+ and the Schwarz triangle.

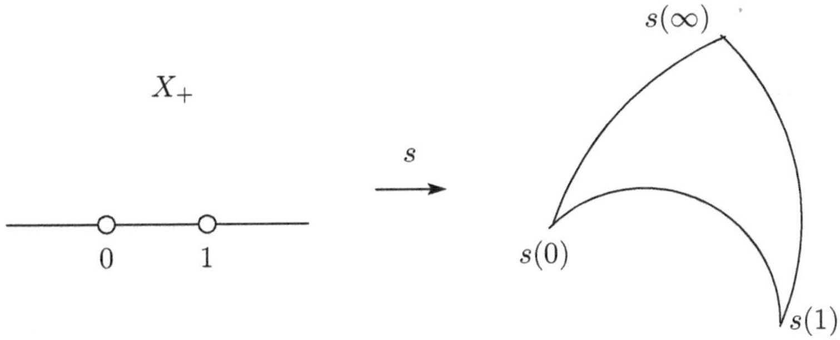

FIGURE 10. Schwarz triangle

4.2. Schwarz map when exponents are real

The global behavior of the Schwarz map can be seen by applying repeatedly the Schwarz reflection principle to the three sides of the Schwarz triangle. For generic parameters, the picture of the reflected triangles gradually becomes chaotic. For special parameters, it can remain neat. For example, if

$$|\mu_0| = \frac{1}{p}, \quad |\mu_1| = \frac{1}{q}, \quad |\mu_\infty| = \frac{1}{r}, \quad p, q, r \in \{2, 3, \dots, \infty\},$$

then the whole image is nice; note that this is not a *necessary* condition. Paint the original triangle black (since it is called the Schwarz triangle), adjacent ones white, and so on. The picture of these black and white triangles thus obtained can be roughly classified into three cases depending on whether

$$\frac{1}{p} + \frac{1}{q} + \frac{1}{r}$$

is bigger than or equal to or less than 1.

- In the first case, the number of the triangles is finite, and those cover the whole sphere. The possible triples ($p \le q \le r$) are

$$(2, 2, r), \quad (2, 3, 3), \quad (2, 3, 4), \quad (2, 3, 5).$$

  The monodromy groups are the dihedral, tetrahedral, octahedral and icosahedral groups, respectively. Figure 11 shows 120 icosahedral triangles.
- In the second case, the three circles generated by the three sides of the triangles pass through a common point. If you send this point to infinity, then the triangles are bounded by (straight) lines, and they cover the whole plane. The possible triples ($p \le q \le r$) are

$$(2, 2, \infty), \quad (3, 3, 3), \quad (2, 4, 4), \quad (2, 3, 6).$$

FIGURE 11. 120 icosahedral triangles and 60 central ones

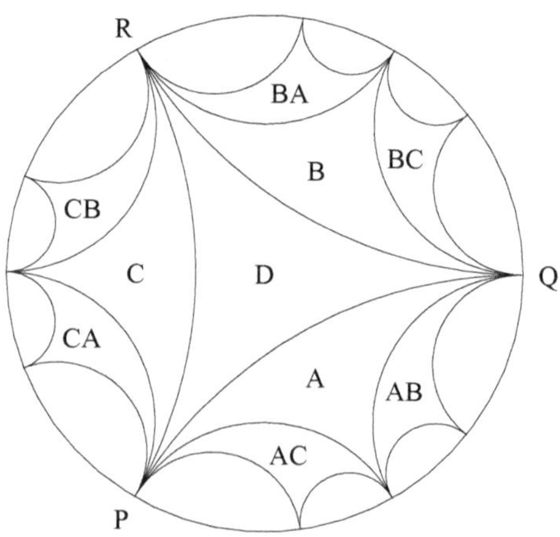

FIGURE 12. (∞, ∞, ∞)

- In the third case, there is a unique circle which is perpendicular to all of the circles generated by the sides of the triangles, and these triangles fill the disc bounded by this circle. In particular, when

$$(p, q, r) = (2, 3, \infty), \quad (\infty, \infty, \infty), \quad (2, 3, 7),$$

you can find in many books nice pictures of the tessellation of these triangles; See Figure 12.

4.3. Schwarz map when the exponents are not real

4.3.1. Schwarz maps with purely imaginary exponents. In this case, the monodromy group is the so-called Schottky group. See [11].

4.3.2. An example with complex exponents. If, for example, we take the parameters as

$$a = \frac{\arccos \frac{1+i}{2}}{2\pi}, \quad b = 1 - a, \quad c = 1,$$

then the monodromy group is generated by

$$\begin{pmatrix} 1 & i \\ 0 & 1 \end{pmatrix} \quad \text{and} \quad \begin{pmatrix} 1 & 0 \\ 1+i & 1 \end{pmatrix},$$

which is the Whitehead-link-complement-group. Though the image of the Schwarz map is chaotic, the image surface of our hyperbolic Schwarz map is a closed surface in hyperbolic 3-space.

5. Hyperbolic spaces

5.1. 2-dimensional models

The upper-half plane

$$\boldsymbol{H}^2 = \{\tau \in \boldsymbol{C} \mid \Im(\tau) > 0\}$$

is called the hyperbolic 2-space. The group of (orientation-preserving) isometric transformations are given by

$$GL_2(\boldsymbol{R}) \ni g = \begin{pmatrix} a & b \\ c & d \end{pmatrix} : \boldsymbol{H}^2 \ni \tau \mapsto \frac{a\tau + b}{c\tau + d} \in \boldsymbol{H}^2.$$

A 'line' is a hemi-circle or a line perpendicular to the boundary line $\Im\tau = 0$. The upper-half plane can be also identified with a Poincaré disc $\{z \in \boldsymbol{C} \mid |z| < 1\}$, by

$$z = \frac{\tau - i}{\tau + i}, \qquad \tau = i\frac{1+z}{1-z}.$$

A 'line' is a hemi-circle perpendicular to the boundary circle $|z| = 1$. See Figure 12.

5.2. 3-dimensional models

The space $\boldsymbol{H}^3 = \mathrm{Her}^+(2)/\boldsymbol{R}^+$ is called the hyperbolic 3-space. The group of (orientation-preserving) isometric transformations are given by

$$GL_2(\boldsymbol{C}) \ni P : \mathrm{Her}^+(2) \ni H \mapsto PH\,{}^t\overline{P} \in \mathrm{Her}^+(2).$$

This space $\boldsymbol{H}^3$ can be identified with the upper half-space $\boldsymbol{C} \times \boldsymbol{R}^+$ as

$$\boldsymbol{C} \times \boldsymbol{R}^+ \ni (z,t) \longmapsto \begin{pmatrix} t^2 + |z|^2 & \bar{z} \\ z & 1 \end{pmatrix} \in \mathrm{Her}^+(2),$$

$$\begin{pmatrix} h & \bar{w} \\ w & k \end{pmatrix} \in \mathrm{Her}^+(2) \longmapsto \boldsymbol{C} \times \boldsymbol{R}^+ \ni \frac{1}{k}\left(w, \sqrt{hk - |w|^2}\right).$$

A 'line' is a hemi-circle or a line perpendicular to the boundary plane $t = 0$. A 'plane' is a hemi-sphere or a plane perpendicular to the boundary plane $t = 0$. The upper half-space can be also identified with a subvariety

$$L_1 = \{x_0^2 - x_1^2 - x_2^2 - x_3^2 = 1\}$$

of the Lorentz-Minkowski 4-space

$$L(+,-,-,-) = \{(x_0, x_1, x_2, x_3) \in \boldsymbol{R}^4 \mid x_0^2 - x_1^2 - x_2^2 - x_3^2 > 0, x_0 > 0\}$$

by

$$\mathrm{Her}^+(2) \ni \begin{pmatrix} h & \bar{w} \\ w & k \end{pmatrix} \longmapsto \frac{1}{2\sqrt{hk - |w|^2}} \left(h + k, w + \bar{w}, \frac{w - \bar{w}}{i}, h - k \right) \in L_1$$

and with the Poincaré ball

$$B_3 = \{(x_1, x_2, x_3) \in \boldsymbol{R}^3 \mid x_1^2 + x_2^2 + x_3^2 < 1\},$$

by

$$L_1 \ni (x_0, x_1, x_2, x_3) \longmapsto \frac{1}{1 + x_0}(x_1, x_2, x_3) \in B_3.$$

A 'line' is a hemi-circle perpendicular to the boundary sphere. A 'plane' is a hemi-sphere perpendicular to the boundary sphere. The Poincaré ball is a space form of constant curvature -1. Through identifications above, each model has also the induced metric that is invariant under respective automorphisms. Its form on the space $\boldsymbol{C} \times \boldsymbol{R}^+$ is $(dzd\bar{z} + dt^2)/t^2$.

We use these models according to convenience.

6. Hyperbolic Schwarz map – fundamental properties

We first transform our equation $E(a, b, c)$ a little. Consider in general an equation of the form

$$u'' + pu' + qu = 0.$$

If we make a change from u to v by $u = fv$, then we have

$$v'' + \left(p + 2\frac{f'}{f}\right) v' + \left(q + p\frac{f'}{f} + \frac{f''}{f}\right) v = 0.$$

Choose f solving $p + 2f'/f = 0$. Then the coefficient of v is given as

$$q - \frac{1}{2}p' - \frac{1}{4}p^2.$$

Perform this transformation to $E(a, b, c)$: By the change of the unknown

$$u \longrightarrow \sqrt{x^c(1 - x)^{a+b+1-c}}\; u,$$

we get the SL-form $E^{\mathrm{SL}}(a, b, c)$,

$$E^{\mathrm{SL}} : u'' - q(x)u = 0, \quad q = -\frac{1}{4}\left\{\frac{1 - \mu_0^2}{x^2} + \frac{1 - \mu_1^2}{(1 - x)^2} + \frac{1 + \mu_\infty^2 - \mu_0^2 - \mu_1^2}{x(1 - x)}\right\},$$

where $\mu_0 = 1-c, \mu_1 = c-a-b, \mu_\infty = a-b$.

Unless otherwise stated, we always take a pair (u_0, u_1) of linearly independent solutions of E^{SL} satisfying $u_0u_1' - u_0'u_1 = 1$, and set

$$U = \begin{pmatrix} u_0 & u_0' \\ u_1 & u_1' \end{pmatrix}.$$

The hyperbolic Schwarz map is defined as

$$\mathscr{S}\colon X \ni x \longmapsto H(x) = U(x)\,{}^t\overline{U}(x) = \begin{pmatrix} |u_0|^2 + |u_0'|^2 & u_0\bar{u}_1 + u_0'\bar{u}_1' \\ \bar{u}_0u_1 + \bar{u}_0'u_1' & |u_1|^2 + |u_1'|^2 \end{pmatrix} \in \boldsymbol{H}^3.$$

6.1. Monodromy group

Let $\{v_0, v_1\}$ be another pair of linearly independent solutions of E^{SL}. Then there is a non-singular matrix, say P, such that $U = PV$ and so that

$$U\,{}^t\overline{U} = PV\,{}^t\overline{V}\,{}^t\overline{P}, \quad \text{where} \quad V = \begin{pmatrix} v_0 & v_0' \\ v_1 & v_1' \end{pmatrix}.$$

Thus the hyperbolic Schwarz map is determined by the system up to orientation preserving automorphisms. The monodromy group $\mathrm{Mon}(a, b, c)$ with respect to U acts naturally on $\boldsymbol{H}^3$ by

$$H \longmapsto MH\,{}^t\overline{M}, \quad M \in \mathrm{Mon}(a, b, c).$$

Note that the hyperbolic Schwarz map to the upper half-space model is given by

$$X \ni x \longmapsto \frac{\left(\bar{u}_0(x)u_1(x) + \bar{u}_0'(x)u_1'(x),\ 1\right)}{|u_1(x)|^2 + |u_1'(x)|^2} \in \boldsymbol{C} \times \boldsymbol{R}^+.$$

6.2. Use of the Schwarz map

Let u and v be solutions of the equation E^{SL} such that $uv' - vu' = 1$. The Schwarz map is defined as $X \ni x \mapsto z = u(x)/v(x) \in Z$; it is convenient to study the hyperbolic Schwarz map by regarding z as variable.

Especially when the inverse of the Schwarz map is single-valued globally, this choice of variable is very useful, because the inverse map is often given explicitly as an automorphic function for the monodromy group acting properly discontinuously on the image of the Schwarz map. The solution U can be written in terms of the inverse of the Schwarz map $x(z)$ and their derivatives:

$$U = i\frac{1}{\sqrt{\dot{x}}} \begin{pmatrix} z\dot{x} & 1 + \frac{z}{2}\frac{\ddot{x}}{\dot{x}} \\ \dot{x} & \frac{1}{2}\frac{\ddot{x}}{\dot{x}} \end{pmatrix}, \tag{6.1}$$

where $\dot{}\ = d/dz$. Here, we summarize the way to show the formula: Since $z'(:= dz/dx) = -1/v^2$ and $\ddot{x} = d^2x/dz^2$, we have

$$v = i\sqrt{\frac{1}{z'}} = i\sqrt{\dot{x}}, \quad u = vz,$$

and

$$v' = \frac{dv}{dx} = \frac{dv}{dz}\frac{dz}{dx} = \frac{i}{2}(\dot{x})^{-3/2}\ddot{x}, \qquad u' = i\frac{1}{\sqrt{\dot{x}}} + z\frac{i}{2}(\dot{x})^{-3/2}\ddot{x}.$$

6.3. Singularities of hyperbolic Schwarz maps

6.3.1. Geometry of surfaces defined by the hyperbolic Schwarz map. A map f from a 2-manifold M into a 3-manifold N is called a front if it is the projection of a smooth map $L : M \to T_1N$ so that $df_p(X)$ is perpendicular to $L(p)$, where T_1N is the tangent sphere bundle of N. A front may have singularity. However, the parallel immersion f_t, $t \in \boldsymbol{R}$, is well-defined, as the set of points that are equidistant from f. It is also a front. A front is said to be flat, if the Gaussian curvature vanishes at any nonsingular point and if, at a singular point, any parallel front f_t is nonsingular for any small value of t. It is known that the flat front f is written as $U\,{}^t\overline{U}$, where $U : M \to SL(2, \boldsymbol{C})$ is an immersion. Thus, in our case where $N = \boldsymbol{H}^3$, the hyperbolic Schwarz map defines a flat front.

6.3.2. Singularities of $\mathscr{S} : X \to \boldsymbol{H}^3$. Since the equation E^{SL} has singularities at 0, 1 and ∞, the corresponding hyperbolic Schwarz map $\mathscr{S}$ has singularities at these points. In terms of flat fronts in $\boldsymbol{H}^3$, they are considered as *ends* of the surface. On the other hand, the map $\mathscr{S}$ may not be an immersion at $x \in X$, even if x is not a singular point of E^{SL}.

In this subsection, we analyze properties of these singular points of the hyperbolic Schwarz maps.

The following criterion is known.

Lemma 6.1 ([5]). (1) *A point $p \in X$ is a singular point of the hyperbolic Schwarz map H if and only if $|q(p)| = 1$.*

(2) *A singular point $x \in X$ of H is equivalent to the cuspidal edge if and only if*

$$q'(x) \neq 0 \quad \textit{and} \quad q^3(x)\bar{q}'(x) - q'(x) \neq 0.$$

(3) *A singular point $x \in X$ of H is equivalent to the swallowtail if and only if $q'(x) \neq 0$,*

$$q^3(x)\bar{q}'(x) - q'(x) = 0, \quad \textit{and} \quad \Re\left\{\frac{1}{q}\left(\left(\frac{q'(x)}{q(x)}\right)' - \frac{1}{2}\left(\frac{q'(x)}{q(x)}\right)^2\right)\right\} \neq 0.$$

We apply Lemma 6.1 to the hypergeometric equation. The coefficient q of the equation E^{SL} is written as

$$q =: \frac{-Q}{4x^2(1-x)^2}, \quad Q = 1 - \mu_0^2 + (\mu_\infty^2 + \mu_0^2 - \mu_1^2 - 1)x + (1 - \mu_\infty^2)x^2. \tag{6.2}$$

Hence $x \in X$ is a singular point if and only if

$$|Q| = 4|x^2(1-x)^2|. \tag{6.3}$$

The condition $q^3(x)\bar{q}'(x) - q'(x) = 0$ is equivalent to

$$Q^3\bar{R}^2 \text{ is real non-positive,} \tag{6.4}$$

where R is given by

$$q' = -\frac{Q'x(1-x) - 2Q(1-2x)}{4x^3(1-x)^3} =: \frac{-R}{4x^3(1-x)^3}.$$

Thus cuspidal edges and swallowtails can be found by solving a system of *algebraic* equations.

6.3.3. At a singular point of the equation $E(a,b,c)$. By using local expressions of solutions around the singularities 0, 1 and ∞, we can prove that if the parameters a, b and c are real, the hyperbolic Schwarz map $\mathscr{S}$ extends to these singular points and to the boundary of $\boldsymbol{H}^3$. Its image is nonsingular and is tangent to the boundary at these point.

7. Hyperbolic Schwarz maps – Examples

When the monodromy group of the equation $E(a,b,c)$ is a finite group or a typical Fuchsian group, we study the singularities of the hyperbolic Schwarz map, and visualize the image surface.

7.1. Finite (polyhedral) monodromy groups

We first recall fundamental facts about the polyhedral groups and their invariants.

7.1.1. Basic data. Let the triple (k_0, k_1, k_∞) be one of

$$(2,2,n)\ (n = 1,2,\dots), \quad (2,3,3), \quad (2,3,4), \quad (2,3,5),$$

in which case, the projective monodromy group is of finite order N:

$$N = \quad 2n, \quad 12, \quad 24, \quad 60,$$

respectively. Note that

$$\frac{2}{N} = \frac{1}{k_0} + \frac{1}{k_1} + \frac{1}{k_\infty} - 1.$$

For each case, there is a triplet $\{R_1, R_2, R_3\}$ of reflections whose mirrors bound a Schwarz triangle (cf. [8]). In the dihedral case $(2,2,n)$, for example, we can take

$$R_1 : z \mapsto \bar{z}, \quad R_2 : z \mapsto e^{2\pi i/n}\bar{z}, \quad R_3 : z \mapsto \frac{1}{\bar{z}}.$$

The monodromy group Mon (a polyhedral group) is the group of even words of these three reflections.

The (single-valued) inverse map

$$s^{-1} : Z \ni z \longmapsto x \in \bar{X} \cong \boldsymbol{P}^1,$$

invariant under the action of Mon, is given as follows. Let $f_0(z), f_1(z)$ and $f_\infty(z)$ be the monic polynomials in z with simple zeros exactly at the images $s(0), s(1)$ and $s(\infty)$, respectively. If $\infty \in Z$ is not in these images, then the degrees of these

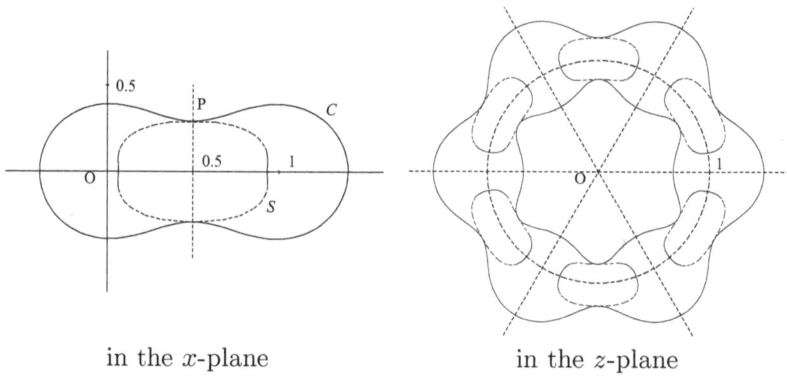

FIGURE 13. The curve $C : |Q| = 4|x(1-x)|^2$, when $(k_0, k_1, k_\infty) = (2, 2, 3)$

polynomials are $N/k_0, N/k_1$ and N/k_∞, respectively; if for instance $\infty \in s(0)$, then the degree of f_0 is $N/k_0 - 1$. Now the inverse map s^{-1} is given by

$$x = A_0 \frac{f_0(z)^{k_0}}{f_\infty(z)^{k_\infty}},$$

where A_0 is a constant; we also have

$$1 - x = A_1 \frac{f_1(z)^{k_1}}{f_\infty(z)^{k_\infty}}, \quad \frac{dx}{dz} = A\frac{f_0(z)^{k_0-1} f_1(z)^{k_1-1}}{f_\infty(z)^{k_\infty+1}},$$

for some constants A_1 and A. Their explicit forms can be found for example in [8]. In the dihedral case $(k_0, k_1, k_\infty) = (2, 2, n)$, for example, we have

$$A_0 = \frac{1}{4}, \quad A_1 = -\frac{1}{4}, \quad A = \frac{n}{4}, f_0 = z^n + 1, \quad f_1 = z^n - 1, \quad f_\infty = z.$$

Note that f_∞ is of degree $1 = 2n/n - 1$, since $\infty \in s(\infty)$, that is, $x(\infty) = \infty$.

7.1.2. Dihedral cases. We consider a dihedral case: $(k_0, k_1, k_\infty) = (2, 2, n)$, $n = 3$. The curve C in the x-plane defined by (6.3): $|Q| = 4|x(1-x)|^2$ is symmetric with respect to the line $\Re(x) = 1/2$ and has a shape of a cocoon (see Figure 13 (left)). We next study the conditions (6.3) and (6.4). We can prove that, on the upper half x-plane, there is a unique point (the intersection P of the curve C and the line $\Re x = 1/2$) such that the image surface has a swallowtail at this point, and has cuspidal edges along $\mathscr{S}(C)$ outside $\mathscr{S}(P)$. We omit the computation.

The curve which gives the self-intersection is tangent to C at P, and crosses the real axis perpendicularly; this is the dotted curve in Figure 13 (left), and is made as follows. Since the curve is symmetric with respect to the line $\Re x = 1/2$, on each level line $\Im x = t$, we take two points x_1 and x_2 ($\Re(x_1 + x_2) = 1$), compute the distance between their images $\mathscr{S}(x_1)$ and $\mathscr{S}(x_2)$, and find the points where the two image points coincide.

We substitute the inverse of the Schwarz map:

$$x = \frac{1}{4}\,\frac{(z^n+1)^2}{z^n}, \qquad n = 3$$

into the expression (6.1) of the hyperbolic Schwarz map, and visualize the image surface in the Poincaré ball model explained in Section 5.2. The upper half x-space corresponds to a fan in the z-plane bounded by the lines with argument $0, \pi/3, 2\pi/3$, and the unit circle (see Figure 13 (right).). The image $s(C)$ consists of two curves; the dotted curves in the figures form the pre-image of the self-intersection.

Let Φ denote the hyperbolic Schwarz map in the z-variable:

$$\Phi := \mathscr{S} \circ s^{-1} : Z \ni z \longmapsto H(z) \in \boldsymbol{H}^3.$$

We visualize the image of the hyperbolic Schwarz map when $n = 3$. Figure 14(upper left) is a view of the image of one fan in the z-plane under Φ (equivalently, the image of upper/lower half x-plane under $\mathscr{S}$). The cuspidal edge traverses the figure from left to right and one swallowtail is visible in the center. The upper right figure is the antipode of the left. Figure 14(below) is a view of the image of six fans dividing the unit z-disk. To draw the images of fans with the same accuracy, we make use of the invariance of the function $x(z)$ under the monodromy groups.

The image of the sixty icosahedral triangles (Figure 11) are shown in Figure 15.

7.2. A Fuchsian monodromy group

We study only the case $(k_0, k_1, k_\infty) = (\infty, \infty, \infty)$.

7.2.1. Singular locus. We find the singular locus of the image when $\mu_0 = \mu_1 = \mu_\infty = 0$. We have

$$Q = 1 - x + x^2, \quad R = (-1 + 2\,x)\left(x^2 - x + 2\right).$$

The singularities lie on the image of the curve

$$C : f := 16|x(1-x)|^4 - |Q|^2 = 0.$$

Note that this curve is symmetric with respect to the line $\Re x = 1/2$.

We can prove that, on the upper half x-plane, there is a unique point P – the intersection of the curve C and the line $\Re(x) = 1/2$ – that the image surface has a swallowtail singularity at P, and has cuspidal edges along $\mathscr{S}(C)$ outside $\mathscr{S}(P)$.

We omit the actual computation.

7.2.2. Lambda function. The inverse of the Schwarz map is a modular function known as the lambda function,

$$\lambda : \boldsymbol{H}^2 = \{z \in \boldsymbol{C} \mid \Im z > 0\} \longrightarrow X.$$

The hyperbolic Schwarz map is expressed in terms of its derivatives. The lambda function can be expressed in terms of the theta functions. Since these functions

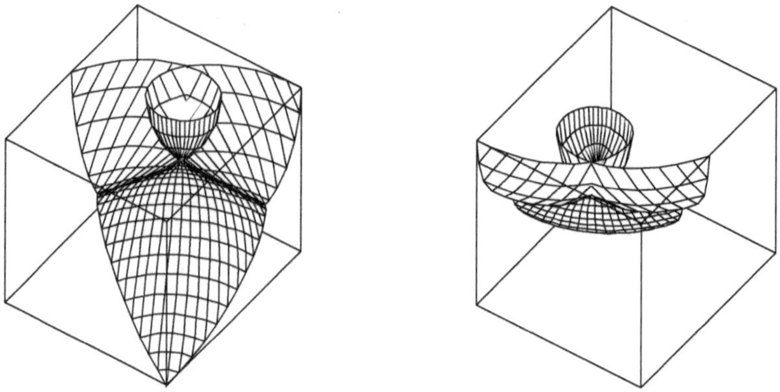

Image of a fan under Φ; a swallowtail can be seen

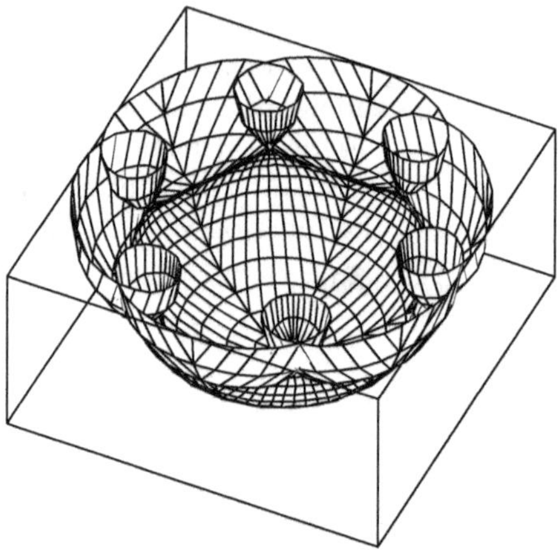

Image of six fans; six swallowtails can be seen

FIGURE 14. Dihedral case

have q-series expansions, which converge very fast, we can accurately compute the image of the hyperbolic Schwarz map.

7.2.3. Visualizing the image surface. The image of the hyperbolic Schwarz map is shown in Figure 17. The first one is the image of the triangle $\{D\}$, and the second one the ten triangles $\{D, A, B, C, AB, BA, AC, CA, BC, CB\}$; these triangles are labeled in Figure 12.

7.3. Parallel family of flat fronts connecting Schwarz and derived Schwarz maps

The parallel front f_t of distance $t \in \boldsymbol{R}$ of the flat front f is defined as

$$\varphi_t(x) = \exp_{\varphi(x)} t\nu(x) = (\cosh t)\varphi(x) + (\sinh t)\nu(x),$$

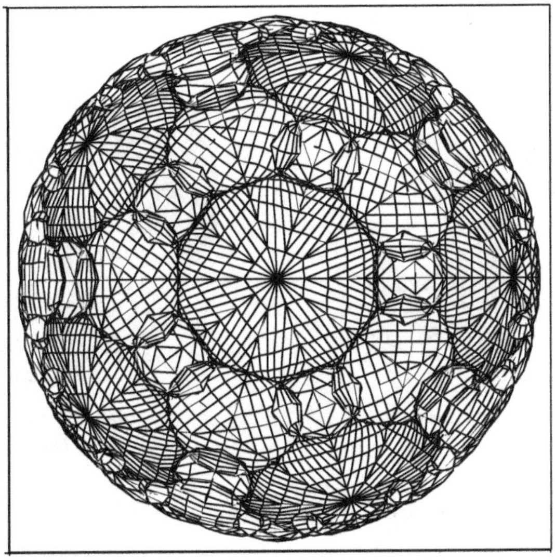

FIGURE 15. Sixty swallowtails can be seen

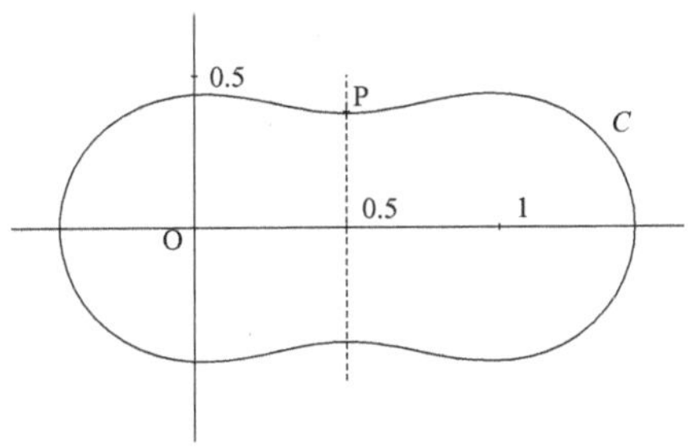

FIGURE 16. The curve $C : |Q| = 4|x(1-x)|^2$, when $(k_0, k_1, k_\infty) = (\infty.\infty, \infty)$

where ν is the unit normal vector field and exp denotes the exponential map of $\boldsymbol{H}^3$. It has another expression as

$$\varphi_t = U_t{}^t\overline{U_t}, \qquad U_t = U \begin{pmatrix} e^{t/2} & 0 \\ 0 & e^{-t/2} \end{pmatrix}.$$

As t tends to plus and minus infinity, the parallel front approaches the hyperbolic Schwarz map and the derived hyperbolic Schwarz map, respectively.

The family of such flat fronts is seen in Figure 18.

7.4. Confluence of swallowtail singularities

The generic singularities of the hyperbolic Schwarz map are cuspidal edges and swallowtail singularities. Their location depends obviously on the parameters (a, b, c).

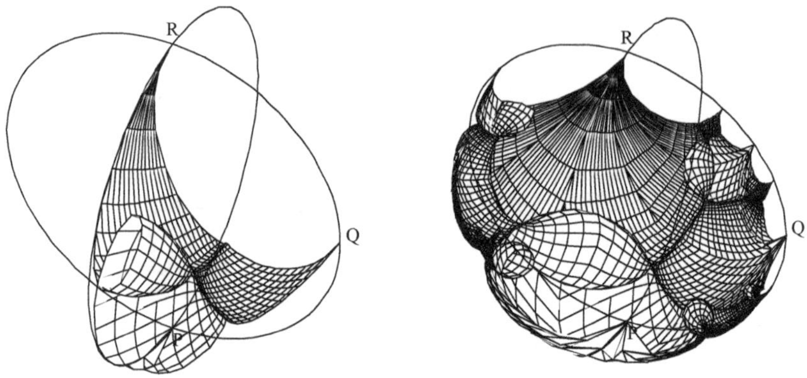

Image of one triangle D

Image of $\{D, A, B, C, AB, BA, AC, CA, BC, CB\}$

FIGURE 17. Images of the hyperbolic Schwarz map when $k_0 = k_1 = k_\infty = \infty$

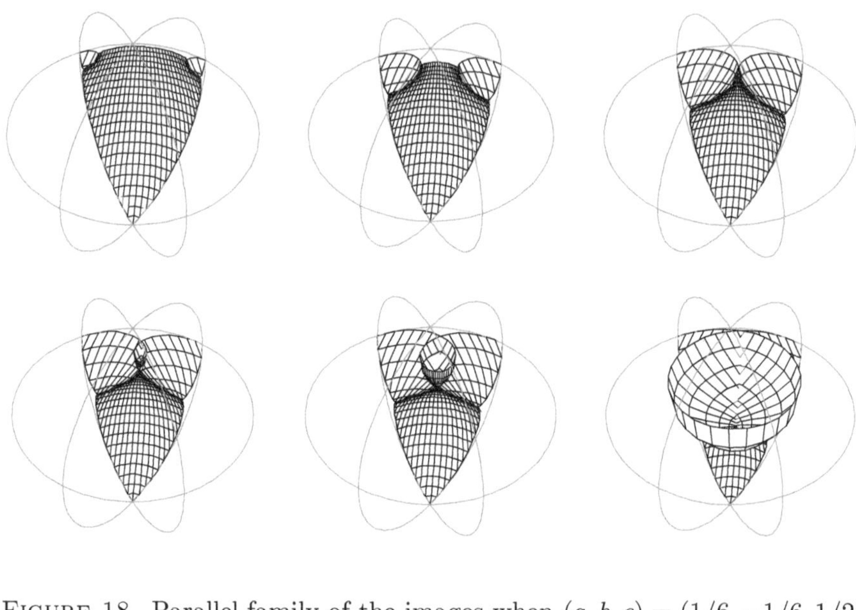

FIGURE 18. Parallel family of the images when $(a, b, c) = (1/6, -1/6, 1/2)$

It is interesting to trace the shapes of the cuspidal edges and to see how creation/elimination of swallowtail singularities occur depending on the parameters.

In Figure 19, for three values of c, we show pictures of the surfaces with parameters $(a, b, c) = (1/2, 1/2, c)$ and the cuspidal edge curves on the surfaces.

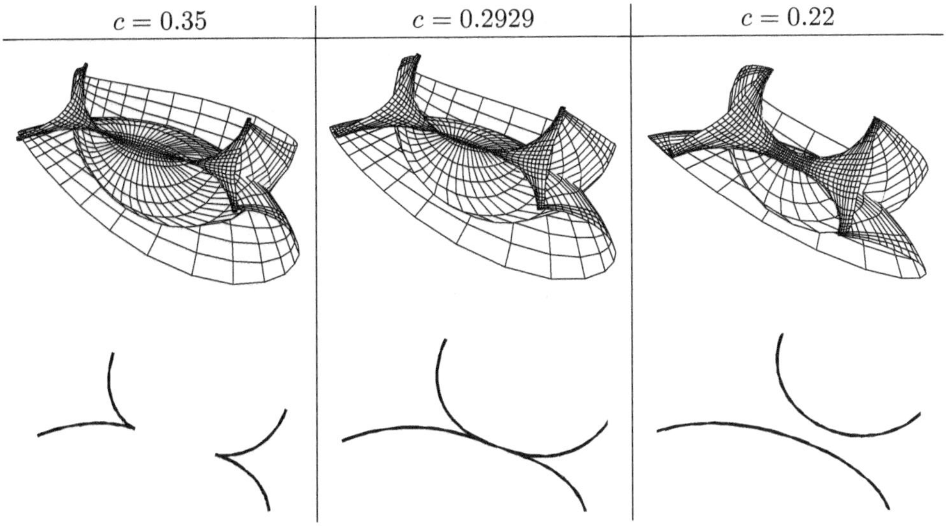

FIGURE 19. Dependence of surfaces and cuspidal edges on parameter c

When $c > 0.2929$, there is a pair of swallowtail singularities on the image surface carried by a pair of cuspidal components of the cuspidal edge curve. When c tends to 0.2929, two singularities come together and kiss, and when $c < 0.2929$, the cuspidal edge curve becomes a pair of nonsingular curves.

We note that the confluence of swallowtail singularities was studied by Arnold [1]; five types $(1, \dots, 5)$ of confluence (bifurcation) as in Figure 20 are known. The confluence shown in Figure 19 corresponds to Type 2. Types 3 and 5 also appear in hyperbolic Schwarz maps, see [7].

References

[1] V. I. Arnold, S. M. Gusein-Zade, and A. N. Varchenko, *Singularities of differentiable maps, Vol.* 1, Monographs in Math. **82**, Birkhäuser, Boston, 1985.

[2] K. Iwasaki, H. Kimura, S. Shimomura and M. Yoshida, *From Gauss to Painlevé – A modern theory of special functions,* Vieweg Verlag, Wiesbaden, 1991.

[3] F. Klein, *Vorlesungen über das Ikosaeder und die auflösung der Gleichungen vom fünften Grade,* Teubner, Leipzig, 1884.

[4] M. Kokubu, M. Umehara, and K. Yamada, "Flat fronts in hyperbolic 3-space", Pacific J. Math. **216**(2004), 149–175.

[5] M. Kokubu, W. Rossman, K. Saji, M. Umehara, and K. Yamada, "Singularities of flat fronts in hyperbolic space", Pacific J. Math. **221**(2005), 303–351.

[6] R. Langevin, G. Levitt and H. Rosenberg, "Classes d'homotopie de surfaces avec rebroussements et queues d'aronde dans $\mathbf{R}^3$", Can. J. Math. **47**(1995), 544–572.

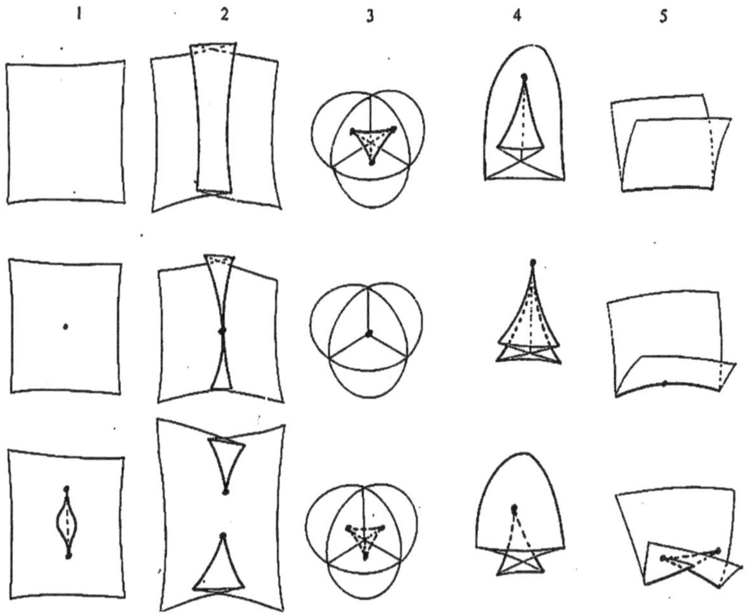

FIGURE 20. Confluence of swallowtail singularities ([6])

[7] M. Noro, T. Sasaki, K. Yamada and M. Yoshida, "Confluence of swallowtail singularites of the hyperbolic Schwarz map defined by the hypergeometric differential equation", Experimental Math. **17**(2008), 191–204.

[8] T. Sasaki, K. Yamada and M. Yoshida, "The hyperbolic Schwarz map for the hypergeometric differential equation", Experimental Math. **17**(2008), 269–282.

[9] T. Sasaki, K. Yamada and M. Yoshida, "Derived Schwarz map of the hypergeometric differential equation and a parallel family of flat fronts", Internat. J. Math. **19**(2008), 847–863.

[10] M. YOSHIDA, *Hypergeometric Functions, My Love,* Vieweg Verlag, 1997.

[11] M. YOSHIDA, "From the power function to the hypergeometric function", Progress in Math., **260**(2007), 407–429.

Takeshi Sasaki
Department of Mathematics, Kobe University, Kobe 657-8501, Japan
e-mail: sasaki@math.kobe-u.ac.jp

Masaaki Yoshida
Faculty of Mathematics, Kyushu University, Fukuoka 810-8560, Japan
e-mail: myoshida@math.kyushu-u.ac.jp

Progress in Mathematics, Vol. 283, 273–281

On the Extendability of Free Multiarrangements

Masahiko Yoshinaga

Abstract. A free multiarrangement of rank k is defined to be extendable if it is obtained from a simple rank $(k+1)$ free arrangement by the natural restriction to a hyperplane (in the sense of Ziegler). Not all free multiarrangements are extendable. We will discuss extendability of free multiarrangements for a special class. We also give two applications. The first is to produce totally non-free arrangements. The second is to give interpolating free arrangements between extended Shi and Catalan arrangements.

Mathematics Subject Classification (2000). Primary 52C35; Secondary 32S22.

Keywords. Multiarrangements, freeness, extendability.

1. Introduction

Let $V = \mathbb{C}^\ell$ be a complex vector space with coordinate $(x_1, \dots, x_\ell)$, let $\mathcal{A} = \{H_1, \dots, H_n\}$ be a central arrangement of hyperplanes. Let us denote by $S = \mathbb{C}[x_1, \dots, x_\ell]$ the polynomial ring and fix $\alpha_i \in V^*$ a defining equation of H_i, i.e., $H_i = \alpha_i^{-1}(0)$. A multiarrangement is a pair $(\mathcal{A}, m)$ of an arrangement $\mathcal{A}$ with a map $m : \mathcal{A} \to \mathbb{Z}_{\geq 0}$, called the multiplicity. We set $Q(\mathcal{A}, m) = \prod_{i=1}^n \alpha_i^{m(H_i)}$ and $|m| = \sum_i m(H_i)$. An arrangement $\mathcal{A}$ can be identified with a multiarrangement with constant multiplicity $m \equiv 1$, which is sometimes called a simple arrangement. Under this notation, the main object in this article is the following module of S-derivations which has contact to each hyperplane in the order m.

Definition 1.1. Let $(\mathcal{A}, m)$ be a multiarrangement, then define

$$D(\mathcal{A}, m) = \{\delta \in \mathrm{Der}_S \,|\delta\alpha_i \in (\alpha_i)^{m(H_i)}, \forall i\}.$$

The module $D(\mathcal{A}, m)$ is obviously a graded S-module. A multiarrangement $(\mathcal{A}, m)$ is said to be free with exponents $(e_1, \dots, e_\ell)$ if $D(\mathcal{A}, m)$ is an S-free module

This work was partially supported by JSPS Postdoctoral Fellowships for Research Abroad.

and there exists a basis $\delta_1, \ldots, \delta_\ell \in D(\mathcal{A}, m)$ such that $\deg \delta_i = e_i$. Note that the degree $\deg \delta$ of a derivation δ is the polynomial degree, that is defined by $\deg(\delta f) = \deg \delta + \deg f - 1$ for a homogeneous polynomial f. An arrangement $\mathcal{A}$ is said to be free if $(\mathcal{A}, 1)$ is free. Here we recall that the freeness is closed under localization. More precisely, let $X \subset V$ be a subset and define $\mathcal{A}_X = \{H \in \mathcal{A} \mid H \supset X\}$. Then the freeness of $(\mathcal{A}, m)$ implies that of $(\mathcal{A}_X, m|_{\mathcal{A}_X})$. (The proof is similar to the simple case, see [10, Theorem 4.37].)

A multiarrangement naturally appears as a restriction of a simple arrangement [22]. Let $\mathcal{A}$ be an arrangement. The arrangement $\mathcal{A}$ determines the restricted arrangement $\mathcal{A}^H = \{H \cap H' \mid H' \in \mathcal{A}, H' \neq H\}$ on $H \in \mathcal{A}$. The restricted arrangement $\mathcal{A}^H$ possesses a natural multiplicity

$$\begin{array}{rccc} m^H: & \mathcal{A}^H & \longrightarrow & \mathbb{Z} \\ & X & \longmapsto & \sharp\{H' \in \mathcal{A} \mid X = H \cap H'\}. \end{array}$$

The freeness of $\mathcal{A}$ and $(\mathcal{A}^H, m^H)$ are connected by the following theorem due to Ziegler.

Theorem 1.2. [22] *If $\mathcal{A}$ is free with exponents $(1, e_2, \ldots, e_\ell)$, then the restriction $(\mathcal{A}^H, m^H)$ is free with exponents $(e_2, \ldots, e_\ell)$.*

Recently freeness of multiarrangements has been extensively studied [3, 4, 12, 15, 16, 17]. The motivation for this article is to ask if a free multiarrangement is obtained as a restriction of a free simple arrangement. Theorem 1.2 leads us to introduce the following notion, which seems to give an important class of free multiarrangements.

Definition 1.3. Let $(\mathcal{A}, m)$ be a free multiarrangement in $\mathbb{K}^\ell$. We say $(\mathcal{A}, m)$ is extendable if it can be obtained as a restriction of a free simple arrangement in $\mathbb{K}^{\ell+1}$.

Example (Non-extendable free multiarrangement). Consider a multiarrangement in $\mathbb{R}^2$,

$$Q(\mathcal{A}, m) = x^3 y^3 (x-y)^1 (x-\alpha y)^1 (x-\beta y)^1,$$

with $\alpha, \beta \neq 0, \pm 1$ and assume α and β are algebraically independent over $\mathbb{Q}$. (Indeed $\alpha\beta \neq 1$ is enough, and in case $\alpha\beta = 1$, it is extendable.) If the slopes α and β are generic, then $(\mathcal{A}, m)$ is free with exponents $(4, 5)$ [17]. So the product of exponents is always ≤ 20. We can prove that it is not extendable. It can be proved as follows (details are left to the reader). The deconing $\overline{\mathcal{A}}$ ([10]) with respect to the hyperplane at infinity of an extension of $(\mathcal{A}, m)$ is an affine line arrangement $\mathbb{R}^2$ having the following defining equations:

$$\begin{aligned} x &= a_1, a_2, a_3, \\ y &= b_1, b_2, b_3, \\ x - y &= c, \\ x - \alpha y &= d, \\ x - \beta y &= e, \end{aligned}$$

where $a_i, b_i, c, d, e \in \mathbb{R}$. The characteristic polynomial $\chi(\overline{\mathcal{A}}, t)$ is of the form $\chi(\overline{\mathcal{A}}, t) = t^2 - 9t + p$, and we can prove that $p > 20$. Thus $\chi(\overline{\mathcal{A}}, t)$ is not factored. It follows from Terao's factorization theorem ([14]) that any extension of $\mathcal{A}$ is not free.

Thus a free multiarrangement $(\mathcal{A}, m)$ is not necessarily extendable in general. In the next section, we focus on some special kinds of multiarrangements.

2. Extendability of locally A_2 arrangements

Definition 2.1. An arrangement $\mathcal{A} = \{H_1, \ldots, H_n\}$ is said to be *locally* A_2 if $|\mathcal{A}_X| \leq 3$ is satisfied for any codimension 2 intersection X. A system of defining equations $\{\alpha_1, \ldots, \alpha_n\}$ of a locally A_2 arrangement $\mathcal{A}$ is called a *positive system* if the following condition is satisfied: Suppose X is a codimension 2 intersection with $|\mathcal{A}_X| = 3$. Setting $\mathcal{A}_X = \{H_i, H_j, H_k\}$. Then one of $\alpha_i = \alpha_j + \alpha_k$, $\alpha_j = \alpha_i + \alpha_k$ or $\alpha_k = \alpha_i + \alpha_j$ holds.

Example. The following are examples of locally A_2 arrangements with positive systems.

(1) Generic in codimension 3. (Equivalently, $|\mathcal{A}_X| = 2$ for any codimension 2 intersection X.) In this case any system of defining equations is a positive system.
(2) Coxeter arrangement of type ADE. In this case, a positive root system is corresponding to a positive system of defining equations.
(3) Subarrangements or direct products of locally A_2 arrangements with positive systems possess the same property. Especially, this class is closed under localization.
(4) (Shi arrangement of type A_2) $Q = xyz(x+y)(x-z)(y-z)(x+y-z)$.

Remark 2.2. Note that a locally A_2 arrangement does not necessarily have a positive system (e. g., $Q = xyz(x+y)(x-z)(y-z)(x+y-2z)$).

We will discuss the extendability for multiarrangements of this class. More precisely, we consider the following concrete extension $E(\mathcal{A}, m)$ of $(\mathcal{A}, m)$ for given locally A_2 arrangement $\mathcal{A}$ with a positive system $(\alpha_H)_H$. Let $(x_1, \ldots, x_\ell, z) \in \mathbb{C}^\ell \times \mathbb{C}$ be a coordinate system of $V \times \mathbb{C}$ and define

$$E(\mathcal{A}, m) = \{z = 0\} \cup \left\{ \alpha_H = kz \,\middle|\, k \in \mathbb{Z}, -\frac{m(H)-1}{2} \leq k \leq \frac{m(H)}{2} \right\}.$$

Then, if we write $H_0 = \{z = 0\}$, it is obvious that $(E(\mathcal{A}, m)^{H_0}, m^{H_0}) = (\mathcal{A}, m)$. Let us define the deconing of $E(\mathcal{A}, m)$ as follows:

$$\mathbf{d}E(\mathcal{A}, m) = \left\{ \alpha_H = k \,\middle|\, k \in \mathbb{Z}, -\frac{m(H)-1}{2} \leq k \leq \frac{m(H)}{2} \right\}.$$

Note that $\mathbf{d}E(\mathcal{A}, m)$ is an affine arrangement in V.

Remark 2.3. The above definition is motivated by that of the extended Catalan and Shi arrangements [8]. Indeed, let $\mathcal{A}$ be a Coxeter arrangement of type ADE. Choose the positive root system as the positive system proposed above. For a given positive integer $k \in \mathbb{Z}_{>0}$, consider constant multiplicities $m = 2k$ and $m = 2k+1$. Then $E(\mathcal{A}, 2k+1)$ is called the extended Catalan arrangement and $E(\mathcal{A}, 2k)$ is called the extended Shi arrangement, which are known to be free [20].

Theorem 2.4. *Let $\mathcal{A}$ be a locally A_2 arrangement with a positive system in $V = \mathbb{C}^\ell$. We fix a positive system $\Phi^+ = \{\alpha_H \mid H \in \mathcal{A}\} \subset V^*$ of defining equations. Let $m : \mathcal{A} \to \mathbb{Z}_{\geq 0}$ be a multiplicity. We assume the following condition:*

(*) *Let $\mathcal{A}_X = \{H_i, H_j, H_k\}$ be a codimension 2 localization with $\alpha_i = \alpha_j + \alpha_k$. If $m(H_i)$ is odd, then at least one of $m(H_j)$ or $m(H_k)$ is odd.*

Then $(\mathcal{A}, m)$ is free, if and only if it is extendable. Indeed, $E(\mathcal{A}, m)$ is a free arrangement.

We will give the proof in the next section. Here we notice an immediate corollary.

Corollary 2.5. *Let $\mathcal{A}$ be a locally A_2 arrangement with a positive system. Suppose that the multiplicity m satisfies either $m(H)$ is odd $\forall H \in \mathcal{A}$ or $m(H)$ is even $\forall H \in \mathcal{A}$. If the multiarrangement $(\mathcal{A}, m)$ is free, then it is extendable.*

Remark 2.6. The condition (*) in Theorem 2.4 is related to the following phenomenon. Consider a multiarrangement $x^2y^2(x+y)^1$. Then (deconing of) our extension $\mathbf{d}E(x^2y^2(x+y)^1)$ is defined by

$$x(x-1)y(y-1)(x+y),$$

which is not free. However another extension

$$x(x-1)y(y-1)(x+y-1)$$

is free. This shows that even $E(\mathcal{A}, m)$ is not free, $(\mathcal{A}, m)$ might have another free extension. The author does not know if the extendability can be proved without assuming condition (*). We look at a little more complicated example.

Example. Let us consider a multiarrangement $x^4y^4z^4(x+y)^5(y+z)^5(x+y+z)^4$. It is known to be free with exponents $(8, 9, 9)$ (see [5, 18] or Proposition 5.2 below). The extension $E(x^4y^4z^4(x+y)^5(y+z)^5(x+y+z)^4)$ is defined by

$$\begin{aligned} x, y, z &= kw \ (k = -1, 0, 1, 2), \\ x+y, y+z &= kw \ (k = -2, -1, 0, 1, 2), \\ x+y+z &= kw \ (k = -1, 0, 1, 2), \\ w &= 0, \end{aligned}$$

which is not free (look at the localization at $x = y = w = 0$ and use Lemma 3.1 (3-ii)). However another extension

$$\begin{aligned} x, y, z &= kw \ (k = -1, 0, 1, 2), \\ x + y, y + z &= kw \ (k = -1, 0, 1, 2, 3), \\ x + y + z &= kw \ (k = 0, 1, 2, 3), \\ w &= 0, \end{aligned}$$

is free.

We can check the following for $\ell = 3$.

Question 2.7. Suppose $\mathcal{A}$ is of type A_ℓ and $(\mathcal{A}, m)$ is free. Then is $(\mathcal{A}, m)$ always extendable?

3. Proof

Proof of Theorem 2.4 is done by induction on the rank ℓ. If $\ell = 2$, then $\mathcal{A}$ is either $|\mathcal{A}| = 2$ or type A_2. Suppose $|\mathcal{A}| = 2$. Then $E(\mathcal{A}, m)$ is obviously free. Suppose $(\mathcal{A}, m)$ is defined by $x^a y^b (x + y)^c$. The next lemma is elementary.

Lemma 3.1. *Assume $a \leq b$. Set $k = a + b + c$ and $\mathcal{E} = E(x^a y^b (x + y)^c)$.*

(1) *If $c < b - a + 1$, then $\chi(\mathcal{E}, t) = (t - 1)(t - b)(t - a - c)$.*
(2) *If $c \geq a + b + 1$, then $\chi(\mathcal{E}, t) = (t - 1)(t - a - b)(t - c)$.*
(3) *$b - a \leq c - 1 < a + b$,*
 (i) *$(a, b, c) \neq$ (even, even, odd), then $\chi(\mathcal{E}, t) = (t - 1)(t - \lfloor k/2 \rfloor)(t - \lceil k/2 \rceil)$.*
 (ii) *$(a, b, c) =$ (even, even, odd), then $\chi(\mathcal{E}, t) = (t - 1)\left((t - \frac{k}{2})^2 + \frac{3}{4}\right)$.*

The next result is due to Wakamiko.

Proposition 3.2. [16] *Let $(\mathcal{A}, m) = x^a y^b (x + y)^c$. Assume $a \leq b$ and set $k = a + b + c$ as above. Since it is rank 2, $(\mathcal{A}, m)$ is always free. The exponents are given as follows:*

(1) *If $c < b - a + 1$, then $\exp(\mathcal{A}, m) = (b, a + c)$.*
(2) *If $c \geq a + b + 1$, then $\exp(\mathcal{A}, m) = (c, a + b)$.*
(3) *$b - a \leq c - 1 < a + b$, then $\exp(\mathcal{A}, m) = (\lfloor k/2 \rfloor, \lceil k/2 \rceil)$.*

In [21], a characterization of freeness for rank 3 arrangements is given. It can be stated as follows.

Proposition 3.3. *For $\ell = 2$, $E(\mathcal{A}, m)$ is free with exponents $(1, d_1, d_2)$ if and only if*

- *$\chi(E(\mathcal{A}, m), t) = (t - 1)(t - d_1)(t - d_2)$ and*
- *$\exp(\mathcal{A}, m) = (d_1, d_2)$.*

Combining these results, we can prove Theorem 2.4 for $\ell = 2$. (Note that the condition (*) in the theorem is the same as saying that Lemma 3.1 (3) (ii) does not occur.)

We now consider the case $\ell \geq 3$. Let us first recall the following result.

Proposition 3.4. [20] *$E(\mathcal{A}, m)$ is free with exponents $(1, e_1, \ldots, e_\ell)$ if and only if $(\mathcal{A}, m)$ is free with exponents $(e_1, \ldots, e_\ell)$ and $E(\mathcal{A}, m)_X$ is free for any positive dimensional intersection $X \subset V$.*

It is easily seen that $E(\mathcal{A}, m)_X = E(\mathcal{A}_X, m|_{\mathcal{A}_X})$. Since the localization $(\mathcal{A}_X, m|_{\mathcal{A}_X})$ of a free multiarrangement $(\mathcal{A}, m)$ is free with rank at most $\ell - 1$, it follows from the inductive hypothesis that $E(\mathcal{A}, m)_X$ is free. Hence Proposition 3.4 shows that $E(\mathcal{A}, m)$ is free. □

4. Totally non-free arrangements

In a recent paper [4] Abe, Terao and Wakefield observed several phenomena concerning multiplicities and freeness of a multiarrangement $(\mathcal{A}, m)$. In particular they proved that generic four-planes $(\mathcal{A}, m)$ defined by $x_1^{m_1} x_2^{m_2} x_3^{m_3} (x_1 + x_2 + x_3)^{m_4}$ will never be free for any positive multiplicity $m : \mathcal{A} \to \mathbb{Z}_{>0}$. Such an arrangement $\mathcal{A}$ is called totally non-free. As an application of extendability techniques, we give a straightforward proof of totally non-freeness for generic arrangements.

Proposition 4.1. *Suppose $\ell = \dim V \geq 3$ and $\mathcal{A}$ is a generic arrangement with $|\mathcal{A}| > \ell$. Let $m : \mathcal{A} \to \mathbb{Z}_{>0}$. Then $(\mathcal{A}, m)$ is not free.*

Proof. Fix a defining equations α_H for each H. As is already noticed, it is a positive system. Since $(\mathcal{A}, m)$ is locally Boolean, $E(\mathcal{A}, m)_X = E(\mathcal{A}_X, m|_{\mathcal{A}_X})$ is free for any nonzero subspace $X \subset V \times \{0\}$. Hence if $(\mathcal{A}, m)$ is free, then Proposition 3.4 shows that $E(\mathcal{A}, m)$ is also free. However, let us consider the restriction to the subspace $X = \{0\} \times \mathbb{C} \subset V \times \mathbb{C}$. Then the localization $E(\mathcal{A}, m)_X$ is isomorphic to $\mathcal{A}$ which is not free. This is a contradiction. □

5. Free interpolations between extended Shi and Catalan arrangements

Let $\mathcal{A}$ be a crystallographic Coxeter arrangement with a fixed positive system Φ^+ of roots. As is already mentioned, $E(\mathcal{A}, 2k+1)$ and $E(\mathcal{A}, 2k)$ are free for any $k \in \mathbb{Z}_{>0}$. Obviously these two families of arrangements are related to each other as

$$\cdots \subset E(\mathcal{A}, 2k-1) \subset E(\mathcal{A}, 2k) \subset E(\mathcal{A}, 2k+1) \subset \cdots .$$

In [19], it is observed that there exist many free arrangements $\mathcal{B}$ such that $E(\mathcal{A}, 2k) \subset \mathcal{B} \subset E(\mathcal{A}, 2k+1)$. The purpose of this section is to give a complete description of free arrangements interpolating these families for type ADE.

Let $m : \mathcal{A} \to \{0,1\}$ be a $\{0,1\}$-valued multiplicity. Any interpolating arrangement can be described as $E(\mathcal{A}, 2k \pm m)$ for some m. We will describe free interpolations by using $\{0,1\}$-valued multiplicity m. Our main result in this section is the following.

Theorem 5.1. *Let $\mathcal{A}$ be an irreducible Coxeter arrangement of type ADE with the Coxeter number h. Fix Φ^+ a positive root system. Let k be a positive integer. Then the following conditions are equivalent.*

(1) $m : \mathcal{A} \to \{0,1\}$ *satisfies the following condition.*
  (1-i) $m^{-1}(1) \subset \mathcal{A}$ *is a free subarrangement with exponents* $(e_1, \dots, e_\ell)$.
  (1-ii) *if* $\alpha_1 = \alpha_2 + \alpha_3 (\alpha_i \in \Phi^+)$ *and* $m(H_1) = 1$, *then at least* $m(H_2) = 1$ *or* $m(H_3) = 1$.

(2) $E(\mathcal{A}, 2k + m)$ *is free with exponents* $(1, kh + e_1, \dots, kh + e_\ell)$.

(3) $E(\mathcal{A}, 2k - m)$ *is free with exponents* $(1, kh - e_1, \dots, kh - e_\ell)$.

Before giving a proof of Theorem 5.1, let us recall a result from [5].

Proposition 5.2. [5, Corollary 12] *Let $\mathcal{A}$ be the Coxeter arrangement with the Coxeter number h, and $m : \mathcal{A} \to \{0,1\}$ be a multiplicity. Let $k \in \mathbb{Z}_{>0}$. Then the following conditions are equivalent.*

- $(\mathcal{A}, m)$ *is free with exponents* $(e_1, \dots, e_\ell)$.
- $(\mathcal{A}, 2k + m)$ *is free with exponents* $(kh + e_1, \dots, kh + e_\ell)$.
- $(\mathcal{A}, 2k - m)$ *is free. with exponents* $(kh - e_1, \dots, kh - e_\ell)$.

First we prove (1)⇒(2). Suppose m satisfies (1-ii). Then the multiplicity $2k+m$ satisfies the condition (*) in Theorem 2.4. Thus the extension $E(\mathcal{A}, 2k+m)$ is free if and only if the multiarrangement $(\mathcal{A}, 2k + m)$ is free. But this is done by the assumption (1-i) and Proposition 5.2.

The implication (1)⇒(3) is similar.

Finally let us prove (2)⇒(1). Suppose $E(\mathcal{A}, 2k+m)$ is free. Then by restricting to H_0, we have (by Theorem 1.2), the multiarrangement $(\mathcal{A}, 2k+m)$ is free. Again from Proposition 5.2, we have $(\mathcal{A}, m)$ is free, in other words, $m^{-1}(1) \subset \mathcal{A}$ is a free subarrangement. Thus we have (1-i). To prove (1-ii), suppose that there exists H_1 such that $\alpha_1 = \alpha_2 + \alpha_3$ and $m(H_1) = 1, m(H_2) = m(H_3) = 0$. Then set $X := H_1 \cap H_2 \cap H_3$, which is a codimension 2 subspace. From Lemma 3.1 (3-ii), the localization $E(\mathcal{A}, 2k + m)_X$ is not free. It is a contradiction. Thus (1-ii) is satisfied. □

Using Terao's factorization theorem, we obtain the following corollary.

Corollary 5.3. *Let $\mathcal{A}$ be a Coxeter arrangement with the Coxeter number h and $m : \mathcal{A} \to \{0,1\}$ be a multiplicity satisfying the condition* (1) *of Theorem* 5.1. *Then*

$$\chi(\mathbf{d}E(\mathcal{A}, 2k \pm m), t) = \prod_{i=1}^{\ell} (t - kh \mp e_i).$$

The above formula implies

$$\chi(\mathbf{d}E(\mathcal{A}, 2k - m), t) = (-1)^{\ell}\chi(\mathbf{d}E(\mathcal{A}, 2k + m), 2kh - t). \tag{5.1}$$

We should note that the formula (5.1) is very similar to the "functional equation" discovered by Postnikov and Stanley [11]. It might be worth asking whether the formula (5.1) holds for any crystallographic Coxeter arrangement $\mathcal{A}$ and any multiplicity $m : \mathcal{A} \to \{0, 1\}$.

Remark 5.4. Recently Abe, Nuida and Numata obtained more general results for type A_ℓ arrangements [2, 9]. Their results suggest that (5.1) holds even for wider class of multiplicities, namely, $m : \mathcal{A} \to \{-1, 0, 1\}$.

References

[1] T. Abe, *The stability of the family of A_2-type arrangements.* J. Math. Kyoto Univ. **46** (2006), no. 3, 617–639.

[2] T. Abe, K. Nuida, Y. Numata, *Signed-eliminable graphs and free multiplicities on the braid arrangement.* J. Lond. Math. Soc. (2) **80** (2009), no. 1, 121–134.

[3] T. Abe, H. Terao, M. Wakefield, *The characteristic polynomial of a multiarrangement.* Adv. in Math. **215** (2007), 825–838.

[4] T. Abe, H. Terao, M. Wakefield, *The Euler multiplicity and the addition-deletion theorems for multiarrangements.* to appear in J. London Math. Soc.

[5] T. Abe, M. Yoshinaga, *Coxeter multiarrangements with quasi-constant multiplicities.* J. Algebra **322** (2009), no. 8, 2839–2847.

[6] C. A. Athanasiadis, *Deformations of Coxeter hyperplane arrangements and their characteristic polynomials.* Arrangements—Tokyo 1998, 1–26, Adv. Stud. Pure Math., 27, Kinokuniya, Tokyo, 2000.

[7] C. A. Athanasiadis, *Generalized Catalan numbers, Weyl groups and arrangements of hyperplanes.* Bull. London Math. Soc. **36** (2004), no. 3, 294–302.

[8] P. H. Edelman, V. Reiner, *Free arrangements and rhombic tilings.* Discrete Comput. Geom. **15** (1996), no. 3, 307–340.

[9] K. Nuida, *A Characterization of Edge-Bicolored Graphs with Generalized Perfect Elimination Orderings.* arXiv:0712.4118

[10] P. Orlik and H. Terao, *Arrangements of hyperplanes.* Grundlehren der Mathematischen Wissenschaften, 300. Springer-Verlag, Berlin, 1992. xviii+325 pp.

[11] A. Postnikov, R. Stanley, *Deformations of Coxeter hyperplane arrangements.* J. Combin. Theory Ser. A **91** (2000), no. 1-2, 544–597.

[12] L. Solomon, H. Terao, *The double Coxeter arrangement.* Comm. Math. Helv. **73** (1998) 237–258.

[13] H. Terao, *Arrangements of hyperplanes and their freeness.* I, II. J. Fac. Sci. Univ. Tokyo Sect. IA Math. **27** (1980), no. 2, 293–320.

[14] H. Terao, *Generalized exponents of a free arrangement of hyperplanes and Shepherd-Todd-Brieskorn formula.* Invent. Math. **63** (1981), no. 1, 159–179.

[15] H. Terao, *Multiderivations of Coxeter arrangements.* Invent. Math. **148** (2002), no. 3, 659–674.

[16] A. Wakamiko, *On the Exponents of 2-Multiarrangements.* Tokyo J. Math. **30** (2007), 99-116.

[17] M. Wakefield, S. Yuzvinsky, *Derivations of an effective divisor on the complex projective line.* Trans. A. M. S. **359** (2007), 4389–4403.

[18] M. Yoshinaga, *The primitive derivation and freeness of multi-Coxeter arrangements.* Proc. Japan Acad., **78**, Ser. A (2002) 116–119.

[19] M. Yoshinaga, *Some characterizations of freeness of hyperplane arrangement.* math.CO/0306228

[20] M. Yoshinaga, *Characterization of a free arrangement and conjecture of Edelman and Reiner.* Invent. Math. **157** (2004), no. 2, 449–454.

[21] M. Yoshinaga, *On the freeness of 3-arrangements.* Bull. London Math. Soc. **37** (2005), no. 1, 126–134.

[22] G. Ziegler, *Multiarrangements of hyperplanes and their freeness.* Singularities (Iowa City, IA, 1986), 345–359, Contemp. Math., 90, Amer. Math. Soc., Providence, RI, 1989.

Acknowledgment

The author would like to thank Takuro Abe and Max Wakefield for useful conversations and comments. He also thanks the referee for pointing out a crucial mistake in the draft and giving many suggestions.

Masahiko Yoshinaga
Department of Mathematics
Faculty of Science, Kobe University
Kobe 657-8501
Japan
e-mail: myoshina@math.kobe-u.ac.jp

Progress in Mathematics, Vol. 283, 283–319

Problem Session

Edited by
Ayşe Altıntaş and Celal Cem Sarıoğlu

Abstract. This article contains the open problems posed at the special session of the CIMPA summer school "Arrangements, Local Systems and Singularities" held at Galatasaray University, İstanbul, 2007.

1. Introduction

The problem session of the conference "Arrangements, Local Systems and Singularities" was held on June 22, 2007 in İstanbul. Here, we present some open problems which were discussed during the session. In order to provide an insight into the relevant subjects, we also give details of historical developments together with some background information where we have thought it to be useful.

We would like to thank Graham Denham, Michael Falk, Daniel Matei, David Mond, Lê Dũng Tráng, A. Muhammed Uludağ, Masahiko Yoshinaga, Fouad ElZein and Sergey Yuzvinky for suggesting these problems and for their help through numerous comments.

2. Problems

2.1. $\mathcal{A}_e$-versal unfoldings of differentiable map germs

Problem 1 (D. Mond). *Let $f : (\mathbb{C}^n, 0) \to (\mathbb{C}^{n+1}, 0)$ be a map germ of finite $\mathcal{A}_e$-codimension and suppose that $(n, n+1)$ are in Mather's range of nice dimensions ([89]), i.e., $n < 7$. Let $\mu_I(f)$ be the rank of the n-th homology group of the image of a stable perturbation of f. Is it true that $\mathcal{A}_e - codim(f) \leq \mu_I(f)$?*

In the theory of isolated hypersurface singularities and more generally of complete intersection singularities, we meet several important results regarding the rank μ of the middle-dimensional homology of the Milnor fibre, the dimension τ of the base of a miniversal deformation of the singularity, and the algebraic codimensions of germs defining those singularities. By the results of Milnor [93], in the case of hypersurface singularities, and Hamm [62], for isolated complete intersection singularities, the Milnor fibre of an n-dimensional isolated complete intersection singularity has the homotopy type of a wedge of n-spheres. The number of such spheres is the Milnor number μ. Greuel [58] proved that $\mu = \tau$ for weighted homogeneous isolated complete intersection singularities, and in [81], Looijenga and Steenbrink showed that the inequality $\mu \geq \tau$ holds for arbitrary isolated complete intersection singularities.

Later, in [86] Marar and Mond gave remarkable results regarding map germs $f : (\mathbb{C}^n, 0) \to (\mathbb{C}^p, 0)$ with $n < p$, including the fact that the above examples are not the only cases in which one can relate algebraic invariants to topological ones. They showed that for a map germ $f : (\mathbb{C}^n, 0) \to (\mathbb{C}^p, 0)$ of finite $\mathcal{A}_e$-codimension (or equivalently, admiting an $\mathcal{A}_e$-versal unfolding) and corank 1, i.e., with $\dim_{\mathbb{C}}(\mathrm{Ker}(df_0(0))) = 1$, the image X_t of a stable perturbation f_t of f plays a similar role to the Milnor fibre of an isolated singularity (see notes below for the definitions of an $\mathcal{A}_e$-versal unfolding and a stable perturbation). Moreover, substantial information on the topology of X_t can be gathered from the multiple point schemes D^k of the map germ f. In the case of $(n,p) = (2,3)$ and $(3,4)$, they obtained a formulae for the Euler characteristic $\chi(X_t)$ of the image X_t. For all map germs listed in [97] (which are all weighted homogeneous), they verified the following equality:

$$\mathcal{A}_e - \mathrm{codim}(f) = \chi(X_t) - 1.$$

The results in that paper were improved by those of [55] where Goryunov and Mond described the topology of X_t by constructing an alternating semi-simplicial resolution **Alt** $\mathbb{Z}_{D^\cdot}$ of the constant sheaf $\mathbb{Z}_{X_t}$ which relates the topology of X_t to that of multiple point spaces of X_t. For the corank 1 case, they reproved and extended Marar's formulae ([84]) for calculating the Euler characteristic of the image X_t. Additionally, in the case where f is weighted homogeneous, they studied its canonical mixed Hodge structures.

In a more general context, Pellikaan and de Jong (unpublished), then de Jong and van Straten [65] and later Mond [96] proved that for map germs $f : (\mathbb{C}^2, 0) \to (\mathbb{C}^3, 0)$ of finite $\mathcal{A}_e$-codimension,

$$\mathcal{A}_e - \mathrm{codim}(f) \leq \chi(X_t) - 1, \text{ with equality if } f \text{ is weighted homogeneous.} \quad (2.1)$$

In [96], Mond also proved another result regarding map germs from $\mathbb{C}^n$ to $\mathbb{C}^{n+1}$: If $f : (\mathbb{C}^n, 0) \to (\mathbb{C}^{n+1}, 0)$ is of finite $\mathcal{A}_e$-codimension and $(n, n+1)$ are nice dimensions, then the image X_t of a stable perturbation of f has the homotopy type of a wedge of n-spheres. The number of spheres in the wedge is an $\mathcal{A}$-invariant of f. Mond called this number the image Milnor number. It is usually denoted by $\mu_I(f)$. Here, $(n, n+1)$ being nice dimensions assures that the map admits a stable perturbation ([89]). We note that Mond's result on the homotopy type of X_t follows from Lê's theorem in [73]: Lê proved that a general hyperplane section of a complex analytic germ of a complete intersection variety S has the homotopy type of a wedge of spheres in real dimension $\dim S - 1$. In our case, X_t is a general hyperplane section of the image of a stable unfolding of f whence one gets the required homotopy type.

As $\mu_I(f)$ is the rank of the n-th homology of the image X_t, (2.1) can be restated as follows:

$$\mathcal{A}_e - \mathrm{codim}(f) \leq \mu_I(f), \text{ with equality if } f \text{ is weighted homogeneous.} \tag{2.2}$$

In order to explain the notion of $\mathcal{A}_e$-codimension, here we include a brief introduction to the theory of differentiable mappings. We note that most of the theory is built the same way for the real case. For general ideas and notation in singularity theory, we suggest Wall's survey article [154] and for a clear exposition of the local theory of singularities and unfoldings Martinet's book [87].

Let $E_{n,p}$ denote the space of germs at $0 \in \mathbb{C}^n$ of holomorphic mappings from $\mathbb{C}^n$ to $\mathbb{C}^p$. Let $x = (x_1, \ldots, x_n)$ and $y = (y_1, \ldots, y_p)$ be the chosen coordinate systems on $\mathbb{C}^n$ and $\mathbb{C}^p$, respectively. The space E_n of $\mathbb{C}$-valued holomorphic germs at $0 \in \mathbb{C}^n$ has a ring structure with the maximal ideal m_n generated by $x_1, \ldots, x_n$. Similarly, the maximal ideal in E_p is $m_p := < y_1, \ldots, y_p >$. We can interpret $E_{n,p}$ as an E_n module via the identification $E_{n,p} = \bigoplus_{i=1}^p E_n$.

Let us consider the group $\mathrm{Diff}(\mathbb{C}^n, 0)$ of local diffeomorphisms at the origin of $\mathbb{C}^n$. If $\phi \in \mathrm{Diff}(\mathbb{C}^n, 0)$ then, by definition,

(i) $\phi(0) = 0$, and
(ii) ϕ defines a diffeomorphism between an open subset U' of a neighbourhood U of $0 \in \mathbb{C}^n$ and $\phi(U')$. Equivalently, $d\phi(0)$ is invertible.

Similar properties hold for a diffeomorphism in $\mathrm{Diff}(\mathbb{C}^p, 0)$.

An action of the group $\mathcal{A} := \mathrm{Diff}(\mathbb{C}^n, 0) \times \mathrm{Diff}(\mathbb{C}^p, 0)$ on $E_{n,p}$ is defined by

$$\begin{array}{rcl} \mathrm{Diff}(\mathbb{C}^n, 0) \times \mathrm{Diff}(\mathbb{C}^p, 0) \times E_{n,p} & \to & E_{n,p} \\ (\phi, \psi, f) & \mapsto & \psi \circ f \circ \phi^{-1}. \end{array} \tag{2.3}$$

This action is in fact an equivalence relation in the usual sense and we say that a map germ $g \in E_{n,p}$ is *$\mathcal{A}$-equivalent* to $f \in E_{n,p}$ if there exist germs of diffeomorphisms $\phi \in \mathrm{Diff}(\mathbb{C}^n, 0)$ and $\psi \in \mathrm{Diff}(\mathbb{C}^p, 0)$ so that $g = \psi \circ f \circ \phi^{-1}$. Since we are allowed to apply coordinate changes on the target space, we can consider only map germs which map $0 \in \mathbb{C}^n$ to $0 \in \mathbb{C}^p$. Let us denote the space of such map germs by $E^0_{n,p}$.

A *d-parameter unfolding* F of $f \in E^0_{n,p}$ is a differentiable map germ

$$\begin{array}{rcl} F : (\mathbb{C}^n \times \mathbb{C}^d, 0) & \to & (\mathbb{C}^p \times \mathbb{C}^d, 0) \\ (x, t) & \mapsto & (f(x,t), t) = (f_t(x), t) \end{array} \tag{2.4}$$

with $f_0(x) = f(x)$ and $t = (t_1, \ldots, t_d)$. Two unfoldings F and G (with the same number of parameters) of f are *isomorphic* if there exist germs of diffeomorphisms $\phi \in \mathrm{Diff}(\mathbb{C}^n \times \mathbb{C}^d, 0)$ and $\psi \in \mathrm{Diff}(\mathbb{C}^p \times \mathbb{C}^d, 0)$ such that $G = \psi \circ F \circ \phi^{-1}$. Note that in this case ϕ and ψ are given by

$$\phi(x,t) = (\tilde{\phi}_t(x), t) \text{ and } \psi(y,t) = (\tilde{\psi}_t(y), t) \tag{2.5}$$

with $\tilde{\phi}_t \in \mathrm{Diff}(\mathbb{C}^n,0)$ and $\tilde{\psi}_t \in \mathrm{Diff}(\mathbb{C}^p,0)$. An unfolding is called *trivial* if it is isomorphic to the constant unfolding $(x,t) \mapsto (f(x),t)$. The following lemma states an infinitesimal criterion for a 1-parameter unfolding to be trivial.

Lemma 2.1 ([87], Lemma 4.1, Ch. XIII). *Let $f_0 : (\mathbb{R}^n,0) \to (\mathbb{R}^p,0)$ be a map germ and $F : (\mathbb{R}^n \times \mathbb{R},0) \to (\mathbb{R}^p \times \mathbb{R},0)$ be a 1-parameter unfolding of f_0. The unfolding F is trivial if and only if there exist two germs of C^∞ vector fields:*

$$X = \frac{\partial}{\partial t} + \sum_{i=1}^{n} X_i(x,t)\frac{\partial}{\partial x_i} \quad on\ (\mathbb{R}^n \times \mathbb{R},0),$$

$$Y = \frac{\partial}{\partial t} + \sum_{i=1}^{p} Y_i(y,t)\frac{\partial}{\partial y_i} \quad on\ (\mathbb{R}^p \times \mathbb{R},0) \tag{2.6}$$

such that $dF \cdot X = Y \circ F$.

Even though this lemma is given for the real case, it can be easily shown that it is also true for complex map germs. If $F : (\mathbb{C}^n \times \mathbb{C},0) \to (\mathbb{C}^p \times \mathbb{C},0)$ is a 1-parameter unfolding of $f \in E^0_{n,p}$ given by $F(x,t) = (f_t(x),t)$ with $t \in \mathbb{C}$, then there exist two germs of vector fields $X = \frac{\partial}{\partial t} + \sum_{i=1}^n X_i(x,t)\frac{\partial}{\partial x_i}$ on $(\mathbb{C}^n \times \mathbb{C},0)$ and $Y = \frac{\partial}{\partial t} + \sum_{i=1}^p Y_i(y,t)\frac{\partial}{\partial y_i}$ on $(\mathbb{C}^p \times \mathbb{C},0)$ satisfying $dF \cdot X = Y \circ F$.

Now, if we put $X_t(x) = \sum_{i=1}^n X_i(x,t)\frac{\partial}{\partial x_i}$ and $Y_t(y) = \sum_{i=1}^n Y_i(y,t)\frac{\partial}{\partial y_i}$, then the triviality condition takes the form $\frac{\partial f_t}{\partial t} = -df_t \cdot X_t + Y_t \circ f_t$. For $t = 0$, we get

$$\frac{\partial f_t}{\partial t}\ |_{t=0} = -df_0 \cdot X_0 + Y_0 \circ f_0 \tag{2.7}$$

with $X_0 \in \theta_{\mathbb{C}^n,0}$ and $Y_0 \in \theta_{\mathbb{C}^p,0}$. Here, $\theta_{\mathbb{C}^n,0}$ (respectively $\theta_{\mathbb{C}^p,0}$) denotes the space of germs of a vector field at $0 \in \mathbb{C}^n$ (respectively at $0 \in \mathbb{C}^p$). For any 1-parameter unfolding $F(x,t) = (f_t(x),t)$, the mapping $\frac{\partial f_t}{\partial t}\ |_{t=0}$ is called the initial speed of F. Clearly, a necessary condition for a 1-parameter unfolding F of $f_0 \in E^0_{n,p}$ to be trivial is that its initial speed has the form (2.7). This motivates the definition of the tangent space at f_0 to $\mathcal{A}f_0$:

$$T\mathcal{A}_e f_0 := \{df_0 \cdot X + Y \circ f_0 \mid X \in \theta_{\mathbb{C}^n,0} \text{ and } Y \in \theta_{\mathbb{C}^p,0}\}. \tag{2.8}$$

By definition, $T\mathcal{A}_e f_0$ is a subspace of the space of vector fields along f_0, $\theta(f_0)$, which consists of vector fields $\xi : \mathbb{C}^n \to T\mathbb{C}^p$ satisfying $pr \circ \xi = f_0$ where $pr : T\mathbb{C}^p \to \mathbb{C}^p$ is the natural projection. Hence, one can define maps

$$\begin{aligned} tf_0 : \theta_{\mathbb{C}^n,0} &\to \theta(f_0) \\ X &\mapsto df_0 \cdot X \end{aligned} \tag{2.9}$$

and

$$\begin{aligned} wf_0 : \theta_{\mathbb{C}^p,0} &\to \theta(f_0) \\ Y &\mapsto Y \circ f_0 \end{aligned} \tag{2.10}$$

and then set $T\mathcal{A}_e f_0 = tf_0(\theta_{\mathbb{C}^n,0}) + wf_0(\theta_{\mathbb{C}^p,0})$. It follows that the deformation space is the quotient

$$N\mathcal{A}_e f_0 = \frac{\theta(f_0)}{tf_0(\theta_{\mathbb{C}^n,0}) + wf_0(\theta_{\mathbb{C}^p,0})}. \tag{2.11}$$

For any $f_0 \in E^0_{n,p}$, $\mathcal{A}_e - \operatorname{codim}(f)$ is defined to be the $\mathbb{C}$-dimension of the space $N\mathcal{A}_e f$.

An unfolding F of f_0 is $\mathcal{A}_e$-*versal* if every other unfolding of f_0 is $\mathcal{A}$-equivalent to an unfolding induced from F. This is the case if and only if its initial speeds $\{\frac{\partial f_t}{\partial t_1}|_{t=0}, \ldots, \frac{\partial f_t}{\partial t_d}|_{t=0}\}$ generate $N\mathcal{A}_e f$ as a $\mathbb{C}$-vector space ([87]). A map germ f_0 is called $\mathcal{A}$-stable if any small perturbation of f_0 is $\mathcal{A}$-equivalent to f_0. A germ f_0 is stable if and only if any $\mathcal{A}_e$-versal unfolding of f is trivial and this is the case if and only if $T\mathcal{A}_e f_0 = \theta(f_0)$ ([92]).

Now consider the case $p = n+1$. Let $F : (\mathbb{C}^n \times \mathbb{C}^d, 0) \to (\mathbb{C}^{n+1} \times \mathbb{C}^d, 0)$, $F(x,t) = (f_t(x), t)$ be an $\mathcal{A}_e$-versal unfolding of f_0 and let $\pi : \mathbb{C}^{n+1} \times \mathbb{C}^d \to \mathbb{C}^d$ be the natural projection. If f_0 is not stable then there exists an analytic set $B \subset \mathbb{C}^d$ called the bifurcation set, such that for $t \in B$ the map f_t is not stable. Provided f_0 does have a stable perturbation then B is a proper analytic subset. In addition, there exists an $\varepsilon > 0$, neighbourhoods U of the origin of $\mathbb{C}^n \times \mathbb{C}^d$ and T of the origin of $\mathbb{C}^d$ and a proper, finite-to-one representative of the unfolding $F : U \to B_\varepsilon \times T$ such that, if $X = F(\bar{U})$ then the stable type stratified mapping $F : F^{-1}(\bar{B}_\varepsilon \times (T - B)) \to X \cap (\bar{B}_\varepsilon \times (T - B))$ is locally trivial over $T - B$ with respect to the stratified submersion $\pi : \bar{B}_\varepsilon \times (T - B) \to T - B$ ([85]). This provides a fibration of the mapping $F : \bar{U} \to \operatorname{image}(F) = X$ whose fibre over a parameter $t \in T - B$ is the stable map $f_t : U_t \to X_t$ where $X_t = X \cap (\bar{B}_\varepsilon \times \{t\})$. It has been already proved that X_t has the homotopy type of a wedge of n spheres for such t ([96], [73]).

Moreover for map germs $f_0 : (\mathbb{C}^n, 0) \to (\mathbb{C}^p, 0)$ of finite $\mathcal{A}_e$-codimension with $n \geq p$, Problem 1 has been answered by Damon and Mond in [32]. In this case, the notion of discriminant of a perturbation takes the place of image in the above discussion. If $F : (\mathbb{C}^{n'}, 0) \to (\mathbb{C}^{p'}, 0)$ is an unfolding of f_0 and f_t is a stable perturbation, then the *discriminant* $D(f_t)$ of f_t is the image of the set of points where the rank of df_t is less than p. By the results of Damon and Mond, $D(f_t) \cap B_\varepsilon$ with $\varepsilon > 0$ sufficiently small, i.e., for any $0 < \varepsilon' \leq \varepsilon$, $D(f_t)$ is stratified transverse to $S_{\varepsilon'}$, is independent up to homeomorphism of the choice of t and ε and it has the homotopy type of a wedge of $(p-1)$-spheres. Furthermore,

$$\mathcal{A}_e - \operatorname{codim}(f_0) \leq \mu_\Delta(f) \text{ , with equality if } f_0 \text{ is weighted homogeneous} \tag{2.12}$$

where $\mu_\Delta(f_0)$, the discriminant Milnor number, is the number of spheres in the wedge.

Damon and Mond proved (2.12) by reducing it to a question about $\mathcal{K}_V$-equivalence for sections of an appropriate variety V and $\mathcal{K}_H$-equivalence for a germ H defining V. These groups are introduced by Damon in [30] as subgroups of the contact group $\mathcal{K}$ ([91]) which is defined to be

$$\mathcal{K} = \{\Phi \in \mathrm{Diff}(\mathbb{C}^n \times \mathbb{C}^p, 0) \mid \exists \varphi \in \mathrm{Diff}(\mathbb{C}^n, 0) \text{ such that } \Phi \circ i = i \circ \varphi \text{ and } \pi \circ \Phi = \varphi \circ \pi\} \tag{2.13}$$

where $i : \mathbb{C}^n \to \mathbb{C}^n \times \mathbb{C}^p$ is the inclusion $i(x) = (x, 0)$ and $\pi : \mathbb{C}^n \times \mathbb{C}^p \to \mathbb{C}^n$ is the projection $\pi(x, y) = x$. Then $\mathcal{K}$ acts on $E_{n,p}$ by

$$(\varphi(x), \Phi \cdot f_0(x)) = \Phi(x, f_0(x)). \tag{2.14}$$

Two germs are $\mathcal{K}$-equivalent if and only if they define isomorphic subspaces. For a given germ $(V, 0) \subset (\mathbb{C}^p, 0)$, the subgroup $\mathcal{K}_V$ is defined as

$$\mathcal{K}_V = \{\Phi \in \mathcal{K} \mid \Phi(\mathbb{C}^n \times V) \subseteq \mathbb{C}^n \times V\}. \tag{2.15}$$

Furthermore, if $H : (\mathbb{C}^p, 0) \to (\mathbb{C}, 0)$ is a map germ, then

$$\mathcal{K}_H = \{\Phi \in \mathcal{K} \mid H \circ \pi_2 \circ \Phi = H \circ \pi_2\} \tag{2.16}$$

where $\pi_2 : \mathbb{C}^n \times \mathbb{C}^p \to \mathbb{C}^p$ is the natural projection. These subgroups also yield equivalence relations. As $\mathcal{K}$-equivalence captures the equivalence of the germs of varieties $f_0^{-1}(0)$, so too the equivalence $\mathcal{K}_V$ captures the equivalence of the germs of varieties $f_0^{-1}(V)$. Tangent spaces with respect to these equivalence relations are defined as follows:

$$T\mathcal{K}_{V,e} f_0 = t f_0(\theta_{\mathbb{C}^n,0}) + f_0^*(\mathrm{Der}(-\log V)) \tag{2.17}$$

and

$$T\mathcal{K}_{H,e} f_0 = t f_0(\theta_{\mathbb{C}^n,0}) + f_0^*(\mathrm{Der}(-\log H)) \tag{2.18}$$

where $\mathrm{Der}(-\log V)$ is the module of vector fields in $\theta_{\mathbb{C}^p,0}$ which are tangent to V and $\mathrm{Der}(-\log H)$ is the module consisting of vector fields $\xi \in \theta_{\mathbb{C}^p,0}$ satisfying $\xi(H) = 0$. In [31], Damon relates the notion of $\mathcal{A}_e$-codimension with $\mathcal{K}_{V,e}$-codimension using the following construction:

By Mather [89], if $f_0 : (\mathbb{C}^n, 0) \to (\mathbb{C}^p, 0)$ is a holomorphic germ of finite singularity type, i.e., finite $\mathcal{K}$-codimension (and/or f admits a stable unfolding), then there is a stable germ $F : (\mathbb{C}^{n'}, 0) \to (\mathbb{C}^{p'}, 0)$ and a germ of an immersion $g_0 : (\mathbb{C}^p, 0) \to (\mathbb{C}^{p'}, 0)$ with g_0 transverse to F such that f_0 is obtained as a fibre product from the following diagram:

$$\begin{array}{ccc} \mathbb{C}^{n'}, 0 & \xrightarrow{F} & \mathbb{C}^{p'}, 0 \\ \uparrow & & \uparrow g_0 \\ \mathbb{C}^n, 0 & \xrightarrow{f_0} & \mathbb{C}^p, 0 \end{array} \tag{2.19}$$

(here F is a stable unfolding of f_0). We set V as the image of F and let H be a defining germ for V. Then, by the results of Damon [31], g_0 has finite $\mathcal{K}_{V,e}$-codimension if and only if f_0 has finite $\mathcal{A}_e$-codimension. Furthermore, the complement $N\mathcal{K}_{V,e} g_0$ to the $\mathcal{K}_{V,e}$-tangent space of g_0 is isomorphic to $N\mathcal{A}_e f_0$. Hence, $\mathcal{A}_e - \mathrm{codim}(f_0)$ is equal to $\mathcal{K}_{V,e}$-codimension of g_0. Also, $\mathcal{K}_{V,e} - \mathrm{codim}(g_0) \leq \mathcal{K}_{H,e} - \mathrm{codim}(g_0)$, with equality if f_0 is weighted homogeneous (which naturally implies g_0 and F are also weighted homogeneous). Note that when $\mathcal{K}_{V,e}$ − codim(g_0) is finite, $\mathcal{K}_{H,e}$ − codim(g_0) is also finite. In this case, the germ g_0 is said to have an isolated instability at $0 \in \mathbb{C}^{p'}$.

If we choose a 1-parameter deformation $G = g_t : (\mathbb{C}^p \times \mathbb{C}, 0) \to (\mathbb{C}^{p'}, 0)$ of g_0, we can recover a 1-parameter unfolding $f : (\mathbb{C}^n \times \mathbb{C}, 0) \to (\mathbb{C}^p \times \mathbb{C}, 0)$, $(x, t) \mapsto (f_t(x), t)$, of f_0 as a fibre product from a diagram similar to (2.19). The crucial step in Damon and Mond's proof is Theorem 5.2 of [32] which states that $N\mathcal{K}_{e/\mathbb{C}}G$, the quotient of $\theta(G)$ by $T\mathcal{K}_{H,e/\mathbb{C}}G = tG(\theta_{\mathbb{C}^n \times \mathbb{C}/\mathbb{C}}) + G^*(\mathrm{Der}(-\mathrm{log}H))$ is free over the ring $\mathcal{O}_{\mathbb{C}}$. The proof of this statement is based on the fact that the discriminant V is a free divisor, i.e., Der$(-\mathrm{log}V)$ is a free $\mathcal{O}_{\mathbb{C}^{p'}}$-module of rank p' (see [123] for the case $p' = 1$ and [80] for the general case). Then, assuming that the isolated point 0 of $\mathcal{K}_H$-instability of g_0 breaks up into the points $y_1, \dots, y_r$ of $\mathcal{K}_H$-instability of g_t, $\sum_{i=1}^r \dim_{\mathbb{C}} N\mathcal{K}_{H,e}(g_t; y_i) = \dim_{\mathbb{C}} N\mathcal{K}_{H,e} g_0$ by Corollary 5.5 of [32]. The proof of (2.12) follows after relating each $\dim_{\mathbb{C}} N\mathcal{K}_{H,e}(g_t; y_i)$ for $g_t(y_i) \notin V$, with the Milnor number of the image of a stable perturbation of f_0 via Lemma 5.6 of [32].

This technical structure for the proof of (2.12) fails in the case of map germs $f : (\mathbb{C}^n, 0) \to (\mathbb{C}^{n+1}, 0)$ only because the image of f is no longer a free divisor.

Problem 2 (D. Mond). *Let $f : (\mathbb{C}^n, 0) \to (\mathbb{C}^p, 0)$ be a map germ of finite $\mathcal{A}_e$-codimension, and suppose that $n \geq p$ and (n, p) is in Mather's range of nice dimensions ([89]). Does the equality $\mu_\Delta(f) = \mathcal{A}_e$-codim$(f)$ imply that f is a quasi-homogeneous map?*

Here, quasihomogeneous means that the singularity is defined by a polynomial which is equivalent to a weighted homogeneous map under a coordinate transformation on the domain. It is easy to observe that if f is a $\mathbb{C}$-valued quasihomogeneous map defining an isolated singularity then its Milnor number and Tjurina number are equal. However, the converse of this statement is not trivial. In [122], Saito gave an intelligent proof for the implication of quasihomogeneity from $\mu = \tau$. The equality $\mu = \tau$ also holds for general quasihomogeneous isolated complete intersection singularities ([58]). The converse of this statement was proved by Greuel, Martin and Pfister [59] for isolated complete intersection curve singularities and by Wahl [153] in the case of isolated complete intersection surface singularities. Finally, Vosegaard proved it for any isolated complete intersection singularity in [151, 152]. By the results of Damon and Mond, in [32], if f is a map germ as described in the problem, and weighted homogeneous, then $\mu_\Delta(f) = \mathcal{A}_e$ − codim(f) (see the notes for Problem 1). This, together with the other similarities between the theory of isolated singularities and the theory of differentiable mappings suggest that a similar implication might exist for such map germs.

2.2. Multiple point spaces for maps

Problem 3 (D. Mond). *Let $f : X \to Y$ be a finite analytic map between two analytic spaces X and Y of dimension n and $n+1$, respectively. Does $\mathrm{Fitt}_k(f_*\mathcal{O}_X)$ give a good structure to the set $M_k(f)$ of multiple points, e.g. a Cohen-Macaulay structure?*

Here, $M_k(f)$ is defined to be the set of points in Y whose preimage consists of k or more points, counted with multiplicity. A good structure for $M_k(f)$ is meant to be an analytic structure such that it is calculable and reduced at ordinary k-tuple points of f (see definition below) and that it commutes with base change and behaves well under deformation, i.e., if F is a deformation of f over base S then $M_k(f)$ is flat over S.

Mond and Pellikaan studied this problem in [98] and suggested that the requested analytic structure should be given by the Fitting ideals of coherent $\mathcal{O}_Y$-module $f_*\mathcal{O}_X$. More precisely, $M_{k+1}(f)$ should be defined by the k−th Fitting ideal $\mathrm{Fitt}_k(f_*\mathcal{O}_X)$ of $f_*\mathcal{O}_X$. Recall that if

$$R^p \xrightarrow{\lambda} R^q \to M \to 0 \tag{2.20}$$

is a presentation of a finite R-module M (where R is a commutative ring with identity), then the ideal $\mathrm{Fitt}_k(M)$ is generated by all $(q-k)\times(q-k)$ minors of the matrix λ, for $q > k \geq q-p$ and $\mathrm{Fitt}_k(M)$ is defined to be equal to R for $k \geq q$ and 0 for $k < q-p$. The condition f being finite guarantees that a presentation of $f_*\mathcal{O}_X$ over the ring $\mathcal{O}_Y$ does exist.

Assuming X is Cohen-Macaulay and Y is smooth with $\dim(Y) = \dim(X)+1$, e.g., $Y = \mathbb{C}^{n+1}$, Mond and Pellikaan constructed an algorithm for obtaining a presentation of $f_*\mathcal{O}_X$ from which all the Fitting ideals can be calculated. Then they showed that for any k, this structure is in fact reduced at ordinary k-tuple points (a point $y \in Y$ is called an ordinary k-tuple point of f if $f^{-1}(y)$ consists of k distinct preimages $x_1,\dots,x_k$, at each of which X is smooth and f is an embedding, such that the tangent spaces at y to the images $f(X,x_i)$ are in general position in T_yY). Furthermore, it commutes with base change. But the flatness condition is only satisfied in special cases: When X is Cohen-Macaulay, $M_1(f)$, i.e., the image of f, and $M_2(k)$, providing it has codimension 2, are also Cohen-Macaulay. As a result, they are flat over the base space of a deformation. Nevertheless, to conclude that $M_3(f)$ is Cohen-Macaulay, it is necessary to assume that X is a Gorenstein space. When $f : (X,x) \to (\mathbb{C}^{n+1},0)$ is a map germ giving $\mathcal{O}_X$ a cyclic structure over $\mathcal{O}_{\mathbb{C}^{n+1}}$, i.e., when $\mathcal{O}_X/f^*m_{\mathbb{C}^{n+1},0} \cong \mathbb{C}[t]/(t^h)$ for a finite $h \in \mathbb{N}$, then for any $1 \leq k \leq n+1$, $M_k(f)$ is Cohen-Macaulay, provided it has codimension k. Finite map germs $f : (\mathbb{C}^n,0) \to (\mathbb{C}^{n+1},0)$ of corank 1 constitute examples for such a case (see [86, 97] for details).

Later, in [70], Kleiman, Lipman and Ulrich studied the case where X and Y are smooth varieties of dimension n and $n+1$ over an algebraically closed field and f is curvilinear, that is for all $x \in X$ the differential $df(x)$ has rank at most $n-2$. They proved that if each component of $M_k(f)$, or equivalently $f^{-1}M_k(f)$, has the minimal possible dimension $n-k$, then $M_k(f)$ and $f^{-1}M_k(f)$ are Cohen-Macaulay. We note that their notation differs from Mond and Pellikaan's: They denoted the multiple point spaces by $N_k(f)$ and the preimage of the multiple point spaces by $M_k(f)$. Their work also includes results about the Hilbert scheme Hilb_f^k of degree k-schemes of the fibres of f and the universal subscheme Univ_f^r of $\mathrm{Hilb}_f^k \times_Y X$ which were obtained by using iteration formulae of [69].

2.3. Invariance of the topological type

Problem 4 (D. Mond). *In [72], Lê and Ramanujam proved the following: Let $f : (\mathbb{C}^{n+1} \times D_\eta, 0 \times D_\eta) \to (\mathbb{C}, 0)$ be a polynomial such that each $f_t : (\mathbb{C}^{n+1}, 0) \to (\mathbb{C}, 0)$ has an isolated singularity at 0 for all t belonging to the disc $D_\eta = \{z \in \mathbb{C} : |z| \leq \eta\}$ and $\mu(f_t)$, the Milnor number of f_t at 0, is independent of t. Then,*

(i) *The fibre-homotopy type of the Milnor fibration of f_t at 0 is independent of t; and if $n \neq 2$,*
(ii) *the diffeomorphism type of the Milnor fibration of f_t at 0 is independent of t; and*
(iii) *the ambient topological type of $V(f_t)$ at 0 is independent of t.*

Does a similar fact hold for μ_Δ or μ_I?

For the definitions of μ_Δ and μ_I, see the notes following Problem 1.

2.4. Representations of quivers

Problem 5 (D. Mond). *The quivers of finite representation type are the Dynkin quivers, i.e., the underlying graph is of type A_n, D_n, E_6, E_7 and E_8. Moreover, if the dimension of the quiver is a root of the associated simple Lie algebra, then there exists an open orbit ([54]). The complement of this open orbit, which is also called a discriminant, is a linear free divisor ([24]). Find a theory of vanishing cycles subject to this theory.*

The notion of free divisor is introduced by Saito in [123]. A reduced divisor $D = V(h) \subset \mathbb{C}^n$ is a free divisor if the module of vector fields on $\mathbb{C}^n$ tangent to D, $\mathrm{Der}(-\log D)$, is a locally free $\mathcal{O}_{\mathbb{C}^n}$-module. It is called a linear free divisor if additionally $\mathrm{Der}(-\log D)$ has a basis consisting of vector fields whose coefficients in the basis $\frac{\partial}{\partial x_1}, \ldots, \frac{\partial}{\partial x_n}$ are linear functions. The theory of free divisors actually lies on the boundary between Singularity Theory and Representation Theory. In [24], Buchweitz and Mond showed that examples of linear free divisors also arose as degeneracies in representation spaces of Dynkin quivers. Later, in [56, 57], more examples were given in the case of quivers without oriented cycles.

Let us recall some definitions: A *quiver* Q is a finite directed graph with a finite set Q_0 of nodes (or vertices), a finite set of arrows Q_1 and two maps $h, t : Q_1 \to Q_0$ that assign to each arrow $\phi \in Q_1$ its head $h\phi$ and its tail $t\phi$ in Q_0. A *representation* V of Q consists of a $\mathbb{k}$-vector space V_a for each $a \in Q_0$ and a $\mathbb{k}$-linear map $V(\phi) : V_{t\phi} \to V_{h\phi}$ for each $\phi \in Q_1$. For a fixed dimension vector

$d = ((d(a))_{a\in Q_0}$, the $\mathbb{k}$-space of representations is

$$\mathrm{Rep}(Q,d) := \prod_{\phi\in Q_1} \mathrm{Hom}_{\mathbb{k}}(V_{t\phi}, V_{h\phi}) \cong \prod_{\phi\in Q_1} \mathrm{Hom}_{\mathbb{k}}(\mathbb{k}^{d(t\phi)}, \mathbb{k}^{d(h\phi)}). \tag{2.21}$$

The group $\mathrm{Gl}(Q,d) := \prod_{a\in Q_0} Gl_{d(a)}(\mathbb{k})$ acts on $\mathrm{Rep}(Q,d)$ by

$$(g_a)_{a\in Q_0} \cdot (V(\phi))_{\phi\in Q_1} = (g_{h\phi} \circ V(\phi) \circ g_{t\phi}^{-1})_{\phi\in Q_1}. \tag{2.22}$$

If V and W are representations of Q, then a morphism of representations $V \to W$ is a set of linear maps $(\psi_a : V(a) \to W(a))_{a\in Q_0}$ satisfying the commutation relations

$$\psi_{h\phi} \circ V(\phi) = W(\phi) \circ \psi_{t\phi} \tag{2.23}$$

for all $\phi \in Q_1$. In [120], Ringel constructed an exact sequence:

$$0 \to \mathrm{Hom}_Q(V,W) \to \mathrm{Hom}_{\mathbb{k}}(V,W) \overset{M_{V,W}}{\to} \mathrm{Hom}_{\mathbb{k}}(tV,hW) \overset{E_{V,W}}{\to} \mathrm{Ext}^1_Q(V,W) \tag{2.24}$$

where

$$\mathrm{Hom}_{\mathbb{k}}(V,W) := \prod_{a\in Q_0} \mathrm{Hom}_{\mathbb{k}}(V(a), W(a)) ,$$

$$\mathrm{Hom}_{\mathbb{k}}(tV,hW) := \prod_{\phi\in Q_1} \mathrm{Hom}_{\mathbb{k}}(V(t\phi), W(h\phi))$$

and the map $M_{V,W}$ sends $(\psi_a)_{a\in Q_0}$ to $(\psi_{h\phi}\circ V(\phi) - W(\phi)\circ\psi_{t\phi})_{\psi\in Q_1}$, $E_{V,W}$ sends $(\gamma_\phi)_{\phi\in Q_1}$ to an extension

$$0 \to W \overset{i}{\to} X \overset{j}{\to} V \to 0$$

where $X(a) = W(a) \oplus V(a)$ for each $a \in Q_0$, i and j are the usual inclusion and projection and for each $\phi \in Q_1$, $X(\phi)$ has matrix

$$\begin{bmatrix} W(a) & -\gamma_\phi \\ 0 & V(a) \end{bmatrix}.$$

Given $V \in \mathrm{Rep}(Q,d)$ and $V' \in \mathrm{Rep}(Q,d')$, the direct sum $V \oplus V'$ belongs to $\mathrm{Rep}(Q, d+d')$. It is the representation with $(V \oplus V')_a = V_a \oplus V'_a$ for $a \in Q_0$ and

$$(V \oplus V')_\phi = \begin{bmatrix} V(\phi) & 0 \\ 0 & V'(\phi) \end{bmatrix}.$$

A representation is *indecomposable* if it is not isomorphic to a direct sum in $\mathrm{Rep}(Q,d)$. Dimension vector d is said to be a *root* if $\mathrm{Rep}(Q,d)$ contains an indecomposable representation, it is a *real root* if $\mathrm{Rep}(Q,d)$ contains exactly one orbit of, necessarily isomorphic, indecomposable representations. If a general representation in Rep(Q,d) is indecomposable, then d is called a *Schur root.*

Gabriel [54] proved the following fundamental result: A connected quiver Q is of finite type, i.e., it has only finitely many indecomposable representations up to isomorphism, if and only if it is a Dynkin quiver. Moreover if d is a positive root of the underlying Dynkin diagram then d is a real Schur root: there is a unique open orbit in $\mathrm{Rep}(Q,d)$ whose points correspond to indecomposable representations. Buchweitz and Mond showed that if d is a real Schur root of the Dynkin quiver Q

then the complement D of the open orbit, the *discriminant*, is a linear free divisor ([24]).

Let V_0 be a smooth point of D, and assume that the orbit $\mathrm{Gl}(Q,d)\cdot V_0$ of V_0 is open in D. Then Lemma 5.2 of [57] states that V_0 is a direct sum of uniquely determined representations V_1 and V_2, both representations have open orbits in their representation spaces and their dimensions d_1 and d_2 are real Schur roots and that, possibly after a permutation of V_1 and V_2, $\mathrm{Ext}^1_Q(V_1,V_2)=0$ and $\mathrm{Ext}^1_Q(V_2,V_1)$ is 1-dimensional. They also show that in this case the general representation $V\in \mathrm{Rep}(Q,d)$ is given by an extension of V_2 by a unique subrepresentation isomorphic to V_1:

$$0\to V_1\to V\to V_2\to 0. \tag{2.25}$$

The extension (2.25) splits *when V moves onto the discriminant.* In this way, it resembles a vanishing cycle in Singularity Theory. The extension (2.25) is determined by a choice of path joining V to a regular point on the discriminant, and vanishes along this path. So, the question is whether it is possible to develop a theory of vanishing cycles for quiver representations.

2.5. Injective maps

Problem 6 (Lê Dũng Tráng). *Let $f:(\mathbb{C}^2,0)\to(\mathbb{C}^3,0)$ be a map germ such that $df(0)=0$. Can f be injective?*

In fact, this problem is in connection with finding a counter-example to Lê's conjecture which originally had the following formulation: Suppose that the normalisation $f:(N,0)\to(S,0)$ of a surface singularity $(S,0)$ is a bijection from a smooth germ $(N,0)$ to $(S,0)$; then S is isomorphic to the total space of an equisingular deformation of an irreducible curve singularity. One can show that for an injective map germ $f:(\mathbb{C}^2,0)\to(\mathbb{C}^3,0)$, the rank of the differential of f at 0 is at least 1 if and only if its image $f(\mathbb{C}^2)$ is the total space of an equisingular deformation of an irreducible plane curve singularity. So, Lê's conjecture can be restated as follows: If the map germ $f:(\mathbb{C}^2,0)\to(\mathbb{C}^3,0)$ is injective then the rank of $df(0)$ is at least 1. Over the last thirty years, many mathematicians have tried to prove this conjecture or to give an example of an injective map with trivial differential at the origin. So far, it is proven only for special cases.

One of the attempts was made by Némethi in [100]. He considered injective analytic map germs $f:(\mathbb{C}^n,0)\to(\mathbb{C}^{n+1},0)$ and claimed that if the image X of such germ is "good" then the rank of $df(0)$ is at least $n-1$ and $(X,0)$ is an equisingular family of plane curves. However, Keilen found a fundamental mistake in Némethi's proof ([67]). Some time later, Keilen and Mond posted Keilen's counter-example on the arXiv (see [68]). But this example does not constitute a counter-example to Némethi's theorem.

In [83], Luengo and Pichon proved the conjecture for the case of cyclic covers over normal surface singularities totally ramified along the zero locus of an analytic function. Then, Fernandez de Bobadilla [18] gave a new proof of Lê's conjecture for the case of singularities containing a smooth curve through the origin which also covers many interesting classes of singularities, in particular the case where the intersection of the two homogeneous forms defining f is a union of lines contained in $f^{-1}(0)$. He also stated a conjecture which corresponds to Lê's and only involves plane curve singularities and their relative polar curves.

2.6. Representations of Lie Groups

Problem 7 (Lê Dũng Tráng). *Let G be a semisimple complex algebraic group with the Lie algebra $\mathfrak{g}$. Let B be the standard Borel subgroup of G and $\mathfrak{b}$ its Lie algebra. Moreover, let $\mathfrak{n}$ denote the nilpotent radical of $\mathfrak{b}$. We consider the following objects: the flag manifold $\mathcal{D}$ associated with G, the quotient $G \times^B \mathfrak{n}$ of $G \times \mathfrak{n}$ by the right action of B and the embedding*

$$\begin{array}{rcl} G \times^B \mathfrak{n} & \to & \mathcal{D} \times \mathfrak{g} \\ g * x & \mapsto & (gB, g \cdot x) \end{array} \tag{2.26}$$

together with its image $\mathcal{Y}$ in $\mathcal{D} \times \mathfrak{g}$ ([135]). We define a map $\varphi : \mathcal{Y} \to \mathfrak{g}$ as the restriction of the natural projection $\mathcal{D} \times \mathfrak{g} \to \mathfrak{g}$ to $\mathcal{Y}$. Take G to be $\mathrm{SL}_{n+1}(\mathbb{C})$ and describe $\varphi^{-1}(x)$ for any nilpotent element x of $\mathfrak{sl}_{n+1}(\mathbb{C})$ in terms of its Jordan form. Solve this problem in case of other Lie algebras.

In the following, we include some definitions and references to the works related to this problem.

Let G be a semisimple (connected) complex algebraic group with $\mathfrak{g} = \mathrm{Lie}(G)$ on which G acts by the adjoint action given by $g \cdot u := gug^{-1}$ for any $g \in G$ and any $u \in \mathfrak{g}$. Let $\mathfrak{h}$ be a Cartan subalgebra and let $\mathcal{W}$ denote the associated Weyl group. Consider a Chevalley-Cartan decomposition of $\mathfrak{g}$: $\mathfrak{g} = \mathfrak{h} \oplus \sum_{\alpha \in R} \mathfrak{g}_\alpha$ where R is a root system of $\mathfrak{g}$ relative to this choice of $\mathfrak{h}$. If Π is the set of primitive roots, then we can decompose R into two sets: the set of positive roots R^+ and the negative roots R^- with respect to Π. The standard Borel subalgebra is defined as $\mathfrak{b} := \mathfrak{h} \oplus \sum_{\alpha \in R^+} \mathfrak{g}_\alpha$ and its nilpotent radical is $\mathfrak{n} := \sum_{\alpha \in R^+} \mathfrak{g}_\alpha$. Let B be the Borel subgroup of G such that $\mathrm{Lie}(B) = \mathfrak{b}$.

Let $G \times^B \mathfrak{n}$ be the quotient of $G \times \mathfrak{n}$ by the right action of B given by $(g, x)b = (gb, b^{-1} \cdot x)$ for $g \in G$, $b \in B$ and $x \in \mathfrak{n}$. By the Killing form we get $G \times^B \mathfrak{n} = T^*(G/B)$ where $\mathcal{D} := G/B$ is the flag manifold. Let $g * x$ denote the class of (g, x) in $G \times^B \mathfrak{n}$. Then $G \times^B \mathfrak{n}$ can be identified with the closed subvariety $\mathcal{Y} := \{(gB, x) \in \mathcal{D} \times \mathfrak{g} \mid x \in g \cdot \mathfrak{n}\}$ via the embedding (2.26) ([135]). The map $f : G \times^B \mathfrak{n} \to \mathfrak{g}$ defined by $g * x \mapsto g \cdot x$ is called the *Springer resolution.* It is proper since G/B is complete and its image is $G \cdot \mathfrak{n} = \mathcal{N}$, the nilpotent variety of $\mathfrak{g}$ ([139]). We have a commutative diagram

$$\begin{array}{ccc} G \times^B \mathfrak{n} & \xrightarrow{\cong} & \mathcal{Y} \\ & \searrow^{f} & \downarrow \varphi \\ & & \mathfrak{g} \end{array} \tag{2.27}$$

where φ is the restriction of the projection map $\mathcal{D} \times \mathfrak{g} \to \mathfrak{g}$, $(gB, x) \mapsto x$.

For a nilpotent element $x \in \mathfrak{n}$, the preimage $\mathcal{D}_x := f^{-1}(x) = \{gB \in \mathcal{D} \mid x \in g \cdot \mathfrak{n}\}$ is called *Springer fibre* above x. This notion has been of interest in many areas, for example, in Representation Theory and Singularity Theory. It is still a mysterious object mainly because it is difficult to give a geometric description which is only known in special cases: If x is a regular element in $\mathfrak{g}$ then $\mathcal{D}_x$ consists of one point. If x is a subregular element then the fibre is a finite union of projective lines which intersect each other transversally and has the configuration of a Dynkin diagram ([140]).

In the case of $G = GL_n(\mathbb{C})$, any $x \in \mathfrak{n}$ has 0 as the only characteristic value. So its Jordan form is defined by a partition λ of n: $\lambda = (\lambda_1, \dots, \lambda_k)$ where λ_i is the length of the i-th Jordan block. An ordered partition can be presented as a Young diagram, an array with k rows of boxes starting on the left with the i-th row containing λ_i boxes. In this way, one can form a bijection between Springer fibres and Young diagrams. Moreover, Spaltenstein [138] and Steinberg [141] showed that there is a bijection between components of $\mathcal{D}_x$ and Young tableaux assigned to λ.

The description of Springer fibres was completely done for the hook and two-row Young diagrams in [53, 150] in the general context. Lorist showed, in the case of Springer fibres of dimension 2, all the irreducible components of the fibre are either the product of two projective lines or are ruled surfaces over a projective line having invariant $e = 2$ ([82]).

In [107], Pagnon solved the problem for a special case: He studied the group $G = SL_n(\mathbb{C})$ considering the generalized map $f_{\mathfrak{p}} : G \times^P \mathfrak{n}_{\mathfrak{p}} \to \bar{\mathcal{O}}_{\mathfrak{p}}$ where P is a parabolic subgroup of G, $\mathfrak{n}_{\mathfrak{p}}$ the nilpotent radical of the Lie algebra $\mathfrak{p}$ of P and $\mathcal{O}_{\mathfrak{p}} \subset \mathfrak{g}$ is the Richardson nilpotent G-orbit induced from $\mathfrak{p}$. He established an estimate for the dimension of $f_{\mathfrak{p}}^{-1}(x)$ where x is a Richardson nilpotent induced by a parabolic subgroup $Q \subset P$. He also showed that there is a bijection between the irreducible components of $\mathcal{O}_{\mathfrak{q}} \cap \mathfrak{n}_{\mathfrak{p}}$ and the components of $f_{\mathfrak{p}}^{-1}(x)$, $x \in \mathcal{O}_{\mathfrak{q}}$ and described the components in some cases using Jordan forms.

2.7. Homotopy type of hyperplane arrangements

Problem 8 (M. Falk). *When does the combinatorial data determine the homotopy type of hyperplane arrangements?*

Let V^ℓ be the ℓ-dimensional vector space over a field $\mathbb{k}$. An *arrangement* $\mathcal{A}$ of hyperplanes is a finite collection of codimension 1 affine subspaces in V^ℓ. The combinatorial data of an arrangement is given by the *intersection lattice* $L(\mathcal{A})$ defined as a set of nonempty intersections of elements of $\mathcal{A}$ ordered by reverse inclusion.

The seminal result in the homotopy theory of arrangements is the calculation of the cohomology algebra of the complement $M(\mathcal{A}) = \mathbb{C}^\ell \setminus \bigcup_{\mathcal{H} \in \mathcal{A}} \mathcal{H}$ introduced by Orlik and Solomon in [105]. Motivated by the work of Arnol'd [6], by using tools established by Brieskorn [23], Deligne [34] and Hattori [63], they defined a graded algebra $A(\mathcal{A})$, which is constructed using only $L(\mathcal{A})$ and known as *Orlik-Solomon algebra.*

Let $\mathcal{A} = \{\mathcal{H}_i | \ i \in [1,n]\}$, and $E(\mathcal{A})$ denote the free graded exterior algebra over $\Bbbk$ generated by 1 and degree 1 elements e_i for $\mathcal{H}_i \in \mathcal{A}$. The Orlik-Solomon algebra $A(\mathcal{A})$, associated with $\mathcal{A}$, is the quotient of $E(\mathcal{A})$ by the homogeneous ideal

$$I(\mathcal{A}) = \{\partial e_C | C \in \mathcal{C}\}, \tag{2.28}$$

where $\mathcal{C}$ is a set of circuits ($C \subseteq \{1, 2, \dots, n\}$ is a *circuit* of $\mathcal{A}$ if it is minimally dependent, i.e., it is a minimal set satisfying $\operatorname{codim}(\bigcap_{i\in C} \mathcal{H}_i) < |C|$), ∂ is a usual boundary operator

$$\partial e_C = \sum_{k\in[1,p]} (-1)^{k-1} e_{i_1} \cdots \hat{e}_{i_k} \cdots e_{i_p} \tag{2.29}$$

for $C = \{i_1, \dots, i_p\}$. The image of e_i in $A(\mathcal{A})$ is denoted by a_i. Then $A(\mathcal{A})$ is a graded-commutative $\Bbbk$-algebra generated by 1 and degree 1 elements a_i, $i \in [1,n]$. According to [105] and [106], $A(\mathcal{A})$ is isomorphic to the cohomology algebra of $M(\mathcal{A})$ with coefficients in $\Bbbk$. The generators a_i correspond to logarithmic 1-forms $d\alpha_i/\alpha_i$, where $\alpha_i : \mathbb{C}^\ell \to \mathbb{C}$ is a linear form with kernel $\mathcal{H}_i$. This gives a presentation of the cohomology ring of the complement of a complex arrangement in terms of generators and relations.

This result from [105] gave rise to a collection of homotopy type conjectures which assert that various homotopy invariants of the complement depend only on $L(\mathcal{A})$. The major positive result in this direction is Randell's lattice-isotopy theorem given in [118]. It states that the homotopy type, indeed the diffeomorphism type of the complement, remains constant through a lattice isotopy that is, a one parameter family of arrangements in which the intersection lattice remains constant.

The rational homotopy type of the complement was studied by Falk [44] and Randell [49]. Their results were reviewed in the survey article [50] by Falk and Randell and later updated in [51]. A presentation of the fundamental group of the complement of a complexified real arrangement was obtained by Randell [50] and Salvetti [126]. The problem for the presentation of the fundamental group of a complement was solved for arbitrary complex arrangements by Arvola [14]. In that case, one can first take a planar section, so that one is working with an affine arrangement in $\mathbb{C}^2$, and then use a braid monodromy technique (see [26, 74, 95]). In [126], Salvetti constructed a finite cell complex of the homotopy type of the complement of a complexified real arrangement. Björner and Ziegler ([17]) used a stratification of V^ℓ compatible with $\mathcal{A}$ to construct a finite regular cell complex of the homotopy type of $M(\mathcal{A})$. In the case of complexified real arrangements it generalizes the work of Salvetti. In [103], Orlik gave a construction of a finite cell complex of the homotopy type of the complement of an arbitrary arrangement of affine subspaces. One can expect to obtain Zariski pairs by constructing cell complexes, but the Zariski pairs obtained by this way are not unions of lines, as in Rybnikov's example. Despite the existence of cell complex constructions, there is still no good criterion to detect nonzero elements in the homotopy groups of the complement or to determine if the complement is a $K(\pi,1)$ space. For more information on $K(\pi,1)$ arrangements it is suggested to read the reference [45].

The fundamental question whether the homotopy type of $M(\mathcal{A})$ is uniquely determined by $L(\mathcal{A})$ was answered in the negative sense by Rybnikov in [121]. By using the MacLane matroid, he constructed two complex arrangements having the same combinatorial type but non-isomorphic fundamental groups. This kind of such pairs is known as *Zariski pairs* (for the original proof of the existence of Zariski pairs see [161]). Once one has an example of a Zariski pair which is distinguished by the fundamental group, it is possible to give infinite families of Zariski pairs using Cremona transformations and covering maps. These techniques have been used by Oka [101], Shimada [130, 131] and Uludağ [147]. The references [7] and [13] are good surveys on Zariski pairs.

The proof proposed by Rybnikov has two steps. Let $G_j := \pi_1(\mathbb{P}^2 \setminus \bigcup \mathcal{H}_{ij})$, $j = 1, 2$. Recall that the homology of the complement of a hyperplane arrangement depends only on the combinatorics. This way one can identify the abelianization of G_1 and G_2 with a combinatorially determined abelian group H. Rybnikov first proved that no isomorphism exists between G_1 and G_2 which induces the identity on H. In particular, this result proves that both line arrangements have different relative topologies. The reason can be outlined as follows: any automorphism of the combinatorics of Rybnikov's arrangement can be obtained from a diffeomorphism of $\mathbb{P}^2$, and hence it produces an automorphism of fundamental groups. Any homeomorphism of pairs $(\mathbb{P}^2, \bigcup \mathcal{H}_{ij})$ induces an automorphism of the combinatorics of $\bigcup \mathcal{H}_{ij}$. Therefore, after a composition one can assume that any homeomorphism of pairs induces the identity on H. The second step is essentially combinatorial but computations are hard to verify. The main point is that one needs to truncate the lower central sequence of G_j such that the quotient K depends only on the combinatorics. Rybnikov proposes to prove that an automorphism of K induces the identity on H (up to sign and automorphism of the combinatorics). This proof is only outlined in [121]. It is worth pointing out that such a result can not be expected for any arrangement. The main difference between relative topology and topology of the complement in terms of isomorphisms of the fundamental group is that homeomorphisms of pairs induce isomorphisms that send meridians to meridians, whereas homeomorphism of the complement can induce any kind of isomorphism. Even if we know that the isomorphism induces the identity on homology, this is not enough to claim that meridians are sent to meridians.

In [11], the authors followed the idea behind Rybnikov's work and provided a detailed proof of his result by using slightly different techniques. In order to prove the first step of Rybnikov's work, they proposed a new approach related to derived series, which is also useful in the study of characteristic varieties and the Alexander invariant. For the second step, they studied combinatorial properties of line arrangements which ensure that any automorphism of the fundamental group of the complement essentially induces the identity on homology (that is analogous to Proposition 4.2 in [121]). For the sake of simplicity, they only presented their progress on line arrangements with double and triple points. However this is an interesting question that can be applied to general line arrangements.

As a result, one can consider the following question: "For which kind of hyperplane arrangements is homotopy type determined by combinatorial data?", e.g. among complexified real arrangements, graphic arrangements, supersolvable arrangements and line arrangements, etc. The second challenge related to this problem is: "Find a general invariant of fundamental groups of the complement of line arrangements that distinguishes the two Rybnikov arrangements, and generalize his construction".

2.8. Resonance varieties

Problem 9 (M. Falk). *Let $\mathcal{A}$ be a braid arrangement of type A_ℓ. Compute the pth resonance variety $\mathcal{R}^p(\mathcal{A}, \Bbbk)$, or, equivalently compute the cohomology of the complex $(A(\mathcal{A}), a)$ for arbitrary $a = \sum_{i\in[1,n]} \lambda_i a_i \in A^1$.*

Resonance varieties provide powerful invariants of the fundamental group (when $p = 1$) and of the cohomology ring of the complement M that other invariants such as Poincaré polynomial can not distinguish. They were first studied implicitly by S. Yuzvinsky in [158] and formally introduced by Falk in [46] as a tool in the classification of Orlik-Solomon algebras $A(\mathcal{A})$. He gave combinatorial conditions that the components of $\mathcal{R}^p(\mathcal{A}, \Bbbk)$ satisfy and conjectured that the components are linear projective spaces; this was proved simultaneously by Cohen and Suciu [27], and Libgober and Yuzvinsky [78].

Now, let us fix some notation and summarize some details for resonance varieties. Let $\Bbbk$ be a field, $\mathcal{A}$ be a central arrangement of n hyperplanes in $\mathbb{C}^\ell$ and $A := A(\mathcal{A}) = \bigoplus_{p=0}^{\ell} A^p$ the Orlik-Solomon algebra of $\mathcal{A}$ over $\Bbbk$. Since A is a quotient of exterior algebra, left-multiplication by a fixed element $a = \sum_{i\in[1,n]} \lambda_i a_i \in A^1 \cong \Bbbk^n$ gives a degree 1 differential on A yielding a cochain complex (A, a):

$$0 \longrightarrow A^0 \longrightarrow A^1 \longrightarrow A^2 \longrightarrow \cdots \longrightarrow A^{\ell-1} \longrightarrow A^\ell \longrightarrow 0. \qquad (2.30)$$

Aomoto studied this complex (A, a) in connection with his work on hypergeometric functions in [5]. The complex was subsequently studied in relation to local system cohomology by Esnault, Schectman and Viehweg in [42]. It is easy to see that (A, a) is exact as long as $\sum_{i\in[1,n]} \lambda_i \neq 0$. In [158], Yuzvinsky showed that in fact (A, a) is generically exact except at the last two positions $\ell - 1$ and ℓ.

The *p-th resonance variety* of $\mathcal{A}$ over $\Bbbk$ consists of points $a_\lambda = (\lambda_1 : \cdots : \lambda_n)$ in $\mathbb{P}(A^1) \cong \mathbb{P}^{n-1}$ corresponding to an element $a = \sum_{i\in[1,n]} \lambda_i a_i \in A^1$ for which (A, a) fails to be exact at A^p. So, for each $p \geq 1$,

$$\mathcal{R}^p(\mathcal{A}, \Bbbk) = \{a_\lambda \in \mathbb{P}^{n-1} | H^p(A, a) \neq 0\}, \qquad (2.31)$$

and there is a filtration by dimension

$$\mathcal{R}^p_k(\mathcal{A}, \Bbbk) = \{a_\lambda \in \mathbb{P}^{n-1} | \dim H^p(A, a) \geq k\}. \qquad (2.32)$$

As noted by Yuzvinsky [158] and Falk [46], $\mathcal{R}^p(\mathcal{A}, \Bbbk)$ is contained in the hyperplane $\Delta_n := \{\sum_{i\in[1,n]} \lambda_i = 0\}$ for all p. Indeed, $a\partial b + \partial(ab) = (\partial a)b = (\sum_{i\in[1,n]} \lambda_i)b$ for

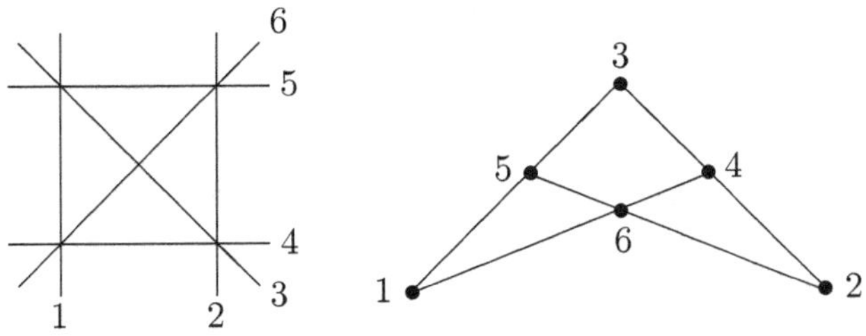

FIGURE 1. The braid arrangement $\mathcal{A}_3$ and its matroid $\mathcal{M}(\mathcal{A}_3)$.

any $b \in A$, and $(\sum_{i\in[1,n]} \lambda_i)\partial$ is a chain contraction of the complex (A, a), which implies that $H^*(A, a) = 0$ if $\sum_{i\in[1,n]} \lambda_i \neq 0$.

A strong motivation and many applications on the topological theory arose initially from the connection with braids. Let $\mathcal{A}_\ell = \{\mathcal{H}_{ij} | \mathcal{H}_{ij} = ker(z_i - z_j),\ 1 \leq i < j \leq \ell\}$ be the arrangement of diagonal hyperplanes in $\mathbb{C}^\ell$, with complement the configuration space $M_\ell := M(\mathcal{A}_\ell)$. In 1962, Fadell and Neuwirth [43] showed that $\pi_1(M_\ell) = P_\ell$, the pure braid group of ℓ strings. Then it was generalized to computation of the fundamental group of the complement of any hyperplane arrangement. Briefly, for an arbitrary hyperplane arrangement in $\mathbb{C}^\ell$, the fundamental group of the complement, $\pi_1(M)$, can be computed algorithmically by using the braid monodromy associated to a generic projection of a generic slice in $\mathbb{C}^2$ (see [14, 26] and references given in those). The end result is a finite presentation with generators x_i corresponding to meridians around the n hyperplanes and commutator relators of the form $x_i^{-1}\alpha_j(x_i)$ where $\alpha_j \in P_n$ are the (pure) braid monodromy generators acting on the meridians via the Artin representation $P_n \hookrightarrow Aut(F_n)$.

In 1960, Arnol'd computed the cohomology ring $H^*(M_\ell, \mathbb{C})$ of the complement of braid ℓ-arrangement as part of his approach to Hilbert's thirteenth problem ([6]). Then, Brieskorn [23] proved that the cohomology ring is isomorphic to the subalgebra of the DeRham complex generated by the logarithmic one-forms, and computed Betti numbers in the case of reflection arrangements. In 1980, Orlik and Solomon gave a simple combinatorial description of the $\Bbbk$-algebra $H^*(M, \Bbbk)$, for any field $\Bbbk$ (for details, see [105, 106, 158, 159]).

Here, we give an example to explain the motive. Let $\mathcal{A}$ be the braid arrangement in $\mathbb{P}^2$ of type $\mathcal{A}_3$. Its defining polynomial is $Q = xyz(x-y)(x-z)(y-z)$ and it is projectively equivalent to Ceva(2) arrangement defined by $(x^2-y^2)(x^2-z^2)(y^2-z^2) = 0$ (see Figure 1).

From the matroid in Figure 1, it is easy to see that the Orlik-Soloman algebra A is the quotient of the exterior algebra E on generators $e_1, \ldots, e_6$ by the ideal $I =< \partial e_{135}, \partial e_{146}, \partial e_{234}, \partial e_{256}, \partial e_{ijkl} >$, where $ijkl$ runs over all four-tuples. Therefore the **nbc** (no broken circuit) Gröbner basis for A contains the following terms,

$$1,$$
$$a_1, a_2, a_3, a_4, a_5, a_6,$$
$$a_{12}, a_{15}, a_{16}, a_{24}, a_{26}, a_{34}, a_{35}, a_{36}, a_{45}, a_{46}, a_{56},$$
$$a_{126}, a_{156}, a_{246}, a_{346}, a_{356}, a_{456}.$$

The Orlik-Solomon algebra A has Hilbert series

$$\pi(A,t) = 1 + 6t + 11t^2 + 6t^3 = (1+t)(1+2t)(1+3t).$$

Fix an element $a = \sum_{i\in[1,6]} \lambda_i a_i \in A^1$, and consider the complex (A, a) in (2.30) for $\ell = 3$. Then

$$H^1(A,a) = \{b \in A^1 | \, ab = 0\}\Big/\{b \in A^1 | \, b = \delta a \text{ for some } \delta \in \Bbbk\}$$

and the first resonance variety of the rank 3 braid arrangement is the projective variety

$$\mathcal{R}^1(\mathcal{A}_3, \Bbbk) = \{a_\lambda \in \Delta_6 | \, \exists b_\mu \in \mathbb{P}^5 \setminus \{a_\lambda\} \text{ s.t. } ab = 0, \text{ where } b = \sum_{i\in[1,6]} \mu_i a_i \in A^1\} \subset \mathbb{P}^5$$

ruled by projective lines. This follows from the fact that if $a_\lambda \in \mathcal{R}^1(\mathcal{A}_3, \Bbbk)$ then there exists $b_\mu \in \mathbb{P}^5 \setminus \{a_\lambda\}$ such that the projective line $a_\lambda * b_\mu$ spanned by a_λ and b_μ lies in $\mathcal{R}^1(\mathcal{A}_3, \Bbbk)$.

Since the set of rank 2 circuits of $\mathcal{A}_3$ is $\mathcal{C} = \{135, 146, 234, 256\}$, we have

$$\begin{aligned}
\partial e_{135} &= e_{35} - e_{15} + e_{13} = (e_3 - e_1)(e_5 - e_3) \in I,\\
\partial e_{146} &= e_{46} - e_{16} + e_{14} = (e_4 - e_1)(e_6 - e_4) \in I,\\
\partial e_{234} &= e_{34} - e_{24} + e_{23} = (e_3 - e_2)(e_4 - e_3) \in I,\\
\partial e_{256} &= e_{56} - e_{26} + e_{25} = (e_5 - e_2)(e_6 - e_5) \in I,
\end{aligned}$$

and therefore $a_{\lambda_1} = (-1:0:1:0:0:0)$, $a_{\lambda_2} = (-1:0:0:1:0:0)$, $a_{\lambda_3} = (0:-1:1:0:0:0)$ and $a_{\lambda_4} = (0:-1:0:0:1:0)$ lie in $\mathcal{R}^1(\mathcal{A}, \Bbbk)$ with respective $b_{\mu_1} = (0:0:-1:0:1:0)$, $b_{\mu_2} = (0:0:0:-1:0:1)$, $b_{\mu_3} = (0:0:-1:1:0:0)$ and $b_{\mu_4} = (0:0:0:0:-1:1)$. Then, one can easily see that the lines

$$\begin{aligned}
L_{135} &: \; z_1 + z_3 + z_5 = z_2 = z_4 = z_6 = 0,\\
L_{146} &: \; z_1 + z_4 + z_6 = z_2 = z_3 = z_5 = 0,\\
L_{234} &: \; z_2 + z_3 + z_4 = z_1 = z_5 = z_6 = 0,\\
L_{156} &: \; z_1 + z_5 + z_6 = z_2 = z_3 = z_4 = 0
\end{aligned}$$

coming from rank 2 circuits lie in $\mathcal{R}^1(\mathcal{A}_3, \Bbbk)$ and that they are local components of $\mathcal{R}^1(\mathcal{A}_3, \Bbbk)$.

On the other hand, we have

$$(e_1 + e_2 - e_4 - e_5)(e_1 + e_2 - e_3 - e_6) = -\partial e_{135} + \partial e_{146} - \partial e_{234} + \partial e_{256} \in I.$$

Then the projective line

$$L_F : z_1 = z_2,\ z_3 = z_6,\ z_4 = z_5,\ z_1 + z_3 + z_5 = 0$$

generated by $a_{\lambda_5} = (1 : 1 : 0 : -1 : -1 : 0)$ and $b_{\mu_5} = (1 : 1 : -1 : 0 : 0 : -1)$ also lies in $\mathcal{R}^1(\mathcal{A}_3, \Bbbk)$ and it is a non-local component corresponding to neighbourly partition $F = \{12|36|45\}$. This is the only neighbourly partition of any rank 3 submatroid of $\mathcal{M}(\mathcal{A}_3)$. Thus, the first resonance variety of the rank 3 braid arrangement of type $\mathcal{A}_3$ is

$$\mathcal{R}^1(\mathcal{A}_3, \Bbbk) = L_{135} \cup L_{136} \cup L_{234} \cup L_{156} \cup L_F.$$

As it is seen from this example, the first resonance variety has a beautiful theory related with rank 2 circuits of arrangement, neighbourly partitions (see [46, 47, 48, 158]) and nets (see [52, 142, 148, 160]).

The second cohomology ring of the complex (A, a) for the rank 3 braid arrangement of type $\mathcal{A}_3$ is

$$H^2(A, a) = \{d \in A^2 |\ ad = 0\} \Big/ \{d \in A^2 |\ \exists c \in A^1 \text{ s.t. } d = ac\},$$

and the second resonance variety is

$$\begin{aligned}\mathcal{R}^2(\mathcal{A}_3, \Bbbk) &= \{a_\lambda \in \Delta_6 |\ H^2(A, a) \neq 0\} \\ &= \{a_\lambda \in \Delta_6 |\ \exists d \in A^2 \text{ s.t. } ad = 0, \text{ and } d \neq ac,\ \forall c \in A^1\} \subset \mathbb{P}^5.\end{aligned}$$

Since

$$\begin{aligned}\partial e_{ijkl} &= e_{jkl} - e_{ikl} + e_{ijl} - e_{ijk} \\ &= (e_j - e_i)(e_k - e_j)(e_l - e_k) \\ &= (e_j - e_i)\partial e_{jkl} \in I.\end{aligned}$$

then all points lying on the projective plane $a_{\lambda_{ij}} * a_{\lambda_{jk}} * a_{\lambda_{kl}}$ spanned by $a_{\lambda_{ij}}$, $a_{\lambda_{jk}}$ and $a_{\lambda_{kl}}$ also lie in $\mathcal{R}^2(\mathcal{A}, \Bbbk)$. Here, $a_{\lambda_{ij}} = a_j - a_i$ and rank 3 circuits $ijkl$ are 1234, 1256 and 3456. For the other four-tuples, either ∂e_{jkl} may lie in I, i.e, may equal 0 in A^2, or its factorization contains an element from A^1. So, it is clear that the set of six points corresponding to linear factors of the boundary of these circuits span Δ_6, i.e., $\mathcal{R}^2(\mathcal{A}_3, \Bbbk) = \Delta_6$.

Similarly the third resonance variety is

$$\begin{aligned}\mathcal{R}^3(\mathcal{A}_3, \Bbbk) &= \{a_\lambda \in \Delta_6 |\ H^3(A, a) \neq 0\} \\ &= \{a_\lambda \in \Delta_6 |\ \exists f \in A^3 \text{ s.t. } af = 0, \text{ and } f \neq ad,\ \forall d \in A^2\} \subset \mathbb{P}^5 \\ &= \Delta_6.\end{aligned}$$

In fact, if $\overline{\mathcal{A}_3}$ is the deconing of $\mathcal{A}_3$ onto the plane $z = 0$ then its Hilbert polynomial is $\pi(\overline{\mathcal{A}_3}, t) = (1 + 2t)(1 + 3t)$ and the β-invariant of the matroid $\mathcal{M}(\mathcal{A}_3)$ is $\pi(\overline{\mathcal{A}_3}, -1) = 2$. Therefore, if a_λ is generic, i.e., if $\sum_{\mathcal{H}_i \in \mathcal{A}_X} \lambda_i \neq 0$ for all irreducible $X \in \mathcal{L}$, with $a_\lambda \in \Delta_6$, then

$$H^p(A(\overline{\mathcal{A}_3}), a) = 0, \text{forall } p < \ell - 1 \tag{2.33}$$

and

$$H^p(A,a) = \begin{cases} 0, & p=0,1, \\ \mathbb{C}^2, & p=2,3. \end{cases} \tag{2.34}$$

This also says that $\mathcal{R}^2(\mathcal{A}_3,\Bbbk) = \mathcal{R}^3(\mathcal{A}_3,\Bbbk) = \Delta_6$.

The braid arrangement $\mathcal{A}_3$ is the smallest arrangement for which the resonance variety contains a non-local component ([46]). The rank ℓ braid arrangement is given by the equation

$$\prod_{1\le i<j\le \ell+1} (z_i - z_j) = 0, \tag{2.35}$$

or equivalently the braid arrangement of type $\mathcal{A}_\ell$ is given by the equation

$$Q(\mathcal{A}_\ell) = \prod_{i=1}^{\ell} z_i \cdot \prod_{1\le i<j\le \ell} (z_i - z_j) = 0. \tag{2.36}$$

Compute the p-th resonance variety $\mathcal{R}^p(\mathcal{A}_\ell,\Bbbk)$ and study its geometry. Results may be a motivation to compute $\mathcal{R}^p(\mathcal{A},\Bbbk)$ for any arrangement $\mathcal{A}$. It might be helpful to remember that if $A = \bigoplus_{i=0}^{\ell} A^i$, then each A^i has grading by the rank i elements of the intersection lattice $\mathcal{L}$, i.e.,

$$A^i = \bigoplus_{X\in\mathcal{L},\ rk(X)=i} A_X,$$

and $A_X = A(\mathcal{A}_X)$ for the subarrangement $\mathcal{A}_X = \{\mathcal{H}|\ X \subset \mathcal{H}\} \subseteq \mathcal{A}$.

The solution of this problem will be a motivation to understand the geometry of resonance varieties.

On the other hand, another most studied case of numerical invariants of an arrangement group $G = \pi_1(M)$ are the ranks of its LCS (lower central series) quotients,

$$\phi_k = \phi_k(G) = \text{rank } \gamma_k(G)/\gamma_{k+1}(G).$$

The impetus came from the work of Kohno [71], who used rational homotopy theory to compute the LCS ranks of the braid arrangement group. Falk and Randell [49], using more direct methods, established the celebrated LCS formula for the broader class of fibre type arrangements, expressing ϕ_k in terms of the exponents of the arrangement $\mathcal{A}$. The LCS formula of Falk and Randell is

$$\prod_{k\ge 1}(1-t^k)^{\phi_k} = \pi(M,-t) = \prod_{i=1}^{\ell}(1-d_i t).$$

Consequently,

$$\phi_k(G) = \sum_{i=1}^{\ell} w_k(d_i).$$

As shown by Papadima and Yuzvinsky in [110], the converse holds for arrangements in $\mathbb{C}^3$: If $\prod_{k\ge 1}(1-t^k)^{\phi_k} = \pi(M,-t)$, then $\mathcal{A}$ is fibre type. The best known fibre type arrangement is the braid arrangement with exponents $\{1,2,...,\ell-1\}$. In that case, the LCS formula was first proved by Kohno [71], and it has been interpreted as a consequence of Koszul duality by Shelton and Yuzvinsky in [129], and has been generalized to hypersolvable arrangements by Jambu and Papadima in [64]. It also has been proved for rational $K(\pi,1)$ arrangements by Papadima and Yuzvinsky in [110], for decomposable arrangements by Papadima and Suciu in [109], and more recently by Lima-Fihlo and Schenck [79] for graphical arrangements.

In [143], Suciu conjectured the resonance LCS formula only for arrangements satisfying the equality $\phi_4 = \theta_4$ between the ranks of LCS quotients and the ranks of Chen groups of arrangements. The formula looks very similar to the LCS formula for fibre type arrangements, with the exponents being replaced by the dimensions of the components of the resonance variety. However, Falk and Randell's classical LCS formula does not imply the Suciu's resonance LCS formula, since fibre type arrangements do not satisfy the equality $\phi_4 = \theta_4$. Schenck and Suciu, in [128], recovered a formula of Falk for ϕ_3 and obtained a formula for ϕ_4, and showed that resonance LCS conjecture is true for graphic arrangement.

If one completely understood the geometry of resonance varieties, then it is expected that the resonance LCS conjecture could be solved.

Problem 10 (G.Denham). *If $\mathcal{L}_\lambda$ is a rank-1 local system on the complement of M of a rank ℓ hyperplane arrangement, is it necessarily the case that*

$$H^i(M,\mathcal{L}_\lambda) \neq 0 \;\Rightarrow\; H^{i+1}(M,\mathcal{L}_\lambda) \neq 0$$

for all $i < \ell$?

Problem 11 (A. Suciu). *Show that, for any arrangement $\mathcal{A}$,*

$$\theta_k(G) = (k-1)\sum_{r\geq 1} h_r \binom{r+k-1}{k} \text{ for large enough } k, \tag{2.37}$$

where h_r is the number of components of $\mathbb{P}\mathcal{R}^1(\mathcal{A})$ of dimension r, and θ_k is the rank of the k-th lower central series quotient of G/G''.

During his lectures, Denham mentioned Problems 10 and 11, and his lecture note [36] ends with them. Problem 11 is first suggested by Alex Suciu in [143]. In order to collect all open problems together, we include these problems as well. For motivation to the problems and notation please read Denham's lecture note [36] and Suciu's paper [143].

2.9. Torsion in the first fundamental group

Problem 12 (M. Falk). *Is $\pi_1(M)$ torsion free for every hyperplane arrangement $\mathcal{A}$?*

One of the fundamental problems in the topological study of polynomial functions $f : (\mathbb{C}^\ell, \mathbf{0}) \to (\mathbb{C}, 0)$ is the computation of the first fundamental group of the complement M to the hypersurface $f^{-1}(0)$. In 1929, Zariski proved in [161] that the fundamental group of the complement of n-lines in general position is abelian. A few years later, van Kampen ([66]) developed a general method for computing the fundamental group of the complement of any curve in $\mathbb{CP}^2$. Since an arrangement $\mathcal{A}$ is a union of hyperplanes, the Zariski-van Kampen method for line arrangements was used as motivation to develop a simple algorithm for the general case.

In the special case when $\mathcal{A}$ is a complexified real arrangement, a presentation of $\pi_1(M)$ was obtained by Randell [117] and Salvetti [125, 126]. Their main result asserts that a presentation may be obtained from the underlying real arrangement of $\mathcal{A}$. In his Ph.D. thesis [14], Arvola gave a presentation for the first fundamental group of the complement of an arbitrary arrangement. A different approach, using the notion of "labyrinth" is adopted by Dung and Vui [40] to arrive at similar presentations of $\pi_1(M)$ for any arrangement $\mathcal{A}$.

In these presentations one first takes a planar section so that one is working with an affine line arrangement in $\mathbb{C}^2$. Then the generators are meridians for each line, and relations come from each intersection point of lines. This method is useful to compute the first fundamental group.

The concept of braid monodromy was introduced by Moishezon [95]. By using Hansen's theory of polynomial covering maps, Cohen and Suciu [26] gave an explicit description of the braid monodromy of a complex arrangement, and showed that the resulting presentation of the first fundamental group is equivalent to the Randell-Arvola presentation.

On the other hand, the most popular such a group is certainly the pure braid group. It appears as the fundamental group of the complement of the braid arrangement. So, $\pi_1(M)$ can be considered as a generalization of the pure braid group, and one can expect to show that many properties of the pure braid group also hold for $\pi_1(M)$. However, the only general known results on this group are its presentations, and some examples of arrangement groups G for which $H^3(G, \mathbb{Q})$ is infinite dimensional (see [14, 88, 12]). Many interesting questions remain, for example, to know whether such a group is torsion free or has finite cohomological dimension.

In 1972, Deligne [34] proved that for a complexification of a real simplicial arrangement, the complement M is aspherical (also expressed by saying that M is a $K(\pi, 1)$ space). That is, the universal cover of M is contractible. By Brieskorn's work [23], the complement M for a complexification of any real reflection arrangement is $K(\pi, 1)$, since all real reflection arrangements are simplicial. Following [144], Falk and Randell introduced in [49] the notion of a fibre-type arrangement and observed that for this class M is aspherical, essentially by the iterated fibration argument of Fadell and Neuwirth ([43]).

If M is aspherical, the cohomology of M is isomorphic to the cohomology of the group $\pi_1(M)$. Since M has cohomological dimension $rk(\mathcal{A}) < \infty$, so does $\pi_1(M)$. In addition, $\pi_1(M)$ has no torsion, and there is a $K(\pi, 1)$ space, $\pi = \pi_1(M)$, with the homotopy type of a finite complex. So, it is natural to ask torsion freeness of $\pi_1(M)$ for any hyperplane arrangement $\mathcal{A}$. The answer is of course affirmative for simplicial arrangements, real reflection arrangements, fibre-type or supersolvable arrangements and for $K(\pi, 1)$ arrangements (see [49, 111]). For non-$K(\pi, 1)$

examples, such as general position arrangements and the examples of Artal, Matei and Cogolludo ([12]), nevertheless have finite cohomological dimension, hence torsion free fundamental groups.

2.10. Torsion in the homology of the Milnor fibre

Problem 13 (M. Falk). *Is the first homology of the Milnor fibre of a hyperplane arrangement $\mathcal{A}$ torsion free?*

Let $\mathcal{A}$ be a hyperplane arrangement in $\mathbb{C}^\ell$ given by the defining polynomial $Q = \prod_{i=1}^n \alpha_i$ which is homogeneous of degree n. In this case, Q can be considered as a map $Q : M \to \mathbb{C}^* = \mathbb{C} \setminus \{0\}$, where M is a complement of the hyperplane arrangement $\mathcal{A}$. It is well known that this map is the projection of a fibre bundle, called the *Milnor fibration* ([93]), and the *Milnor fibre* $F = Q^{-1}(1)$ should be of interest. In [118], it was shown that the Milnor fibration is constant in a lattice-isotopic family, so that the Milnor fibre is indeed an invariant of lattice-isotopy. Because of this, (analogous to the definition made in the theory of knots) two arrangements are said to be topologically equivalent (or have the same topological type) if they are lattice isotopic. Thus, the Milnor fibre and fibration are invariants of topological type ([118]).

Early results concerning the Milnor fibre of an arrangement (often in the general context of plane curves) appear in the work of Libgober [74, 75, 76, 77] and Randell [116], particularly with respect to Alexander invariants. Libgober's works gave considerable information about the homology of the Milnor fibre by means of the number of singularities appearing in the arrangement together with their types and position, and the number of lines. In the article [116], it was observed that an Alexander polynomial is equal to the characteristic polynomial of the monodromy on the Milnor fibre. In [25], Cohen and Suciu used covering space theory and homology with local coefficients to study the Milnor fibre of a homogeneous polynomial and, applying these techniques to hyperplane arrangements, they obtained an explicit algorithm for computing the Betti numbers of the associated Milnor fibre, as well as the dimensions of the eigenspaces of the algebraic monodromy, for arbitrary real central arrangements in $\mathbb{C}^3$. They also obtained combinatorial formulas for these invariants of the Milnor fibre of a generic arrangement of arbitrary dimension, recovering a result due to Orlik and Randell [104]. A few years later, by using the group presentation and Fox calculus, Cohen and Suciu formed twisted chain complexes whose homology gave that of the Milnor fibre ([28]). They provided an effective method supported by several explicit examples.

Later, in [37], Dimca and Némethi posed the problem of finding a homogeneous polynomial $f : (\mathbb{C}^{\ell+1}, \mathbf{0}) \to (\mathbb{C}, 0)$ such that the homology of the complement of the hypersurface defined by f is torsion free, but the homology of the Milnor fibre F of f has torsion. Cohen, Denham and Suciu proved in [29] that this was indeed possible and showed by constructing that, for each prime p, there is a polynomial $f_p : \mathbb{C}^3 \to \mathbb{C}$ with p-torsion in the homology of the Milnor fibre $F_p = f_p^{-1}(1)$. More precisely, they considered the polynomial

$$f_p := f_p(x,y,z) = \begin{cases} xy(x^p - y^p)^2(x^p - z^p)(y^p - z^p), & \text{if } p \text{ is odd,} \\ x^2y(x^2 - y^2)^3(x^2 - z^2)^2(y^2 - z^2), & \text{if } p = 2, \end{cases} \tag{2.38}$$

which defines a multi-arrangement of hyperplanes. As is well-known, for any hyperplane (multi)-arrangement, the homology groups $H_*(M,\mathbb{Z})$ of the complement M are finitely generated and torsion free (see [35, 119]). They computed the first homology of the Milnor fibre $F_p = f_p^{-1}(1)$ as

$$H_1(F_p,\mathbb{Z}) = \begin{cases} \mathbb{Z}^{3p+1} \oplus \mathbb{Z}_p \oplus T, & \text{if } p \text{ is odd,} \\ \mathbb{Z}^{3p+1} \oplus \mathbb{Z}_2 \oplus \mathbb{Z}_2, & \text{if } p = 2, \end{cases} \tag{2.39}$$

where T is a finite abelian group satisfying $T \otimes \mathbb{Z}_q = 0$ for every prime q such that $q \nmid 2(2p+1)$, and showed that it has p-torsion.

2.11. Stability of hyperplane arrangements

Problem 14 (M. Yoshinaga). *Let $a_1 \le a_2$, $b_1 \le b_2$ and $c_1 \le c_2$ be given positive integers and $\mathcal{A}(k)$ denote the arrangement in $\mathbb{C}^3$ defined by*

$$Q(\mathcal{A}(k)) := z \cdot \prod_{p=a_1-k}^{a_1+k} (x - pz) \cdot \prod_{q=b_1-k}^{b_1+k} (y - qz) \cdot \prod_{r=c_1-k}^{c_1+k} (x + y - rz) = 0. \tag{2.40}$$

Study the structure of $D(\mathcal{A}(k))$, particularly, dependence of k.

A hyperplane arrangement $\mathcal{A}$ is a finite collection of codimension 1 affine hyperplanes of a fixed ℓ-dimensional $\Bbbk$-vector space $V \simeq \Bbbk^\ell$. An arrangement $\mathcal{A}$ is *central* if each hyperplane is a vector subspace of V. Let $\{X_1, X_2, \ldots, X_\ell\}$ be a basis for the dual space V^* and put $S := \mathrm{Sym}(V^*) \simeq \Bbbk[X_1, X_2, \ldots, X_\ell]$. For each hyperplane $\mathcal{H} \in \mathcal{A}$, let us fix a nonzero linear form $\alpha_\mathcal{H} \in V^*$ such that its kernel is $\mathcal{H}$, and put $Q(\mathcal{A}) := \prod_{\mathcal{H}\in\mathcal{A}} \alpha_\mathcal{H}$. For an arrangement $\mathcal{A}$, we can define the S-module

$$\begin{aligned} D(\mathcal{A}) &:= \{\theta \in \mathrm{Der}_\Bbbk(S) \mid \theta(\alpha_\mathcal{H}) \in S \cdot \alpha_\mathcal{H}, \forall \mathcal{H} \in \mathcal{A}\} \\ &= \{\theta \in \mathrm{Der}_\Bbbk(S) \mid \theta(Q(\mathcal{A})) \in S \cdot Q(\mathcal{A})\}, \end{aligned} \tag{2.41}$$

where $\mathrm{Der}_\Bbbk(S)$ is the S-module of $\Bbbk$-linear derivations of S.

The module $D(\mathcal{A})$ is called the module of logarithmic vector fields (with respect to $\mathcal{A}$). A nonzero element $\theta = \sum_{i\in[1,\ell]} f_i \frac{\partial}{\partial X_i}$ is said to be *homogeneous* of degree p if $f_i \in S_p$ for $i \in [1,\ell]$. An arrangement $\mathcal{A}$ is *free* if $D(\mathcal{A})$ is a free module. When $\mathcal{A}$ is free, there exists a homogeneous basis $\{\theta_1, \theta_2, \ldots, \theta_\ell\}$ for $D(\mathcal{A})$. Then the *exponents* of a free arrangement $\mathcal{A}$ are defined by

$$\exp(\mathcal{A}) := \{\deg(\theta_1), \deg(\theta_2), \ldots, \deg(\theta_\ell)\} \tag{2.42}$$

and $\exp(\mathcal{A})$ do not depend on the choice of a basis.

On the other hand, the characteristic polynomial $\chi(\mathcal{A},t)$ and the Poincaré polynomial $\pi(\mathcal{A},t)$ are related as follows:

$$\chi(\mathcal{A},t) = t^\ell \pi(\mathcal{A}, -1/t). \tag{2.43}$$

These polynomials are important concepts in the theory of hyperplane arrangements. Actually, there a lot of combinatorial and geometric interpretations of the characteristic polynomial (for details, see [106]). The Poincaré polynomial can be used to calculate the Chern polynomial.

Theorem 2.2 ([99], **Theorem 4.1).** *For a polynomial $F(t) \in \mathbb{Z}[t]$, let $\overline{F(t)}$ denote the class of $F(t)$ in $\mathbb{Z}[t]/(t^\ell)$. Let $\mathcal{A}$ be a central ℓ-arrangement and assume $\widetilde{D(\mathcal{A})}$ is a sheafification of $D(\mathcal{A})$. Then it holds that*

$$c_t(\widetilde{D(\mathcal{A})}) = \overline{\pi(\mathcal{A}, -t)}.$$

In particular, if $\ell = 3$ and $\chi_0(\mathcal{A}, t) = t^2 - c_1 t + c_2$ then for any central 3-arrangement $\mathcal{A}$ it holds that

$$c_t(\widetilde{D(\mathcal{A})}) = (1 - c_1 t + c_2 t^2)(1 - t).$$

In dimension 3, fix a basis $\{X, Y, Z\}$ for V^* in such a way that the hyperplane $\{Z = 0\}$ is an element of $\mathcal{A}$. The module of reduced logarithmic vector fields $D_0(\mathcal{A})$ is defined by

$$D_0(\mathcal{A}) := \{\theta \in D(\mathcal{A}) \mid \theta(Z) = 0\}.$$

Note that for any central arrangement $\mathcal{A}$, there exists a derivation

$$\theta_{\mathbf{E}} := X\frac{\partial}{\partial X} + Y\frac{\partial}{\partial Y} + Z\frac{\partial}{\partial Z} \in D(\mathcal{A}),$$

which is called Euler derivation. It is obvious that $D_0(\mathcal{A}) = D(\mathcal{A})/(S \cdot \theta_E)$. Moreover, in the notation of Theorem 2.2, it holds that

$$c_t(\widetilde{D_0(\mathcal{A})}) = 1 - c_1 t + c_2 t^2.$$

These results, together with [136, 145], show that the characteristic polynomial of an arrangement $\mathcal{A}$ is determined by algebraic structures of $D(\mathcal{A})$. It is therefore natural to ask whether the combinatorial behaviour of characteristic polynomials is controlled by algebraic structures of logarithmic vector fields ([41, 114, 156]). In our example, since it is known that ([15])

$$\chi(\mathcal{A}(k+1), t) = \chi(\mathcal{A}(k), t-3), \text{ for } k \gg 0,$$

the corresponding modules $D(\mathcal{A}(k))$ and $D(\mathcal{A}(k+1))$ should be closely related ([4, 124, 144, 146, 155]).

Next, we review some definitions and results on the stability of vector bundles on projective spaces.

A torsion free sheaf E on the projective space $\mathbb{P}^n_{\Bbbk}$ is said to be *stable* if for any coherent subsheaf $F \subset E$ with $0 < rank(F) < rank(E)$ we have

$$slope(F) = \frac{c_F}{rank(F)} < \frac{c_E}{rank(E)} = slope(E)$$

and it is *semistable* if

$$slope(F) = \frac{c_F}{rank(F)} \leq \frac{c_E}{rank(E)} = slope(E).$$

Moreover, E is *unstable* if it is not stable. A 3-arrangement $\mathcal{A}$ is called *stable* (resp. *semistable*) if the reflexive sheaf $\widetilde{D_0(\mathcal{A})}$ is a stable (resp. semistable) torsion free sheaf on $\mathbb{P}(V)$.

Lemma 2.3 (**[102], Lemma 1.2.4, Ch. II).** *A torsion free sheaf E on the projective space $\mathbb{P}^n_{\Bbbk}$ is stable if and only if $E \otimes \mathcal{O}_{\mathbb{P}^n}(d)$ is stable for some $d \in \mathbb{Z}$.*

For a rank 2 vector bundle E on $\mathbb{P}^n_{\Bbbk}$ $(n \geq 2)$, there exists a unique integer d_E such that $c_1(E \otimes \mathcal{O}_{\mathbb{P}^n}(d_E)) \in \{0, -1\}$. Here, $E \otimes \mathcal{O}_{\mathbb{P}^n}(d_E)$ is called the normalisation of E and denoted by E_{norm}. The normalised Chern polynomial of E is then the Chern polynomial of E_{norm}.

Lemma 2.4 (**[102], Lemma 1.2.5, Ch. II).** *Let E be a rank 2 bundle on $\mathbb{P}^n_{\Bbbk}$ $(n \geq 2)$. Then E is stable if and only if $H^0(\mathbb{P}^n_{\Bbbk}, E_{norm}) = 0$. Moreover, if $c_1(E)$ is even, then E is semistable if and only if $H^0(\mathbb{P}^n_{\Bbbk}, E_{norm}(-1)) = 0$.*

The stability of normal crossing arrangements was studied by Dolgachev and Kapranov in [39]. The following criterion for the stability of arrangements was given by Schenck in [127].

Theorem 2.5 (**[127], Theorem 4.5).** *Let $\mathcal{A}$ be an arrangement of d lines in $\mathbb{P}^2$, $\mathcal{H}_0$ be a line in $\mathcal{A}$, and let us put $\mathcal{A}' := \mathcal{A} \setminus \{\mathcal{H}_0\}$. Then the following hold:*

(i) *If d is odd, then $\mathcal{A}$ is stable if $\mathcal{A}'$ is stable and $|\mathcal{A} \cap \mathcal{H}_0| > (d+1)/2$.*
(ii) *If d is odd, then $\mathcal{A}$ is semistable if $\mathcal{A}'$ is semistable and $|\mathcal{A} \cap \mathcal{H}_0| > (d-1)/2$.*
(iii) *If d is even, then $\mathcal{A}$ is stable if $\mathcal{A}'$ is semistable and $|\mathcal{A} \cap \mathcal{H}_0| > d/2$.*

Under the light of this information, Abe [1] studied the stability of the family of arrangements which are not normal crossings and gave a necessary and sufficient combinatorial condition for the stability and the freeness of A_2-type arrangements (the A_2-type arrangement is a special case of the Coxeter arrangement of A_2-type). Moreover, he determined explicitly when the normalisation of the sheafification of its module of reduced logarithmic vector fields is isomorphic to $T_{\mathbb{P}^2}(-2)$ and gave a partial answer to the 3-shift problem, which is the conjecture on the root system posed by Yoshinaga (see Remark 3.3 in [1]). These results pose a problem whether the stability of arrangements is determined by the combinatorial structure of arrangement. This is similar to the Terao conjecture, which asserts the freeness of arrangements depends only on the combinatorics of arrangements. Since Abe got answers in a positive sense for the stability problem, in his next paper [2], he introduced B_2-type arrangements as a generalization of the classical Coxeter arrangement of type B_2, and considered the stability and the freeness of it. He showed their (semi)stability is determined by its combinatorics. Moreover, he gave a partial answer to the 4-shift problem. Roughly speaking, it asserts the interesting combinatorial behavior of a family of B_2-type arrangements $\{\mathcal{A}(k)\}$ is governed by the geometric property of the sheaf of logarithmic vector fields$\{\widetilde{D_0(\mathcal{A}(k))}\}$, i.e., there exist isomorphisms

$$\widetilde{D_0(\mathcal{A}(k+1))} \simeq \widetilde{D_0(\mathcal{A}(k))} \otimes \mathcal{O}(-4), \quad (k \gg 0). \tag{2.44}$$

Since the shift and the Coxeter number of the root system of type B_2 are both 4, this conjecture is called the 4-shift problem. For details on the 4-shift problem, see Remark 2.3 in [2].

2.12. Profinite completion of fundamental groups

Problem 15 (D. Matei). *Is there any pair of plane curves $\mathcal{C}_1, \mathcal{C}_2 \subset \mathbb{P}^2$ such that if $G_i = \pi_1(\mathbb{P}^2 \setminus \mathcal{C}_i)$, $i = 1, 2$, then $\widehat{G}_1 \cong \widehat{G}_2$, but $G_1 \not\cong G_2$?*

The *profinite completion* $\widehat{G}$ of a group G is the inverse limit of the directed system of finite quotients of G, i.e.,

$$\widehat{G} := \varprojlim_{N \lhd G} G/N \tag{2.45}$$

such that G/N is finite. If G is a *residually finite group* (i.e., if for every nontrivial element g in G there is a homomorphism h from G to a finite group such that $h(g) \neq 1$) then the natural map $G \to \widehat{G}$ is injective. In [60], Grothendieck discovered a remarkably close connection between the representation theory of a finitely generated group and its profinite completion: if $A \neq 0$ is a commutative ring and $u : G_1 \to G_2$ is a homomorphism of finitely generated groups, then $\widehat{u} : \widehat{G}_1 \to \widehat{G}_2$ is an isomorphism if and only if the restriction functor $u_A^* : \mathrm{Rep}_A(G_2) \to \mathrm{Rep}_A(G_1)$ is an equivalence of categories, where $\mathrm{Rep}_A(G)$ is the category of finitely presented A-modules with a G-action.

Grothendieck investigated circumstances under which $\widehat{u} : \widehat{G}_1 \to \widehat{G}_2$ being an isomorphism implies that u is an isomorphism of the original groups. This led him to pose the celebrated problem:

Is there any pair G_1, G_2 of finitely presented, residually finite groups such that $\widehat{G}_1 \cong \widehat{G}_2$ but $G_1 \not\cong G_2$?

A negative solution to the corresponding problem for finitely generated groups was given by Platanov and Tavgen in [112, 113]. The methods used in [112] subsequently inspired Bass and Lubotzky's construction of finitely generated linear groups that are super-rigid but are not of arithmetic type ([16]). They discovered a host of other finitely generated, residually finite groups such that $\widehat{G}_1 \cong \widehat{G}_2$ but $G_1 \not\cong G_2$. All of these examples are based on a fibre product construction and it seems that none of them is finitely presentable. Indeed, as the authors of [16] note, "a result of Grunewald ([61], Prob. B) suggests that such fibre products are rarely finitely presentable."

In [115], L. Pyber constructed continuously many pairs of 4-generator groups G_1, G_2 such that $\widehat{u} : \widehat{G}_1 \to \widehat{G}_2$ is an isomorphism but $u : G_1 \to G_2$ is not. Once again, these groups are not finitely presented.

For further readings about Grothendieck's problems, the following references are suggested: Bridson [19, 20, 21, 22], Grunewald [61], Platanow-Tavgen [112, 113] and Pyber [115].

A pair of $(\mathcal{C}_1, \mathcal{C}_2)$ of plane curves of the same degree is called a *Zariski pair* if there exist tubular neighbourhoods $T_1 \subset \mathbb{P}^2$ of $\mathcal{C}_1$ and $T_2 \subset \mathbb{P}^2$ of $\mathcal{C}_2$ such that $(T_1, \mathcal{C}_1)$ and $(T_2, \mathcal{C}_2)$ are diffeomorphic while $(\mathbb{P}^2, \mathcal{C}_1)$ and $(\mathbb{P}^2, \mathcal{C}_2)$ are not homeomorphic. The first example of Zariski pairs was studied by Zariski [161] in order to show that an equisingular family of plane curves need not be connected. Later, Zariski pairs were studied by several authors. Since then, many examples of Zariski pairs have been given.

A Zariski pair $(\mathcal{C}_1, \mathcal{C}_2)$ is called an *arithmetic Zariski pair* if $\mathcal{C}_1$ and $\mathcal{C}_2$ are conjugate. A difficulty in constructing examples of arithmetic Zariski pairs comes from the fact that, if $\mathcal{C}_1$ and $\mathcal{C}_2$ are conjugate, then $\pi_1(P^2 \setminus \mathcal{C}_1)$ and $\pi_1(P^2 \setminus \mathcal{C}_2)$ have the same profinite completions. Artal Bartalo, Carmona Ruber, and Cogolludo Agustín [9, 10] constructed an arithmetic Zariski pair in degree 12. They distinguished $(\mathbb{P}^2, \mathcal{C}_1)$ and $(\mathbb{P}^2, \mathcal{C}_2)$ by means of the braid monodromy. Later, in [132], Shimada introduced an invariant $N_\mathcal{C}$ of the homeomorphism type of $(\mathbb{P}^2, \mathcal{C})$ for plane curves $\mathcal{C}$ of even degree. By means of this invariant, he presented some examples of arithmetic Zariski pairs in degree 6. Let us explain shortly his examples by Dynkin types. A *Dynkin type* is a finite formal sum

$$R = \sum_{\ell \geq 1} a_\ell A_\ell + \sum_{m \geq 4} b_m B_m + \sum_{n=6}^{8} e_n E_n, \tag{2.46}$$

where a_ℓ, d_m, e_n are non-negative integers and almost all of them are zero. The rank of the Dynkin type R is defined by

$$\operatorname{rank}(R) = \sum_{\ell \geq 1} a_\ell \ell + \sum_{m \geq 4} b_m m + \sum_{n=6}^{8} e_n n. \tag{2.47}$$

An *ADE-sextic* is a plane curve of degree 6 with only simple singularities. The type R of ADE-sextic $\mathcal{C}$ is the Dynkin type of the singularities of $\mathcal{C}$. Then $\operatorname{rank}(R)$ is equal to the total Milnor number of $\mathcal{C}$. Hence, it is at most 19. If $\mathcal{C}$ is an ADE-sextic, then the minimal resolution $X_\mathcal{C}$ of the double covering $Y_\mathcal{C} \to \mathbb{P}^2$ that branches exactly along $\mathcal{C}$ is a $K3$-surface. When $\mathcal{C}$ is a maximizing sextic, i.e., $\operatorname{rank}(R) = 19$, then the invariant $N_\mathcal{C}$ of the pair $(\mathbb{P}^2, \mathcal{C})$ coincides with the transcendental lattice of $X_\mathcal{C}$. Combining this result with Artal-Bartalo, Carmona-Ruber, Cogolludo-Agustín [8], Degtyarev [33] and the result in [134], Shimada proved the existence of arithmetic Zariski pairs of maximizing sextics for each of Dynkin types $A_{16} + A_2 + A_1$, $A_{16} + A_3$, $A_{18} + A_1$ and $A_{10} + A_9$ (for proofs see [132]). Also in [133], he gave more examples of arithmetic Zariski pairs of maximizing sextics (see Table 5.1 in [133]).

References

[1] T. Abe, *The stability of the family of A_2−type arrangements.* J. Math. Kyoto Univ. **46** (3) (2006), 617–636.

[2] T. Abe, *The stability of the family of B_2−type arrangements.* Hokkaido University Preprint Series 790 (2006).

[3] T. Abe and M. Yoshinaga, *Splitting criterion for reflexive sheaves.* Proc. of the Amer. Math. Soc. **136** (2008), 1887–1892.

[4] T. Abe and M. Yoshinaga, *Coxeter multiarrangements with quasi-constant multiplicities.* preprint arXiv:0708.3228.

[5] K. Aomoto, *Un théorème du type de Matsushima-Murakami concernant l'intégrale des functions multiformes.* J. Math. Pures Appl. **52** (1973), 1–11.

[6] V. I. Arnol'd, *The cohomology ring of the colored braid group.* Mat. Zametki, **5** (1960), 229–231.

[7] E. Artal Bartolo and J. Carmona Ruber, *Zariski pairs, fundamental groups and Alexander polynomials.* J. of Math. Soc. Japan **50** (1998), 521–543.

[8] E. Artal Bartolo, J. Carmona Ruber and J. I. Cogolludo Agustín, *On sextic curves with big Milnor number.* In Trends in singularities, Trends Math., 1–29, Birkhäuser, Basel, 2002.

[9] E. Artal Bartolo, J. Carmona Ruber and J. I. Cogolludo Agustín, *Braid monodromy and topology of plane curves.* Duke Math. J. **118** (2) (2003), 261–278.

[10] E. Artal Bartolo, J. Carmona Ruber and J. I. Cogolludo Agustín, *Effective invariants of braid monodromy.* Trans. Amer. Math. Soc. **359** (2007), 165–183.

[11] E. Artal Bartolo, J. Carmona Ruber, J. I. Cogolludo Agustin and M. M. Buzunáriz, *Invariants of combinatorial line arrangements and Rybnikov's example.* Compositio Mathematica **141** (2005), 1578–1588.

[12] E. Artal Bartolo, J. I. Cogolludo Agustin and D. Matei, *Arrangements of hypersurfaces and Bestvina-Brady groups.* In: Mini-Workshop: Topology of closed one-forms and cohomology jumping loci, Oberwolfach Rep. **4** no. 3 (2007), 2321–2360.

[13] E. Artal Bartolo, J. I. Cogolludo Agustin and H. Tokunaga, *A survey on Zariski pairs.* Seminario Matemático, García de galdeano, Universidad de Zaragoza, 2006.

[14] W. Arvola, *The fundamental group of the complement of an arrangement of complex hyperplanes.* Ph.D. Thesis, University of Wisconsin-Madison, 1990.

[15] C. A. Athanasiadis, *Generalized Catalan numbers, Weyl groups and arrangements of hyperplanes.* Bull. London Math. Soc. **36** no. 3 (2004), 294–302.

[16] H. Bass and A. Lubotzky, *Nonarithmetic superrigid groups: counterexamples to Platonov's conjecture.* Ann. of Math. **151** (2000), 1151–1173.

[17] A. Björner and G. Ziegler, *Combinatorial stratification of complex arrangements.* J. Amer. Math. Soc. **5** no. 1 (1992), 105–149.

[18] J. Fernandez de Bobadilla, *A reformulation of Lê's conjecture.* Indag. Math. (N.S.) **17** (3) (2006), 345–352.

[19] M. R. Bridson, *Decision problems and profinite completions of groups.* Preprint (2008).

[20] M. R. Bridson, *Direct factors of profinite completions and deciability.* J. Group Theory (2008), to appear.

[21] M. R. Bridson, *The Schur multiplier, profinite completions and deciability.* preprint, University of Oxford, April 2008.

[22] M. R. Bridson and F. J. Grunewald, *Grothendieck's problems concerning profinite completions and representations of groups.* Ann. of Math. **160** (2004), 359–373.

[23] E. Brieskorn, *Sur les groupes de tresses.* In: Séminaire Bourbaki 1971/72. Lecture Notes in Math. **317**, 21–44, Springer, Berlin, Heidelberg, New York, 1973.

[24] R.-O. Buchweitz and D. Mond, *Linear free divisors and quiver representations.* In: Singularities and Computer Algebra, C. Lossen and G. Pfister (Eds), London Math. Soc. Lecture Notes in Math. **324**, 41–77, Cambridge University Press, 2006.

[25] D. Cohen and A. Suciu, *On Milnor fibrations of arrangements.* J. London Math. Soc. **51** (1) (1995), 105–119.

[26] D. Cohen and A. Suciu, *The braid monodromy of plane algebraic curves and hyperplane arrangements.* Comment. Math. Helvetici **72** (1997), 285–315.

[27] D. Cohen and A. Suciu, *Characteristic varieties of arrangements.* Math. Proc. Cambridge Phil. Soc. **127** (1999), 33–53.

[28] D. Cohen and A. Suciu, *Alexander invariants of complex hyperplane arrangements.* Trans. Amer. Math. Soc. **351** (1999), 4043–4067.

[29] D. Cohen, G. Denham and A. Suciu, *Torsion in Milnor fiber homology.* Algebraic and Geometric Topology **3** (2003), 511–535.

[30] J. Damon, *Deformations of sections of singularities and Gorenstein surface singularities.* Amer. J. Math. **109** (1987), 695–722.

[31] J. Damon, *$\mathcal{A}$-equivalence and the equivalence of sections of images and discriminants.* In: D. Mond and J. Montaldi (Eds.), Singularity Theory and Applications, Warwick 1989, Lecture Notes in Math. **1462**, Springer, Berlin, Heidelberg, New York, 1991.

[32] J. Damon and D. Mond, *$\mathcal{A}$-Codimension and the vanishing topology of discriminants.* Invent. math. **106** (1991), 217–242.

[33] A. Degtyarev, *On deformations of singular plane sextics.* J. Algebraic Geom. **17** (2008), 101–135.

[34] P. Deligne, *Les immeubles des groupes de tresses génralisés.* Invent. Math. **17** (1972), 273–302.

[35] G. Denham, *The Orlik-Solomon complex and Milnor fiber homology.* Topology Appl. **118** (1-2) (2002), 45–63.

[36] G. Denham, *Homological aspects of hyperplane arrangements.* In: F. ElZein, A. Suciu, M. Tosun, A.M. Uludağ and S. Yuzvinsky (Eds.), Lecture notes of CIMPA summer school: Arrangements, Local Systems and Singularities. Progress in Mathematics, Birkhäuser, Basel, 2010.

[37] A. Dimca and A. Némethi, *Hypersurface complements, Alexander modules and monodromy.* Proceedings of the 7th workshop on Real and Complex Singularities, Sao Carlos, 2002; M. Ruas and T. Gaffney Eds, Contemp. Math., Amer. Math. Soc. (2004), 19–43.

[38] A. Dimca and S. Yuzvinsky *Lectures on Orlik-Solomon algebras.* In: F. ElZein, A. Suciu, M. Tosun, A.M. Uludağ and S. Yuzvinsky (Eds.), Lecture notes of CIMPA summer school: Arrangements, Local Systems and Singularities. Progress in Mathematics, Birkhäuser, Basel, 2010.

[39] I. Dolgachev and M. Kapranov, *Arrangements of hyperplanes and vector bundle on $\mathbb{P}^n$.* Duke Math. J. **71** no. 3 (1993), 633–664.

[40] N. V. Dung and H. H. Vui, *The fundamental group of complex hyperplane arrangements.* Acta Math. Vietnam **20** (1995), 31–41.

[41] P. H. Edelman and V. Reiner, *Free arrangements and rhombic tilings.* Discrete Comput. Geom. **15** no. 3 (1996), 307–340.

[42] H. Esnault, V. Schechtman and E. Viehweg, *Cohomology of local systems on the complement of hyperplanes.* Invent. Math. **109** (1992), 557–561.

[43] E. Fadell and L. Neuwirth, *Configuration spaces.* Mathematica Scandinavica **10** (1962), 111–118.

[44] M. Falk, *The minimal model of the complement of an arrangement of hyperplanes.* Trans. Amer. Math. Soc. **309** (1988), 543–556.

[45] M. Falk, $K(\pi,1)$ *arrangements.* Topology, **34** (1995), 141–154.

[46] M. Falk, *Arrangements and cohomology.* Annals of Combinatorics **1** (1997), 135–157.

[47] M. Falk, *The line geometry of resonance varieties.* arXiv:math/0405210v2.

[48] M. Falk, *Geometry and combinatorics of resonance weights.* In: F. ElZein, A. Suciu, M. Tosun, A.M. Uludağ and S. Yuzvinsky (Eds.), Lecture notes of CIMPA summer school: Arrangements, Local Systems and Singularities. Progress in Mathematics, Birkhäuser, Basel, 2010.

[49] M. Falk and R. Randell, *The lower central series of a fiber-type arrangement.* Invent. Math. **82** (1985), 77–88.

[50] M. Falk and R. Randell, *On the homotopy theory of arrangements.* In: Complex Analytic Singularities. Adv. Studies in Pure Math. **8**, 101–124, North Holland, 1986.

[51] M. Falk and R. Randell, *On the homotopy theory of arrangements II.* In: M. Falk, H. Terao (Eds.) Arrangements – Tokyo 1998, Adv. Stud. Pure Math. **27**, 93–25, Kinokoniyo, Tokyo, 2000.

[52] M. Falk and S. Yuzvinsky, *Multinets, resonance varieties, and pencils of plane curves.* Composito Math. **143** (4) (2007), 1069–1088.

[53] F.Y.C. Fung, *On the topology of components of some Springer fibers and their relation to Kazhdan-Lusztig theory.* Adv. Math. **178** (2) (2003), 244–276.

[54] P. Gabriel, *Unzerlegbare Darstellungen I.* Manuscripta Math. **6** (1972), 71–103.

[55] V. Goryunov and D. Mond, *Vanishing cohomology of singularities of mappings.* Compositio Math. **89** (1) (1993), 45–80.

[56] M. Granger, D. Mond, A. Nieto and M. Schulze, *Linear Free Divisors.* submitted.

[57] M. Granger and D. Mond, *Linear Free Divisors and Quivers.* to be published by RIMS Kôkyûroku.

[58] G.-M. Greuel, *Dualität in der lokalen Kohomologie isolierter Singularitäten.* Math. Ann. **250** (1980), 157–173.

[59] G.-M. Greuel, B. Martin and G. Pfister, *Numerische Charakterisierung Quasihomogener Gorenstein-Kurvensingularitäten.* Math. Nachr. **124** (1985), 123–131.

[60] A. Grothendieck, *Représentations linéaries et compactification profinite des proupes discrets.* Manuscripta Math. **2** (1970), 375–396.

[61] F. J. Grunewald, *On some groups which can not be finitely presented.* J. London Math. Soc. **17** (2) (1978), 427–436.

[62] H. Hamm, *Lokale topologische Eigenschaften komplexer Räume.* Math. Ann. **191** (1971), 235–252.

[63] A. Hattori, *Topology of* $\mathbb{C}^n$ *minus a finite number of affine hyperplanes in general position.* J. Math.Invent. math. In: Séminaire Bourbaki 1971/72. Lecture Notes in Math. **317**, 21–44, Springer, Berlin, Heidelberg, New York, 1973.

[64] M. Jambu and S. Papadima, *A generalization of fiber type arrangements and a new deformation method.* Topology **37** (1998), 1135–1164.

[65] T. de Jong and D. van Straten, *Disentanglements.* In: Mond, D., Montaldi, J. (Eds) Singularity theory and its applications, Warwick 1989, Lecture Notes in Math. **1462**, Springer, Berlin, 1991.

[66] E. R. van Kampen, *On the fundamental group of an algebraic curve.* Amer. J. Math. **55** (1933), 255–260.

[67] T. Keilen, *A claim on the rank of an injective map.* Master Thesis, University of Warwick, 1993.

[68] T. Keilen and D. Mond, *Injective analytic maps - a counterexample to the proof.* arXiv:math.AG/0409426.

[69] S. Kleiman, *Multiple-point formulas: I. Iteration.* Acta Math. **147** (1981), 13–49.

[70] S. Kleiman, J. Lipman and B. Ulrich, *The multiple-point schemes of a finite curvilinear map of codimension one.* Ark. Mat. **34** (2) (1996), 285–326.

[71] T. Kohno, *Série de Poincaré-Koszul associée aux groupes de tresses pures.* Invent. Math. **82** (1985), 57–75.

[72] Lê Dũng Tráng and C. P. Ramanujam, *The invariance of Milnor's number implies the invariance of the topological type.* Amer. J. Math. **98** (1976), 67–78.

[73] Lê Dũng Tráng, *Sur les cycles évanouissants des espaces analytiques.* C. R. Acad. Sc. **288** no. 4 (1979), A283–A285.

[74] A. Libgober, *On the homotopy type of the complement to the plane algebraic curves.* J. für Reine und Angewandte Mathematische **367** (1986), 103–114.

[75] A. Libgober, *Fundamental groups of the complements to plane singular curves.* Procedings of Symposia in Pure Mathematics **46** (1987), 29–45.

[76] A. Libgober, *Invariants of plane algebraic curves via representations of braid groups.* Inventiones Mathematicae **95** (1989), 25–30.

[77] A. Libgober, *On the homology of finite Abelian coverings.* Topology and Its Applications **43** (1992), 157–166.

[78] A. Libgober and S. Yuzvinsky, *Cohomology of the Orlik-Solomon algebras and local systems.* Composito Math. **121** (2000), 337–361.

[79] P. Lima-Filho and H. Schenck, *The holonomy Lie algebras and the LCS formula for subarrangements of* A_n. preprint, 2008.

[80] E. J. N. Looijenga, *Isolated singular points on complete intersections.* London Math. Soc. Lecture Notes in Math. **77**, Cambridge, London: Cambridge Univ. Press, 1984.

[81] E. J. N. Looijenga and J.H.M. Steenbrink, *Milnor numbers and Tjurina numbers of complete intersections.* Math. Ann. **271** (1985), 121–124.

[82] P. Lorist, *The geometry of $\mathcal{B}_x$.* Indag. Math. **48** (1986), 423–442.

[83] I. Luengo and A. Pichon, *Lê's conjecture for cyclic covers.* In: J.-P. Brasselet, T. Suwa (Eds.), Singularities Franco-Faponaises, Sémin. Congr., vol. 10, 163–190, Soc. Math. France, Paris, 2005.

[84] W. L. Marar, *The Euler characteristic of the disentanglement of the image of a corank 1 map-germ.* In: D. Mond, D and J. Montaldi (Eds.), Singularity Theory and Applications, Warwick 1989, Lecture Notes in Math. **1462**, Springer, Berlin, Heidelberg, New York, 1991.

[85] W. L. Marar, *Mapping fibrations and multiple point schemes.* Ph.D. Thesis, University of Warwick, 1989.

[86] W. L. Marar and D. Mond, *Multiple point schemes for corank 1 maps.* J. London Math. Soc. **39** (2) (1989), 553–567.

[87] J. Martinet, *Singularities of Smooth Functions and Maps.* Cambridge Univ. Press, 1982.

[88] D. Matei and A. I. Suciu, *Counting homomorphisms onto finite solvable groups,* Journal of Algebra **286** no. 1 (2005), 161–186.

[89] J. Mather, *Stability of C^∞-mappings VI: the nice dimensions.* Proc. Liverpool Singularities Symposium Vol. 1, Lecture Notes in Math. **192**, 207–255, Springer, Berlin, 1971.

[90] J. Mather, *Stability of C^∞-mappings: IV. Classification of stable germs by R-algebras.* Inst. Hautes tudes Sci. Publ. Math. **37** (1969), 223–248.

[91] J. Mather, *Stability of C^∞-mappings. III. Finitely determined map germs.* Inst. Hautes tudes Sci. Publ. Math. **35** (1968), 279–308.

[92] J. Mather, *Stability of C^∞-mappings : II. Infinitesimal stability implies stability.* Ann. of Math., **89** (1969), 259–291.

[93] J. Milnor, *Singular points on complex hypersurfaces.* Ann. Math. Stud. vol. 61, Princeton, 1968.

[94] J. Milnor and P. Orlik, *Isolated singularities defined by weighted homogeneous polynomials.* Topology **9** (1970), 385–393.

[95] B. Moishezon, *Stable branch curves and braid monodromies* . Lecture Notes in Math. **862**, 21–44, Springer, Berlin, Heidelberg, New York, 1981.

[96] D. Mond, *Vanishing cycles for analytic maps.* In: D. Mond, D and J. Montaldi (Eds.), Singularity Theory and Applications, Warwick 1989, Lecture Notes in Math. **1462**, Springer, Berlin, Heidelberg, New York, 1991.

[97] D. Mond, *Some remarks on the geometry and classification of germs of maps from surfaces to 3-spaces.* Topology **26** (1987), 361–383.

[98] D. Mond and R. Pellikaan, *Fitting ideals and multiple points of analytic mappings.* Algebraic geometry and complex analysis (Pátzcuaro, 1987), 107–161, Lecture Notes in Math. **1414**, Springer, Berlin, 1989.

[99] M. Mustată and H. Schenck, *The module of logarithmic p-forms of a locally free arrangement.* J. Algebra **241** no. 2 (2001), 699–719.

[100] A. Némethi, *Injective analytic mappings.* Duke Math. J. **69** (2) (1993), 335–347.

[101] M. Oka, *Two transforms of plane curves and their fundamental groups.* J. Math. Sci. Univ. Tokyo **3** no. 2 (1996), 399–443.

[102] C. Okonek, M. Schneider and H. Spindler, *Vector bundles on complex projective spaces.* Progress in Mathematics **3**. Birkhäuser, Boston, Mass. 1980.

[103] P. Orlik, *Complements of subspace arrangements.* J. Alg. Geom. **1** (1992), 147–156.

[104] P. Orlik and R. Randell, *The Milnor fiber of a generic arrangement.* Arxiv for matematik, **31** (1993), 71–81.

[105] P. Orlik and L. Solomon, *Combinatorics and topology of complements of hyperplanes.* Invent. Math. **56** (1980), 167–189.

[106] P. Orlik and H. Terao, *Arrangements of hyperplanes.* Lecture Notes in Math. **300**, Springer, Berlin, Heidelberg, New York, 1992.

[107] N. G. J. Pagnon, *Generic fibers of the generalized Springer resolution of type A".* Adv. Math. **194** (2) (2005), 437–462.

[108] S. Papadima, *Higher Homotopy Groups of Complements of Complex Hyperplane Arrangements.* Advances in Mathematics **165** (2002), 71–100.

[109] S. Papadima and A. Suciu, *When does the associated graded Lie algebra of an arrangement group decompose?.* Commentarii Mathematici Helvetici **81** (2006), 859–875.

[110] S. Papadima and S. Yuzvinky, *On rational $K[\pi,1]$ spaces and Koszul algebras.* J. Pure Appl. Alg. **144** (1999), 156–167.

[111] L. Paris, *On the fundamental group of the complement of a complex hyperplane arrangement.* In: M. Falk and H. Terao (Eds.), Arrangements - Tokyo 1998, Adv. Stud. Pure Math. **27**, 257–272, Kinokuniya, Tokyo, 2000.

[112] V. Platonov and O. I. Tavgen, *Grothendieck's problem on profinite completions of groups.* Soviet Math. Dokl. **33** (1986), 822–825.

[113] V. Platonov and O. I. Tavgen, *Grothendieck's problem on profinite completions and representations of groups.* K theory **4** (1990), 45–47.

[114] A. Postnikov and R. Stanley, *Deformations of Coxeter hyperplane arrangements.* J. Combin. Theory Ser. A **91** (1-2) (2000), 544–597.

[115] L. Pyber, *Groups of intermediate subgroups growth and a problem of Grothendieck.* Duke Math. J. **121** (2004), 169–188.

[116] R. Randell, *Milnor fibraions of lattice isotopic arrangements.* Proceedings of Symposia in Pure Mathematics **40** (1983), 415–419.

[117] R. Randell, *The fundamental group of the complement of a union of complex hyperplanes.* Invent. Math. **69** (1982), 103–108. Correction: Invent. Math. **80** (1985), 467–468.

[118] R. Randell, *Lattice-isotopic arrangements are topologicaly isomorphic.* Proceedings of the American Mathematical Society **107** (1989), 555–559.

[119] R. Randell, *Homotopy and group cohomology of arrangements*, Topology Appl. **78** (1997), 201-213.

[120] C. M. Ringel, *Representations of K species and bimodules.* J. Alg. **41** (1976), 269–302.

[121] G. Rybnikov, *On the fundamental group of a complex hyperplane arrangement.* DIMACS Technical Reports **107**, 1994; arXiv:math/9805056v1.

[122] K. Saito, *Quasihomogene isolierte Singularitäten von Hyperflächen.* Invent. Math. **14** (1971), 123–142.

[123] K. Saito, *Theory of logarithmic differential forms and logarithmic vector Şelds.* J. Fac. Sci. Univ. Tokyo Sect. Math. **27** (1980), 265–291.

[124] K. Saito, *On a linear structure of the quotient variety by a finite reflexion group.* Publ. Res. Inst. Math. Sci. **29** (4) (1993), 535–579.

[125] M. Salvetti, *Arrangements of lines and monodromy of plane curves.* In: Algebraic Geometry. Lecture Notes in Math. **862**, 107–192, Springer, 1981.

[126] M. Salvetti, *Topology of the complement of real hyperplanes in $\mathbb{C}^N$.* Invent. Math. **88** (1987), 603–618.

[127] H. Schenck, *Elementary modifications on line configurations in $\mathbb{P}^2$.* Comment. Math. Helv. **78** (3) (2003), 447–462.

[128] H. K. Schenck, A. I Suciu, *Lower central series and free resolutions of hyperplane arrangements.* Transactions of the American Mathematical Society, Vol. **354** no. 9 (2002), 3409–3433.

[129] B. Shelton and S. Yuzvinsky, *Koszul algebras from graphs and hyperplane arrangements.* J. London Math. Soc. **56** (2) (1997), 477–490.

[130] I. Shimada, *A note on Zariski pairs.* Composito Math. **104** (2) (1996), 125–133.

[131] I. Shimada, *Fundamental groups of complements to singular plane curves.* Amer. J. Math. **119** (1) (1997), 127–157.

[132] I. Shimada, *On arithmetic Zariski pairs in degree 6.* Adv. Geom. **8** (2) (2008), 205–225.

[133] I. Shimada, *Non-homeomorphic conjugate complex varieties.* arXiv:math/0701115v2.

[134] I. Shimada, *Transcendental lattices and supersingular reduction lattices of a singular $K3$ surface.* Preprint, Trans. Amer. Math. Soc., 2008.

[135] P. Slodowy, *Four lectures on simple groups and singularities.* Commun. Math. Inst. Rijksuniv. Utr. 11, 1980.

[136] L. Solomon and H. Terao, *A formula for the characteristic polynomial of an arrangement.* Adv. in Math. 64 (3) (1987), 305–325.

[137] L. Solomon and H. Terao, *The double Coxeter arrangement.* Comm. Math. Helv. **73** (1998) 237–258.

[138] N. Spaltenstein, *The fixed point set of a unipotent transformation on the flag manifold.* Indag. Math. **38** (1976), 452–456.

[139] T. A. Springer, *The unipotent variety of a semisimple group.* in: S. Abhyankar (Ed.), Proc. of the Bombay Colloqu. in Algebraic Geometry, 373–391, Oxford Univ. Press, London, 1969.

[140] R. Steinberg, *Conjugacy Classes in Algebraic Groups.* Lecture Notes in Math. **366**, Springer, Berlin, 1974.

[141] R. Steinberg, *An occurrence of the Robinson-Schensted correspondence.* J. Algebra **113** (2) (1988), 523–528.

[142] J. Stipins, *Old and new examples of k-nets in* $\mathbb{P}^2$. arXiv:math.AG/0701046.

[143] A. Suciu, *Fundamental groups of line arrangements: Enumerative aspects*, in: E. Priveto (ed.), Advances in algebraic geometry motivated by physics. Contemporary Math. **276**, Amer. Math. Soc. (2001), 43–79.

[144] H. Terao, *Arrangements of hyperplanes and their freness.* I. Journal of the Faculty of Science of the University of Tokyo **27** (1980), 293–320.

[145] H. Terao, *Generalized exponents of a free arrangement of hyperplanes and Shepherd-Todd-Brieskorn formula.* Invent. Math. **63** (1) (1981), 159–179.

[146] H. Terao, *Multiderivations of Coxeter arrangements.* Invent. Math. **148** (3) (2002), 659–674.

[147] A.M. Uludağ, *More Zariski pairs and finite fundamental groups of curve complements.* Manuscripta Math. **106** (2001), 271–277.

[148] G. Urzua, *On line arrangements with applications to 3-nets.* arXiv:0704.0469v3.

[149] A. Wakamiko, *On the Exponents of 2-Multiarrangements.* Tokyo J. Math. **30** (2007), 99–116.

[150] J.A. Vargas, *Fixed points under the action of unipotent elements of* SL_n *in the flag variety.* Bol. Soc. Mat. Mexicana, **24** (1979), 1–14.

[151] H. Vosegaard, *A characterization of quasi-homogeneous purely elliptic complete intersection singularities.* Compositio Math. 124 (2000), 111–121.

[152] H. Vosegaard, *A characterization of quasi-homogeneous complete intersection singularities.* J. Algebraic Geom. **11** (3) (2002), 581–597.

[153] J. Wahl, *A characterization of quasi-homogeneous Gorenstein surface singularities.* Compositio Math. **55** (1985), 269–288.

[154] C.T.C. Wall, *Finite determinacy of smooth map-germs.* Bull. London Math. Soc. **13** (1981), 481–539.

[155] M. Yoshinaga, *The primitive derivation and freeness of multi-Coxeter arrangements.* Proc. Japan Acad., **78**, Ser. A (2002) 116–119.

[156] M. Yoshinaga, *Characterization of a free arrangement and conjecture of Edelman and Reiner.* Invent. Math. **157** (2) (2004), 449–454.

[157] M. Yoshinaga, *On the freeness of 3-arrangements.* Bull. London Math. Soc. **37** (1) (2005), 126–134.

[158] S. Yuzvinsky, *Cohomology of Brieskorn-Orlik-Solomon algebras.* Comm. Algebra **23** (1995), 5339–5354.

[159] S. Yuzvinsky, *Orlik-Solomon algebras in algebra and topology.* Russian Math. Surveys **56** (2) (2001), 293–364.

[160] S. Yuzvinsky, *Realization of finite abelian groups by nets in* $\mathbb{P}^2$. Communications in Algebra **140** (6) (2004), 1614–1624.

[161] O. Zariski, *On the problem of existence of algeraic functions of two variables possessing a given branch curve.* Amer. J. Math. **51** (1929), 305–328.

[162] G. Ziegler, *Multiarrangements of hyperplanes and their freeness.* Singularities (Iowa City, IA, 1986), 345–359, Contemp. Math., 90, Amer. Math. Soc., Providence, RI, 1989.

Ayşe Altıntaş
Mathematics Institute
University of Warwick
Coventry, CV4 7AL
UK
e-mail: `A.Altintas@warwick.ac.uk`

Celal Cem Sarıoğlu
Department of Mathematics
Dokuz Eylül University
Tınaztepe Campus
Faculty of Arts and Sciences
35160 Buca, İzmir
Turkey
e-mail: `celalcem.sarioglu@deu.edu.tr`

Progress in Mathematics (PM)

Edited by
Hyman Bass, University of Michigan, USA
Joseph Oesterlé, Institut Henri Poincaré, Université Paris VI, France
Alan Weinstein, University of California, Berkeley, USA

BIRKHÄUSER

Progress in Mathematics is a series of books intended for professional mathematicians and scientists, encompassing all areas of pure mathematics. This distinguished series, which began in 1979, includes research level monographs, polished notes arising from seminars or lecture series, graduate level textbooks, and proceedings of focused and refereed conferences. It is designed as a vehicle for reporting ongoing research as well as expositions of particular subject areas.

PM 283: El Zein, F. / Suciu, A.I. / Tosun, M. / Uludağ, A. M. / Yuzvinsky, S. (Eds.)
Arrangements, Local Systems and Singularities (2009). ISBN 978-3-0346-0208-2

PM 282: Bogomolov, F. / Tschinkel. Y. (Eds.)
Cohomological and Geometric Approaches to Rationality Problems. New Perspectives (2010). ISBN 978-0-8176-4933-3

PM 281: Kantorovitz, S.
Topics in Operator Semigroups (2010). ISBN 978-0-8176-4931-9

PM 280: Alonso, M.E. / Arrondo, E. / Mallavibarrena, R. / Sols, I. (Eds.)
Liaison, Schottky Problem and Invariant Theory. Remembering Federico Gaeta (2010). ISBN 978-0-0346-0200-6

PM 279: Ceyhan, Ö. / Manin, Y.I. / Marcolli, M. (Eds.)
Arithmetic and Geometry Around Quantization (2010). ISBN 978-0-8176-4830-5

PM 278: Campbell, H.E.A. / Helminck, A.G. / Kraft, H. / Wehlau, D. (Eds.)
Symmetry and Spaces. In Honor of Gerry Schwarz (2010). ISBN 978-0-8176-4874-9

PM 277: Browning, T.
Quantitative Arithmetic of Projective Varieties (2009). ISBN 978-3-0346-0128-3

PM 276: Bartocci, C. / Bruzzo, U. / Hernández Ruipérez, D.
Fourier–Mukai and Nahm Transforms in Geometry and Mathematical Physics (2009). ISBN 978-0-8176-3246-5

PM 275: Juhl, A.
Families of Conformally Covariant Differential Operators, Q-Curvature and Holography (2009). ISBN 978-3-7643-9899-6

PM 274: Puig, L.
Frobenius Categories versus Brauer Blocks. The Grothendieck Group of the Frobenius Category of a Brauer Block (2009). ISBN 978-3-7643-9997-9

PM 273: Snaith, V.P.
Stable Homotopy Around the Arf-Kervaire Invariant (2009). ISBN 978-3-7643-9903-0

PM 272: Barreira, L.
Dimension and Recurrence in Hyperbolic Dynamics (2008). *Winner of the Ferran Sunyer i Balaguer Prize 2008.* ISBN 978-3-7643-8881-2

PM 271: Galicki, K. / Simanca, S. (Eds.)
Riemannian Topology and Geometric Structures on Manifolds (2008). ISBN 978-0-8176-4742-1

PM 270: Tschinkel, Y. / Zarhin, Y. (Eds.)
Algebra, Arithmetic, and Geometry. In Honor of Y.I. Manin. Volume II (2009, to appear). ISBN 978-0-8176-4746-9

PM 269: Tschinkel, Y. / Zarhin, Y. (Eds.)
Algebra, Arithmetic, and Geometry. In Honor of Y.I. Manin. Volume I (2009, to appear). ISBN 978-0-8176-4744-5

PM 268: Gromoll, D. / Walschap, G.
Metric Foliations and Curvature (2008, to appear). ISBN 978-3-7643-8714-3

PM 267: Barge, J. / Lannes, J.
Suites de Sturm, indice de Maslov et périodicité de Bott (2008). ISBN 978-3-7643-8709-9

PM 266: Pigola, S. / Rigoli, M. / Setti, A.G.
Vanishing and Finiteness Results in Geometric Analysis. In Memory of Alexander Reznikov (2008). ISBN 978-3-7643-8641-2

Progress in Mathematics (PM)

BIRKHÄUSER

Edited by
Hyman Bass, University of Michigan, USA
Joseph Oesterlé, Institut Henri Poincaré, Université Paris VI, France
Alan Weinstein, University of California, Berkeley, USA

Progress in Mathematics is a series of books intended for professional mathematicians and scientists, encompassing all areas of pure mathematics. This distinguished series, which began in 1979, includes research level monographs, polished notes arising from seminars or lecture series, graduate level textbooks, and proceedings of focused and refereed conferences. It is designed as a vehicle for reporting ongoing research as well as expositions of particular subject areas.

PM 265: Kapranov, M. / Kolyada, S. / Manin, Yu.I. / Moree, P. / Potyagailo, L. (Eds.)
Geometry and Dynamics of Groups and Spaces (2007).
ISBN 978-3-7643-8607-8

PM 264: Miró-Roig, R.M.
Determinantal Ideals (2007)
Winner of the Ferran Sunyer i Balaguer Prize 2007.
ISBN 978-3-7643-8534-7

PM 263: Todorcevic, S.
Walks on Ordinals and Their Characteristics (2007).
ISBN 978-3-7643-8528-6

PM 262: Fargues, L. / Genestier, A. / Lafforgue, V.
L'isomorphisme entre les tours de Lubin-Tate et de Drinfield.
ISBN 978-3-7643-8455-5

PM 261: Stout, E.L.
Polynomial Convexity (2007)
ISBN 978-0-8176-4537-3

PM 260: Holzapfel, R.-P. / Uludağ, A.M. / Yoshida, M. (Eds.)
Arithmetic and Geometry Around Hypergeometric Functions (2007)
ISBN 978-3-7643-8283-4

PM 259: Capogna, L. / Danielli, D. / Pauls, S.D. / Tyson, J.T.
An Introduction to the Heisenberg Group and the Sub-Riemannian Isoperimetric Problem (2007)
ISBN 978-3-7643-8132-5

PM 258: Gan, W.T. / Kudla, S.S. / Tschinkel, Y. (Eds.). Eisenstein Series and Applications (2008).
ISBN 978-0-8176-4496-3

PM 257: Lakshmibai, V. / Littelmann, P. / Seshadri, C.S. (Eds.)
Schubert Varieties (due 2009)
ISBN 978-0-8176-4153-5

PM 256: Borthwick, D.
Spectral Theory of Infinite-Volume Hyperbolic Surfaces (2007). ISBN 978-0-8176-4524-3

PM 255: Kobayashi, T. / Schmid, W. / Yang, J.-H. (Eds.)
Representation Theory and Automorphic Forms (2007). ISBN 978-0-8176-4505-2

PM 254: Ma, X. / Marinescu, G.
Holomorphic Morse Inequalities and Bergman Kernels (2007)
Winner of the Ferran Sunyer i Balaguer Prize 2006.
ISBN 978-3-7643-8096-0

PM 253: Ginzburg, V. (Ed.)
Algebraic Geometry and Number Theory (2006).
ISBN 978-0-8176-4471-7

PM 252: Maeda, Y. / Michor, P. / Ochiai, T. / Yoshioka, A. (Eds.)
From Geometry to Quantum Mechanics (2007).
ISBN 978-0-8176-4512-8

PM 251: Boutet de Monvel, A. / Buchholz, D. / Iagolnitzer, D. / Moschella, U. (Eds.)
Rigorous Quantum Field Theory (2006). ISBN 978-3-7643-7433-4

PM 250: Unterberger, A.
The Fourfold Way in Real Analysis (2006). ISBN 978-3-7643-7544-7

PM 249: Kock, J. / Vainsencher, I.
An Invitation to Quantum Cohomology. Kontsevich's Formula for Rational Plane Curves (2006).
ISBN 978-0-8176-4456-7

PM 248: Bartholdi, L. / Ceccherini-Silberstein, T. / Smirnova-Nagnibeda, T. / Zuk, A.
(Eds.) Infinite Groups: Geometric, Combinatorial and Dynamical Aspects (2006)
ISBN 978-3-7643-7446-4

PM 247: Baues, H.-J.
The Algebra of Secondary Cohomology Operations (2006)
ISBN 978-3-7643-7448-8

PM 246: Dragomir, S. / Tomassini, G., Differential Geometry and Analysis on CR Manifolds (2006).
ISBN 978-0-8176-4388-1

PM 245: Fels, G. / Huckleberry, A.T. / Wolf, J.A.
Cycle Spaces of Flag Domains. A Complex Geometric Viewpoint (2006). ISBN 978-0-8176-4391-1

PM 244: Etingof, P. / Retakh, V. / Singer, I.M. (Eds.)
The Unity of Mathematics (2006)
ISBN 978-0-8176-4076-7

PM 243: Bernstein, J. / Hinich, V. / Melnikov, A. (Eds.)
Studies in Lie Theory (2006)
ISBN 0-8176-4342-3

PM 242: Dufour, J.-P. / Zung, N.T.
Poisson Structures and their Normal Forms (2005)
ISBN 978-3-7643-7334-4

PM 241: Seade, J.
On the Topology of Isolated Singularities in Analytic Spaces
Winner of the Ferran Sunyer i Balaguer Prize 2005
ISBN 978-3-7643-7322-1

FSC
www.fsc.org
MIX
Papier | Fördert gute Waldnutzung
FSC® C083411

Kolibri 360 GmbH
Ferdinand-Jühlke-Straße 7
99095 Erfurt, Deutschland
produktsicherheit@kolibri360.de